Cell Biology: Theory, Concepts and Applications

Cell Biology: Theory, Concepts and Applications

Edited by Samantha Granger

SYRAWOOD
PUBLISHING HOUSE

New York

Published by Syrawood Publishing House,
750 Third Avenue, 9th Floor,
New York, NY 10017, USA
www.syrawoodpublishinghouse.com

Cell Biology: Theory, Concepts and Applications
Edited by Samantha Granger

International Standard Book Number: 978-1-68286-744-0 (Hardback)

Cataloging-in-Publication Data

Cell biology : theory, concepts and applications / edited by Samantha Granger.
    p. cm.
Includes bibliographical references and index.
ISBN 978-1-68286-744-0
1. Cytology. 2. Biology. I. Granger, Samantha.
QH581.2 .C45 2019
571.6--dc23

TABLE OF CONTENTS

PREFACE

Over the recent decade, advancements and applications have progressed exponentially. This has led to the increased interest in this field and projects are being conducted to enhance knowledge. The main objective of this book is to present some of the critical challenges and provide insights into possible solutions. This book will answer the varied questions that arise in the field and also provide an increased scope for furthering studies.

The cell is the structural, functional and biological unit of living organisms. Cell organelles are parts of the cell that execute specific functional roles. Some of the important organelles in the cell are the centrosome, cell membrane, cytoplasm, endoplasmic reticulum, mitochondrion, ribosomes, etc. Cell biology is a branch of biology that is concerned with the study of the cell and cell organelles. It focuses on the study of eukaryotic cells, prokaryotic cells and their signaling pathways. Cells can be observed under the microscope using the techniques of optical microscopy, electron microscopy, fluorescence electron microscopy, etc. Cell biology plays a crucial role in varied fields of biology like molecular biology, biochemistry, immunology, genetics, etc. This book is a compilation of chapters that discuss the most vital concepts and emerging trends in the field of cell biology. It aims to shed light on some of the unexplored aspects and the recent researches in this field. It will serve as a valuable source of reference for students as well as experts.

I hope that this book, with its visionary approach, will be a valuable addition and will promote interest among readers. Each of the authors has provided their extraordinary competence in their specific fields by providing different perspectives as they come from diverse nations and regions. I thank them for their contributions.

Editor

Early changes in the metabolic profile of activated CD8$^+$ T cells

Clemens Cammann[1][*], Alexander Rath[2], Udo Reichl[2], Holger Lingel[3], Monika Brunner-Weinzierl[3], Luca Simeoni[1], Burkhart Schraven[1,4] and Jonathan A. Lindquist[1,5]

Abstract

Background: Antigenic stimulation of the T cell receptor (TCR) initiates a change from a resting state into an activated one, which ultimately results in proliferation and the acquisition of effector functions. To accomplish this task, T cells require dramatic changes in metabolism. Therefore, we investigated changes of metabolic intermediates indicating for crucial metabolic pathways reflecting the status of T cells. Moreover we analyzed possible regulatory molecules required for the initiation of the metabolic changes.

Results: We found that proliferation inducing conditions result in an increase in key glycolytic metabolites, whereas the citric acid cycle remains unaffected. The upregulation of glycolysis led to a strong lactate production, which depends upon AKT/PKB, but not mTOR. The observed upregulation of lactate dehydrogenase results in increased lactate production, which we found to be dependent on IL-2 and to be required for proliferation. Additionally we observed upregulation of Glucose-transporter 1 (GLUT1) and glucose uptake upon stimulation, which were surprisingly not influenced by AKT inhibition.

Conclusions: Our findings suggest that AKT plays a central role in upregulating glycolysis via induction of lactate dehydrogenase expression, but has no impact on glucose uptake of T cells. Furthermore, under apoptosis inducing conditions, T cells are not able to upregulate glycolysis and induce lactate production. In addition maintaining high glycolytic rates strongly depends on IL-2 production.

Keywords: T-cell activation, Aerobic glycolysis, AKT/PKB, Lactate

Background

T cells play a central role in the immune system and are crucial for the adoptive immune response. Activation of T cells by specific antigens leads to proliferation, differentiation into effector cells, and cytokine production.

A variety of stimuli, including soluble or immobilized antibodies (Abs) that recognize the T cell receptor (TCR), peptide-loaded APCs, or MHC-I tetramers carrying high- or low-affinity peptides, have been used to study T cell responses. It was previously shown that different stimuli lead to either proliferation or apoptosis of thymocytes [1] and mature T cells [2]. However, it is poorly understood how triggering of the same receptor with ligands of different affinity can induce these different outcomes. Since it is known that thymocytes which cannot fulfill their energy demands undergo apoptosis [3] we hypothesized that changes in the metabolic profiles in activated T cells might contribute to cell fate specification.

Stimulation of T cells leads to a change from a quiescent resting state into an activated state, which is characterized by an extensive cell growth, proliferation, and the production of effector proteins, such as cytokines. In the resting state, T lymphocytes maintain their basal energy demands primarily through a mixed usage of glucose and glutamine [3]. However, to meet the increased energy demands following activation, glucose metabolism increases as a source of energy and providing precursor molecules for cellular biosynthesis [4]. Unlike hepatocytes and myocytes, lymphocytes do not have large internal glycogen stores. This makes them highly dependent on extracellular glucose. Glucose uptake in T cells is mediated by the glucose-transporter 1 (GLUT1). It was previously shown that upregulation of GLUT1

* Correspondence: clemens.cammann@med.ovgu.de
[1]Institute of Molecular and Clinical Immunology, Otto-von-Guericke-University, Magdeburg, Germany
Full list of author information is available at the end of the article

expression depends on co-stimulation via CD28 [5, 6]. Co-stimulation is also responsible for the activation of PI3K/AKT, which is thought to be involved in the expression of GLUT1 at the cell surface [7]. However it was shown recently that AKT does not appear to be required for the upregulation of GLUTI and for the increase in glucose uptake upon T cell stimulation [8].

Another important regulator of cellular metabolism is the adenosine-monophosphate kinase (AMPK), which promotes ATP conservation and production through the upregulation of glycolysis, fatty acid oxidation, and the inhibition of ATP-consuming pathways such as protein synthesis, fatty acid synthesis, gluconeogenesis, and glycogen synthesis. AMPK can be activated by an increase in the AMP:ATP ratio followed by phosphorylation through LKB1, a serine/threonine kinase [9–11]. In addition it is known that triggering of the TCR activates AMPK in an AMP-independent, but Ca^{2+}-calmodulin-dependent kinase kinase 2 (CAMKK2)-dependent manner, which was shown to activate AMPK independent of AMP levels [12, 13].

We demonstrate here that stimulation of murine $CD8^+$ T cells with MHC-I tetramers carrying the high affinity OVA-peptide SIINFEKL leads to the transient activation of AMPK followed by an increase in the glycolytic rate and production of lactate, to counter the increased demand for ATP after activation.

Furthermore, we show that the inhibition of lactate production leads to a decreased proliferation. Additionally we confirmed that AKT is required for the glycolytic change in CD8+ T cells whereas mTOR is dispensable. Investigation of later time points revealed a connection between CTLA4 upregulation and downregulation of IL2 production accompanied by subsequent downregulation of lactate production.

Results

Antibody stimulation induces ATP depletion, whereas tetramers do not

In our experimental system we activated OT-I T cells using either cross-linked soluble CD3 and CD8 monoclonal antibodies (mAbs) (antibody stimulation) or cross-linked H-2K^b molecules loaded with the SIINFEKL peptide (tetramer stimulation). As described in our previous study [2], antibody stimulation leads to a strong and transient activation of signaling molecules downstream of the TCR and induces apoptosis. In contrast, tetramer stimulation shows a weak, but sustained activation of signaling molecules that induces T cell proliferation.

We first assessed whether the concentration of ATP changes upon T cell stimulation by measuring the level of intracellular ATP. Figure 1a shows that within 4 h after stimulation with soluble antibodies, there is a

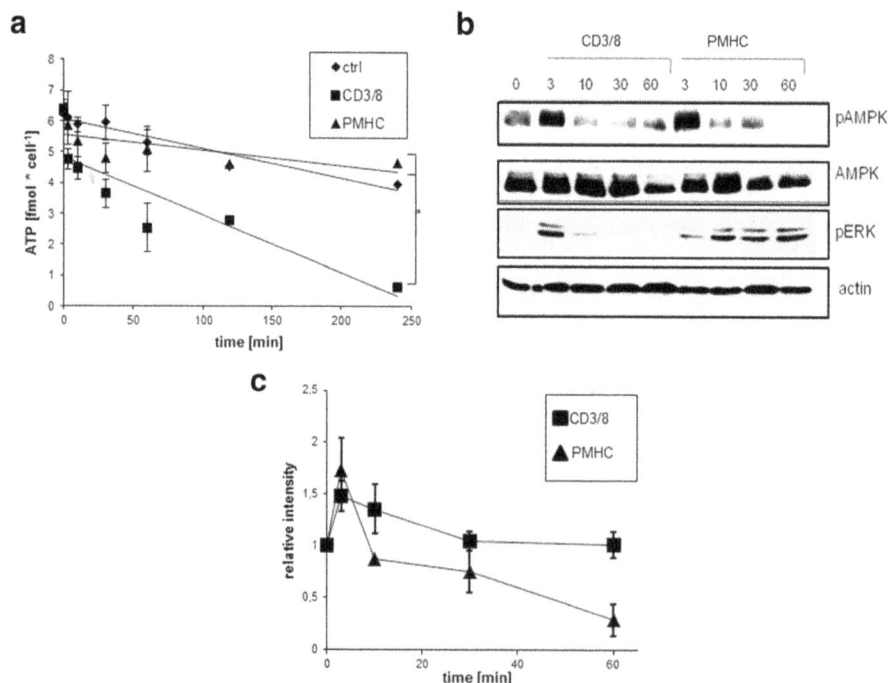

Fig. 1 Stimulation of CD8$^+$ T cells with antibodies leads to rapid ATP consumption. Purified CD8$^+$ T cells were treated with either soluble CD3/CD8 mAbs or OT-I tetramers (PMHC) for the indicated time periods. **a** Cellular ATP production was analyzed using the ATP-Assay Kit. $n = 6$ (**b**) AMPK activation was determined by Western blotting. Phospho-ERK staining was included to show effective stimulation. Total AMPK and actin are included as loading controls. **c** Quantification of AMPK phosphorylation within 60 min after stimulation, $n = 3$

noticeable decrease in the ATP concentration, which is not seen upon tetramer stimulation. Moreover, we confirmed this result by analyzing various metabolites including ATP by mass spectrometry (MS) coupled to high-performance anion-exchange chromatography (data not shown). The observed decrease in the un-stimulated control is due to the fact that purified mouse T cells die when left in culture without a stimulus.

We next analyzed the activation of the metabolic regu-lator AMPK (Fig. 1b), which becomes phosphorylated by LKB1 if the AMP:ATP ratio is increased. Here we show that AMPK is activated immediately by both stimuli. At later time points, antibody-stimulated T cells continue to show a sustained activation of AMPK, whereas tetramer-stimulated cells do not. This led us to conclude that the higher ATP levels observed after tetramer stimulation, result in AMPK inactivation. In contrast,

lower ATP levels observed upon antibody stimulation correlate with sustained AMPK activation.

Tetramer stimulation actives glycolysis, leading to lactate production

It has been previously shown that stimulation of primary human T cells results in the upregulation of glucose up-take and glycolysis [14]. To analyze whether this holds true in our system, we investigated the generation of metabolites for glycolysis and the citric acid cycle. We found that 2-4 h after tetramer stimulation, the concen-trations of the glycolytic metabolites fructose-1,6-bisphosphate and 3-phosphoglycerate were significantly increased (Fig. 2a). The concentrations of the end prod-uct pyruvate were also slightly, but not significantly in-creased. For metabolites of the citric acid cycle, we observed a trend towards increased ketoglutarate, mal-ate, and fumarate upon tetramer stimulation (Fig. 2b). In

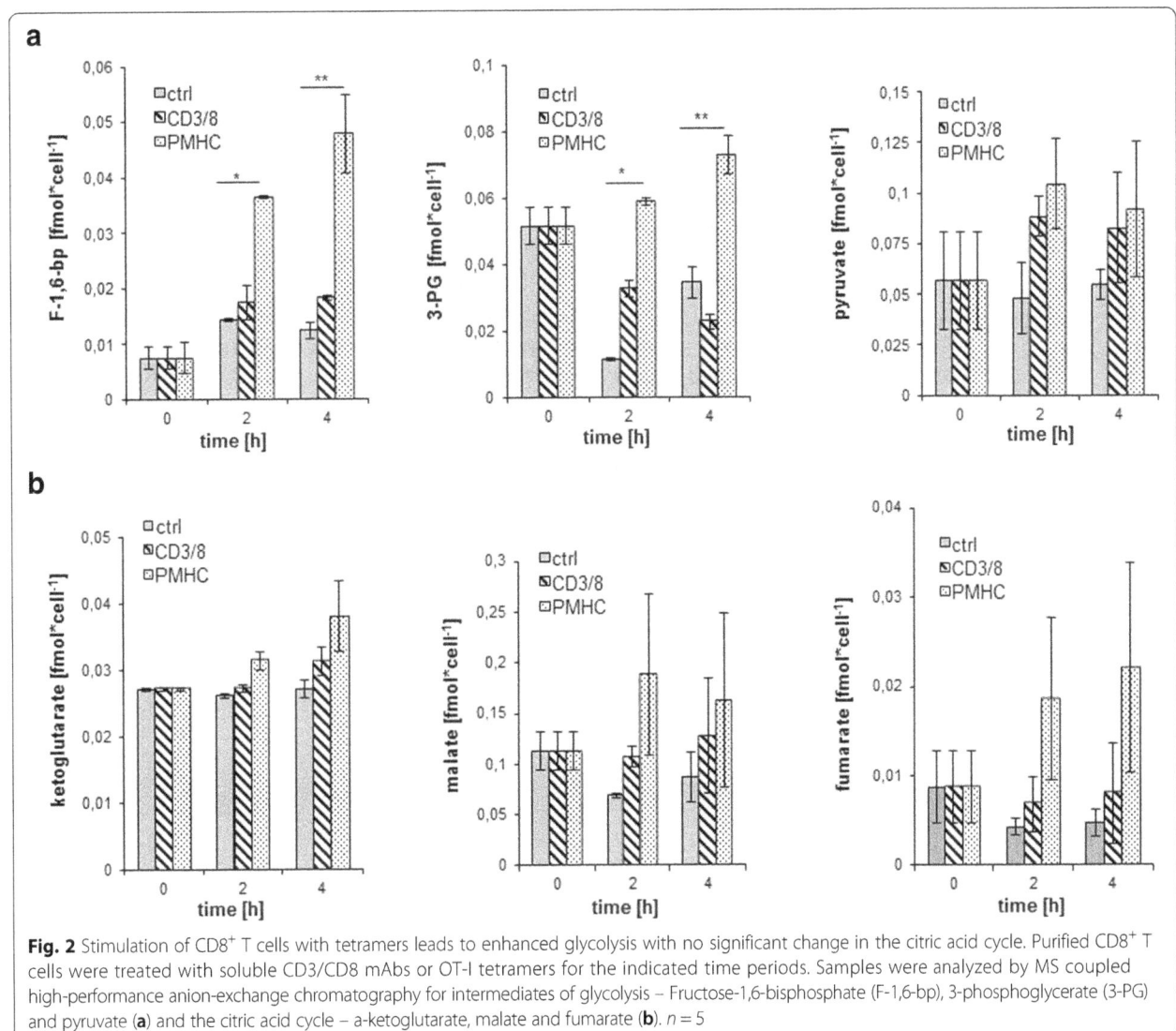

Fig. 2 Stimulation of CD8$^+$ T cells with tetramers leads to enhanced glycolysis with no significant change in the citric acid cycle. Purified CD8$^+$ T cells were treated with soluble CD3/CD8 mAbs or OT-I tetramers for the indicated time periods. Samples were analyzed by MS coupled high-performance anion-exchange chromatography for intermediates of glycolysis – Fructose-1,6-bisphosphate (F-1,6-bp), 3-phosphoglycerate (3-PG) and pyruvate (a) and the citric acid cycle – a-ketoglutarate, malate and fumarate (b). $n = 5$

contrast to tetramer stimulation, changes in glycolysis and citric acid cycle were not observed after antibody stimulation.

Since we did not observe changes in metabolism upon antibody stimulation, we focused our investigation on the analysis of metabolic parameters upon tetramer stimulation. The observed increase in glycolysis without significant changes in the citric acid cycle lead us to hypothesize that pyruvate produced by glycolysis is converted into lactate (i.e. aerobic glycolysis). To test this hypothesis, we directly measured lactate production (Fig. 3a). We observed a high lactate production after stimulation with tetramers, whereas there was no lactate production after antibody stimulation. When we analyzed later timepoints, we found that lactate production occurs at a high rate during the first 48 h following stimulation. At later timepoints we observed a decrease in lactate production. Since T cells die within 24 h upon antibody stimulation, we used the low affinity peptide Q4H7 as a "negative" control, as this induces survival, but not proliferation [1].

Proliferation of activated T cells requires both lactate production and functional electron transport chains

To determine if lactate production is required to maintain T cell function we added oxamate, which inhibits lactate dehydrogenase and hence effectively blocks lactate production. We found a decreased proliferation upon the addition of oxamate, thus indicating the necessity of lactate production for T cell proliferation (Fig. 4a). Although we observed no significant changes in the citric acid cycle, the addition of rotenone, which blocks ATP production by interfering with complex I of the electron transport chain in mitochondria, also caused a complete abrogation of proliferation (Fig. 4a). To control the function of the inhibitors rotenone and oxamate we analyzed lactate production. In the presence of oxamate, as expected, we observed a decrease in lactate

production, whereas the addition of rotenone showed no effect on lactate produced upon stimulation (Fig. 4b). This led us to the conclusion that even if T cells shift their metabolism towards glycolysis upon stimulation, the TCA cycle is still required. To assess the toxicity of rotenone and oxamate we analyzed the cells for apoptosis (Additional file 1: Figure S1D), where no toxic effects could be observed.

Since lactate production rapidly decreases 48 h after stimulation, we hypothesized that there must be a switch that shuts off lactate. IL-2 is an autocrine growth factor which drives proliferation at later stages of T cell activation. We tested the hypothesis that IL-2 is involved in switching off lactate production. Therefore we applied a neutralizing IL-2 antibody to our cultures in order to prevent the binding of IL-2 to its receptor. To confirm that our neutralizing IL-2 antibody was effective and not toxic, we analyzed Stat5 phosphorylation 24 h after stimulation (Additional file 1: Figure S1C) and apoptosis (Additional file 1: Figure S1D). Surprisingly, adding of the neutralizing IL-2 antibody to the culture leads to decreased lactate production (Fig. 4c). Moreover the addition of exogenous IL-2 to the stimulated cells fostered the production of lactate (Fig. 4c). This led us to the conclusion that IL-2 is needed to maintain lactate production and does not act as inhibitor of lactate production. We next assessed lactate production in OT1-T cells from CTLA4$^{-/-}$ mice upon stimulation. Interestingly the decrease in lactate production could not be observed in these cells (Fig. 5a). Moreover the loss of CTLA4 leads to a sustained production of IL-2 (Fig. 5b). Additionally, this could be confirmed by adding exogenous CTL4 to the cells, which had an impact on IL-2 receptor expression (Fig. 5c). These results indicate that CTLA4 plays a critical role in modulating IL-2 and lactate production.

AKT regulates lactate dehydrogenase expression

To further analyze whether, in addition to AMPK, other signaling molecules take part in the regulation of

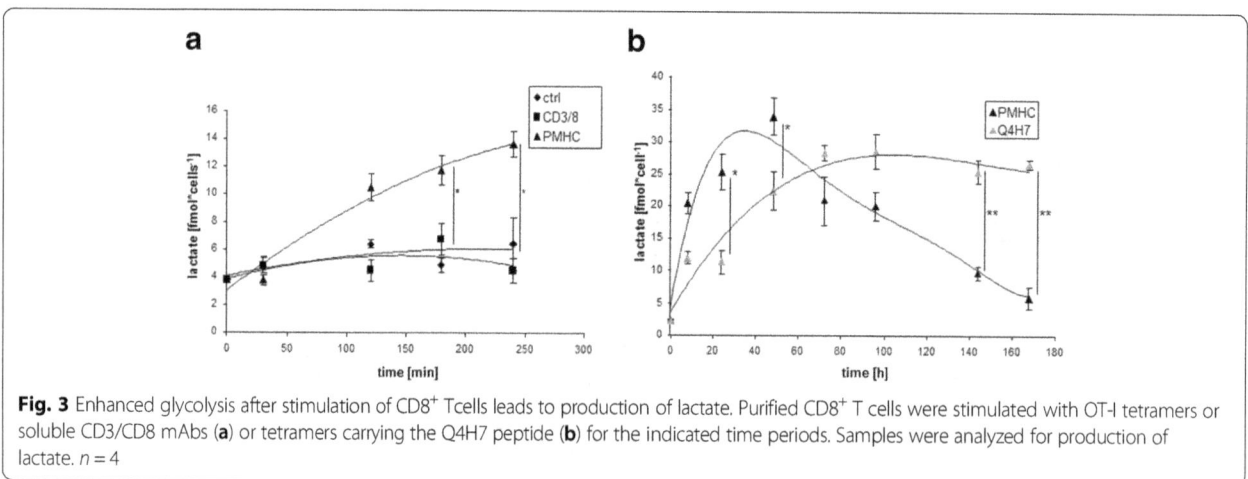

Fig. 3 Enhanced glycolysis after stimulation of CD8$^+$ Tcells leads to production of lactate. Purified CD8$^+$ T cells were stimulated with OT-I tetramers or soluble CD3/CD8 mAbs (**a**) or tetramers carrying the Q4H7 peptide (**b**) for the indicated time periods. Samples were analyzed for production of lactate. $n = 4$

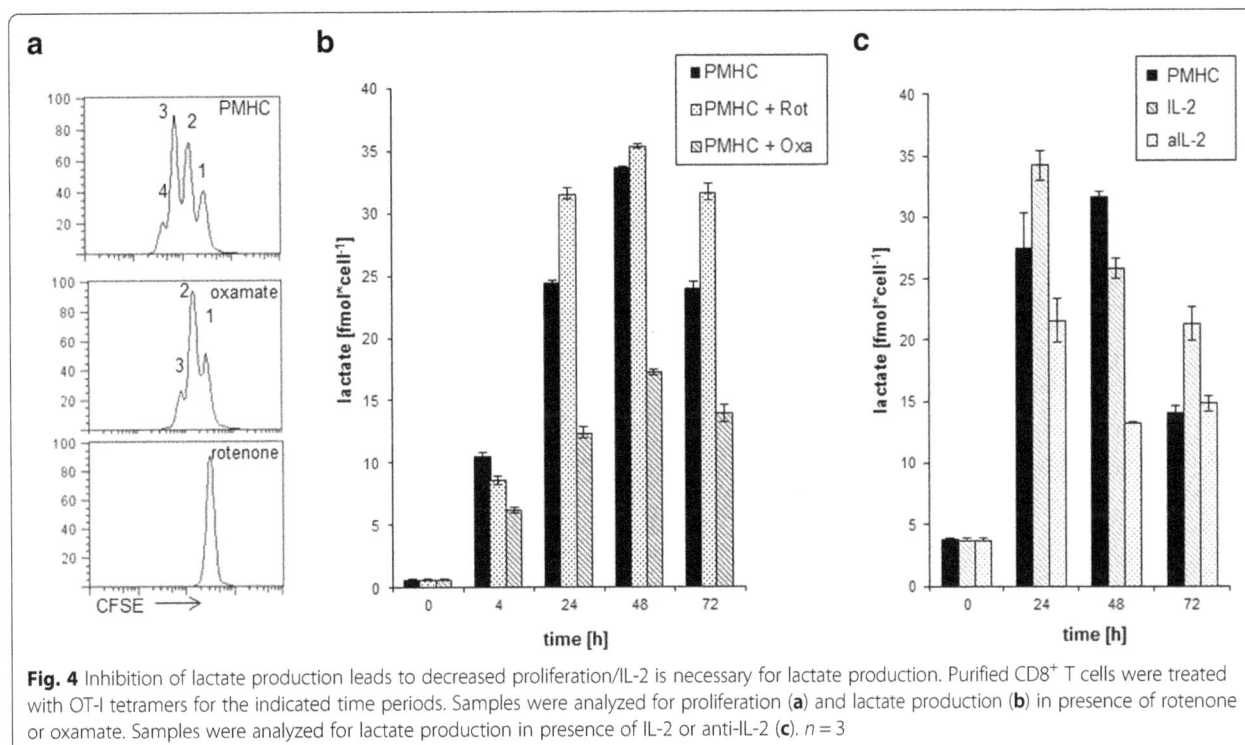

Fig. 4 Inhibition of lactate production leads to decreased proliferation/IL-2 is necessary for lactate production. Purified CD8$^+$ T cells were treated with OT-I tetramers for the indicated time periods. Samples were analyzed for proliferation (**a**) and lactate production (**b**) in presence of rotenone or oxamate. Samples were analyzed for lactate production in presence of IL-2 or anti-IL-2 (**c**). $n = 3$

metabolism, we investigate AKT and mTOR, which are known to be involved in nutrient sensing in T cells [15]. We applied inhibitors of these enzymes to analyze their relative contribution to lactate production. We observed that both inhibitors block the phosphorylation of p70S6K, which is a downstream target of both mTOR and AKT (Additional file 1: Figure S1A and B) thus confirming that both inhibitors work in our system. We found that the inhibition of the mTOR-complex with rapamycin had no effect on lactate production, whereas the inhibition of AKT leads to a complete abrogation of lactate production (Fig. 6a).

To investigate if the inhibition of lactate production upon AKT inhibition is a direct effect, we analyzed the lactate dehydrogenase (LDH), the enzyme responsible for the conversion of pyruvate to lactate. We observed an increase in total LDH levels after tetramer stimulation which was strongly reduced in the presence of the AKT inhibitor (Fig. 6b).

Another proposed target of AKT in T cell metabolism is GLUT1, which is upregulated upon stimulation to ensure increased glucose uptake. We observe an upregulation of GLUT1 after tetramer stimulation (Fig. 6c and d) However, when we applied the AKT inhibitor we saw no effect on GLUT1 upregulation. We next analyzed the glucose uptake after tetramer stimulation (Fig. 6e). Here, we selectively analyzed the glucose uptake in living cells by including an Annexin V/PI staining to exclude dead and apoptotic cells which do not actively uptake glucose.

We confirmed that, upon tetramer stimulation, glucose uptake is increased, but the addition of the AKT inhibitor showed no effect. Hence, we conclude that AKT is not required for the increase of glucose uptake in CD8$^+$ T cells. As a negative control we added apigenin which prevents the uptake of the fluorescent glucose.

Discussion

It was previously described for OT-I mouse T cells that stimulation with soluble antibodies results in apoptosis whereas more physiological stimuli like the stimulation with OT-I tetramers lead to proliferation and differentiation of T cells reflecting the in vivo situation. The reason why these different stimuli have different outcomes on the T cell fate remains still unclear. Therefore we hypothesized that the depletion of the intracellular ATP stores following antibody stimulation could be responsible for the induction of apoptosis. It was previously shown for tumor cell lines that if intracellular ATP concentrations fall below a certain limit, apoptosis is induced [16, 17]. Furthermore several studies showed that apoptotic processes require ATP to transfer apoptotic signals into the nucleus as well as for chromatin condensation and nuclear fragmentation [18–20]. In this study we observed a rapid consumption of ATP upon antibody stimulation, whereas ATP levels remain constant after tetramer stimulation. The question is whether the drop in ATP levels are the cause or a consequence of the resulting apoptosis. First consider that the changes in

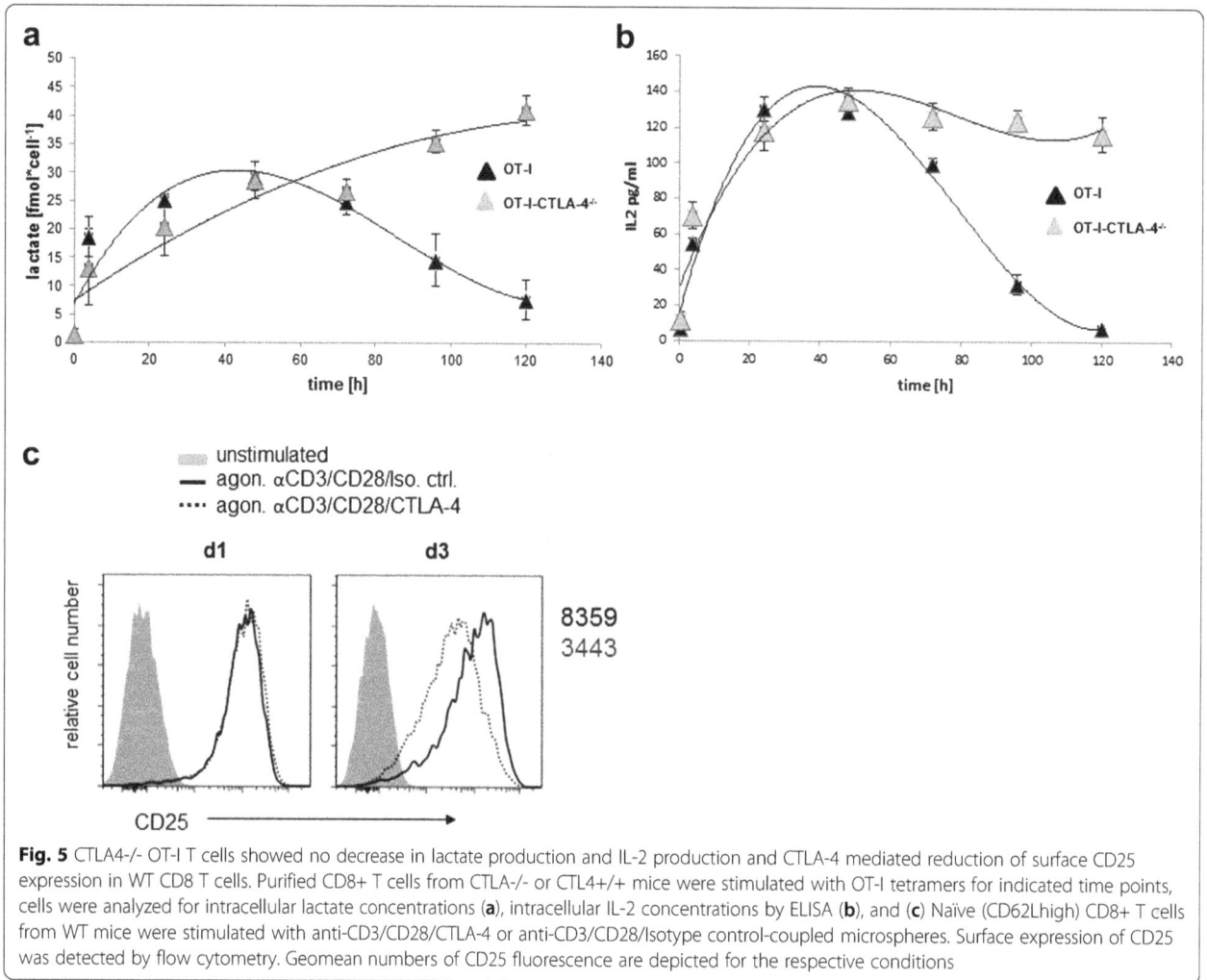

Fig. 5 CTLA4-/- OT-I T cells showed no decrease in lactate production and IL-2 production and CTLA-4 mediated reduction of surface CD25 expression in WT CD8 T cells. Purified CD8+ T cells from CTLA-/- or CTL4+/+ mice were stimulated with OT-I tetramers for indicated time points, cells were analyzed for intracellular lactate concentrations (**a**), intracellular IL-2 concentrations by ELISA (**b**), and (**c**) Naïve (CD62Lhigh) CD8+ T cells from WT mice were stimulated with anti-CD3/CD28/CTLA-4 or anti-CD3/CD28/Isotype control-coupled microspheres. Surface expression of CD25 was detected by flow cytometry. Geomean numbers of CD25 fluorescence are depicted for the respective conditions

ATP levels were observed already within the first hour upon stimulation. From this one may assume that the dramatic phosphorylation which occurs upon antibody stimulation leads to a high consumption of ATP [21]. This in turn results in the induction of apoptosis, which further reinforces the reduction in ATP levels. Therefore our results suggest that following tetramer stimulation, T cells switch on metabolic programs to generate ATP in order to counterbalance their ATP consumption.

We next investigated the activation of AMPK, as this is the major energy sensor in cells and its activation is closely linked to intracellular ATP levels. We observed an initial activation of AMPK upon both tetramer and antibody stimulation. This initial activation was previously reported to be induced by the activation of CAMKK2 upon TCR triggering [13]. Since AMPK is further activated by a high AMP:ATP ratio, the decreased ATP levels observed upon antibody stimulation forces AMPK to remain active. In contrast, upon tetramer stimulation, T cells maintain high levels of ATP, thus inactivating AMPK. This led us to the conclusion that

tetramer stimulation induces additional changes in the metabolic profiles, since the initial activation of AMPK is not sufficient to shift T cell metabolism towards glycolysis, which is needed to maintain proliferation.

Several previous studies [5, 8, 14] analysed the upregulation of glycolysis by monitoring glucose consumption from the media, direct glucose uptake by the cell, or the upregulation of GLUT1. Here we used a new method to analyze 25 metabolites and nucleotides of glycolysis and citric acid cycle to further investigate the shift towards glycolysis [22]. In our experiments with tetramer stimulation, we clearly observed a significant increase in glycolytic metabolites compared to either the antibody stimulated or unstimulated cells. We also observed small changes in the metabolites of the citric acid cycle, however these were not statistically significant. This led us to the conclusion that aerobic glycolysis is the major energy-producing process after activation.

These observations were confirmed by the fact that, after tetramer stimulation, T cells produce high amounts of lactate, which is an indication for increased aerobic

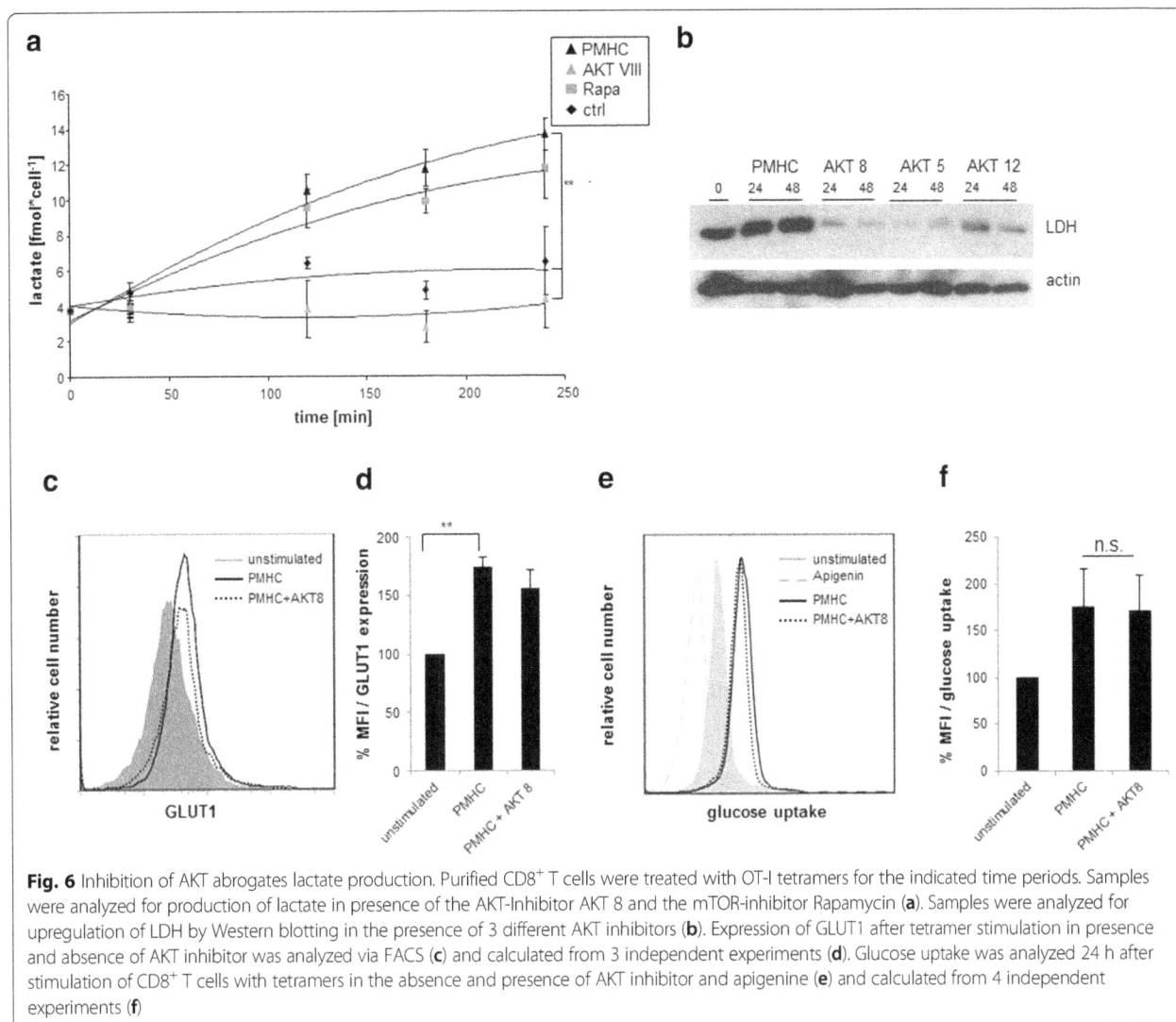

Fig. 6 Inhibition of AKT abrogates lactate production. Purified CD8$^+$ T cells were treated with OT-I tetramers for the indicated time periods. Samples were analyzed for production of lactate in presence of the AKT-Inhibitor AKT 8 and the mTOR-inhibitor Rapamycin (**a**). Samples were analyzed for upregulation of LDH by Western blotting in the presence of 3 different AKT inhibitors (**b**). Expression of GLUT1 after tetramer stimulation in presence and absence of AKT inhibitor was analyzed via FACS (**c**) and calculated from 3 independent experiments (**d**). Glucose uptake was analyzed 24 h after stimulation of CD8$^+$ T cells with tetramers in the absence and presence of AKT inhibitor and apigenine (**e**) and calculated from 4 independent experiments (**f**)

glycolysis, also known as the "Warburg effect". The fact that lactate production is essential for T cell proliferation was confirmed by the observation that addition of oxamate, a lactate dehydrogenase inhibitor, lead to decreased proliferation. Nevertheless, T cells still require the citric acid cycle to maintain proliferation, as shown by inhibiting the ATP production with rotenone. This inhibition is rather indirect because rotenone blocks complex I of the electron transport chain, disrupting the proton gradient at the mitochondrial membrane, which in the end abolishes the ATP synthase reaction. In a recent study comparing metablic features of T cell subsets, the authors observed striking differences between CD8+ T cells, which showed more glycolytic metabolism, compare to CD4+ T cells, which show higher rates of mitochondrial oxidative metabolism and a greater maximal respiratory capacity [23]. However activation and proliferation of both cell types were similar sensitive to the addition of rotenone. This supports the idea that the

TCA cycle is not only required for the generation of ATP, but is also required to deliver substrates for biosynthetic processes like the generation of nucleotides. Furthermore it has been shown that during oxidative phosphorylation, reactive oxygen species are generated which played a critical role in T cell activation [24].

Surprisingly we observed that after 48 h of activation, lactate production decreases. This observation led us to hypothesize that there is a switch in T cell metabolism to shut down aerobic glycolysis. Our hypothesis was that with the transition from antigen-driven proliferation to cytokine-driven proliferation, IL-2 might also be responsible for switching off lactate production and directing pyruvate into the TCA cycle to generate precursors for biosynthesis. When we tested this hypothesis we found that IL-2 is required to maintain lactate production and appears to play no role in switching off lactate. This observation contradicts a previous study were the authors showed that the removal of IL-2 within the first 20 h of stimulation had no effect on lactate production in CD4+ primary human T cells [5]. On the other hand it is known that IL-2 can induce negative regulators of T cell activation like CTLA4, which downmodulates T cell responses in order to prevent an overreaction of the immune system [5, 25]. Interestingly, when we assessed lactate production in OT1-T cells from CTLA4$^{-/-}$ mice, the decrease in lactate production could not be observed indicating that CTLA4 plays a critical role in modulating the switch of glycolytic end products. Furthermore this could be correlated to the IL-2 production in these cells. Under normal conditions the expression of IL-2 ultimately leads to the upregulation of CTLA4 [26, 27]. This in turn leads to feedback inhibition of IL-2 production, which would explain the observed decrease in lactate production [28]. Furthermore the addition of exogenous CTLA4 to the cells leads to reduced CD25 expression, which confirms the role of CTLA4 in regulating IL-2. The possible corresponding ligand triggering surface CTLA4 in CD8+ T cells was shown to be CD80, which is also present on activated T cells serving as a T cell-T cell interaction partner [29]. Indeed, CTLA4 was recently shown to inhibit glycolysis in CD4+ Tcells, thus supporting our observations [30].

In a recent study, it was shown that the upregulation of BCL-6 represses genes encoding molecules involved in aerobic glycolysis that are upregulated during the effector phase of the immune response [29]. A connection between the expression of CTLA4 and BCL-6 has been discussed in the generation of follicular helper T cells [31], but if it is a common feature in modulating T cell metabolism remains unclear and has to be addressed in further experiments.

Since it is known that the PI3K/AKT/mTOR pathway is required for metabolism, we analyzed the contribution of AKT and mTOR, both involved in signaling processes known to regulate cellular metabolism [6, 32, 33]. Inhibition of mTOR with rapamycin showed no effect on lactate production, while the inhibition of AKT completely abrogated lactate production. Additionally, we show for the first time that activation of AKT upregulates the expression of LDH, the enzyme which converts pyruvate to lactate. This observation was confirmed by specific functionally different AKT inhibitors which all abrogated the expression of LDH.

Moreover, we observed an increase in GLUT1 expression and glucose uptake upon tetramer stimulation which was also described before in primary human T cells [5, 6]. In contrast to these previous studies, the upregulation of GLUT1 and glucose uptake in OT1 CD8+ T cells was not AKT-dependent. These contradictory results were also described in a recent study by Macintyre and coworkers [8], showing that inhibition of AKT had no effect on glucose uptake after stimulation of P14

TCR tg T cells with gp33 peptide. Thus, we confirm these previous results showing that AKT has no impact on the uptake of glucose [8]. However we additionally show that AKT is required for the upregulation of lactate dehydrogenase expression. Therefore, we suggest that GLUT1 upregulation and glucose uptake are regulated in an AKT-independent manner whereas lactate production strongly requires the activation of AKT. It was shown that phosphoinositol-dependent protein kinase 1 (PDK1), an upstream activator of AKT, is responsible for upregulation of glucose uptake independent of the PI3K/AKT pathway [8]. Since the role of PDK1 was assessed in T cell blasts in the presence of high IL-2 concentrations, there is a strong temporal separation from our system. We clearly observe the upregulation of LDH in parallel with the activation of AKT within 48 h upon stimulation. This leads to the hypothesis that AKT activation upon stimulation induces upregulation of LDH whereas IL-2 production induces an AKT independent upregulation of glucose uptake via PDK1.

Surprisingly the inhibition of mTOR by rapamycin had no effect on lactate production, since it was described that mTOR is one of the crucial players in nutrient sensing in mammalian cells. Previous studies revealed an important role of mTOR in the regulation of differentiation into CD4+ T cell subsets like Th1, Th2 and Th17 cells [34, 35]. Furthermore we observed no activation of HIF1α (data not shown) which was proposed to be a link between mTOR and upregulation of glycolytic enzymes [36]. Therefore we could conclude that mTOR activation is not required for upregulation of glycolysis in CD8+ T cells.

Conclusion

In summary, we show that antibody stimulation did not induce a significant increase in glycolytic metabolites and hence no shift towards glycolysis. The observed drop in ATP levels might be the cause and the consequence of the resulting apoptosis. Another possibility is that apoptotic processes which where induced by antibody stimulation, inhibit the metabolic reprogramming of T cells.

Tetramer stimulation of CD8$^+$ T cells leads to the induction of aerobic glycolysis in order to fulfill the increased energy demands following T cell activation, which is IL-2 dependent and counteracted by the upregulation of CTLA4 . We show that AKT plays a major role in aerobic glycolysis via upregulation of LDH. Further experiments are required to identify other possible targets of AKT and to identify other players regulating T cell metabolism. The study by MacIntyre and coworkers [8] suggests that PDK1 is the master regulator of T cell metabolism by phosphorylating members of the AGC kinase family, like RSK, PKCs, and SGKs independent of the PI3K/AKT pathway. Recently a new molecule was identified called lymphocyte expansion molecule (LEM, [32]), which was shown to promote antigen specific CD8+ T cell expansion, effector function, and memory cell generation. LEM was observed to regulate the protein complexes of the oxidative phosphorylation pathway in the inner membrane of the mitochondria thereby upregulating the generation of reactive oxygen species which play a crucial role in T cell proliferation. Nevertheless the whole mechanism how this modulation effects T cell behaviour remains unclear.

The activation of the metabolic regulator AMPK appears to be not required for the shift of metabolism towards enhanced glycolysis since it is switched off shortly after stimulation. A recent study suggested that AMPK may be involved in the generation of memory CD8[+] T cells. It was shown that addition of the drug metformin, a strong activator of AMPK, leads to reduced proliferation and an enhanced development of memory CD8[+] T cells [37]. In contrast to CD4+ T cells mTOR appears to be dispensable for upregulation of glycolysis and induction of proliferation in CD8+ T cells. Therefore inhibition of AKT or an upregulation of CTLA4 could be possible targets for preventing the overreaction and exhaustion of T cells as seen in chronic viral infections, like HIV or HCV. Until now some metabolic inhibitors have been shown to suppress T cell responses in EAE, asthma, and graft versus host disease [38–40]. This leads to the conclusion that changes in T cell metabolism can alter T cell expansion and differentiation making metabolic regulation a powerful target for treatment of a large variety of diseases like immune-related diseases characterized by hyperactive T cells.

Methods

Mice, cell purification, stimulation

OT-I TCR transgenic (tg) mice were maintained in pathogen free conditions. All experiments were performed with samples taken from euthanized animals in accordance with the German National Guidelines for the Use of Experimental Animals (Animal Protection Act, Tierschutzgesetz, TierSchG). Animals were handled in accordance with the European Communities Council Directive 86/609/EEC. All possible efforts were made to minimize animal suffering and the number of animals used. Splenic CD8[+] T cells from OT-I TCR tg mice were purified and stimulated as previously described [2]. CTLA4-/-/OT-I mice were kindly provided by M. Brunner-Weinzierl. Crosslinking of CTLA-4 (CD152) on C57BL/6 CD8+ T cells was performed using latex microspheres coated with antibodies. In brief, 107 microspheres/ml were suspended in PBS with CD3 (0.75 µg/ml), CD28 (2.5 µg/ml), CTLA-4 or a hamster isotype control antibody (A19-3, 8 µg/ml) and incubated for 1 h at 37 °C, followed

by washing in PBS and blocking with complete media. CD8+ T cells (1.5×10^6/ml) were stimulated at a ratio of 1:1 with antibody-coupled microspheres in the presence of IL-12 (5 ng/µl). Specificity of crosslinking of CD152 with antibody-coupled microspheres was controlled by stimulating naive CD8+ T cells (CD8+ CD62Lhigh) of OT-1 CTLA4-/- mice with CD3, CD28, and CTLA-4 or CD3, CD28, and isotype control-coupled microspheres.

Human T cells

Approval for these studies was obtained from the Ethics Committee of the Medical Faculty at the Otto-von-Guericke University, Magdeburg, Germany. Informed consent was obtained in accordance with the Declaration of Helsinki. Peripheral blood mononuclear cells were isolated as previously described [40]. The Jurkat E6.1 T cell line was maintained in RPMI 1640 medium [BioChrom] supplemented with 10 % heat inactivated Fetal Bovine Serum (FBS) [PAN] and cultured at 37 °C and 5 % CO_2 in humidified atmosphere.

Metabolic analysis

CD8[+] T cells were cultured in RPMI 1640 medium containing 10 % FCS (PAN Biotech), 100 U/ml penicillin, 100 µg/ml streptomycin (all from Biochrom AG), and 50 µM 2-ME in 48-well plates at a concentration of 4 x 10^6 cells/well. Cells were left unstimulated or stimulated with either CD3/CD8 mAbs or OT-I tetramers at 37 °C. Cells were harvested after 2 h and 4 h. The cells were immediately resuspended in 600 µl ice cold Methanol/Chloroform (2:1). The subsequent metabolic extraction, measurements, and analysis were performed as previously described [41].

Lactate and IL2 production

Stimulated OT-I T cells were harvested at indicated timepoints and analyzed for lactate production with Lactate assay kit (Biocat) measuring colorimetric changes at 450 nm according manufacturers instruction and the production of IL2 with ELISA MAX™ Standard Kit (Biolegend) using a microplate reader.

Glucose uptake

OT-I T cells were stimulated for 24 h with OT-1 tetramers and then analyzed with glucose uptake cell-based assay kit (Cayman Chemical Company). Cells were harvested and starved for 15 min in PBS at 37C°. Afterwards fluorescent labelled glucose (150 µg/ml, 2-NBDG) was added to the cells for 15 min. After subsequent washing, the cells were additionally stained with pacific blue - Annexin V (Biolegend) and PI (Biolegend) and analyzed via FACS.

Inhibitors and Western blotting

CD8[+] T cells were either left untreated or stimulated with CD3/CD8 mAbs or OT-I tetramers. The inhibitors AKT V (1 µM) AKT-VIII (2 µM) AKT XII (5 µM;all Calbiochem) or Rapamycin (2 mM) were directly added to the samples after stimulation. Anti-AMPKα, anti-phospho(p)AMPKα (Thr172), Anti-p-ERK1/2, anti-pS6K (all Cell Signaling) and anti-β-actin (clone AC-15) (Sigma-Aldrich) were used for Western blotting.

Flow cytometry

CD8[+] T cells were stimulated with agonistic antibody coated microspheres for the indicated time, washed once and stained with CD25-AF647 [Biolegend] for 15 min at 4 °C. After one washing, samples were analyzed on FACSCantoII using FACSDiva software [BD].

Statistics

All experimental results were analyzed for statistical relevance with ANOVA, $p < 0.05$ was considered significant.

Additional file

> **Additional file 1: Figure S1.** Purified CD8+ T cells were treated with OT-I-streptamers for the indicated time periods. Samples were analyzed by Western blotting using the indicated Abs to determine the inhibition of mTOR (A) and AKT (B). Purified CD8+ T cells were treated with OT-I-streptamers for 24 h in presence or absence of aIL-2. Samples were analyzed for pSTAT5 activation to determine the function of aIL-2 antibody (C) Toxicity was assessed for the inhibitors rotenone and oxamate and the aIL-2 antibody by AnnexinV PI staining 24 h after stimulation (D). (TIF 189 kb)

Abbreviations

AKT/PKB, Proteinkinase B; AMPK, AMP kinase; ATP, adenosine triphosphate; CTLA4, Cytotoxic T-lymphocyte-associated Protein 4; GLUT1, Glucose transporter 1; IL-2, Interleukin 2; LDH, Lactate dehydrogenase; LKB1, Liverkinase B1; mTOR, Mammalian target of rapamycin; TCA cycle, Tricarboxylic acid cycle; TCR, T cell receptor

Acknowledgments

The authors would like to thank Dr. Tilo Beyer for critical reading of the manuscript, Katja Ehrecke and Anja Polanetzki for technical assistance, Dr. Ursula Bommhardt for providing AKT inhibitor and rapamycin, and Prof. Werner Hoffmann for helpful discussion. U.R., B.S., and J.A.L. are members of the Magdeburg Center for Systems Biology (MaCS) and B.S. and J.A.L. are also members of the SYBILLA consortium [European Union 7th Frame Program]. L.S., B.S., M.B-W and J.A.L. are member of the CRC854.

Funding

This work was supported in part by grants from the German Research Society (DFG) [JL 1031/1-3] and CRC 854 TP13 to JL, TP B14 to MBW, and TP B19 to LS/BS, the German Ministry of Education and Research (BMBF) FOR-SYS program [0313922], the State of Sachsen-Anhalt (Dynamic Systems) [XD3639HP/0306], and the European Union 7th Frame Program (SYBILLA) [HEALTH-F4-2008-201106].

Authors' contributions

CC designed and performed experiments, analyzed and interpreted results, and wrote the manuscript; A.R. and H.L. performed experiments and analyzed data; UR and MB-W provided vital equipment; LS designed experiments and analyzed results, BS supervised the work, interpreted results, and edited the manuscript; and JAL directed the study, designed experiments, analyzed and interpreted results, and wrote the manuscript. All authors have read and approved the manuscript.

Competing interests

The authors declare that they have no competing interests.

Author details

[1]Institute of Molecular and Clinical Immunology, Otto-von-Guericke-University, Magdeburg, Germany. [2]Max-Planck-Institute for Dynamics of Complex Technical Systems, Magdeburg, Germany. [3]Department of Experimental Pediatrics, Otto-von-Guericke-University, Magdeburg, Germany. [4]Department of Immune Control, Helmholtz Centre for Infection Research, Braunschweig, Germany. [5]Department of Nephrology and Hypertension, Diabetes and Endocrinology, Otto-von-Guericke University, Magdeburg, Germany.

References

1. Daniels MA, Teixeiro E, Gill J, Hausmann B, Roubaty D, Holmberg K, et al. Thymic selection threshold defined by compartmentalization of Ras/MAPK signalling. Nature. 2006;444:724–9.
2. Wang X, Simeoni L, Lindquist JA, Saez-Rodriguez J, Ambach A, Gilles ED, et al. Dynamics of proximal signaling events after TCR/CD8-mediated induction of proliferation or apoptosis in mature CD8+ T cells. J Immunol. 2008;180:6703–12.
3. Brand K, Williams JF, Weidemann MJ. Glucose and glutamine metabolism in rat thymocytes. Biochem. 1984;221:471–5.
4. Roos D, Loos JA. Changes in the carbohydrate metabolism of mitogenically stimulated human peripheral lymphocytes. II. Relative importance of glycolysis and oxidative phosphorylation on phytohaemagglutinin stimulation. Exp Cell Res. 1973;77:127–35.
5. Frauwirth KA, Riley JL, Harris MH, Parry RV, Rathmell JC, Plas DR, et al. The CD28 signaling pathway regulates glucose metabolism. Immunity. 2002;16:769–77.
6. Rathmell JC, Elstrom RL, Cinalli RM, Thompson CB. Activated Akt promotes increased resting T cell size, CD28-independent T cell growth, and development of autoimmunity and lymphoma. Eur J Immunol. 2002;33: 2223–32.
7. Jacobs SR, Herman CE, Maciver NJ, Wofford JA, Wieman HL, Hammen JJ, Rathmell JC. Glucose uptake is limiting in T cell activation and requires CD28-mediated Akt-dependent and independent pathways. J Immunol. 2008;180:4476–86.
8. Macintyre AN, Finlay D, Preston G, Sinclair LV, Waugh CM, Tamas P, et al. Protein kinase B controls transcriptional programs that direct cytotoxic T cell fate but is dispensable for T cell metabolism. Immunity. 2011;34:224–36.
9. Hardie DG, Salt IP, Hawley SA, Davies SP. AMP-activated protein kinase: an ultrasensitive system for monitoring cellular energy charge. Biochem J. 1999;338:717–22.
10. Woods A, Johnstone SR, Dickerson K, Leiper FC, Fryer LGD, Neumann D, et al. LKB1 is the upstream kinase in the AMP-activated protein kinase cascade. Curr Biol. 2003;13:2004–8.
11. Cao Y, Li H, Liu H, Zheng C, Ji H, Liu X. The serine/threonine kinase LKB1 controls thymocyte survival through regulation of AMPK activation and Bcl-XL expression. Cell Res. 2010;20:99–108.
12. Hurley RL, Anderson KA, Franzone JM, Kemp BE, Means AR, Witters LA. The Ca2+/calmodulin-dependent protein kinase kinases are AMP-activated protein kinase kinases. J Biol Chem. 2005;280:29060–6.
13. Tamás P, Hawley SA, Clarke RG, Mustard KJ, Green K, Hardie DG, Cantrell DA. Regulation of the energy sensor AMP-activated protein kinase by antigen receptor and Ca2+ in T lymphocytes. J Exp Med. 2006;203:1665–70.
14. Bauer DE, Harris MH, Plas DR, Lum JJ, Hammerman PS, Rathmell JC, et al. Cytokine stimulation of aerobic glycolysis in hematopoietic cells exceeds proliferative demand. FASEB J. 2004;18:1303–5.
15. Salmond RJ, Emery J, Okkenhaug K, Zamoyska R. MAPK, phosphatidylinositol 3-kinase, and mammalian target of rapamycin pathways converge at the level of ribosomal protein S6 phosphorylation to control metabolic signaling in CD8 T cells. J Immunol. 2009;183:7388 97.

16. Garland JM, Halestrap A. Energy metabolism during apoptosis. Bcl-2 promotes survival in hematopoietic cells induced to apoptose by growth factor withdrawal by stabilizing a form of metabolic arrest. J Biol Chem. 2004;272:4680–8.

17. Izyumov DS, Avetisyan AV, Pletjushkina OY, Sakharov DV, Wirtz KW, Chernyak BV, Skulachev VP. "Wages of fear": transient threefold decrease in intracellular ATP level imposes apoptosis. Biochim Biophys Acta. 2004;1658:141–7.

18. Tsujimoto Y. Apoptosis and necrosis: intracellular ATP level as a determinant for cell death modes. Cell Death Differ. 1997;4:429–34.

19. Yasuhara N, Eguchi Y, Tachibana T, Imamoto N, Yoneda Y, Tsujimoto Y. Essential role of active nuclear transport in apoptosis. Genes Cells. 1997;2:55–64.

20. Kass GE, Eriksson JE, Weis M, Orrenius S, Chow SC. Chromatin condensation during apoptosis requires ATP. Biochem J. 1996;318:749–52.

21. Leist M, Single B, Castoldi AF, Kühnle S, Nicotera P. Intracellular adenosine triphosphate (ATP) concentration: a switch in the decision between apoptosis and necrosis. J Exp Med. 1997;185:1481–6.

22. Ritter JB, Genzel Y, Reichl U. Simultaneous extraction of several metabolites of energy metabolism and related substances in mammalian cells: optimization using experimental design. Anal Biochem. 2008;373:349–69.

23. Cao Y, Rathmell J. C; Macintyre, A N. Metabolic reprogramming towards aerobic glycolysis correlates with greater proliferative ability and resistance to metabolic inhibition in CD8 versus CD4 T cells. PLoS One. 2014;9:e104104.

24. Sena LA, Chandel NS. Physiological roles of mitochondrial reactive oxygen species. Mol Cell. 2012;48:158–67.

25. Parry RV, Chemnitz JM, Frauwirth KA, Lanfranco AR, Braunstein I, Kobayashi SV, et al. CTLA-4 and PD-1 receptors inhibit T-cell activation by distinct mechanisms. Mol Cell Biol. 2005;25:9543–53.

26. Alegre ML, Noel PJ, Eisfelder BJ, Chuang E, Clark MR, Reiner SL, Thompson CB. Regulation of surface and intracellular expression of CTLA4 on mouse T cells. J Immunol. 1996;157:4762–70.

27. Wang XB, Zheng CY, Giscombe R, Lefvert AK. Regulation of surface and intracellular expression of CTLA-4 on human peripheral T cells. Scand J Immunol. 2001;54:453–8.

28. Carreno BM, Bennett F, Chau TA, Ling V, Luxenberg D, Jussif J, et al. TLA-4 (CD152) can inhibit T cell activation by two different mechanisms depending on its level of cell surface expression. J Immunol. 2000;165:1352–6.

29. Thaventhiran JE, Hoffmann A, Magiera L, de la Roche M, Lingel H, Brunner-Weinzierl M, Fearon DT. Activation of the Hippo pathway by CTLA-4 regulates the expression of Blimp-1 in the CD8+ T cell. Proc Natl Acad Sci U S A. 2012;109:E2223–9.

30. Patsoukis N, Bardhan K, Chatterjee P, Sari D, Liu B, Bell LN, Karoly ED, Freeman GJ, Petkova V, Seth P, Li L, Boussiotis VA. PD-1 alters T-cell metabolic reprogramming by inhibiting glycolysis and promoting lipolysis and fatty acid oxidation. Nat Commun 2015:6:6692.

31. Oestreich KJ, Read KA, Gilbertson SE, Hough KP, McDonald PW, Krishnamoorthy V, Weinmann AS. Bcl-6 directly represses the gene program of the glycolysis pathway. Nat Immunol. 2014;15:957–64.

32. Okoye I, Wang L, Pallmer K, Richter K, Ichimura T, Haas R, et al. T cell metabolism. The protein LEM promotes CD8+ T cell immunity through effects on mitochondrial respiration. Science. 2015;348:995–1001.

33. Gottlob K, Majewski N, Kennedy S, Kandel E, Robey RB, Hay N. Inhibition of early apoptotic events by Akt/PKB is dependent on the first committed step of glycolysis and mitochondrial hexokinase. Genes Dev. 2001;15:1406–18.

34. Ciofani M, Zúñiga-Pflücker JC. Notch promotes survival of pre-T cells at the beta-selection checkpoint by regulating cellular metabolism. Nat Immunol. 2005;6:881–8.

35. Michalek RD, Gerriets VA, Jacobs SR, Macintyre AN, Maciver NJ, Mason EF, et al. Cutting edge: distinct glycolytic and lipid oxidative metabolic programs are essential for effector and regulatory CD4+ T cell subsets. J Immunol. 2011;186:3299–303.

36. Shi LZ, Wang R, Huang G, Vogel P, Neale G, Green DR, Chi H. HIF1alpha-dependent glycolytic pathway orchestrates a metabolic checkpoint for the differentiation of TH17 and Treg cells. J Exp Med. 2011;208:1367–76.

37. Pearce EL, Walsh MC, Cejas PJ, Harms GM, Shen H, Wang L-S, et al. Enhancing CD8 T-cell memory by modulating fatty acid metabolism. Nature. 2009;460:103–7.

38. Dang EV, Barbi J, Yang H-Y, Jinasena D, Yu H, Zheng Y, et al. Control of T(H)17/T(reg) balance by hypoxia-inducible factor 1. Cell. 2011;146:772–84.

39. Ostroukhova M, Goplen N, Karim MZ, Michalec L, Guo L, Liang Q, Alam R. The role of low-level lactate production in airway inflammation in asthma. Am J Physiol Lung Cell Mol Physiol. 2012;302:L300–7.

40. Gatza E, Wahl DR, Opipari AW, Sundberg TB, Reddy P, Liu C, et al. Manipulating the bioenergetics of alloreactive T cells causes their selective apoptosis and arrests graft-versus-host disease. Sci Transl Med. 2011;3:67ra8.

41. Ritter JB, Wahl AS, Freund S, Genzel Y, Reichl U. Metabolic effects of influenza virus infection in cultured animal cells: Intra- and extracellular metabolite profiling. BMC Syst Biol. 2010;4:61.

2

High resolution imaging reveals heterogeneity in chromatin states between cells that is not inherited through cell division

David Dickerson[1], Marek Gierliński[1], Vijender Singh[1], Etsushi Kitamura[1], Graeme Ball[1], Tomoyuki U. Tanaka[1] and Tom Owen-Hughes[1,2]* (iD)

Abstract

Background: Genomes of eukaryotes exist as chromatin, and it is known that different chromatin states can influence gene regulation. Chromatin is not a static structure, but is known to be dynamic and vary between cells. In order to monitor the organisation of chromatin in live cells we have engineered fluorescent fusion proteins which recognize specific operator sequences to tag pairs of syntenic gene loci. The separation of these loci was then tracked in three dimensions over time using fluorescence microscopy.

Results: We established a work flow for measuring the distance between two fluorescently tagged, syntenic gene loci with a mean measurement error of 63 nm. In general, physical separation was observed to increase with increasing genomic separations. However, the extent to which chromatin is compressed varies for different genomic regions. No correlation was observed between compaction and the distribution of chromatin markers from genomic datasets or with contacts identified using capture based approaches. Variation in spatial separation was also observed within cells over time and between cells. Differences in the conformation of individual loci can persist for minutes in individual cells. Separation of reporter loci was found to be similar in related and unrelated daughter cell pairs.

Conclusions: The directly observed physical separation of reporter loci in live cells is highly dynamic both over time and from cell to cell. However, consistent differences in separation are observed over some chromosomal regions that do not correlate with factors known to influence chromatin states. We conclude that as yet unidentified parameters influence chromatin configuration. We also find that while heterogeneity in chromatin states can be maintained for minutes between cells, it is not inherited through cell division. This may contribute to cell-to-cell transcriptional heterogeneity.

Keywords: Chromatin structure, Fluorescence microscopy, Live cell imaging, Epigenetic inheritance

Abbreviations: ChIP, Chromatin immunoprecipitation; FOV, Field of view; K-S test, Kolmogorov-Smirnov test; LMS, Laser milled slide; MSDC, Mean square distance change; OMX, Optical microscope experimental; PSF, Point spread function; RI, Refractive index; TAD, Topologically associated domain

* Correspondence: t.a.owenhughes@dundee.ac.uk
[1]Centre for Gene Regulation and Expression, College of Life Sciences, University of Dundee, Dundee DD1 5EH, UK
[2]Wellcome Trust Building, University of Dundee, Dow Street, Dundee DD1 5EH, UK

Background

Chromatin is a DNA-protein complex which provides cells with a framework for important packaging and regulatory functions. Biochemical reconstitution has provided profound insight into the structure of the nucleosome [1], the basic unit of chromatin organisation. Biophysical studies have also revealed that arrays of nucleosomes spontaneously reorganise to form chromatin fibres with a diameter of approximately 30 nm under the appropriate ionic conditions [2–4]. Proposed nucleosome arrangements for such fibres include the 1-start solenoidal and 2-start supercoiled models, as well as combinations of the two and less ordered structures [5–7]. However, studies of native chromatin provide evidence for well organised 30 nm fibres in only a few specialised cases [2, 8, 9]. Growing evidence from close-to-native-state methodologies favours the existence of relatively disordered arrays of nucleosomes in both mitotic [10–13] and interphase chromosomes [8, 12–15].

On a larger scale, studies in a variety of organisms have indicated that chromosomes are arranged into chromosomal territories [16–19]. These territories have been characterised as associations of megabase-scale topologically associated domains (TADs), which are thought to result from complex physical interactions between various regions of genomes [20–26] and this concept has been supported by Chromosome Conformation Capture strategies such as Hi-C and 5C [27]. Hi-C based approaches provide important insights into chromosome organisation, but many are subject to complications arising from a reliance on cross-linking as well as the difficulty in generating temporal information regarding the chromosomal interactions.

Chromatin organisation and mobility has also been studied in vivo using fluorescent tagging of genomic loci and analysing the cells via microscopy [14, 28–36]. These approaches make feasible the measurement of native chromatin characteristics such as compaction ratio, flexibility, and diffusive behaviour. Previously, comparison of genomic and physical separation in fixed cells has shown that squared inter-probe distances are related to genomic separation [37, 38]. The extent of folding has been observed to vary in different regions of metazoan chromosomes and between cell types [39]. Changes in compaction have also been observed to occur during differentiation at some loci [40] but not at others [41].

A great deal of effort has gone into the development of polymer models to describe chromatin structure. Random Walk, or a Self-avoiding Walk models were initially used to describe non-looping chromatin fibres [42]. Most recently, the diffusive properties of fluorescently tagged loci have been observed to be consistent with a rouse-like polymer [14]. Fractal models explain some of the observed properties of chromatin with organisation that is self-similar at different scales [24, 43] however, this is not fully supported by Hi-C data [44]. More recently, models including looping and polymer melt geometries have gained prevalence [42, 45, 46]. Looping models account for data from sources including 3C technologies and fluorescence in situ hybridisation, which indicate a non-linear relationship between spatial and genomic distance [46–48]. Looping interactions have the potential to juxtapose important regulatory regions as appropriate over time. Polymer Melt models currently hold widespread support given that they chromatin is modelled as relatively disordered arrays of nucleosomes rather than folded fibres consistent with cryo-EM and small angle X-ray scattering of native chromatin [11, 49]. The Strings and Binders Switch model, which is largely based on 3C data is also attractive in that it accounts for looping while simultaneously predicting nucleosomal DNA to be the predominant fibre [50].

Improvements in optics, image acquisition electronics, and live imaging techniques, together with the ability to label specific loci using fluorescent fusion proteins [51] enable the study of the dynamic nature of chromatin organization in cells in three dimensions over time with greater precision than has been possible previously. In this study we introduce distinct fluorescent tags flanking a range of genomic regions and track the motion of the labelled reporters using an OMX Blaze microscope. We describe a work flow that enables 3D live cell tracking with a mean measurement error of 63 nm. We find that within individual yeast cells the separation of operator sequences exhibits substantial variation over time. Genomic loci are able to reorganize extensively below a threshold of approximately 70 kb. However above this, there is a transition to independent motion constrained by the nuclear environment. Within a clonal population of cells the mean conformations of reporter loci vary significantly and can persist over time frames of 1–10 min. By comparing chromatin states in related mother-daughter cell pairs we observe no evidence for inheritance of chromatin conformation.

Results

A system to measure chromatin compaction in live cells

In order to assay chromatin organization in vivo, we generated seven strains with fluorescently tagged chromosomal loci flanking various lengths of genomic DNA (Fig. 1a). The fluorescent repressor operator system (FROS) we adopted involves flanking different sides of reporter loci with arrays of 224 tet operators and 256 lac operators. These were then visualised though their interaction with mCherry TetR and GFP LacI (Fig. 1b). In order to mitigate the potential for arrays of repressor-

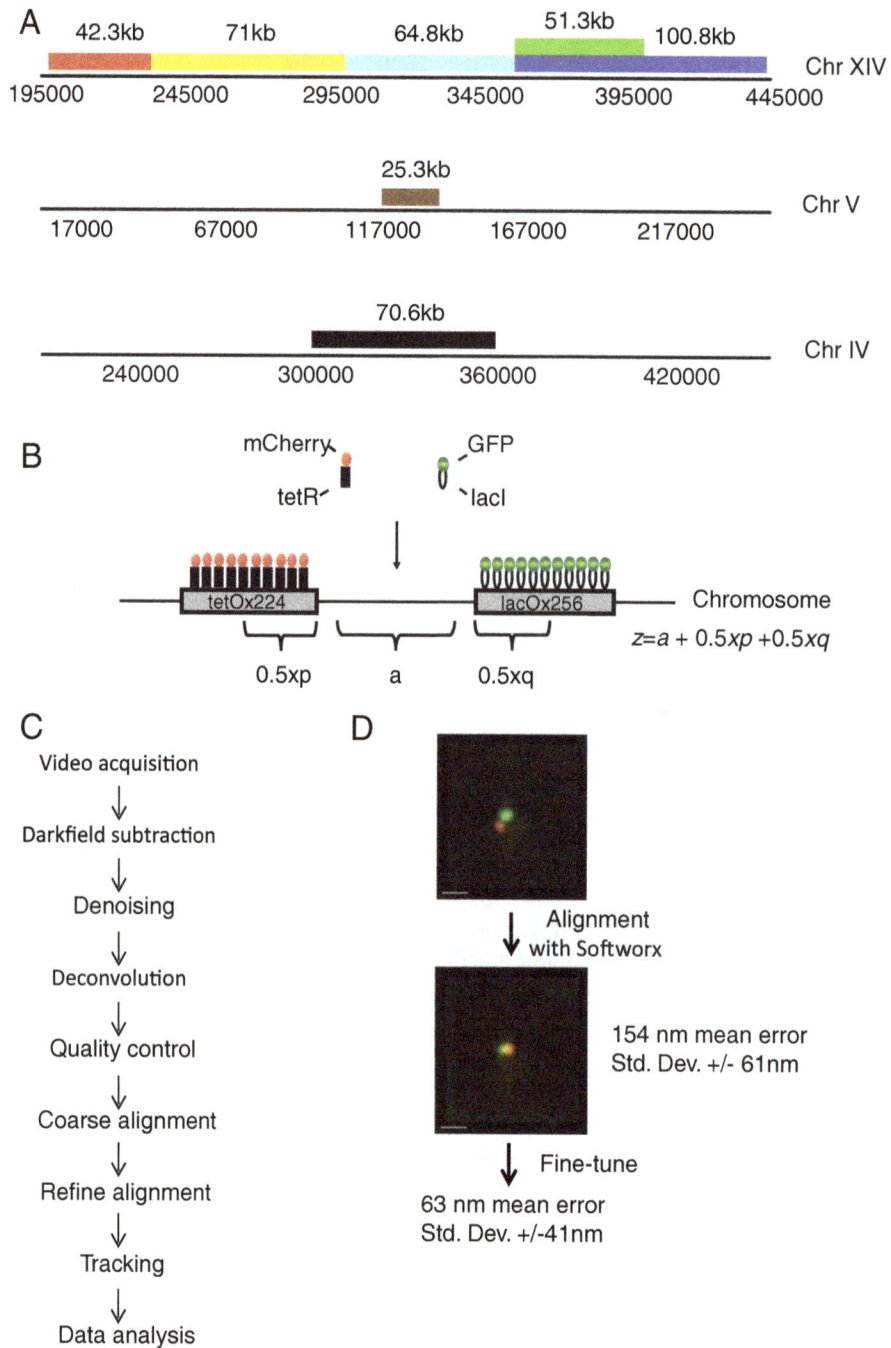

Fig. 1 Establishment of a work-flow for live cell 4D imaging. **a** Seven sample strains were generated which introduced lac ×256 and tet ×224 operator arrays flanking regions on Chrs XIV, IV and V. **b** Operator arrays were detected using fluorescent tagged repressor proteins as indicated. A naming convention was adopted that includes the endogenous genomic distance (**a**) as well as half the distance of each operator array (0.5x_p, 0.5x_q). Not shown: An additional strain was generated with a single tetO ×224 array on Chr XIV which expressed both tetR-GFP and tetR-mCherry to produce colocalising green and red spots. This strain was used for channel alignment and mean measurement error estimation purposes. **c** Summary of the work flow for image processing. **d** Two stage channel alignment was found to improve standard error from 154 nm to 63 nm

bound operators to generate heterochromatin low concentrations of tetracycline were included in media to reduce tetR binding and a lacI point mutant was used [52].

As the fluorescent intensity of foci is likely to be centred at the midpoint of the lac or tet operator arrays, we adopted the naming convention of describing the genomic separation (z) present in these strains as $z = a + 0.5x_p + 0.5x_q$ where a is the length of intragenic spacer and x_p and x_q are the lengths of the two flanking operators (Fig. 1b). Therefore, the strain with 60.6 kb of yeast genomic DNA flanked by lac and tet operator arrays was referred to as having 71 kb separation (60.6 kb genomic DNA + 10.4 kb of flanking operator array DNA), and so on. An additional strain was generated with colocalising green and red spots on Chr. XIV, for use as a control for channel alignment and measurement error.

Live cell 3-dimensional videos of these 8 strains were generated using the OMX microscope. The workflow used is summarised in Fig. 1c. Briefly, video acquisition was performed with CMOS cameras. To remove noise a dark field subtraction step was included as described in Materials and Methods. Subsequently, a second level of denoising was performed using ND-SAFIR using settings described by the Sedat Lab [53]. Deconvolution was performed using Softworx software. Quality control was performed as described in the Materials and Methods section. As the green and red channels are directed to different cameras on this system, channel alignment is critical to minimize translational, rotational, and scale errors. Initially we followed an established method which utilizes the imaging of multi-wavelength fluorescent beads to perform a coarse channel alignment. To improve the resolution that could be obtained in vivo we adopted a two-step channel alignment procedure. Firstly, coarse alignment was performed using beads or an etched slide and the Softworx alignment software. This was then refined using a strain in which tet operators are bound by both tetR-GFP and tetR-mCherry. The mean deviations of the centres of the red and green foci in three planes were used to generate a vector which was then applied to all red-channel frames. This reduced the mean measurement error from 110 nm to 63 nm (Fig. 1d). This reduction is likely due to the fact that the vector generated in the colocalising strain factors in differences in refractive indices between the objective lens and the subject being viewed (media and cells). A histogram of measurements from the coarse- and fine-tune-aligned colocalising strain is presented in Additional file 1: Figure S1. As the signal to noise ratios (SNRs) of the fluorescent foci of all two colour operator strains were similar to those of the colocalising strain, and as all satisfied identical quality control criteria, we consider it reasonable to assume that the mean measurement error (63 nm) is applicable to all the measurements described below.

Non-linear relationship between physical distance and genomic separation

Using the workflow described above it was possible to measure the distance between two fluorescently tagged

loci over time in several strains. Spot distance behaviours from all videos are presented in Additional file 1: Figure S2. Distance measurements for strains with varying genomic separations in G1 of the cell cycle are presented as histograms in Fig. 2a. When the distributions obtained from all strains are plotted in boxplot format, it is apparent that for the longer genomic separations there is a progressive but non-linear increase in the physical distance (Fig. 2b), similar to that previously reported [54].

The physical separation distance can be normalised for the genomic separation and expressed as compaction (Fig. 2c). This shows that the 42, 64 and 100 kb strains are more compact than the other strains (Fig. 2c). This suggests that locus specific effects may influence chromatin compaction in addition to the genomic separation. One potential explanation for the increased compaction of the 64 kb locus is that it participates in a more extensive network of looping interactions. Chromatin capture analysis has been used extensively to monitor looping interactions. As 4C and Micro-C data have been collected for the whole yeast genome under growth conditions comparable to those we have used, it is possible to investigate the frequency of interactions observed across the chromosomal regions we studied. 4C [55] and micro-C [56] interactions are plotted across a region of chromosome XIV (Fig. 3a–c, e). The highest density of 4C interactions falls within the 71 kb strain (Fig. 3c), which is not anomalously compact (Fig. 2b). The more compact 64 kb locus is not shown to have an increased density of interacting loci. Similarly, total contacts and boundaries detected by micro-C do not correlate with the compaction observed by imaging (Fig. 3b, e).

With rich data describing the distributions of many different chromatin features being available for budding yeast, we sought to determine whether any other factors correlate with the compaction observed at the loci we have studied. Chromatin immunoprecipitation (ChIP) enrichments for 18 different chromatin features including histone modifications, histone H3 occupancy, general transcription factors and RNA polymerase, were plotted across chromosomes XIV and IV (Additional file 1: Figure S3; S5). None of these factors correlate well with the higher compaction observed in the 42, 64 and 100 kb strains. The distributions of RNA pol II, Sir2, Histone H3 ChIP, and Histone H3K4 monomethylation are shown as examples (Fig. 3d–g).

Anisotropy is increased for large separation distances

During imaging, it was noticeable that the relative orientations of the tagged loci in a subset of cells were markedly constrained. To analyse this quantitatively we developed a test which assigns a statistic, D (see Materials and Methods), which quantifies the anisotropy of

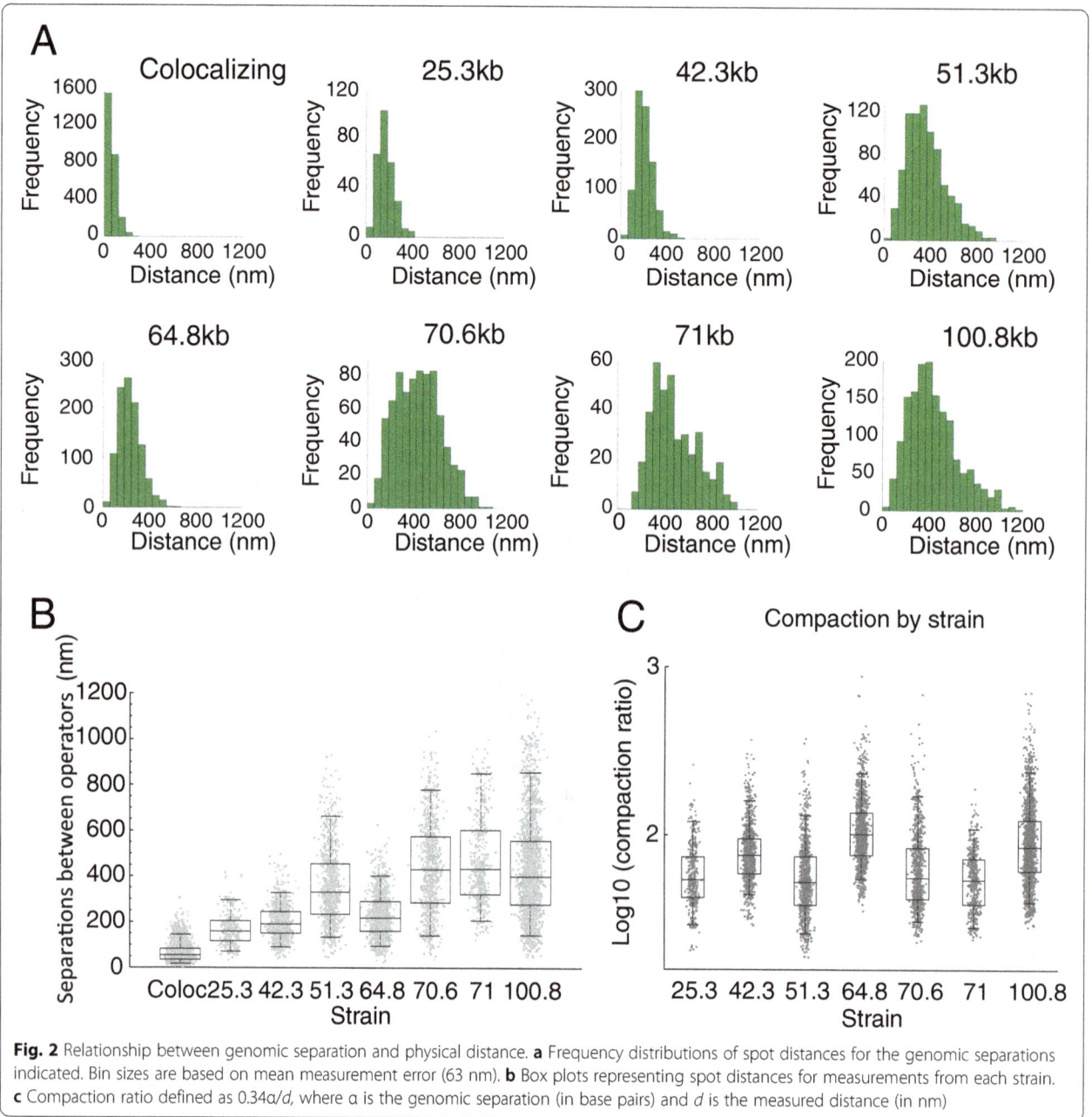

Fig. 2 Relationship between genomic separation and physical distance. **a** Frequency distributions of spot distances for the genomic separations indicated. Bin sizes are based on mean measurement error (63 nm). **b** Box plots representing spot distances for measurements from each strain. **c** Compaction ratio defined as 0.34a/d, where a is the genomic separation (in base pairs) and d is the measured distance (in nm)

the spatial orientation of the two-spot system. The higher D, the greater the degree of anisotropy. Lower D values were generally observed for genomic separations below 70 kb (Fig. 4b) and anisotropy is correlated with the mean distance (Fig. 4c). This could be an indication that for small physical distances that relative motion of foci is less constrained. As physical distance increases, the nuclear environment acts as a constraint restricting the extent of relative motion.

Analysis of independence of motion

If two loci are rigidly coupled, then their motion is anticipated to correlate over time. In contrast, if two loci are distant and elastically connected, their motion is expected to be independent. We estimate the spatial correlation of the two dots by the RV coefficient, a multivariate extension of the Pearson's correlation coefficient [57]. There is a weak trend for independence of motion to increase with increasing genomic distance (Fig. 4d). The exceptions to this trend are the relatively compact 64 and 42 kb genomic separations which adopt a more compact state (Fig. 2c). The increased spatial correlation at short separations could arise from a more rigid coupling across shorter intervening genomic separations, or as a result of concerted motion of chromatin within local territories within the nucleus.

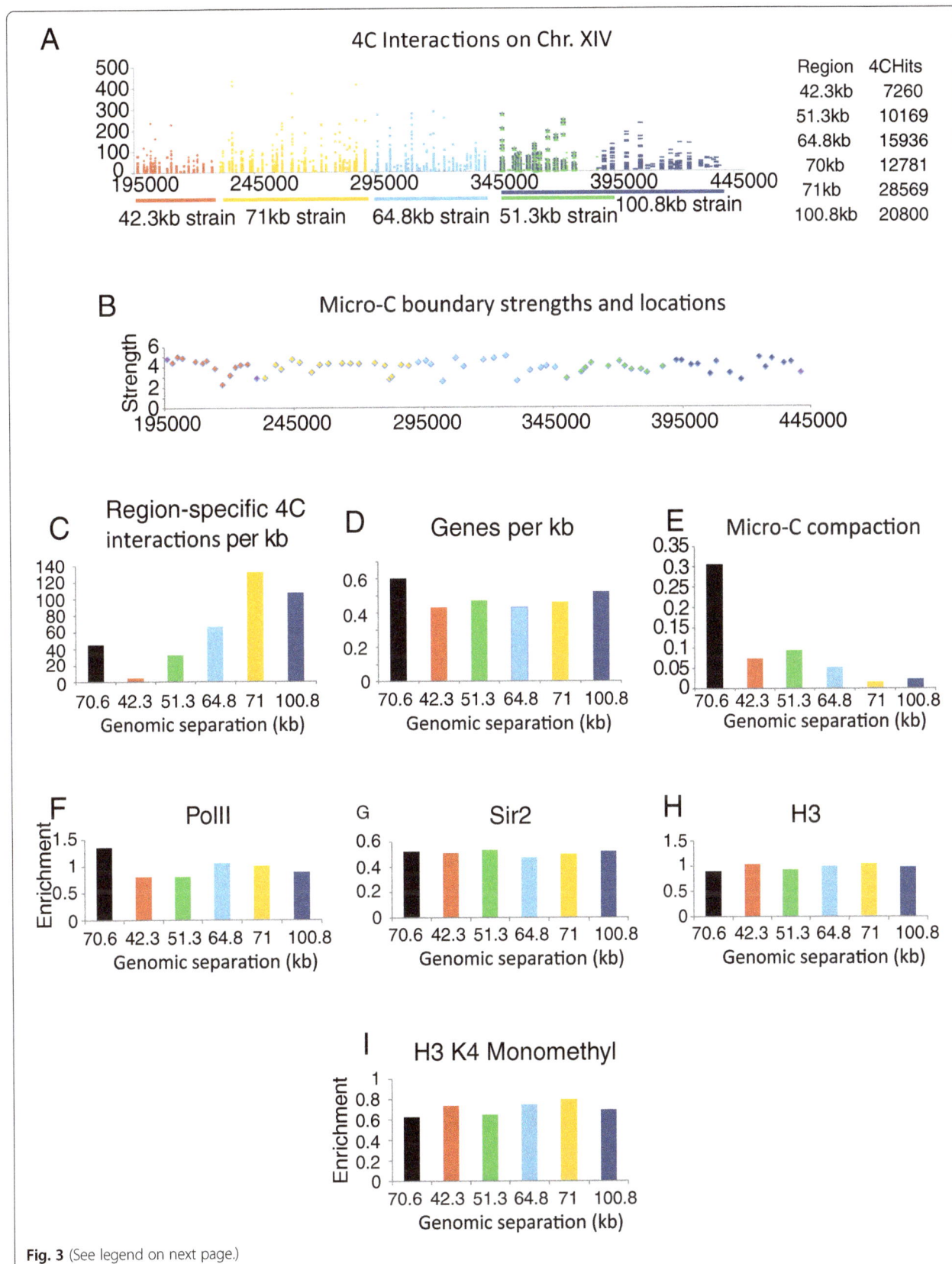

Fig. 3 (See legend on next page.)

(See figure on previous page.)
Fig. 3 Capture contacts and chromatin composition do not correlate with compaction. **a** Distribution of Hi-C contacts [95] across a region of chromosome XIV. The different genomic separations are colour coded along with the contacts measured across each region. The total number of contacts identified within each region is indicated to the right. **b** Strengths and locations of boundaries as determined by Micro-C [56]. The same colour coding is used to identify boundaries of a given strain. The number of region-specific 4C interactions per Kb (**c**) [95], number of genes per kb (**d**), compaction measured by micro-C (multiplied by −1) (**e**) [56], mean enrichment for PolII (**f**), mean enrichment for Sir2 (**g**) [96], mean enrichment for histone H3 (**h**) [96] and mean enrichment for histone H3 monomethylated at lysine 4 (**i**) [91] are plotted for each strain with the genomic separation between operators indicated. No factors were identified that correlate well with compaction. The distribution of additional factors across this region is plotted in Additional file 1: Figure S3)

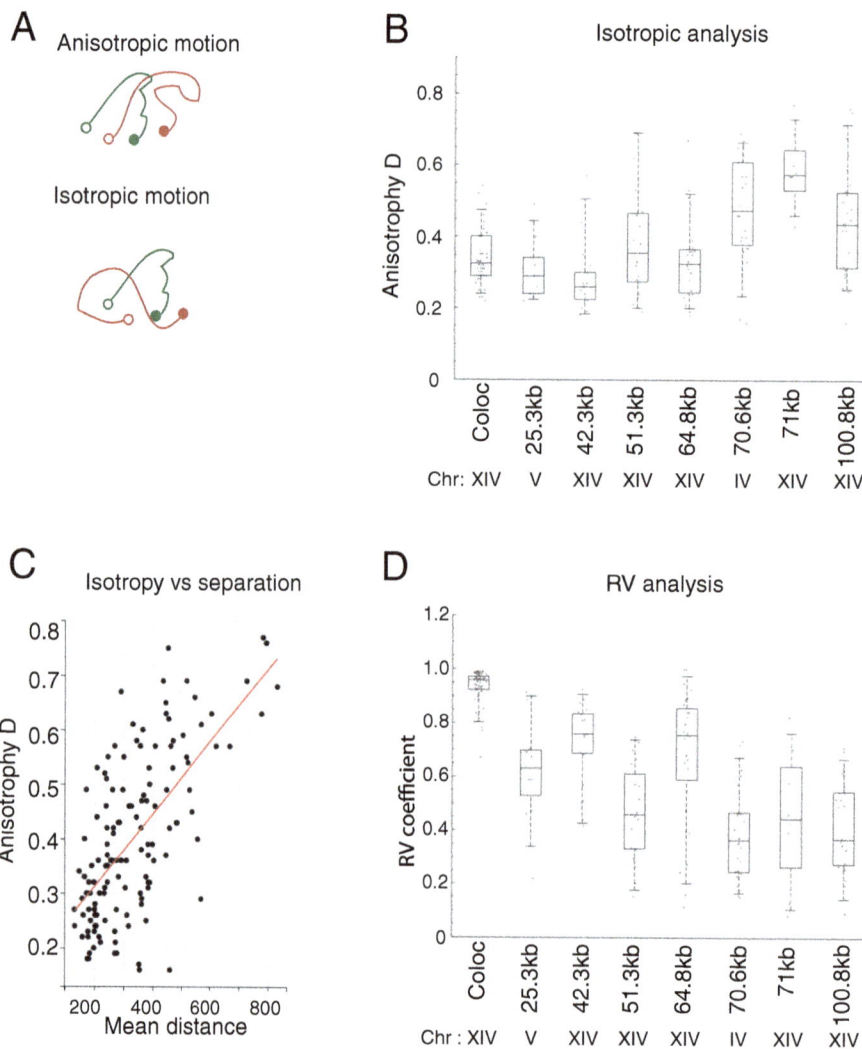

Fig. 4 Motion analysis for strains with different genomic separations. **a** When the relative orientation of two particles is maintained as they proceed along their trajectories (indicated as lines) their motion is considered isotropic (*bottom*), when orientation changes over time motion is anisotropic (*top*). A statistic D was calculated as a measure of anisotropy (Materials and Methods; Fig. 4). Larger D values correspond to greater anisotropy. D is observed to generally increase with genomic separation (**b**). In **c** the mean D is plotted against the mean distance ($R^2 = 0.43$). **d** To assess the correlation of the direction of motion of the two operators flanking a locus, comparisons of the vectors describing the motion of each operator were made at each time point. An RV coefficient was calculated for this purpose. The general trend observed was a decrease of motion correlation as the genomic separation between the operators increased. However, for the 64 kb and 42 kb strains which adopt a more compact configuration, the trend was not observed

Heterogeneity in chromatin configuration between single cells persists for minutes

As many videos, each of which consisting of up to 100 time points, were acquired during the characterisation of each strain, it was possible to compare fluctuations in the distance between operator sequences observed within and between individual cells within a given strain. In all cases cell-to-cell variation was observed, and the scale with which the means varied ranged from below 2-fold to over 4-fold for the longer genomic separations (Fig. 5a–f). From visual inspection of individual distance versus time traces it is clear that in some cases the separation distance remains relatively constant but distinct between cells (Additional file 1: Figure S2.2 fff). In other cases, a single transition is observed during the course of a movie (Additional file 1: Figure S2.2 uu), while in other cells a

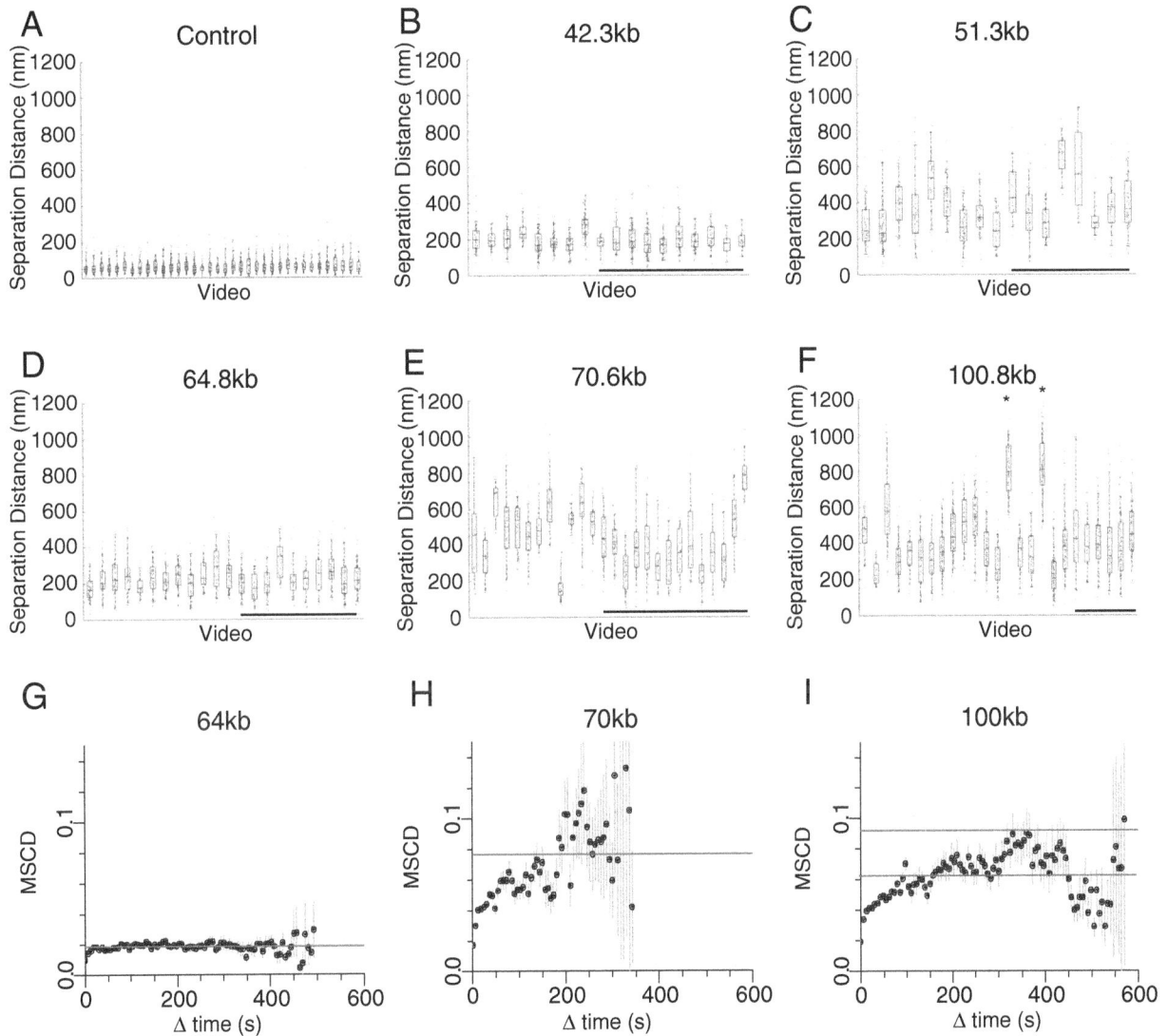

Fig. 5 Variations in the motion of reporter loci in single cells. **a–f** Box plots of distances for individual cells. Non-underlined boxes indicate data collected from videos of approximately one minute in length during which Z stacks were collected approximately once per second. Underlined boxes were collected over 10 min with each Z stack collected at 6 s intervals. Variation in median distance between cells is observed, especially for the 51, 70 and 100 kb separations, in some cases over durations of up to 10 min. The the mean square change in distance (MSCD) as a function of the time interval was calculated for the 64, 70 and 100 kb strains where distance was sufficiently distinct from the 63 nm measurement error (**g-i**). To extend the range of time intervals that could be used, data from videos in which distance was measured at short intervals and at longer intervals were superposed and binned in 6 s time intervals. Error bars represent standard error. In each case a trend is observed for MSCD to increase reaching a plateau after 30–150 s. The red line shows the MSCD calculated by bootstrapping across random time points from random videos for a given strain (a cell-independent and time-independent estimate). The green line (in panel **i**) shows the same quantity found excluding data from two videos with unusually large distances (marked with asterisks in panel **f**). These observations suggest that over longer time intervals, the MSCD observed within cells approaches that observed between cells

series of rapid fluctuations in distance is observed over short time intervals (Additional file 1: Figure S2.2 ii). Differences in mean distance were observed over both 1 min. and 10 min. (Fig. 5a–f). To assess the time scale over which distance varies, the mean square change in distance (MSCD) was plotted for different time intervals. This can be interpreted as one-dimensional mean square displacement, where changes in distance between the spots is squared rather than simply changes in position squared are plotted (Fig. 5g–l). These curves indicate that MSCD typically increases over time intervals up to 30 s. However, beyond 150 s, a plateau is reached at which little additional change in separation distance is observed. The magnitude of this plateau for MSCD varies considerably between strains with the value in the 64 kb strain being 3–4 fold less than observed in the 70 and 100 kb strains.

If the observed changes in distance occur as a result of vibrational motion that reaches a maximum by 150 s, then we would expect the distance after long time intervals within one cell to be comparable with the variation observed between cells. To test this, time independent distance changes between randomly selected time points in different cells were calculated by bootstrapping.

This time-independent and cell-independent MSCD is plotted as a red line in Fig. 5g–l. For the 64 and 70 kb strains the MSCD measured between cells is similar to the maximum MSCD observed within cells after time intervals of greater than 150 s. However, this is not the case for the 100 kb strain, as the between-cell MSCD is greater than that measured within cells (Fig. 5l). This is likely due to the unusually large spot distance in two cells in this strain (marked with asterisks in Fig. 5f). When the time-independent measurement between cells is recalculated omitting these two cells, the mean is a better fit to the value observed at time intervals of greater than 150 s within cells (green line; Fig. 5g). Large differences in mean distance were also observed in the 70 kb strain, even in videos acquired over 10 min (Fig. 5e). However, in this case these differences in separation are evenly distributed around the mean. The observation that the mean separation distance between the same loci can be distinct in specific cells for several minutes raises the possibility that there may be mechanisms acting to maintain different conformations in a subset of cells.

Differences in chromatin conformation are not inherited through cell divisions

Within cultures of Saccharomyces cerevisiae, following cell division, mother and daughter cells remain associated and can readily be identified as larger mother cells associated with a smaller daughter cell (Fig. 6a). The ease with which related siblings can be identified cytologically provides an opportunity to investigate whether chromatin state is inherited through cell divisions.

Representative histograms of paired mother and daughter cells are shown for the 70.6 kb strain in Fig. 6b. From this it is clear that chromosome conformation varies in several mother daughter pairs indicating that in these cases separation distance was not conserved through cell division. The mean spot distances observed in mother and daughter cells are similar for each strain (Fig. 5c). When changes in compaction are measured for paired mother daughter cells and compared to that in randomly selected unrelated mother daughter pairs across all strains, the means and standard distributions are similar (Fig. 6d). In addition the mean distance changes observed in comparisons between unrelated mother and daughter cells are similar (Fig. 6d), indicating that the reduced volume of the nuclei of daughter cells does not affect the variation in distance.

Discussion

Recent developments in fluorescent microscopy provide new opportunities to study the organisation of native chromatin in live cells. Here we describe a workflow that enables live cell 3-D two-channel measurements to be made with a mean measurement accuracy of 63 nm. A key step in this workflow was the adoption of a channel alignment control that takes into consideration the optical properties of the experimental sample. Although our application of this approach was based in yeast cells, similar two colour alignment using two versions of a fluorescent reporter is applicable to many cell based systems. It is likely that improvements in several aspects of the work flow will further reduce measurement error. As the greatest error component is along the axial dimension, approaches such as confocal microscopy, multiphoton microscopy, or optical astigmatism would all likely decrease mean measurement error.

We use this system to study the configuration of chromatin across a series of chromosomal loci with distinct genomic separations. In all cases the physical distance between the fluorescent reporters is observed to fluctuate over time. In addition, a trend is observed for distance to increase at larger genomic intervals. The scale of the measurements is comparable to data obtained previously using FISH, 2D [54] and more recently 3D data [34]. Information of this type can be used to test different physical models describing chromatin fibres.

At longer distances above roughly 300 nm (corresponding to a genomic separation of approximately 70 kb), we observe that the two reporter loci have a higher tendency to remain in the same relative orientation with respect to one another. Structure within the nucleus may restrict diffusive motion on this scale. At the same time there is greater correlation of motion of

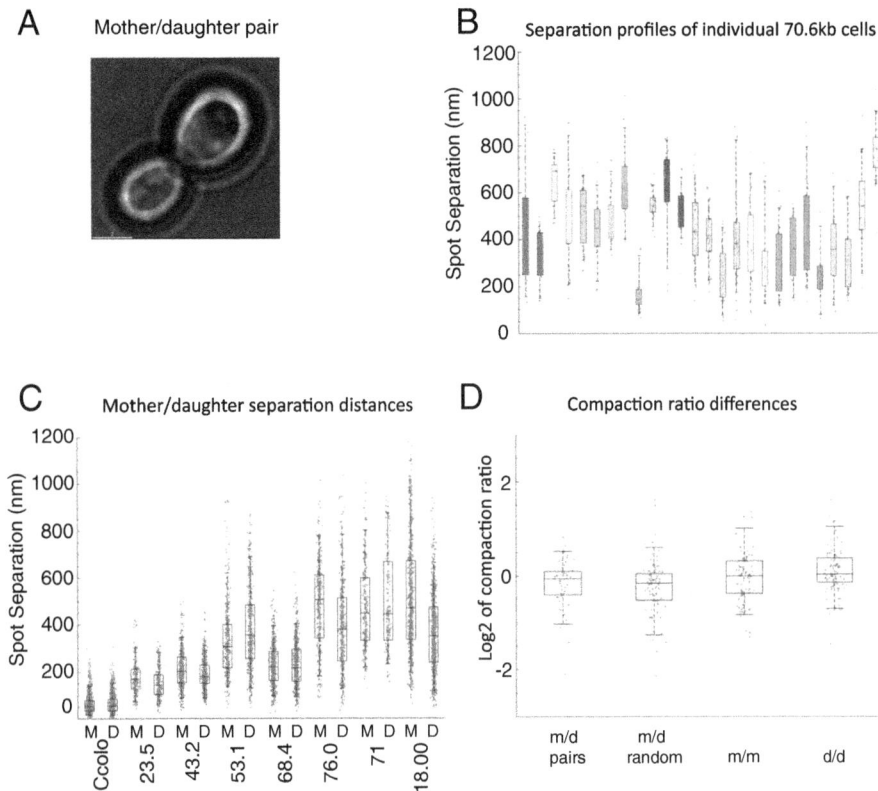

Fig. 6 Chromatin conformation is not inherited. Mother and daughter cells can readily be identified in *Saccharomyces cerevisiae* cultures as physically attached pairs (**a**). Each mother-daughter pair shares the same parent providing an opportunity to assess whether chromatin configuration is conserved through cell division. Movies obtained in individual cells from the 70 kb strain are plotted with mother-daughter pairs in the same colour in (**b**), in each case with data from the mother to the left. Unpaired cells are in grey. In some cases it is clear that mothers and daughters have differing spot distances. The distance box plots for all mothers and daughters of each strain are plotted in (**c**). In order to compare changes in distance between a large sample of related and unrelated mother and daughter cell pairs, changes in compaction were calculated using movies from each genomic separation (**d**). Compaction is calculated by dividing genomic separation by the measured distance between operators. The change in compaction between different populations was then calculated as a ratio indicated on the y-axis as 'Log2 of compaction ratio'. This comparison could be made between related m/d pairs and unrelated m/d pairs selected at random. In both cases, the distributions of the differences in compaction are similar and not statistically significant (Mann-Whitney p-value = 0.181). This indicates that the separation of the reporter loci studied here is not inherited. Little change in separation is observed when the comparison is made between unrelated mother/mother and daughter/daughter pairs. This suggests that the difference in the volume of mother and daughter cell nuclei has little effect on the separation of these operator tagged loci

loci at closer proximity than ~300 nm. This could arise as a result a physical coupling between the two operators. It is also possible that the local chromatin environment extending to 300 nm, a form of sub chromosomal territory or domain, undergoes some localised flow-like motion that accounts for the correlated motion. On scales greater than 300 nm the density of intervening chromatin may act to exclude free diffusion and restrict isotropy. At these larger distances, the motion of the two operators becomes independent as indicated by the reduced RV coefficient. The differences in motion observed over different ranges may relate to previously reported effects of nuclear exclusion on chromosome organisation [58].

The measured separation distance did not vary consistently with genomic separation across all the strains studied. The 42, 64 and 100 kb strains show higher levels of compaction than the other strains (Fig. 2c). When the diffusive motion of the lac and tet operators flanking these loci is studied independently, the mean square displacement curves are similar (Additional file 1: Figure S4). This suggests that the differences in compaction do not result from constraints to the motion of either of the reporter sequences. Another potential explanation for the reduced separation distances observed in the 42, 64 and 100 kb strains could be that chromatin is arranged in a more compact state over these genomic loci. To investigate this we took advantage of the large number of previously published genomic datasets available in budding yeast and searched for factors that correlated with chromatin compaction across the different loci we have studied. Amongst the

18 factors selected (including histone occupancy, post-translational modifications and measures of transcriptional activity) none show a strong correlation with compaction (Fig. 3; Additional file 1: Figure S3). In addition, the loci within these strains are not adjacent to loci such as centromeres, telomeres or the rDNA locus that have previously been observed to influence subnuclear motion [38].

Looping interactions that affect the separation of reporter loci provide an attractive explanation for the variations in the observed separation distance both between strains and over time. Hi-C approaches are widely used to detect looping interactions. Like microscopy-based approaches they detect heterogeneity in chromatin conformation between cells [59] and can be used to model chromosome architecture [55]. However, the distributions of cross-links obtained by Hi-C, and high resolution micro-C, do not correlate with the higher compaction observed in the 42, 64 and 100 kb strains (Fig. 3). This is perhaps not surprising as although these loci are relatively compact, the distance between operators is typically 200 nm which is beyond the range likely to enhance DNA ligation, the readout for capture based approaches. Chromosomal loci can potentially be constrained through interactions with any relatively immobile object within the nucleus. Such interactions may bring heterologous DNA sequences into closer proximity but still out with the range required to enhance ligation. Factors such as these are likely to contribute to the previously noted discord between Hi-C and imaging based chromatin measurements [60]. Where chromatin is especially well ordered there is a greater chance that imaging and Hi-C approaches converge. An example of this is provided by the mating type loci on budding yeast chromosome III [61].

Changes in the association of loci with relatively immobile bodies within the nucleus that affect the distance between reporter loci provide an attractive means of accounting for some of the heterogeneity we have observed. Such interactions could be stable over differing time scales. Transient interactions could account for the rapid variation in distance observed at some loci (Additional file 1: Figure S2). Where interactions are more stable they could contribute to variation in the mean distance observed in different cells. A diverse range of factors could act to influence the localisation of a given chromosomal locus and this could explain why no obvious correlation between genomic features and separation distance was identified.

The use of Saccharomyces cerevisiae makes it relatively easy to identify pairs of cells that share a common mother. Comparing the conformation of chromatin between related and unrelated mother daughter pairs, chromatin conformation was not observed to be conserved in related cells. However, it is possible that this will not be the case for all loci. In budding yeast it is well established that expression of genes at the mating type loci [62] and within subtelomeric regions [63] can be inherited. Furthermore the nuclear localisation of these regions is distinct [61, 64]. The chromosomal regions we studied do not include these regions which may be exceptions within the context of Saccharomyces cerevisiae genes. Higher eukaryotes possess additional chromatin features such as HP1 proteins and polycomb that are more likely to influence both inheritance and nuclear localisation [65, 66]. Consistent with this, inducible decompaction of reporter loci in mouse embryonic stem cells has been observed to be sufficient to cause a change in subnuclear localisation that persists through cell divisions [67].

Although we do not observe any evidence for the inheritance of chromatin configuration through cell division, we do observe individual cells that have distinct chromatin configurations that persist for times periods of up to 10 min. This suggests that alternate chromatin configurations can be maintained in individual cells. It's possible that this heterogeneity may affect the ability of cells to respond to environmental stimuli. There is good evidence indicating that the subnuclear localisation of genes can play an important role in their regulation. Tethering genes to the nuclear periphery is known to favour establishment of silent heterochromatin [68, 69], many genes have been observed to transiently associate with the nuclear pore during activation [70]. In mammalian cells changes in the localisation of genes has been observed to correlate with changes in transcription [71–73]. If the conformation of loci has a similar influence on gene regulation, then it could contribute to the heterogeneity in transcriptional responses that have been observed in single yeast cells [74–76]. This heterogeneity potentially provides an advantage for individual cells in being able to respond rapidly to an environmental change. However, unlike the changes occurring during the development of multicellular organisms there is not a need for such changes to be inherited. Instead, if a cell is well placed to adapt to a new environment it may be best to restore heterogeneity in subsequent generations providing capability to respond rapidly to a diverse range of future challenges. As we do not observe evidence for the inheritance of chromatin configuration at this level, it is possible that the processes of DNA replication and cell division provide an opportunity to reset the configuration to the spectrum of states observed within the population. In this way non inherited heterogeneity may serve an important biological function. In multicellular organisms there is a need both for the flexibility to respond to environmental change and the precision required for tissue development. This may involve a different balance between inherited and non-inherited states.

Conclusions

A workflow has been established to study the separation of fluorescently tagged reporter loci in live cells. The mean separation of reporter loci was observed to increase with increasing intervening genomic sequence. However, this increase is non-linear indicating that different regions of the genome are in different configurations. No genomic features were identified that correlate with the observed separation of loci suggesting that as yet uncharacterised factors influence chromosome organisation. Separation is observed to vary within cells over time and between cells. This heterogeneity may contribute to heterogeneity in the transcriptional response at the level of single cells. Distinct chromatin configurations were however, not observed to be inherited through cell division.

Methods

Plasmids and strains

The plasmids used in this study are summarized in Additional file 2: Table S1. Plasmids pT1196 [77], pFA6a-mCherry-natMX6 [78], pAT253corrected [52], pLAU43 and pLAU44 [79], pRS416 [80], pAFS59 [32], pAFS135 [81], pRS306tetO224 [82], pYiplac204-Gal1Pro-MDN1 [83], pGVH30 [54], pFA6KanMX6 [84] and pAG25 and pAG32 [85] have been described previously. pDD2244 (tetR-GFP-tetR-mCherry::ADE2) was generated in 4 cloning steps from pAT253corrected (Taddei), pT1196, and pKS391. pDD2245 (GFP-lacI**-TetR-mCherry::ADE2) was generated in 4 cloning steps from pKS391, pGVH30, pAG32, and pAT253corrected. pDD2246 (tetO-UBP10::TRP1) was generated from pLAU44, an NdeI-UPB10 integration site-HindIII PCR product, and an AatII-TRP1-NdeI PCR product fragment in 2 cloning steps. pDD202 was generated in 5 cloning steps from pRS306tetO224, pRS416 and 3 PCR products. pDD206 was generated in 4 cloning steps from pYCG_YLR106c, pRS416, and pFA6KanMX6. pDD207 was generated in 4 cloning steps from pYiplac204-Gal1Pro-MDN1, pRS416, and pAFS59. The lac operator array plasmids pDD249, pDD251, pDD253, and pDD254 used to generate yeast strains DD1471-1475 were constructed by cloning the appropriate genomic integration target sequence from Chr XIV adjacent to the lac operator array in a pLAU43 lacOx240 clone which had previously been modified with a SalI-URA3-SalI PCR product fragment. The tet operator array plasmids pDD2246, pDD250, pDD252, pDD255, pDD256 and p2577 used to generate strains DD1471-1475 were constructed by cloning a AatII-TRP1-NdeI PCR product fragment into the pLAU44 tetOx240 plasmid and then cloning the appropriate genomic integration

target sequence from Chr XIV adjacent to the tet operator array. All plasmid sequences are available upon request. All plasmids were verified by multiple restriction digests as well as sequencing of crucial regions.

The *S. cerevisiae* strains in this study are summarized in Additional file 2: Table S2 and illustrated in Fig. 1a. The tet and lac operator arrays for all Chr XIV strains were integrated between convergent genes. The terminators of these genes were duplicated and flank the insertion sites such that all genes retain their wild type terminators. Additional file 2: Table S2 columns 5' and 3' indicate the pairs of convergent genes where the insertions took place. DD1407 generation: WT yeast strain K699 (W303) was transformed first with pDD2244 (tetR-GFP-TetR-mCherry::ADE2) linearized with PciI, and then with pDD2246 linearized with PfoI. DD1413 was generated by successively cloning in linearized pAFS135, pDD2248, pDD202, pDD206, and pDD207. It was verified by PCR using primer pairs 1988 + 1952, 2061 + 2062, 2044 + 2529, and 2051 + 2058, which flank the appropriate integration sites at the URA3 locus. Strains DD1471-1475 were generated using plasmids pDD2244 and pDD2245, and the appropriate lacO and tetO array plasmids pDD246, 2247, 2248, and pDD249-256, and were verified by PCR using primers 2452–2468. Strain DD1336 was generated by cloning linearized pDD2247 into T6002 and was verified by PCR using primers 2067–2074.

Cell culture

Tetracycline was added to all cell cultures to diminish the affinity of the tet repressor DNA binding protein for its DNA binding site, as per Dubarry [52]. Optimal concentration resulting in 94 % maximum fluorescence intensity was determined via concentration series and measured on the OMX. The colocalising strain and most sample strains were streaked to YPAD and cultured overnight, propagated in liquid culture for 8 h, cultured overnight in 75%SC/25%YPA + 2%dextrose + 20 ng/ml tetracycline, washed in the culture media and placed on ice. In all cases cells were adhered to concanavalin A-treated 35 mm glass-bottomed MatTek culture dishes for 10 min at 22.0 °C, then allowed to temperature-equilibrate in the microscope enclosure at 23.5 °C for an additional 10 min.

Image acquisition

All imaging was performed with a GE|OMX Blaze® microscope. Immersion oil with refractive index 1.514 was used in all cases. Typical video acquisition included 5 μm stack height, 250 nm step size, 21 images per stack, 128 × 128 field of view (FOV), 50–100 time points, excitation 3–8 msec, ND 31–100 for mCherry and ND 5–10 for GFP, and sequential channel acquisition. The microscope enclosure is maintained at 23.5° Celsius.

Videos and associated tracking data are available at url: http://dx.doi.org/10.17867/10000102.

Channel alignment protocol and image processing

Channel alignment parameters for an initial coarse alignment were generated using single stacks of twenty-nine 1024x1024 FOV images with step size of 125 nm of a Tetraspeck 100 nm bead slide or laser milled slide (LMS) using the red and green channel cameras. Coarse alignment lateral parameters were calculated using Softworx software, which included a translation, magnification, and rotation. Coarse axial alignment offset was determined manually using the Softworx Measure Chromatic Correction function. Channel alignment fine tuning parameters were calculated using live cell 3-D tracking data from the colocalising strain, DD1407, which has single red and green colocalising spots of dimensions assumed to be smaller than the PSF. 50–100 time points of 17 128×128 FOV images/stack were generated with 3-8 msec exposures generated sequentially from red and green channels. Videos were split into individual channels and saved in 4-byte float format. Dark-field images, generated previously by taking the mean pixel intensity of 1000 images at set exposure times, were then subtracted from the individual video channels to correct for noise arising from the CMOS cameras. Videos were denoised with the ND-SAFIR denoising software [86] using the Sedat Lab settings iter = 5, p = 3, sampling = 2, noise = Gaussian, adapt = 0, island = 4, and np = 8 [53], and then deconvolved via Softworx using a ratio (conservative) method tailored to an idealized objective of the model used in our OMX Blaze, and saved in 4-byte float format. Individual channel files were fused into a single red plus green video file and coarse alignment was performed with Softworx using the bead slide or LMS alignment offset parameters. The spots were tracked using Imaris, the x, y, and z offsets between the different channels were determined for each time point using Excel spreadsheet, and the means of these offsets were calculated from data from multiple videos. Fine-tuning alignment was performed in Excel by subtracting these mean offsets from the red channel x, y, and z spot coordinates, which resulted in a final translation. After the fine tuning alignment had been performed on the colocalising strain videos, the mean distance between the red and green spots for all time points from all videos was determined via Pythagorean Theorem to be 63 nm, with a standard deviation of 37 nm. The tracking data is available at https://idr-demo.openmicroscopy.org/webclient/annotation/1645869.

Quality control

Plotting z coordinate versus error indicated that once tagged loci diffused to within 1 µm of the top or bottom of a stack, error increased (not shown), and for this reason these data points were eliminated from both control and sample data sets. Plotting maximum spot intensity or contrast versus error, followed by LOESS smoothing, revealed correlations which allowed for elimination of data points with high error (not shown). Based on this GFP and mCherry contrast thresholds were both set at 12, while minimum intensity threshold for mCherry and GFP was set at 25 and 16 respectively. Use of these thresholding values removed images where the positions of foci were not sufficiently well defined to obtain high resolution locations. Using these threshold values, we were able to generate live cell 3-D 2-channel videos with 250 nm step size, 5 µm stack size, 21 images per stack, and up to 100 time points. As experimental video data were subjected to the same thresholding protocol, it is assumed that the resulting error was also 63 nm.

Relative orientation anisotropy

We assessed the degree of anisotropy of the spatial orientation of each pair of marked loci by mapping them to UV-space. The Cartesian co-ordinates $r_i = (x_i, y_i, z_i)$ of the distance vector between the ends of a locus were mapped into a unit square,

$$u_i = \frac{1}{2\pi} \tan^{-1} \frac{y_i}{x_i}, \qquad v_i = \frac{|r_i| - z_i}{2|r_i|}. \qquad (1)$$

If the original vectors r_i are isotropic (uniformly distributed on a sphere $\mathbb{S}^2$), then the transformed coordinates (u_i, v_i) are uniformly distributed over $[0, 1]^2$. We applied a two-dimensional Kolmogorov-Smirnov test [87, 88] to all (u_i, v_i) across each video. The test statistic, D, measures the degree of anisotropy in the r_i distribution.

RV analysis

RV coefficient analysis is similar to Pearson's correlation coefficient analysis, but is a multivariate generalization rather than bivariate [57]. When used to compare the positions of two genomic loci in 3D over time, the RV coefficient indicates relative independence of motion of the two loci. Larger magnitudes of the coefficient correspond to a greater tendency for the two loci to track together.

Mean square displacement analysis

Mean square displacement analysis was performed for individual fluorescent loci using accepted methods [89].

The OMX Blaze microscope

The GE|OMX Blaze® microscope is fitted with an Olympus UPlanSApo 60× 1.42NA oil objective, a Piezo stage, a BGR standard filter set (DAPI 436/31, FITC 528/48,

A568 609/37, cy5 683/40), conventional widefield solid-state 461–189 nm and 563–588 nm LED solid state light sources, 3 back-illuminated 15 bit scientific CMOS cameras (PCO AG, Germany) with 1024 × 1024 chip size, and a temperature control chamber set to 23.5 C. The instrument is controlled by proprietary GE software.

Additional files

Additional file 1: Figure S1. Two step channel alignment improves resolution. Channel alignment has traditionally been performed using multispectral beads of dimensions smaller than the PSF. Stacks of 100 nm Tetraspeck bead images in separate channels were analysed using Softworx alignment software to calculate rotational, translational, tilt, and magnification offsets. When colocalising strain videos were aligned following this protocol the mean measured distance between the tagged loci was calculated to be 110 nm. Including the fine-tuning step reduced the mean error to 63 nm. Error bars are standard deviation. **Figure S2.** Distance verses time plots. For individual videos acquired in this study, spot separation distance in um is plotted against time in seconds. Graphs are grouped by strain as indicated. Each video is assigned an alphabetical identifier which corresponds. **Figure S3.** Chromatin composition at reporter loci. High-resolution ChIP profiling enrichments for the chromatin constituents indicated across the loci studied. Rpb1, PolII Ser2P, PolII Ser5P, Pcf11, Spt4, Spt5, Spt6, Spt16, and TFIIB data were generated by [90]. Histone H3 K36 monomethylation, K36 dimethylation, and K79 trimethylation data were generated by [91]. Histone H2B K123 ubiquitylation data were generated by [92]. Ino80 data were generated by [93]. **Figure S4.** MSD curves for operators flanking each reporter locus. Mean square displacement is plotted for the fluorescently tagged operator sequences flanking each locus used. Data for the GFP tagged lacI are shown in green and mCherry tetR in red. Data from movies taken over different time scales is combined. The grey line indicates the profile anticipated for anomalous diffusion, $MSD(\Delta t) \propto \Delta t^{1/2}$ [89]. The green and red lines show the best-fitting lines of Brownian diffusion, $MSD(\Delta t) \propto \Delta t$. **Figure S5.** PolII enrichment at tetO and lacO array integration sites on Chr XIV. Enrichment of PolII subunit Rpb3 along the relevant loci is shown in green [94]. Integration sites of arrays of bacterial repressor binding sites are indicated by down carats ('V'). Locations of open reading frames, as well as their position on Watson or Crick strands, are indicated by blue and purple arrows. (A) All lacO and tetO arrays on Chr XIV were integrated between convergent genes and the terminators of the convergent genes were duplicated such that each copy flanked the insertion site. In this way all genes retained wild type copies of their terminators after insertion. (B) PolII enrichments at the integration sites of the 70.6 kb strain. TetO array was integrated between YDL089w and YDL088c, and lacO array was integrated between YDL055c and YDL054c. (C) PolII enrichments at the integration sites of the 25.3 kb strain. This strain is flanked by the ura3-1 point mutant and wild type URA3. (PDF 5372 kb)

Additional file 2: Plasmids and strains used in this study. (PDF 460 kb)

Acknowledgements

We thank Markus Posch, Bavishna Balagopal, Paul Appleton, Graeme Ball, and Samuel Swift and other members of the Dundee Imaging Facility for their assistance with the microscopes and analysis. We thank the Angela Taddei Lab, the David Sherratt Lab, the Susan Gasser Lab, the Ken Sawin Lab, the Kim Nasmyth Lab, and the Aaron Straight Lab for generously providing plasmids. We thank the J. Salamero Lab for providing the ND-SAFIR software, and the John Sedat Lab for providing the Priism software.

Funding

This work was supported by BBSRC grant BB/K008676/1 Wellcome Senior Research Fellowship [095062], MRC Next Generation Optical Microscopy Award (Ref: MR/K015869/1), Wellcome Principal Research Fellowship [096535], and Wellcome Trust strategic award 097945/B/11/Z.

Authors' contributions

DRD conducted all the experiments. DRD and TOH designed the study. DRD, MG, and VS wrote Python and Pearl scripts for the statistical analysis. DRD, TOH, and MG interpreted the data. DRD and GB designed the SNR analysis strategy. EK assisted in clone identification. All authors read and approved the final manuscript.

Competing interests

The authors declare that they have no competing interests.

References

1. Luger K, Mader AW, Richmond RK, Sargent DF, Richmond TJ. Crystal structure of the nucleosome core particle at 2.8 A resolution. Nature. 1997; 389(6648):251–60.
2. Thoma F, Koller T, Klug A. Involvement of histone H1 in the organization of the nucleosome and of the salt-dependent superstructures of chromatin. J Cell Biol. 1979;83(2 Pt 1):403–27.
3. Bednar J, Horowitz RA, Grigoryev SA, Carruthers LM, Hansen JC, Koster AJ, Woodcock CL. Nucleosomes, linker DNA, and linker histone form a unique structural motif that directs the higher-order folding and compaction of chromatin. Proc Natl Acad Sci U S A. 1998;95(24):14173–8.
4. Woodcock CL, Dimitrov S. Higher-order structure of chromatin and chromosomes. Curr Opin Genet Dev. 2001;11(2):130–5.
5. Dorigo B, Schalch T, Kulangara A, Duda S, Schroeder RR, Richmond TJ. Nucleosome arrays reveal the two-start organization of the chromatin fiber. Science. 2004;306(5701):1571–3.
6. Robinson PJ, Rhodes D. Structure of the '30 nm' chromatin fibre: a key role for the linker histone. Curr Opin Struct Biol. 2006;16(3):336–43.
7. Grigoryev SA, Arya G, Correll S, Woodcock CL, Schlick T. Evidence for heteromorphic chromatin fibers from analysis of nucleosome interactions. Proc Natl Acad Sci U S A. 2009;106(32):13317–22.
8. Fussner E, Djuric U, Strauss M, Hotta A, Perez-Iratxeta C, Lanner F, Dilworth FJ, Ellis J, Bazett-Jones DP. Constitutive heterochromatin reorganization during somatic cell reprogramming. Embo J. 2011;30(9):1778–89.
9. Woodcock CL. Chromatin fibers observed in situ in frozen hydrated sections. Native fiber diameter is not correlated with nucleosome repeat length. J Cell Biol. 1994;125(1):11–9.
10. McDowall AW, Smith JM, Dubochet J. Cryo-electron microscopy of vitrified chromosomes in situ. Embo J. 1986;5(6):1395–402.
11. Eltsov M, Maclellan KM, Maeshima K, Frangakis AS, Dubochet J. Analysis of cryo-electron microscopy images does not support the existence of 30-nm chromatin fibers in mitotic chromosomes in situ. Proc Natl Acad Sci U S A. 2008;105(50):19732–7.
12. Maeshima K, Hihara S, Eltsov M. Chromatin structure: does the 30-nm fibre exist in vivo? Curr Opin Cell Biol. 2010;22(3):291–7.
13. Nishino Y, Eltsov M, Joti Y, Ito K, Takata H, Takahashi Y, Hihara S, Frangakis AS, Imamoto N, Ishikawa T, et al. Human mitotic chromosomes consist predominantly of irregularly folded nucleosome fibres without a 30-nm chromatin structure. Embo J. 2012;31(7):1644–53.
14. Hajjoul H, Mathon J, Ranchon H, Goiffon I, Mozziconacci J, Albert B, Carrivain P, Victor JM, Gadal O, Bystricky K, et al. High-throughput chromatin motion tracking in living yeast reveals the flexibility of the fiber throughout the genome. Genome Res. 2013;23(11):1829–38.
15. Bouchet-Marquis C, Dubochet J, Fakan S. Cryoelectron microscopy of vitrified sections: a new challenge for the analysis of functional nuclear architecture. Histochem Cell Biol. 2006;125(1–2):43–51.
16. Abney JR, Cutler B, Fillbach ML, Axelrod D, Scalettar BA. Chromatin dynamics in interphase nuclei and its implications for nuclear structure. Journal of Cell Biology. 1997;137(7):1459–68.
17. Mahy NL, Perry PE, Gilchrist S, Baldock RA, Bickmore WA. Spatial organization of active and inactive genes and noncoding DNA within chromosome territories. J Cell Biol. 2002;157(4):579–89.
18. Rabl C. Uber Zellteilung. Morph Jb. 1885;10:214–330.
19. Baddeley D, Chagin VO, Schermelleh L, Martin S, Pombo A, Carlton PM, Gahl A, Domaing P, Birk U, Leonhardt H, et al. Measurement of replication structures at the nanometer scale using super-resolution light microscopy. Nucleic Acids Res. 2009;38(2):e8.
20. Cook PR. Molecular biology - The organization of replication and transcription. Science. 1999;284(5421):1790–5.

21. Cremer T, Cremer M, Huebner B, Strickfaden H, Smeets D, Popken J, Sterr M, Markaki Y, Rippe K, Cremer C. The 4D nucleome: Evidence for a dynamic nuclear landscape based on co-aligned active and inactive nuclear compartments. Febs Letters. 2015;589(20):2931–43.

22. Kreth G, Finsterle J, von Hase J, Cremer M, Cremer C. Radial arrangement of chromosome territories in human cell nuclei: A computer model approach based on gene density indicates a probabilistic global positioning code. Biophys J. 2004;86(5):2803–12.

23. Pombo A, Branco MR. Functional organisation of the genome during interphase. Curr Opin Genet Dev. 2007;17(5):451–5.

24. Lieberman-Aiden E, van Berkum NL, Williams L, Imakaev M, Ragoczy T, Telling A, Amit I, Lajoie BR, Sabo PJ, Dorschner MO, et al. Comprehensive mapping of long-range interactions reveals folding principles of the human genome. Science. 2009;326(5950):289–93.

25. Lanctot C, Cheutin T, Cremer M, Cavalli G, Cremer T. Dynamic genome architecture in the nuclear space: regulation of gene expression in three dimensions. Nat Rev Genet. 2007;8(2):104–15.

26. Misteli T. Beyond the sequence: cellular organization of genome function. Cell. 2007;128(4):787–800.

27. de Wit E, de Laat W. A decade of 3C technologies: insights into nuclear organization. Genes Dev. 2012;26(1):11–24.

28. Lassadi I, Bystricky K. Tracking of single and multiple genomic loci in living yeast cells. Methods Mol Biol. 2011;745:499–522.

29. Li C, Vagin VV, Lee S, Xu J, Ma S, Xi H, Seitz H, Horwich MD, Syrzycka M, Honda BM, et al. Collapse of germline piRNAs in the absence of Argonaute3 reveals somatic piRNAs in flies. Cell. 2009;137(3):509–21.

30. Robinett CC, Straight A, Li G, Willhelm C, Sudlow G, Murray A, Belmont AS. In vivo localization of DNA sequences and visualization of large-scale chromatin organization using lac operator/repressor recognition. J Cell Biol. 1996;135(6 Pt 2):1685–700.

31. Michaelis C, Ciosk R, Nasmyth K. Cohesins: chromosomal proteins that prevent premature separation of sister chromatids. Cell. 1997;91(1):35–45.

32. Straight AF, Belmont AS, Robinett CC, Murray AW. GFP tagging of budding yeast chromosomes reveals that protein-protein interactions can mediate sister chromatid cohesion. Curr Biol. 1996;6(12):1599–608.

33. Normanno D, Boudarene L, Dugast-Darzacq C, Chen J, Richter C, Proux F, Benichou O, Voituriez R, Darzacq X, Dahan M. Probing the target search of DNA-binding proteins in mammalian cells using TetR as model searcher. Nat Commun. 2015;6:7357.

34. Lassadi I, Kamgoue A, Goiffon I, Tanguy-le-Gac N, Bystricky K. Differential chromosome conformations as hallmarks of cellular identity revealed by mathematical polymer modeling. PLoS Comput Biol. 2015;11(6), e1004306.

35. Marshall WF, Straight A, Marko JF, Swedlow J, Dernburg A, Belmont A, Murray AW, Agard DA, Sedat JW. Interphase chromosomes undergo constrained diffusional motion in living cells. Curr Biol. 1997;7(12):930–9.

36. Vazquez J, Belmont AS, Sedat JW. Multiple regimes of constrained chromosome motion are regulated in the interphase Drosophila nucleus. Curr Biol. 2001;11(16):1227–39.

37. van den Engh G, Sachs R, Trask BJ. Estimating genomic distance from DNA sequence location in cell nuclei by a random walk model. Science. 1992; 257(5075):1410–2.

38. Bystricky K, Laroche T, van Houwe G, Blaszczyk M, Gasser SM. Chromosome looping in yeast: telomere pairing and coordinated movement reflect anchoring efficiency and territorial organization. J Cell Biol. 2005;168(3):375–87.

39. Yokota H, Singer MJ, van den Engh GJ, Trask BJ. Regional differences in the compaction of chromatin in human G0/G1 interphase nuclei. Chromosome Res. 1997;5(3):157–66.

40. Chambeyron S, Bickmore WA. Chromatin decondensation and nuclear reorganization of the HoxB locus upon induction of transcription. Genes Dev. 2004;18(10):1119–30.

41. Garrick D, De Gobbi M, Samara V, Rugless M, Holland M, Ayyub H, Lower K, Sloane-Stanley J, Gray N, Koch C, et al. The role of the polycomb complex in silencing alpha-globin gene expression in nonerythroid cells. Blood. 2008; 112(9):3889–99.

42. Tark-Dame M, van Driel R, Heermann DW. Chromatin folding–from biology to polymer models and back. J Cell Sci. 2011;124(Pt 6):839–45.

43. Bancaud A, Lavelle C, Huet S, Ellenberg J. A fractal model for nuclear organization: current evidence and biological implications. Nucleic Acids Res. 2012;40(18):8783 92.

44. Sanborn AL, Rao SS, Huang SC, Durand NC, Huntley MH, Jewett AI, Bochkov ID, Chinnappan D, Cutkosky A, Li J, et al. Chromatin extrusion explains key features of loop and domain formation in wild-type and engineered genomes. Proc Natl Acad Sci U S A. 2015;112(47):E6456–65.

45. Luger K, Dechassa ML, Tremethick DJ. New insights into nucleosome and chromatin structure: an ordered state or a disordered affair? Nat Rev Mol Cell Biol. 2012;13(7):436–47.

46. Mateos-Langerak J, Bohn M, de Leeuw W, Giromus O, Manders EM, Verschure PJ, Indemans MH, Gierman HJ, Heermann DW, van Driel R, et al. Spatially confined folding of chromatin in the interphase nucleus. Proc Natl Acad Sci U S A. 2009;106(10):3812–7.

47. Bickmore WA, van Steensel B. Genome architecture: domain organization of interphase chromosomes. Cell. 2013;152(6):1270–84.

48. Bohn M, Heermann DW. Diffusion-driven looping provides a consistent framework for chromatin organization. PLoS One. 2010;5(8), e12218.

49. Joti Y, Hikima T, Nishino Y, Kamada F, Hihara S, Takata H, Ishikawa T, Maeshima K. Chromosomes without a 30-nm chromatin fiber. Nucleus. 2012;3(5):404–10.

50. Barbieri M, Chotalia M, Fraser J, Lavitas LM, Dostie J, Pombo A, Nicodemi M. Complexity of chromatin folding is captured by the strings and binders switch model. Proc Natl Acad Sci U S A. 2012; 109(40):16173–8.

51. Lord SJ, Lee HL, Moerner WE. Single-molecule spectroscopy and imaging of biomolecules in living cells. Anal Chem. 2010;82(6):2192–203.

52. Dubarry M, Loiodice I, Chen CL, Thermes C, Taddei A. Tight protein-DNA interactions favor gene silencing. Genes Dev. 2011;25(13):1365–70.

53. Carlton PM, Boulanger J, Kervrann C, Sibarita JB, Salamero J, Gordon-Messer S, Bressan D, Haber JE, Haase S, Shao L, et al. Fast live simultaneous multiwavelength four-dimensional optical microscopy. Proc Natl Acad Sci U S A. 2010;107(37):16016–22.

54. Bystricky K, Heun P, Gehlen L, Langowski J, Gasser SM. Long-range compaction and flexibility of interphase chromatin in budding yeast analyzed by high-resolution imaging techniques. Proc Natl Acad Sci U S A. 2004;101(47):16495–500.

55. Duan Z, Andronescu M, Schutz K, McIlwain S, Kim YJ, Lee C, Shendure J, Fields S, Blau CA, Noble WS. A three-dimensional model of the yeast genome. Nature. 2010;465(7296):363–7.

56. Hsieh TH, Weiner A, Lajoie B, Dekker J, Friedman N, Rando OJ. Mapping Nucleosome Resolution Chromosome Folding in Yeast by Micro-C. Cell. 2015;162(1):108–19.

57. Robert P, Escoufier Y. A Unifying Tool for Linear Multivariate Statistical Methods: The RV- Coefficient. JRSS Series C. 1976;25(3):257–65.

58. Tjong H, Gong K, Chen L, Alber F. Physical tethering and volume exclusion determine higher-order genome organization in budding yeast. Genome Res. 2012;22(7):1295–305.

59. Nagano T, Lubling Y, Stevens TJ, Schoenfelder S, Yaffe E, Dean W, Laue ED, Tanay A, Fraser P. Single-cell Hi-C reveals cell-to-cell variability in chromosome structure. Nature. 2013;502(7469):59.

60. Williamson I, Berlivet S, Eskeland R, Boyle S, Illingworth RS, Paquette D, Dostie J, Bickmore WA. Spatial genome organization: contrasting views from chromosome conformation capture and fluorescence in situ hybridization. Genes Dev. 2014;28(24):2778–91.

61. Belton JM, Lajoie BR, Audibert S, Cantaloube S, Lassadi I, Goiffon I, Bau D, Marti-Renom MA, Bystricky K, Dekker J. The conformation of yeast chromosome III is mating type dependent and controlled by the recombination enhancer. Cell Rep. 2015;13(9):1855–67.

62. Pillus L, Rine J. Epigenetic inheritance of transcriptional states in S-cerevisiae. Cell. 1989;59(4):637–47.

63. Gottschling DE, Aparicio OM, Billington BL, Zakian VA. Position effect at S. cerevisiae telomeres: reversible repression of Pol II transcription. Cell. 1990; 63(4):751–62.

64. Therizols P, Duong T, Dujon B, Zimmer C, Fabre E. Chromosome arm length and nuclear constraints determine the dynamic relationship of yeast subtelomeres. Proc Natl Acad Sci U S A. 2010;107(5):2025–30.

65. Gonzalez-Sandoval A, Towbin BD, Kalck V, Cabianca DS, Gaidatzis D, Hauer MH, Geng LQ, Wang L, Yang T, Wang XH, et al. Perinuclear anchoring of H3K9-methylated chromatin stabilizes induced cell fate in C. Elegans embryos. Cell. 2015;163(6):1333–47.

66. Eskeland R, Leeb M, Grimes GR, Kress C, Boyle S, Sproul D, Gilbert N, Fan Y, Skoultchi AI, Wutz A, et al. Ring1B compacts chromatin structure and represses gene expression independent of histone ubiquitination. Mol Cell. 2010;38(3):452–64.

67. Therizols P, Illingworth RS, Courilleau C, Boyle S, Wood AJ, Bickmore WA. Chromatin decondensation is sufficient to alter nuclear organization in embryonic stem cells. Science. 2014;346(6214):1238–42.

68. Maillet L, Boscheron C, Gotta M, Marcand S, Gilson E, Gasser SM. Evidence for silencing compartments within the yeast nucleus: A role for telomere proximity and SIR protein concentration in silencer-mediated repression. Genes Dev. 1996;10(14):1796–811.

69. Finlan LE, Sproul D, Thomson I, Boyle S, Kerr E, Perry P, Ylstra B, Chubb JR, Bickmore WA. Recruitment to the nuclear periphery can alter expression of genes in human cells. PLoS Genet. 2008;4, e1000039. doi:10.1371/journal.pgen.1000039.

70. Taddei A, Van Houwe G, Hediger F, Kalck V, Cubizolles F, Schober H, Gasser SM. Nuclear pore association confers optimal expression levels for an inducible yeast gene. Nature. 2006;441(7094):774–8.

71. Osborne CS, Chakalova L, Brown KE, Carter D, Horton A, Debrand E, Goyenechea B, Mitchell JA, Lopes S, Reik W, et al. Active genes dynamically colocalize to shared sites of ongoing transcription. Nat Genet. 2004;36(10):1065–71.

72. Brown JM, Green J, das Neves RP, Wallace HAC, Smith AJH, Hughes J, Gray N, Taylor S, Wood WG, Higgs DR, et al. Association between active genes occurs at nuclear speckles and is modulated by chromatin environment. J Cell Biol. 2008;182(6):1083–97.

73. Morey C, Da Silva NR, Perry P, Bickmore WA. Nuclear reorganisation and chromatin decondensation are conserved, but distinct, mechanisms linked to Hox gene activation. Development. 2007;134(5):909–19.

74. Schwabe A, Bruggeman FJ. Single yeast cells vary in transcription activity not in delay time after a metabolic shift. Nat Commun. 2014;5:4798.

75. Zenklusen D, Larson DR, Singer RH. Single-RNA counting reveals alternative modes of gene expression in yeast. Nat Struct Mol Biol. 2008;15(12):1263–71.

76. Raser JM, O'Shea EK. Control of stochasticity in eukaryotic gene expression. Science. 2004;304(5678):1811–4.

77. Renshaw MJ, Ward JJ, Kanemaki M, Natsume K, Nedelec FJ, Tanaka TU. Condensins promote chromosome recoiling during early anaphase to complete sister chromatid separation. Dev Cell. 2010;19(2):232–44.

78. Snaith HA, Samejima I, Sawin KE. Multistep and multimode cortical anchoring of tea1p at cell tips in fission yeast. Embo J. 2005;24(21):3690–9.

79. Lau IF, Filipe SR, Soballe B, Okstad OA, Barre FX, Sherratt DJ. Spatial and temporal organization of replicating Escherichia coli chromosomes. Mol Microbiol. 2003;49(3):731–43.

80. Sikorski RS, Hieter P. A system of shuttle vectors and yeast host strains designed for efficient manipulation of DNA in Saccharomyces cerevisiae. Genetics. 1989;122(1):19–27.

81. Edwards S, Li CM, Levy DL, Brown J, Snow PM, Campbell JL. Saccharomyces cerevisiae DNA polymerase epsilon and polymerase sigma interact physically and functionally, suggesting a role for polymerase epsilon in sister chromatid cohesion. Mol Cell Biol. 2003;23(8):2733–48.

82. Hsu JM, Huang J, Meluh PB, Laurent BC. The yeast RSC chromatin-remodeling complex is required for kinetochore function in chromosome segregation. Mol Cell Biol. 2003;23(9):3202–15.

83. Mason PB, Struhl K. The FACT complex travels with elongating RNA polymerase II and is important for the fidelity of transcriptional initiation in vivo. Mol Cell Biol. 2003;23(22):8323–33.

84. Wach A, Brachat A, Pohlmann R, Philippsen P. New heterologous modules for classical or PCR-based gene disruptions in Saccharomyces cerevisiae. Yeast. 1994;10(13):1793–808.

85. Goldstein AL, McCusker JH. Three new dominant drug resistance cassettes for gene disruption in Saccharomyces cerevisiae. Yeast. 1999;15(14):1541–53.

86. Boulanger J, Kervrann C, Bouthemy P, Elbau P, Sibarita JB, Salamero J. Patch-based nonlocal functional for denoising fluorescence microscopy image sequences. IEEE Trans Med Imaging. 2010;29(2):442–54.

87. Fasano GF. A: A Multidimensional Version of the Kolmogorov-Smirnov Test. MNRAS. 1987;225:155–70.

88. Peacock JA. Two-dimensional goodness-of-fit testing in astronomy. MNRAS. 1983;202(3):615–27.

89. Qian H, Sheetz MP, Elson EL. Single particle tracking. Analysis of diffusion and flow in two-dimensional systems. Biophys J. 1991;60(4):910–21.

90. Mayer A, Heidemann M, Lidschreiber M, Schreieck A, Sun M, Hintermair C, Kremmer E, Eick D, Cramer P. CTD tyrosine phosphorylation impairs termination factor recruitment to RNA polymerase II. Science. 2012;336(6089):1723–5.

91. Weiner A, Hsieh TH, Appleboim A, Chen HV, Rahat A, Amit I, Rando OJ, Friedman N. High-resolution chromatin dynamics during a yeast stress response. Mol Cell. 2015;58(2):371–86.

92. Bonnet J, Wang CY, Baptista T, Vincent SD, Hsiao WC, Stierle M, Kao CF, Tora L, Devys D. The SAGA coactivator complex acts on the whole transcribed genome and is required for RNA polymerase II transcription. Genes Dev. 2014;28(18):1999–2012.

93. Yen K, Vinayachandran V, Batta K, Koerber RT, Pugh BF. Genome-wide nucleosome specificity and directionality of chromatin remodelers. Cell. 2012;149(7):1461–73.

94. Mayer A, Lidschreiber M, Siebert M, Leike K, Soding J, Cramer P. Uniform transitions of the general RNA polymerase II transcription complex. Nat Struct Mol Biol. 2010;17(10):1272–8.

95. Duan X, Yang Y, Chen YH, Arenz J, Rangi GK, Zhao X, Ye H. Architecture of the Smc5/6 Complex of Saccharomyces cerevisiae Reveals a Unique Interaction between the Nse5-6 Subcomplex and the Hinge Regions of Smc5 and Smc6. J Biol Chem. 2009;284(13):8507–15.

96. Thurtle DM, Rine J. The molecular topography of silenced chromatin in Saccharomyces cerevisiae. Genes Dev. 2014;28(3):245–58.

3

Nuclear envelope structural defect underlies the main cause of aneuploidy in ovarian carcinogenesis

Callinice D. Capo-chichi[1,2,3], Toni M. Yeasky[1,2], Elizabeth R. Smith[1,2] and Xiang-Xi Xu[1,2]*

Abstract

Background: The Cancer Atlas project has shown that p53 is the only commonly (96 %) mutated gene found in high-grade serous epithelial ovarian cancer, the major histological subtype. Another general genetic change is extensive aneuploidy caused by chromosomal numerical instability, which is thought to promote malignant transformation. Conventionally, aneuploidy is thought to be the result of mitotic errors and chromosomal nondisjunction during mitosis. Previously, we found that ovarian cancer cells often lost or reduced nuclear lamina proteins lamin A/C, and suppression of lamin A/C in cultured ovarian epithelial cells leads to aneuploidy. Following up, we investigated the mechanisms of lamin A/C-suppression in promoting aneuploidy and synergy with p53 inactivation.

Results: We found that suppression of lamin A/C by siRNA in human ovarian surface epithelial cells led to frequent nuclear protrusions and formation of micronuclei. Lamin A/C-suppressed cells also often underwent mitotic failure and furrow regression to form tetraploid cells, which frequently underwent aberrant multiple polar mitosis to form aneuploid cells. In ovarian surface epithelial cells isolated from p53 null mice, transient suppression of lamin A/C produced massive aneuploidy with complex karyotypes, and the cells formed malignant tumors when implanted in mice.

Conclusions: Based on the results, we conclude that a nuclear envelope structural defect, such as the loss or reduction of lamin A/C proteins, leads to aneuploidy by both the formation of tetraploid intermediates following mitotic failure, and the reduction of chromosome (s) following nuclear budding and subsequent loss of micronuclei. We suggest that the nuclear envelope defect, rather than chromosomal unequal distribution during cytokinesis, is the main cause of aneuploidy in ovarian cancer development.

Keywords: Nuclear envelope, Lamin A/C, Aneuploidy, Polyploidy, Ovarian cancer, Carcinomas

Background

Recently, the Cancer Atlas project [1] determined that mutation in p53 gene is the only common somatic genetic change (96%) found in high-grade serous epithelial ovarian cancer [1, 2], the most common histological subtype. However, inactivation of p53 in ovarian epithelial cells in mouse models has not demonstrated a clear path for epithelial tumorigenesis [3, 4], even in aged mice

following transplantation of p53 mutant ovaries [5]. Thus, the etiology and mechanism of epithelial ovarian cancer is not yet satisfactorily understood.

Another common genetic change in ovarian carcinomas revealed from the cancer genomic study is extensive aneuploidy [1]. The connection of abnormal chromosomes with cancer was first recognized over one hundred years ago by Boveri [6, 7]. Generally, aneuploidy is thought to be the result of mitotic errors and chromosomal nondisjunction during mitosis [8–10]. The majority of human ovarian cancer cells are aneuploid and possess a hyperdiploid (>46) to subtetraploid (<96) chromosome number [11]. Although a correlation between aneuploidy and malignancy has been recognized,

* Correspondence: xxu2@med.miami.edu
[1]Sylvester Comprehensive Cancer Center/University of Miami, Miami, Florida 33136, USA
[2]Department of Cell Biology, University of Miami Miller School of Medicine, Miami, FL 33136, USA
Full list of author information is available at the end of the article

the causes and significance of aneuploidy in cancer remain unsettled [7, 12–17]. Several mechanisms have been noted for the origination of aneuploidy [8, 10, 13, 18]. Genes that cause mitotic failure account for the majority of cases, and chromosomal non-disjunction is thought to cause unequal distribution of chromosomes in daughter cells [7, 10]. Tetraploid cells are believed to form following mitotic failure, and aneuploid cells are produced in subsequent mitotic events [8, 9, 13, 19]. Centrosome amplification also leads to multipolar cytokinesis and aneuploidy [18].

One unique view is that chromosome instability and aneuploidy may provide an unbalanced global expression profile of increases and decreases in gene dosages that create the cancer cell properties [12]. The general interpretation is that chromosome instability and aneuploidy promote the accelerated loss and gain of specific tumor suppressor genes and oncogenes, respectively, leading to selection of mutant cells with a growth advantage and subsequent malignant transformation [14, 16, 19, 20]. Two possible routes, a progressive shift up pathway and a tetraploid intermediate following drift down pathway, may convert a diploid normal cell to an aneuploid cancer cell. Cells with an optimal chromosome composition may have growth advantage, be selected, and become neoplastic.

Enlarged and deformed nuclei are characteristics of cancer cells, and the aberrant nuclear morphology correlates with malignancy and is a diagnostic and prognostic indicator, referred to as "nuclear grade" [21–26]. The increase in chromosome number over normal cells accounts for the larger nuclear size in cancer. Based purely on the nuclear morphology of cells sampled, the PAP test (or PAP smear), invented by Dr. Papanicolaou in the 1930s, is able to make diagnostic and prognostic prediction of the degree of malignancy of uterine and cervical cancers [27]. Changes in the nuclear matrix and/or nuclear envelope have been postulated, and deformation of nuclear morphology was shown to associate with oncogenic signaling [21, 28–31].

Shape of the nucleus is determined by structural proteins of the nuclear envelope lamina, which has been well studied [32–36]. Lamin A/C, but not lamin B1, is critical for the maintenance of a smooth and oval shaped nucleus [37]. Phosphorylation of lamin mediates reversible disassembly and re-formation of nuclear envelope in mitosis [38, 39]. Mutations or loss-of-function in several nuclear envelope structure proteins, including emerin, Man1, Baf, and lamin in *C. elegans*, cause similar nuclear and mitotic phenotypes such as an enlarged and deformed nucleus, defective chromosome segregation, and the formation of chromatin bridges between divided nuclei, suggesting a critical role for the nuclear envelope in cytokinesis and mitosis [33, 34, 38–41]. However, lamin A/C is dispensable for mitosis in mammalian cells since deletion of lamin A/C does not impair development in mice [42]. Nevertheless, lamin A/C is known to affect mitosis [43, 44], and a role in nuclear envelope formation likely influences the process of cytokinesis, though the redundancy of the three lamin genes present in mammals may reduce the impact of a single gene. Thus, roles of nuclear envelope proteins in maintaining the nuclear structure and mediating cytokinesis/mitosis are conserved across species. Lamin A/C expression is absent in embryonic stem cells and early embryos, and is progressively expressed in nearly all tissues in later developmental stages [45, 46]. The cell types that seem to lack lamin A/C, such as embryonic carcinoma cells and some cells of the spleen, thymus, bone marrow and intestine in the adult mouse, may fall into the "stem cell" category [45, 46].

Loss or reduction of lamin A/C expression is often found in cancer cells [47], including leukemia [48, 49], colon [50], prostate [29], lung [51], breast [52], and gastric cancers [53, 54]. Our earlier study also found that lamin A/C expression is lost in about 60% of serous ovarian carcinomas, in which the mRNA is often present despite the loss of protein [55]. AKT and cell cycle associated phosphorylation of lamin A/C lead to this protein degradation [56–58]. One report concluded that lamin A/C proteins are increased in ovarian cancer when normal ovarian tissues (instead of ovarian surface epithelia) were used as controls [59]. However, the ovarian epithelial cells of the surface layer were found to be strongly positive, whereas the stromal cells were largely low for lamin A/C [59]. Thus, the correct interpretation of the result should be that 39% of ovarian cancer cases are positive for lamin A/C, and lamin A/C proteins are lost or greatly reduced in 61% of ovarian cancers. Another report identified lamin C as a marker that is reduced/lost in malignant ovarian cancer but not in borderline tumors based on results from 2-dimensional gel electrophoresis [60]. Thus, the published studies generally support our report that lamin A/C proteins are lost in over half of ovarian cancer. Previously, we found that suppression of lamin A/C caused aneuploidy in human ovarian surface epithelial cells [55]. Here, we further investigated the mechanisms and consequences of the development of aneuploidy and tumor development following the loss of lamin A/C using both human and mouse ovarian surface epithelial cells.

Results
Previously using siRNA to suppress lamin A/C expression in human ovarian surface epithelial (HOSE) cells, we reported that loss of lamin A/C proteins led to a deformed nuclear morphology, polyploidy, and aneuploidy [55].

Following up the previous findings, we explored the mechanisms for the development of aneuploidy upon lamin A/C suppression. The HOSE cells were transfected with histone H2B-GFP that marks the nuclear DNA to monitor the behavior of the cells in cultures, as described previously [28]. Similar to that previously reported, about 50% of HOSE cells expressed GFP after transfection, and the signals persisted for at least 2 weeks over the length of the experiments. We observed that at any given time, 30 to 60% of the lamin A/C-suppressed HOSE cells

exhibited an aberrant nuclear morphology in about 200 cells analyzed. If allowed to follow the cells in cultures over several hours, essentially all cells displayed some degree of nuclear deformation, compared to about 5% in controls that were transfected with scrambled siRNA oligonucleotides.

One apparent feature of nuclear deformation was nuclear herniation, or budding (Fig. 1a, b). In around 20% of over 200 cases examined, the nuclear body had an extremely extended herniation, and the nuclear materials

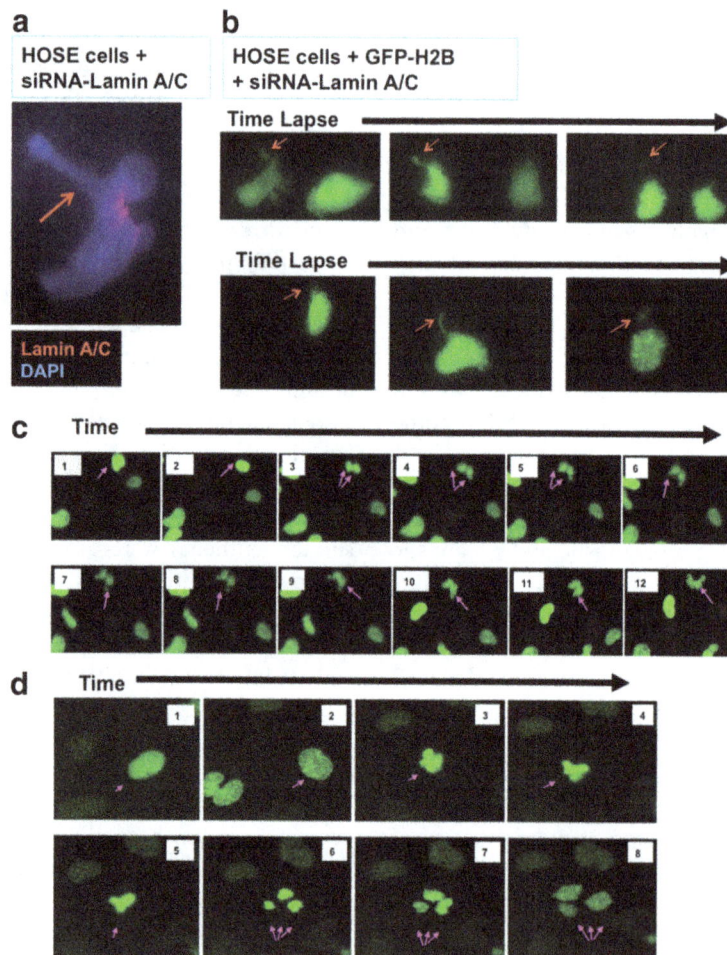

Fig. 1 Nuclear protrusion and herniation, mitotic failure, and aberrant cytokinesis in lamin A/C-suppressed HOSE cells. Lamin A and C were suppressed by siRNA (Santa Cruz biotechnology Inc, Santa Cruz, CA) in primary human ovarian epithelial (HOSE) cells, using lipofectamine 2000 according to the manufacturer protocol (Invitrogen, CA). **a** At day 3, the lamin A/C-suppressed cells were analyzed using immunofluorescence microscopy. An example of a cell undergoing nuclear protrusion and herniation is shown (arrow), for nuclear staining with DAPI (*blue*) and lamin A/C (*red*). **b** The lamin A/C-suppressed primary HOSE cells that were previously transfected with GFP-histone H2B were monitored for nuclear changes by time-lapse video fluorescence micros-copy over a 48-h period. Time-lapse video microscopy was performed 12 h after lamin A/C suppression. Cells were seeded in 24 well falcon plate and transfected the next day with scramble siRNA (control) or lamin A/C siRNA in serum reduced Opti-MEM media. Image acquisition was performed every 5 min for 48 h with a 40× dry objective lens on Nikon Eclipse TE 300 microscope linked to a Roper Scientific photometrics 12-bit range Camera using Meta imaging series (MetaVue) software. Stacked images were assembled with MetaVue software to make the movies. Two examples of sequential time-lapse images 15 min apart are shown. Arrows indicate the nuclear herniation and the formation of micronuclei, which gradually faded. Presumably, aneuploidy was resulted in the remaining nuclei. **c** Sequential time-lapse images (images frame #1 to 12) 15 min apart are shown for a 3-h segment of a video of a mitotic failure of a dividing cell. Arrows indicate the nuclei first underwent DNA condensation, separation, and then fused back to form presumably tetraploid nuclei. **d** Sequential time-lapse images (frame #1 to 8) 15 min apart show an aberrant mitotic process. Arrows indicate the nuclei undergoing a tripolar division. Likely, aneuploid cells were formed

often broke off to produce a micronucleus (Fig. 1a, b). When observed over a 6-h time-lapse, nearly all lamin A/C-suppressed cells underwent nuclear herniation and released micronuclei, as shown in two examples (Fig. 1b). Arrows indicate the nuclear protrusion and the formation of micronuclei, which gradually faded, presumably being degraded by the cellular proteolytic machinery. Consequently, aneuploid cells resulted, containing the remaining nuclei. In comparison, such phenomenon was rare in lamin A/C-positive HOSE cells, and about 4% of over 200 control cells showed formation of micronuclei. Since lamin A/C is frequently reduced or lost in ovarian cancer cells [55], we reason that the lamin A/C-deficient ovarian cancer cells may develop aneuploidy by such a mechanism: nuclear protrusions and formation of micronuclei. The observations that cancer cell often undergo transient nuclear envelope rupture in gap phases [61] and collapse of micronuclei produced [62] support this idea.

Another feature of Lamin A/C-suppressed HOSE cells was frequent mitotic failure and the formation of tetraploid cells (Fig. 1c). In the analysis of 48-h time-lapse video, around a third (22 out of 80 mitotic events observed) of mitosis in GFP-histone H2B-labeled, siRNA-lamin A/C-transfected HOSE cells resulted in tetraploidy, as an example shown (Fig. 1c). Typically in such events, a cell was first observed to undergo nuclear condensation (Fig. 1c, frame #1–3) and subsequent attempt in cytokinesis (Fig. 1c, frame #4–8). However, the forming daughter nucleus failed to separate and then fused (Fig. 1c, frame #9–11), and a presumed tetraploid cell resulted (Fig. 1c, frame #12). In controls HOSE cells, mitotic failure was rare, none among 50 mitoses observed. In mammalian cells that have 3 lamin genes (lmna encoding lamin A/C, lmnb1, and lmnb2), lamin A/C is not essential for mitosis. However, lamin is essential for mitosis in C. elegans, which has only one lamin gene [43]. Nevertheless, mutations in lamin A/C interfere with mitosis and cell cycle progression in mammalian cells [43, 44]. Thus, our current finding that HOSE cells have an increased rate of mitotic failure seems compatible with the function of lamin proteins in cytokinesis during the formation of daughter nuclei in the dividing cells.

Additionally, aberrant mitosis such as tripolar cell division was frequent in lamin A/C-suppressed cells, as an example shown (Fig. 1d). The cells that underwent 3-way mitosis generally had larger nuclei, suggesting that they were tetrapoid cells, and thus aneuploid cells were generated. In the examination of 80 mitotic events of the lamin A/C-suppressed, GFP-Histone H2B-labeled HOSE cells, 6 events of aberrant tripolar mitosis were recognized.

The analysis of the cell behavior following lamin A/C suppression has been repeated over the course of 2 years, and the results were consistent from 4 independent preparations of HOSE cells. The above results led us to suggest

that in lamin A/C-deficient cells, nuclear herniation to form micronuclei, mitotic failure to form tetraploid, and tripolar mitosis are mechanisms leading to the development of aneuploidy. Previously we found that lamin A/C-suppressed cells were not able to continue proliferation in culture, likely because of aneuploidy and activation of p53 check points [55].

We subsequently used primary mouse ovarian surface epithelial (MOSE) cells to further investigate the mechanism and consequence of lamin A/C-suppression. We reasoned that use of MOSE cells would allow us to introduce p53 genetic mutation to mimic the genotype in human ovarian cancer, and to bypass growth arrest following lamin A/C suppression. Primary ovarian surface epithelial cells were prepared from wildtype and p53 knockout mice, and were transfected with scrambled (control) or lamin A/C specific siRNA. The transfection efficiency ranged from 80 to 90% in various experiments based on uptake of labeled cy3-siRNA oligonucleotides. Lamin A protein was significantly reduced upon transfection with targeting siRNA as visualized by Western blots (Fig. 2a). Using immunofluorescence microscopy, it was estimated that around 80% of the MOSE cells had greatly reduced lamin A (Fig. 2b, c). Here, a lamin A - specific antibody was used, since it was found specific to the mouse protein, which is not properly recognized by several other available lamin A/C antibodies tested. The lamin A/C-suppressed MOSE cells also exhibited frequent nuclear herniation, mitotic failure, and aberrant mitosis (Fig. 2b, c), similar to those observed in HOSE cells upon suppression of lamin A/C expression.

We used flow cytometry to analyze cellular DNA content of the cells following siRNA suppression of lamin A/C. Comparing to the control cells (Fig. 2d) that have distinctive G1 (2n) and G2 (4n) peaks, p53 (-/-) MOSE cells showed a slightly higher fraction of polyploid (8n) cells (Fig. 2e). The lamin A/C-siRNA suppressed cells had a distinctive profile (Fig. 2f): the G1 peak separated into two (or more) main populations, which likely indicated the presence of a sub 2n fraction because of loss of one or few chromosomes by nuclear protrusion and the formation of micronuclei that was degraded. The G2 fraction was also reduced in lamin A/C-suppressed cells, likely because a cell cycle checkpoint was activated, as shown previously for HOSE cells [55]. In the p53 null and lamin A/C-suppressed cells, cell populations with various DNA content distributed continuously from 2n to 8n, suggesting the development of massive aneuploidy in these cells (Fig. 2g). Because of the presence of extensive aneuploidy, the profiles of these flow cytometry results were not suitable for analysis using a general flow cytometry program that does not account for aneuploidy.

Both the wildtype and the lamin A/C-suppressed MOSE cells had only limited life span in culture, and became senescent and deteriorated within 1–2 months.

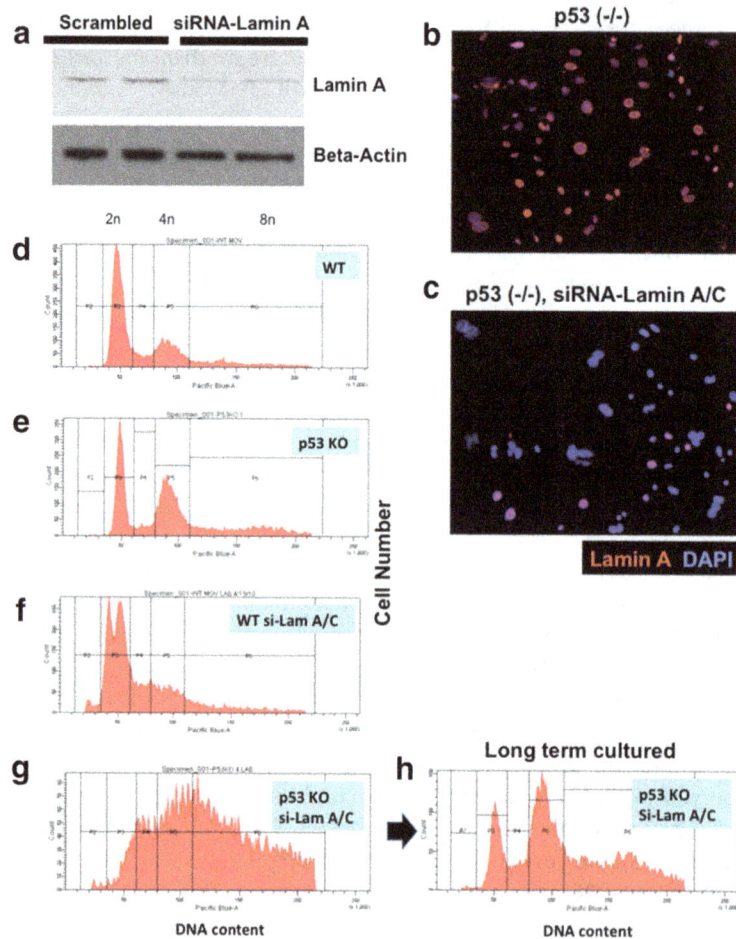

Fig. 2 Lamin A/C suppression in primary mouse ovarian surface epithelial (MOSE) cells results in aneuploidy and polyploidy, synergistic with p53 deletion. Primary wildtype (WT) and p53 knockout (KO) MOSE cells were transfected with control (scrambled) or siRNA (si-Lam A) to suppress lamin A/C suppression. **a** At day 3, the lamin A/C-suppressed cells were analyzed by Western blot for the presence of lamin A protein. Duplicate experiments are shown for control (scrambled siRNA) and siRNA specific to mouse lamin A/C. **b** The cells were analyzed by immunofluorescence microscopy for the expression of lamin A. An example of p53 (-/-) MOSE cells treated with control siRNA is shown. **c** In comparison, staining was reduced in cells treated with siRNA-lamin A/C. **d** Flow cytometry was performed 3 days after siRNA transfection. Cells were collected following trypsin digestion, washed with PBS, and cell pellets were resuspended in ice cold ethanol/PBS (70% v/v) with gentle agitation. The fixed cells were kept at − 20°C until ready to use. Prior to flow cytometric analysis, cells were centrifuged at 1200 rpm for 5 min and washed twice with PBS before resuspension in 0.5 mm vybrant *violet* dye. Cells were then incubated at 37°C for 30 min before flow cytometric analysis for DNA content. Flow cytometry profile for wildtype (WT) cells treated with control siRNA is shown. **e** p53 knockout cells; **f** WT cells treated with siRNA-lamin A/C; **g** p53 knockout cells treated with siRNA-lamin A/C. **h** Flow cytometry profile of the p53 knockout, siRNA-lamin A/C-treated MOSE cells following longer-term (2 months) culturing

However, both the p53-deficient and the and the lamin A/C-suppressed p53-deficient MOSE cells continued to grow in culture. Following 4 weeks in culture, the original p53-deficient and Lamin A/C-suppressed MOSE cells with a wildly variable distributed chromosome number (Fig. 2g) converted into a more defined cellular chromosomal number distribution (Fig. 2h). We interpret that certain clones with optimal karyotypes from the original populations had growth advantage in culture and became the dominating cell populations.

Indeed, chromosome analysis of metaphase spreads indicated aneuploidy and wide range of chromosomal number distribution in the lamin A/C-suppressed p53-deficient MOSE cells, such as 56, 60, 63, 67, 80, 81, 82, 84, 89, and 94 chromosomes, determined in 10 randomly selected metaphase spreads. Two of the examples are shown (Fig. 3a, b). Chromosome identification in two samples revealed complex karyotypes in the lamin A/C-suppressed p53-deficient MOSE cells (Fig. 3c, d), and a marker chromosome was observed in one sample (Fig. 3c). For comparison, metaphases from p53 knockout MOSE cells (without prior lamin A/C-siRNA treatment) were found to be largely near diploid (40 chromosomes) to tetraploid (80 chromosomes), and

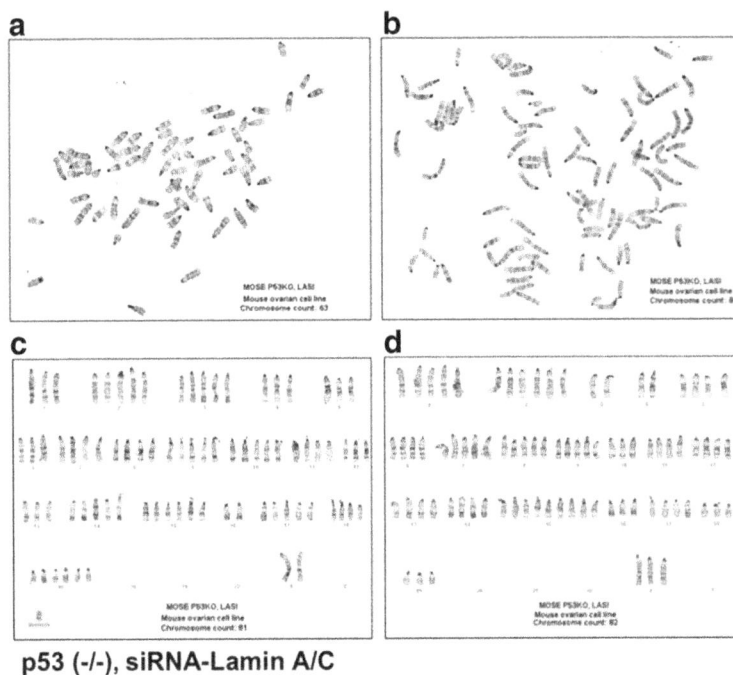

Fig. 3 p53 inactivation and lamin A/C suppression result in aneuploidy and complex karyotypes. Primary p53 knockout MOSE cells were transfected with control or siRNA (si-Lam A) to suppress lamin A/C expression. The cells were maintained and passaged for 2 months in culture, and then subjected to chromosome analysis. Chromosome number counting and cytogenetic analysis were performed in 50 metaphase spreads for each cell preparation. At least 10 chromosome spreads from each preparation were randomly selected and estimated for chromosome number, and 2 appropriate samples were used for karyotyping. **a** and **b**, 2 representative examples of chromosome spreads from p53 (-/-) and siRNA-lamin A/C-treated MOSE cells are shown. **c** and **d**, 2 examples of karyotyping from p53 (-/-) and siRNA-lamin A/C-treated cells are shown

karyotyping by the cytogenetic core facility indicated that obvious structural abnormalities were not observed, but subtle abnormalities cannot be ruled out (quoted from the facility report).

When the MOSE cells were implanted into nude mice, both p53 (-/-) and lamin A/C-suppressed p53 (-/-) MOSE cells were tumorigenic (Fig. 4). Tumors formed in 5 of 6 nude mice when p53 (-/-) MOSE cells were implanted; and lamin A/C-suppressed p53 (-/-) MOSE cells formed tumors in 6 out of 6 mice tested. The tumors derived from the lamin A/C-suppressed p53 (-/-) MOSE cells had unique malignant features (Fig. 4a): the tumor cells often presented as small nodules invaded into muscle fibers (Fig. 4b, c). The tumor cells also exhibited a higher variation in nuclear sizes. In contrast, tumors derived from p53 (-/-) MOSE cells grew as a single mass with a more uniform nuclear morphology and size (Fig. 4g, h). Thus, a transient suppression of lamin A/C and generation of aneuploidy enable the growth of tumors with an increased degree of malignancy. Nevertheless, when implanted into immune competent female littermates from which the MOSE cells were prepared, neither p53 (-/-) nor lamin A/C-suppressed p53 (-/-) MOSE cells were able to produce significant or persistent tumors, indicating these MOSE cells were unable to escape the host immune surveillance in the development of tumors.

When the tumors were analyzed by immunohistochemistry, the tumors derived from lamin A/C-suppressed p53 (-/-) MOSE cells showed lamin A/C staining ranged from high (Fig. 4d), low (Fig. 4e), to mixed (Fig. 4f). Thus, it appears that some lamin A/C expression was recovered following the prior transc1ent suppression by siRNA. Interestingly, tumors derived from p53 (-/-) MOSE cells uniformly had low lamin A/C staining, in 3 out of 3 tumors analyzed as shown by a representative example (Fig. 4i). Thus, unexpectedly, the expression of lamin A/C is lower in tumors from p53 (-/-) MOSE cells without than with prior lamin A/C suppression. The results suggest a preference in growth and tumor development of a low laimin A/C cell population in the p53 (-/-) MOSE cells. The idea that a reduced lamin A/C expression may contribute to tumor development following p53 inactivation will need to be further verified.

Discussion

Previously, we found that lamin A/C expression is commonly lost in ovarian cancer, and suppression of lamin A/C in ovarian surface epithelial cells led to the

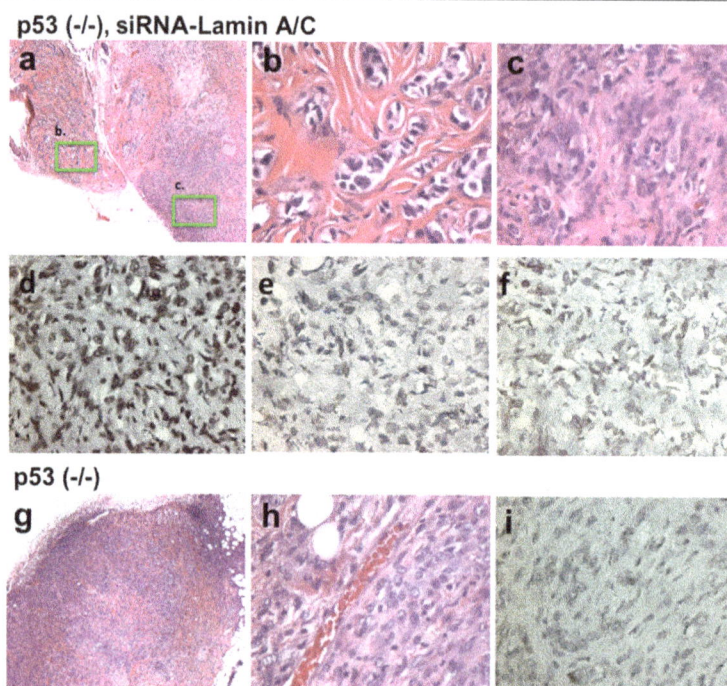

Fig. 4 p53 inactivation and lamin A/C suppression lead to malignant tumors. Primary p53 knockout MOSE cells were transfected with control or siRNA (si-Lam A) and cultured for about 2 months. When the lamin A/C-suppressed and p53 (-/-) MOSE cells were implanted in nude mice subcutaneously, invasive tumors developed in 4 weeks (**a**). Two areas of the tumor are shown in higher magnification (**b**, **c**). Three examples of tumors formed from lamin A/C-suppressed p53 (-/-) MOSE cells were stained with lamin A/C, as shown in (**d**), (**e**), and (**f**). **g** Tumors formed from p53 (-/-) MOSE cells (not treated with lamin A/C-siRNA)) were compared, and a higher magnification (**h**) is shown. **i** An example of lamin A/C immunostaining is shown for a tumor derived for p53 (-/-) MOSE cells

formation of aneuploidy, especially in p53 inactivated cells [55]. In following up the previous finding, here we determined the mechanisms for the development of aneuploidy when lamin A/C is eliminated. We found that when lamin A/C was suppressed, both the HOSE and the MOSE cells often failed in completing cytokinesis, and tetraploid cells were formed. Also, the aberrant nuclei underwent nuclear protrusion to form micronuclei, presumably a mechanism for progressive reduction of chromosome number to select for a growth permissive karyotype (s). Indeed, we observed that the lamin A/C-siRNA treated mouse p53 (-/-) MOSE cells exhibit complex karyotypes that resemble those of ovarian cancer cells.

The observation of frequent mitotic failure in lamin A/C-depleted cells is consistent with the requirement of lamin in cytokinesis in lower organisms such as *C. elegans* and *Drosophila* [33, 34], which have only one lamin gene. Possibly, the redundancy of the three lamin genes in mammals is the reason that the absence of a single lamin isoform would not generally cause cytokinesis failure, but rather increase the frequency of such events, as we have observed here.

The experimental results described here support a hypothesis that nuclear envelope defects (loss of lamin A/C proteins) may be the common cause of chromosomal numerical instability and aneuploidy in ovarian cancer (Fig. 5a-c). The idea explains both nuclear morphological deformation and aneuploidy, two prominent hallmarks of ovarian cancer. Generally, chromosomal disjunction is thought to be the cause of aneuploidy [7, 9, 10]. However, the results reported here leads us to a provocative hypothesis that nuclear envelope defect, such as loss of lamin A/C, rather than chromosomal disjunction (Fig. 5a), may be the main cause of aneuploidy in ovarian cancer (Fig. 5c). We reason that lamin A/C-deficient cells frequently fail to complete cytokinesis. We speculate that this is cause by the failure to properly form new nuclear envelope to encase the two new daughter nuclei, and the dividing nucleus undergoes furrow regression to produce tetraploid intermediates. Subsequently, aneuploid cells are generated by tripolar division. Formation of micronuclei at G-phases is another mechanism for the loss of individual chromosomes [61–64].

Conclusions

Loss of lamin A/C appears to only increase the frequency of such mitotic failure in mammalian cells, though lamin is essential for cytokinesis in C. elegans, which has only one lamin gene [40]. Aneuploid cells

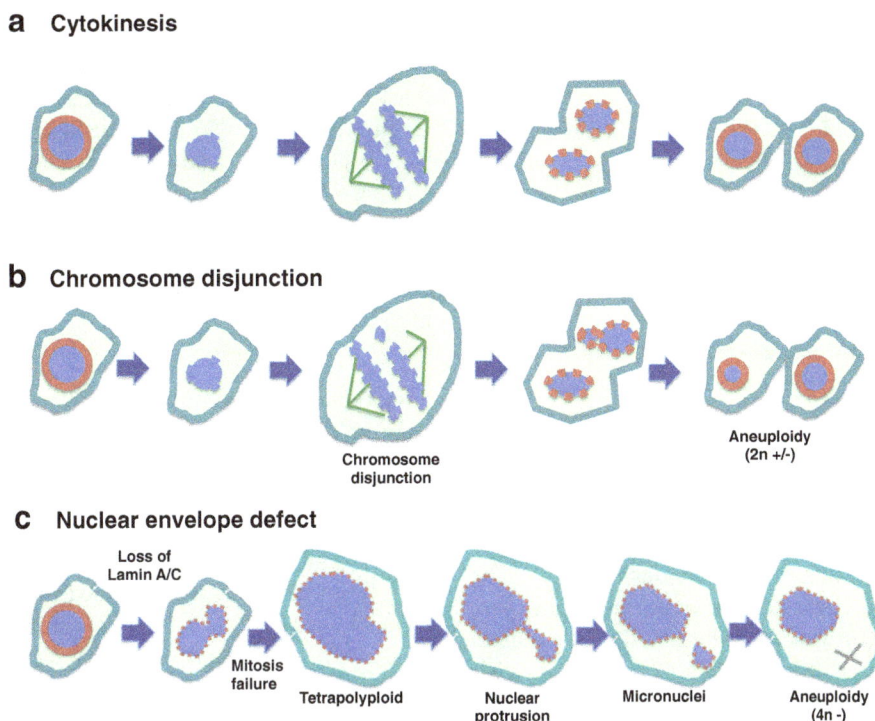

Fig. 5 Working model: nuclear envelope defect is the main cause of aneuploidy in carcinogenesis. **a** Depiction of normal cytokinesis: at the start of M phase, the nuclear envelope dissolves, chromatin undergoes condensation, chromosomes pair and then separate, two new nuclear envelopes form, and cytokinesis is completed. **b** Chromosomal Disjunction: during chromosomal separation, one or more chromosomes are not attached. As a result, the two daughter cells have unequal distribution of chromosomes following cytokinesis. This mechanism is generally thought to be the main cause of aneuploidy. **c** Nuclear envelope defect causes aneuploidy: We reason that loss of a nuclear envelope structural component such as lamin A/C results in a misshapen nucleus. Additionally, the lamin A/C-deficient cells frequently fail to complete cytokinesis. Thus, tetraploid cells and subsequently aneuploid cells are generated. Formation of micronuclei at G-phases is another mechanism for the loss of individual chromosomes. Thus, we propose that the nuclear envelope defect is the main cause of aneuploidy in ovarian cancer development

may be growth retarded and undergo cell growth arrest or death [65, 66]. p53 mutation may allow the cells to survive and undergo clonal selection [67]. Most aneuploid cells generated from transient loss of lamin A/C likely would die, but ultimately, a population of cells with a unique chromosomal composition is selected and expanded to form cancer. Thus, our results advocate a concept that a deformed nuclear envelope is the main source of chromosomal instability of the cancer cells, and is the cause rather than a consequence of neoplastic transformation.

Methods
Reagents
Tissue culture flasks (Falcon), tissue culture media, trypsin, and 100× antibiotic-antimycotic solution (Cellgro, Mediatech, Inc) were purchased from Fisher Scientific Inc (Springfield, NJ). Triazol reagent and transfection reagent were purchased from Invitrogen Inc (Carlsbad, CA). For Western blot detection, Super Signal West Dura Extended Duration Substrate (PIERCE, Rockford, IL) was used. For immunofluorescence microscopy, Alexa Fluor 488 and 596 conjugated secondary antibodies and Hoechst 33342

nuclear counter staining dye were purchased from Molecular Probes Inc (Eugene, Oregon). Primary antibodies, anti-lamin A (H-102, rabbit polyclonal IgG), were purchased from Santa Cruz Biotechnology Inc (Santa Cruz, CA).

Mouse models and xenograft tumor assay
The p53 knockout mice were purchased from Taconic (Hudson, NY) [68]. The mouse colony was kept in the C57BL/6 background and heterozygous pairs were bred to produce homozygous or wild type mice, which were used to prepare ovarian surface epithelial cells for experiments. The mouse colony was maintained and genotyped as described previously [69, 70].

Immunodeficient Scid mice were purchased from Jackson lab (Bar Harbor, ME). The mice were used for xenograft assays to test tumor development from MOSE cells by implanting 5×10^6 cells subcutaneously in the immune deficient nude mice. The cells were tested to ensure free of microorganism contamination before xenograft assay. The mice were monitored daily post implantation to observe tumor development, up to 2 months. At the end of the experiments or when tumors with significant size (less than

10% of body weight) were observed, the mice were euthanized and the tumors were dissected and subjected to histology analysis.

Cell cultures

Primary human ovarian surface epithelial (HOSE) were isolated and provided by Dr. Andrew K. Godwin (Fox Chase Cancer Center). HOSE cells were cultured in media containing 6 g/l of HEPES, 15% FBS, 1× antibiotic-antimycotic, and insulin, as reported previously [55]. Ovarian cancer cells were cultured in RPMI-1640 media supplemented with 10% FBS and 1× antibiotic-antimycotic. All cells were maintained at 37 °C in a humidified atmosphere of 5% CO_2.

Primary mouse ovarian surface epithelial (MOSE) cells were isolated from 2 to 3 ovaries of p53 (-/-) or wild type BL6 mice of 3 to 6 months of age by collagenase digestion for an hour, as described previously [69, 70]. The cells released were harvested for culturing, and were found to be more than 90% epithelial origin as characterized by cytokeratin staining [70]. These primary cells were used for experiments following a brief culture and expansion of 4 to 7 days.

Small interfering RNA (siRNA) transfection

Lamin A/C expression in human cells (HOSE) was silenced using siRNA reagent (Catalog # sc-35776) purchased from Santa Cruz Biotech, Inc (Santa Cruz, CA). siRNA specific for mouse lamin A/C (Lmna (ID 16905) Trilencer-27 Mouse siRNA) was purchased from OriGene Inc. The siRNA is a mixture of 3 to 5 RNA oligonucleotides of 19–27 base long with sequences specifically targeting human lamin A/C, as described previously [55]. The siRNA oligonucleotides were transfected into HOSE or MOSE cells using Lipofectamine 2000 according to the manufacturer's protocol (Invitrogen, CA). Cells were analyzed 72 h after transfection for Western blot and immunofluorescence microscopy analysis. Time-lapse video microscopy was performed 12 h after lamin A/C suppression. Cells were seeded in 24 well Falcon plate and transfected the next day with scrambled siRNA (control) or lamin A siRNA in serum reduced Opti-MEM media. The media was changed 10 h after transfection with phenol red free filming media containing 15% FBS, HEPES, glutamine and antibiotic for ovarian epithelial cells. Time-lapse image acquisition was performed every 5 min for 48 h with a 40× dry objective lens on Nikon Eclipse TE 300 microscope linked to a Roper Scientific photometrics 12-bit range Camera. Image acquisition was done using Meta imaging series (MetaVue) software. Stacked images were assembled with MetaVue software to make the movie.

Flow cytometric analysis

MOSE cells were seeded in T75 flasks and transfected the next day with siRNA in serum reduced Opti-MEM media. For fixed cells analysis, cells were re-suspended in ice-cold ethanol/PBS (70% v/v) by gentle agitation. The fixed cells were kept at – 20 °C until assayed. Prior to flow cytometric analysis, cells were centrifuged at 1,200 rpm for 5 min and washed twice with PBS before re-suspension in 0.5 mm vybrant violet dye. Cells were then incubated at 37 °C for 30 min before flow cytometric analysis for DNA content, as described previously [28, 55].

Immunofluorescence microscopy, time-lapse video, and immunohistochemistry

Briefly for immunofluorescence microscopy, adhered cells on 4-well chambered glasses were washed twice with PBS at room temperature, fixed with 4% paraformaldehyde for 15 min, and permeablized with 0.5% Triton X-100 for 5 min. Then, the cells were washed three times with PBS, blocked with 3% BSA in PBS containing 0.1% Tween-20 for 30 min, and incubated for 1 h at 37 °C with primary antibodies that were diluted (1/200) in 1% BSA in PBS containing 0.1% Tween- 20. AlexaFluor 488-conjugated (green fluorescence) or AlexaFluor 594-conjugated (red fluorescence) secondary antibodies were used. Nuclei and chromosomes were stained with Hoechst 33342 solution (1 M). Cells were washed three times, then mounted and sealed in anti-fade reagent containing 100 mM of n-propyl gallate (pH 7.4), 90% glycerol in PBS. Immunofluorescence stainings were viewed with 60× or 100× oil objective lens on Nikon Eclipse TE 300 microscope linked to a Roper Scientific photometrics 12-bit range camera. Image acquisition was done using Meta imaging series (MetaVue) software. Images were merged using MetaVue software.

For time-lapse video microscopy, the acquisition of sequential image of cells expressing histone H2B-GFP was made every 5 min for up to 24 h as described previously [28].

Karyotyping and chromosome analysis

Chromosome number counting and cytogenetic analysis were performed by the Cytogenetics & Molecular Diagnostic Laboratory of the University of Miami core facility. The cells were growth arrested at metaphase by incubation with colcemid for 8 h. For each cell preparation, 50 metaphase spreads were obtained and subjected to G-bands with 400 banding resolution. At least 10 chromosome spread samples from each preparation were randomly selected and estimated for chromosome number, and a few samples were used for chromosome identification by an experienced cytogenetist and certified by the facility director.

Abbreviations

BSA: Bovine serum albumin; GFP: Green fluorescence protein; HOSE: Human ovarian surface epithelial; MOPS: 3-Morpholinopropane-1-sulfonic acid, 3-(N-Morpholino) propanesulfonic acid and 3-Morpholinopropanesulfonic acid; MOSE: Mouse ovarian surface epithelial; PBS: Phosphate-buffered saline; PMSF: Phenylmethylsulfonyl fluoride; SDS: Sodium dodecyl sulfate; siRNA: Small interfering RNA; TBS: Tris-buffered saline; TBST: Tris-buffered saline with Tween-20

Acknowledgement

Our lab alumni and students including Wensi Tao, Linlin Gao, Santas Rosario, Ziqi Wang, Justin Correa, Jessica Clark, and Anthony Guerrero have worked on this project and contributed to the basis of the current work. We also thank our colleague Dr. Robert Moore for materials and technical advice in the course of the experiments. We acknowledge the excellent technical assistance from Flow Cytometry Core and Cytogenetic Core at the University of Miami Miller School of Medicine. Primary HOSE cells were provided by the lab of Dr. Andrew K. Godwin (Fox Chase Cancer Center). We thank our colleagues, Drs. Sophia George and Robert Moore, for advice and discussion during the course of the experiments and the preparation of the manuscript.

Funding

These studies were supported by funds from concept awards BC097189 and BC076832 from Department of Defense (USA). Grants R01 CA095071, R01 CA099471, and CA79716 to X.X. Xu from NCI, NIH also contributed to the studies. The work is also partially supported by a pilot grant from the University of Miami cFAR grant P30AI073961 from NIH.

Authors' contributions

CDC and XXX made the initial observation and conceived the main hypothesis of the study. CDC performed the majority of the experiments included in this manuscript, including cell culture study of expression by Western blot, flow cytometry, immunofluorescence microscopy, and cell growth assay. ERS contributed to cell culture and immunofluorescence microscopy work. TMY assisted tumor and mouse model experiments. CDC prepared the first draft of the manuscript, XXX made revisions, and all authors were involved in editing and writing. All authors participated in discussion and refining of the ideas and design of the experiments. All read and approved the final manuscript.

Competing interests

The authors declare that they have no competing interests.

Author details

[1]Sylvester Comprehensive Cancer Center/University of Miami, Miami, Florida 33136, USA. [2]Department of Cell Biology, University of Miami Miller School of Medicine, Miami, FL 33136, USA. [3]Institute of Biomedical Sciences, Laboratory of Biochemistry and Molecular Biology, University of Abomey-Calavi, Abomey Calavi, Benin.

References

1. Cancer Genome Atlas Research Network. Integrated genomic analyses of ovarian carcinoma. Nature. 2011;474:609–15. Erratum in: Nature 2012; 490: 298.
2. Berchuck A, Kohler MF, Marks JR, Wiseman R, Boyd J, Bast Jr RC. The p53 tumor suppressor gene frequently is altered in gynecologic cancers. Am J Obstet Gynecol. 1994;70:246–52.
3. Flesken-Nikitin A, Choi KC, Eng JP, Shmidt EN, Nikitin AY. Induction of carcinogenesis by concurrent inactivation of p53 and Rb1 in the mouse ovarian surface epithelium. Cancer Res. 2003;63:3459–63.
4. Orsulic S, Li Y, Soslow RA, Vitale-Cross LA, Gutkind JS, Varmus HE. Induction of ovarian cancer by defined multiple genetic changes in a mouse model system. Cancer Cell. 2002;1(1):53–62.
5. Chen CM, Chang JL, Behringer RR. Tumor formation in p53 mutant ovaries transplanted into wild-type female hosts. Oncogene. 2004;23:7722–5.
6. Boveri T. Zur Frage der Enstehung maligner Tumoren. Jena: Gustav Fischer Verlag; 1914.
7. Holland AJ, Cleveland DW. Boveri revisited: chromosomal instability, aneuploidy and tumorigenesis. Nat Rev Mol Cell Biol. 2009;10:478–4787.
8. King RW. When 2 + 2 = 5: the origins and fates of aneuploid and tetraploid cells. Biochim Biophys Acta. 2008;1786:4–14.
9. Shi Q, King RW. Chromosome nondisjunction yields tetraploid rather than aneuploid cells in human cell lines. Nature. 2005;437:1038–42.
10. Jefford CE, Irminger-Finger I. Mechanisms of chromosome instability in cancers. Crit Rev Oncol Hematol. 2006;59:1–14.
11. Roschke AV, Tonon G, Gehlhaus KS, McTyre N, Bussey KJ, Lababidi S, Scudiero DA, Weinstein JN, Kirsch IR. Karyotypic complexity of the NCI-60 drug-screening panel. Cancer Res. 2003;63:8634–47.
12. Duesberg P. Does aneuploidy or mutation start cancer? Science. 2005;307:41.
13. Ganem NJ, Storchova Z, Pellman D. Tetraploidy, aneuploidy and cancer. Curr Opin Genet Dev. 2007;17:157–62.
14. Micho F, Iwasa Y, Vogelstein B, Lengauer C, Nowak MA. Can chromosomal instability initiate tumorigenesis? Semin Cancer Biol. 2005;15:43–9.
15. Rajagopalan H, Lengauer C. Aneuploidy and cancer. Nature. 2004;432:338–41.
16. Storchova Z, Pellman D. From polyploidy to aneuploidy, genome instability and cancer. Nat Rev Mol Cell Biol. 2004;5:45–54.
17. Weaver BA, Cleveland DW. Does aneuploidy cause cancer? Curr Opin Cell Biol. 2006;18:658–67.
18. Fukasawa K. Oncogenes and tumour suppressors take on centrosomes. Nat Rev Cancer. 2007;7:911–24.
19. Margolis RL. Tetraploidy and tumor development. Cancer Cell. 2005;8:353–4.
20. Pihan G, Doxsey SJ. Mutations and aneuploidy: co-conspirators in cancer? Cancer Cell. 2003;4:89–94.
21. Boyd J, Pienta KJ, Getzenberg RH, Coffey DS, Barrett JC. Preneoplastic alterations in nuclear morphology that accompany loss of tumor suppressor phenotype. J Natl Cancer Inst. 1991;83:862–6.
22. Zink D, Fischer AH, Nickerson JA. Nuclear structure in cancer cells. Nat Rev Cancer. 2004;4:677–87.
23. Hsu CY, Kurman RJ, Vang R, Wang TL, Baak J, Shih IM. Nuclear size distinguishes low- from high-grade ovarian serous carcinoma and predicts outcome. Hum Pathol. 2005;36:1049–54.
24. Palmer JE, Sant Cassia LJ, Irwin CJ, Morris AG, Rollason TP. The prognostic value of nuclear morphometric analysis in serous ovarian carcinoma. Int J Gynecol Cancer. 2008;18:692–701.
25. Partin AW, Walsh AC, Pitcock RV, Mohler JL, Epstein JI, Coffey DS. A comparison of nuclear morphometry and Gleason grade as a predictor of prognosis in stage A2 prostate cancer: a critical analysis. J Urol. 1989;142:1254–8.
26. Pienta KJ, Coffey DS. Correlation of nuclear morphometry with progression of breast cancer. Cancer. 1991;68:2012–6.
27. Papanicolaou GN. A new procedure for staining vaginal smears. Science. 1942;95:438–9.
28. Capo-chichi CD, Cai KQ, Testa JR, Godwin AK, Xu XX. Loss of GATA6 leads to nuclear deformation and aneuploidy in ovarian cancer. Mol Cell Biol. 2009; 29:4766–77.
29. Debes JD, Sebo TJ, Heemers HV, Kipp BR, Haugen DL, Lohse CM, Tindall DJ. p300 modulates nuclear morphology in prostate cancer. Cancer Res. 2005; 65:708–12.
30. Fischer AH, Taysavang P, Jhiang SM. Nuclear envelope irregularity is induced by RET/PTC during interphase. Am J Pathol. 2003;163:1091–100.
31. Nickerson JA. Nuclear dreams: the malignant alteration of nuclear architecture. J Cell Biochem. 1998;70:172–80.
32. Dechat T, Pfleghaar K, Sengupta K, Shimi T, Shumaker DK, Solimando L, Goldman RD. Nuclear lamins: major factors in the structural organization and function of the nucleus and chromatin. Genes Dev. 2008;22:832–53.
33. Gorjánácz M, Jaedicke A, Mattaj IW. What can Caenorhabditis elegans tell us about the nuclear envelope? FEBS Lett. 2007;581:2794–801.
34. Margalit A, Liu J, Fridkin A, Wilson KL, Gruenbaum Y. A lamin-dependent pathway that regulates nuclear organization, cell cycle progression and germ cell development. Novartis Found Symp. 2005;264:231–40. discussion 240–5.
35. Wilson KL. The nuclear envelope, muscular dystrophy and gene expression. Trends Cell Biol. 2000;10:125–9.
36. Wilson KL, Berk JM. The nuclear envelope at a glance. J Cell Sci. 2010;123: 1973–8.
37. Lammerding J, Fong LG, Ji JY, Reue K, Stewart CL, Young SG, Lee RT. Lamins A and C but not lamin B1 regulate nuclear mechanics. J Biol Chem. 2006; 281:25768–80.
38. Liu B, Wang J, Chan KM, Tjia WM, Deng W, Guan X, Huang JD, Li KM, Chau PY, Chen DJ, Pei D, Pendas AM, Cadiñanos J, López-Otín C, Tse HF, Hutchison C, Chen J, Cao Y, Cheah KS, Tryggvason K, Zhou Z. Genomic instability in laminopathy-based premature aging. Nat Med. 2005;11:780 5.

39. Heald R, McKeon F. Mutations of phosphorylation sites in lamin A that prevent nuclear lamina disassembly in mitosis. Cell. 1990;61:579–89.

40. Liu J, Rolef Ben-Shahar T, Riemer D, Treinin M, Spann P, Weber K, Fire A, Gruenbaum Y. Essential roles for Caenorhabditis elegans lamin gene in nuclear organization, cell cycle progression, and spatial organization of nuclear pore complexes. Mol Biol Cell. 2000;11:3937–47.

41. Liu J, Lee KK, Segura-Totten M, Neufeld E, Wilson KL, Gruenbaum Y. MAN1 and emerin have overlapping function (s) essential for chromosome segregation and cell division in Caenorhabditis elegans. Proc Natl Acad Sci U S A. 2003;100:4598–603.

42. Sullivan T, Escalante-Alcalde D, Bhatt H, Anver M, Bhat N, Nagashima K, Stewart CL, Burke B. Loss of A-type lamin expression compromises nuclear envelope integrity leading to muscular dystrophy. J Cell Biol. 1999;147:913–20.

43. Cao K, Capell BC, Erdos MR, Djabali K, Collins FS. A lamin A protein isoform overexpressed in Hutchinson-Gilford progeria syndrome interferes with mitosis in progeria and normal cells. Proc Natl Acad Sci U S A. 2007;104: 4949–54.

44. Dechat T, Shimi T, Adam SA, Rusinol AE, Andres DA, Spielmann HP, Sinensky MS, Goldman RD. Alterations in mitosis and cell cycle progression caused by a mutant lamin A known to accelerate human aging. Proc Natl Acad Sci U S A. 2007;104:955–60.

45. Lin F, Worman HJ. Expression of nuclear lamins in human tissues and cancer cell lines and transcription from the promoters of the lamin A/C and B1 genes. Exp Cell Res. 1997;236:378–84.

46. Röber RA, Weber K, Osborn M. Differential timing of nuclear lamin A/C expression in the various organs of the mouse embryo and the young animal: a developmental study. Development. 1989;105:365–78.

47. Foster CR, Przyborski SA, Wilson RG, Hutchison CJ. Lamins as cancer biomarkers. Biochem Soc Trans. 2010;38:297–300.

48. Agrelo R, Setien F, Espada J, Artiga MJ, Rodriguez M, Pérez-Rosado A, Sanchez-Aguilera A, Fraga MF, Piris MA, Esteller M. Inactivation of the lamin A/C gene by CpG island promoter hypermethylation in hematologic malignancies, and its association with poor survival in nodal diffuse large B-cell lymphoma. J Clin Oncol. 2005;23:3940–7.

49. Stadelmann B, Khandjian E, Hirt A, Lüthy A, Weil R, Wagner HP. Repression of nuclear lamin A and C gene expression in human acute lymphoblastic leukemia and non-Hodgkin's lymphoma cells. Leuk Res. 1990;14:815–21.

50. Willis ND, Cox TR, Rahman-Casañs SF, Smits K, Przyborski SA, van den Brandt P, Van Engeland M, Weijenberg M, Wilson RG, De Bruïne A, Hutchison CJ. Lamin A/C is a risk biomarker in colorectal cancer. PLoS ONE. 2008;3, e2988.

51. Machiels BM, Broers JL, Raymond Y, De Ley L, Kuijpers HJ, Caberg NE, Ramaekers FC. Abnormal A-type lamin organization in a human lung carcinoma cell line. Eur J Cell Biol. 1995;67:328–35.

52. Capo-chichi CD, Cai KQ, Smedberg J, Ganjei-Azar P, Godwin AK, Xu XX. Loss of A-type lamin expression compromises nuclear envelope integrity in breast cancer. Chin J Cancer. 2011;30:415–25.

53. Moss SF, Krivosheyev V, De Souza A, Chin K, Gaetz HP, Chaudhary N, Worman HJ, Holt PR. Decreased and aberrant nuclear lamin expression in gastrointestinal tract neoplasms. Gut. 1999;45:723–9.

54. Wu Z, Wu L, Weng D, Xu D, Geng J, Zhao F. Reduced expression of lamin A/C correlates with poor histological differentiation and prognosis in primary gastric carcinoma. J Exp Clin Cancer Res. 2009;28:8.

55. Capo-chichi CD, Cai KQ, Simpkins F, Ganjei-Azar P, Godwin AK, Xu XX. Nuclear envelope structural defects cause chromosomal numerical instability and aneuploidy in ovarian cancer. BMC Med. 2011;9:28.

56. Naeem AS, Zhu Y, Di WL, Marmiroli S, O'Shaughnessy RF. AKT1-mediated Lamin A/C degradation is required for nuclear degradation and normal epidermal terminal differentiation. Cell Death Differ. 2015;22:2123–32.

57. Kochin V, Shimi T, Torvaldson E, Adam SA, Goldman A, Pack CG, Melo-Cardenas J, Imanishi SY, Goldman RD, Eriksson JE. Interphase phosphorylation of lamin A. J Cell Sci. 2014;127:2683–96.

58. Bertacchini J, Beretti F, Cenni V, Guida M, Gibellini F, Mediani L, Marin O, Maraldi NM, De Pol A, Lattanzi G, Cocco L, Marmiroli S. The protein kinase Akt/PKB regulates both prelamin A degradation and Lmna gene expression. FASEB J. 2013;27:2145–55.

59. Hudson ME, Pozdnyakova I, Haines K, Mor G, Snyder M. Identification of differentially expressed proteins in ovarian cancer using high-density protein microarrays. Proc Natl Acad Sci U S A. 2007;104:17494–9.

60. Alaiya AA, Franzén B, Fujioka K, Moberger B, Schedvins K, Silfversvärd C, Linder S, Auer G. Phenotypic analysis of ovarian carcinoma: polypeptide expression in benign, borderline and malignant tumors. Int J Cancer. 1997; 73:678–83.

61. Vargas JD, Hatch EM, Anderson DJ, Hetzer MW. Transient nuclear envelope rupturing during interphase in human cancer cells. Nucleus. 2012;3:88–100.

62. Hatch EM, Fischer AH, Deerinck TJ, Hetzer MW. Catastrophic nuclear envelope collapse in cancer cell micronuclei. Cell. 2013;154:47–60.

63. Aristei C, Stracci F, Guerrieri P, Anselmo P, Armellini R, Rulli A, Barberini F, Latini P, Menghini AR. Frequency of sister chromatid exchanges and micronuclei monitored over time in patients with early-stage breast cancer: results of an observational study. Cancer Genet Cytogenet. 2009;192:24–9.

64. Shimizu N, Itoh N, Utiyama H, Wahl GM. Selective entrapment of extrachromosomally amplified DNA by nuclear budding and micronucleation during S phase. J Cell Biol. 1998;140:1307–20.

65. Torres EM, Williams BR, Amon A. Aneuploidy: cells losing their balance. Genetics. 2008;179:737–46.

66. Williams BR, Prabhu VR, Hunter KE, Glazier CM, Whittaker CA, Housman DE, Amon A. Aneuploidy affects proliferation and spontaneous immortalization in mammalian cells. Science. 2008;322:703–9.

67. Thompson SL, Compton DA. Proliferation of aneuploid human cells is limited by a p53-dependent mechanism. J Cell Biol. 2010;188:369–1.

68. Jacks T, Remington L, Williams BO, Schmitt EM, Halachmi S, Bronson RT, Weinberg RA. Tumor spectrum analysis in p53-mutant mice. Curr Biol. 1994; 4:1–7.

69. Cai KQ, Wang Y, Smith ER, Smedberg JL, Yang DH, Yang WL, Xu XX. Global deletion of Trp53 reverts ovarian tumor phenotype of the germ cell-deficient white spotting variant (Wv) mice. Neoplasia. 2015;17:89–100.

70. Wang Y, Cai KQ, Smith ER, Yeasky TM, Moore R, Ganjei-Azar P, Klein-Szanto AJ, Godwin AK, Hamilton TC, Xu XX. Follicle Depletion Provides a Permissive Environment for Ovarian Carcinogenesis. Mol Cell Biol. 2016;36:2418–30.

4

Dental pulp pluripotent-like stem cells (DPPSC), a new stem cell population with chromosomal stability and osteogenic capacity for biomaterials evaluation

Raquel Núñez-Toldrà[1,2], Ester Martínez-Sarrà[1,2], Carlos Gil-Recio[1,2], Miguel Ángel Carrasco[3], Ashraf Al Madhoun[4], Sheyla Montori[1,2] and Maher Atari[1,2,5*]

Abstract

Background: Biomaterials are widely used to regenerate or substitute bone tissue. In order to evaluate their potential use for clinical applications, these need to be tested and evaluated in vitro with cell culture models. Frequently, immortalized osteoblastic cell lines are used in these studies. However, their uncontrolled proliferation rate, phenotypic changes or aberrations in mitotic processes limits their use in long-term investigations. Recently, we described a new pluripotent-like subpopulation of dental pulp stem cells derived from the third molars (DPPSC) that shows genetic stability and shares some pluripotent characteristics with embryonic stem cells. In this study we aim to describe the use of DPPSC to test biomaterials, since we believe that the biomaterial cues will be more critical in order to enhance the differentiation of pluripotent stem cells.

Methods: The capacity of DPPSC to differentiate into osteogenic lineage was compared with human sarcoma osteogenic cell line (SAOS-2). Collagen and titanium were used to assess the cell behavior in commonly used biomaterials. The analyses were performed by flow cytometry, alkaline phosphatase and mineralization stains, RT-PCR, immunohistochemistry, scanning electron microscopy, Western blot and enzymatic activity. Moreover, the genetic stability was evaluated and compared before and after differentiation by short-comparative genomic hybridization (sCGH).

Results: DPPSC showed excellent differentiation into osteogenic lineages expressing bone-related markers similar to SAOS-2. When cells were cultured on biomaterials, DPPSC showed higher initial adhesion levels. Nevertheless, their osteogenic differentiation showed similar trend among both cell types. Interestingly, only DPPSC maintained a normal chromosomal dosage before and after differentiation on 2D monolayer and on biomaterials.

Conclusions: Taken together, these results promote the use of DPPSC as a new pluripotent-like cell model to evaluate the biocompatibility and the differentiation capacity of biomaterials used in bone regeneration.

Keywords: Dental pulp, Stem cells, Pluripotency, Genetic stability, Osteogenic differentiation, Biomaterials

* Correspondence: matari@uic.es
[1]Regenerative Medicine Research Institute, Universitat Internacional de Catalunya, Barcelona, Spain
[2]Chair of Regenerative Implantology MIS-UIC, Barcelona, Spain
Full list of author information is available at the end of the article

Background

The increase in life expectancy has been associated with a rise in the number of bone-grafting procedures for diseases such as osteoporosis, arthritis, tumors and trauma; placing an even larger demand on the healthcare system to replace and restore bone loss [1]. Recently, a great deal of efforts has focused in the field of bone tissue engineering, and, particularly, in the area of stem cell biology and how to modulate their behavior through environmental cues [1].

Biomaterials have been shown to allow the guidance of stem cells in vitro as well as in vivo. In order to assess their biocompatibility as well as their ability to differentiate cells into specific lineages, these need to be tested in an in vitro cell culture model. For this purpose, many established cell lines and models have emerged to address the surge in research in this field [2]. For instance, for bone related biomaterials, most studies have examined their osteogenic potential using immature osteoblasts, immortalized cell lines or mesenchymal stem cells among others. Primary cells, such as lineage-specific osteoblasts, can be isolated and cultivated relatively easily; however, they have a limited lifespan [3]. Immortalized cell lines, such as the human sarcoma osteogenic cell line (SAOS-2) have been frequently used in applied biology since they are from human origin while providing unlimited number of cells [2, 4]. Nevertheless, these cell lines, due to their cancer origin, usually possess phenotype changes between passages, aberrations in mitotic processes and lack of growth inhibition, which limits their use in long-term investigations [5]. On the other hand, mesenchymal stem cells (MSC), which can be isolated from many adult tissues, are an attractive cell source for tissue engineering. These cells are self-renewable with a high proliferative capacity and possess a multi-lineage differentiation potential [6]. However, long-term MSC culture conditions, for their maintenance and expansion, cause morphological and immune-phenotypical changes which lead to cell senescence and alternations in their differentiation potential. For example, morphological abnormalities, cellular enlargement, miss expression of specific surface markers and an ultimate growth inhibition are associated with MSC that are cultured beyond passage 12 [7]. Therefore, there is a need to find a cell type with genetic stability and stemness characteristics to be used to evaluate biomaterials in cell therapy applications.

The dental pulp is an accessible niche housing neural crest-derived stem cells. This niche contains several populations of dental pulp stem cells (DPSC), with different properties. The first characterized population was the dental pulp mesenchymal stem cells (DPMSC), with multi-potential capability. Recently, a new stem cell population from the human dental pulp of third molars has been isolated. These cells, named dental pulp pluripotent-like stem cells (DPPSC) have particular culture conditions and, unlike DPMSC, express pluripotency markers until late passages and are able to differentiate into cells from the three germ layers (endoderm, mesoderm and ectoderm) [8, 9]. We therefore consider that DPPSC could be a promising cell population that can be used to evaluate the biological properties of biomaterials. The use of differentiated cells, e.g. osteoblast cells, limits the relevance of the biomaterial since it is already expected that the biomaterial will allow expression of osteogenic markers. For this purpose, the use of pluripotent cells that can potentially differentiate into any lineage can be properly guided by the biomaterial and hence demonstrating the efficiency of the biomaterial [10]. While this is the main purpose of pluripotent stem cells, up to date, embryonic stem cells and induced pluripotent stem cells (iPSC) have limited applications for biomaterials testing due to ethical reasons or low efficient transfections [10]. Hence, DPPSC might be used in order to overcome the current limitations of specific cell lineages or other pluripotent stem cells.

One of the key objectives of bone tissue engineering is the enhancement of stem cell mediated osteogenic differentiation under three-dimensional (3D) scaffold conditions to mimic engineering of clinically applicable bone constructs [1, 11]. In this way, scaffolds provide suitable support for cellular infiltration, migration, as well as, proper cell proliferation and differentiation [12]. Scaffolds are manufactured from several biomaterials including metals, ceramics, synthetic polymers or natural polymers. Currently, the components of the extracellular matrix play an important role as natural substrates for in vitro cells in cultures [13]. In this sense, collagen is regarded as an ideal scaffold for tissue engineering, as it provides support to connective tissues [14–16]. Collagen type I based materials are extensively used for basic cell culture applications, as well as in the fields of bioreactor technology and tissue engineering [17–19]. On the other hand, titanium and titanium alloys are primarily used in bone implant materials. Titanium has been widely used in medical practice, showing excellent biocompatibility and safety [20]. Hence, tissue compatibility, osseointegration and functional maintenance of functions are fundamental criteria for the long-term success of endosseous dental implants [21].

Therefore, the main purpose of this study was to assess the biocompatibility and the osteogenic capacity of DPPSC in the presence of different types of biomaterials used in bone regeneration studies, such as metals or natural scaffolds.

Methods

Patient selection and ethics statement

The third molars of healthy patients were extracted for orthodontic reasons and 6 different patients of different sexes and ages (14-21 years old) were selected.

The procedure and all experiments of this study were performed in accordance with the guidelines on human stem cell research issued which was approved by the Committee on Bioethics of the Universitat Internacional de Catalunya, with the study code: BIO-ELB-2013-03.

Isolation of DPPSC and DPMSC from third molars

In this study, DPPSC and DPMSC were isolated from the same dental pulps as previously described [22]. Briefly, the molars were cleaned using gauze soaked in 70% ethanol previously the extraction of the dental pulp. Then, the dental pulp tissues were disaggregated by digestion with collagenase type I (3 mg/ml; Sigma) for 60 min at 37 °C. After washing twice with DPBS (Sigma), isolated cells were cultured in two different mediums and densities in order to separate DPPSC from DPMSC. DPPSC and DPMSC were maintained and expanded under different culture conditions until passage 15.

Culture of DPPSC

The culture medium for DPPSC consisted of 60% DMEM-low glucose (Life Technologies) and 40% MCDB-201 (Sigma) supplemented with 1X SITE (Sigma), 1X LA-BSA (Sigma), 10^{-4}M ascorbic acid 2-phosphate (Sigma), 1% penicillin/streptomycin (Life Technologies), 2% FBS (Biochrom), 10 ng/ml hPDGF-BB (R&D Systems) and 10 ng/ml EGF (R&D Systems). Flasks were pre-coated with 100 ng/ml fibronectin for one hour at 37 °C in 5% CO_2 incubator. During the 2 weeks of primary culture, the medium was changed every 4 days. Cells were passaged when they were at 30% confluence by adding 0.25% trypsin-EDTA (Life Technologies) and then they were cultured at a density of 100 cells/cm^2.

Culture of DPMSC

The culture medium for DPMSC consisted of DMEM-high glucose (Life Technologies), 10% FBS (Biochrom) and 1% penicillin/streptomycin (Life Technologies). The medium was changed every 4 days during the first 2 weeks of primary culture. To propagate DPMSC, cells were detached at 80% confluence by the addition of 0.25% trypsin-EDTA (Life Technologies) and reseeded at a density of 2x10^3 cells/cm^2.

Culture of SAOS-2

The commercially available SAOS-2 cells at passage 10 (Sigma) were seeded at density of 10^{3}cells/cm^2 in DMEM-high glucose (Life Technologies) supplemented with 10% FBS (Biochrom), 2 mM L-glutamine and 1% penicillin/streptomycin (Life Technologies), at 37 °C in a 5% CO_2 incubator. The medium was changed every 3 days. After reaching 90% confluence, cells were detached by the addition of 0.25% trypsin-EDTA (Life Technologies).

Osteogenic differentiation on 2D

DPPSC isolated from three different patients at passages 1, 5 and 10 were osteogenetically stimulated for 21 days. The osteogenic medium contained α – MEM (Gibco) containing 10% heat-inactivated FBS (Biochrom), 10 mM β-glycerol phosphate (Sigma), 50 μM L-ascorbic acid (Sigma), 0.01 μM Dexamethasone (Sigma) and 1% penicillin/streptomycin (Life Technologies) solution. Cells were cultured on 24 well culture plates at a cell density of 5×10^3 cells/cm^2. The cell line SAOS-2 were used as a control and seeded at the same density. The medium was changed every 2 days.

Osteogenic differentiation on biomaterials

In order to evaluate if DPPSC are appropriate to test the osteogenic capacity of well-known biomaterials, DPPSC from 2 of the same donors, used also in 2D differentiations, were differentiated on biomaterials. The chosen biomaterials, based on the extensive previous research using different types of cells, were collagen I based cell carriers (CCC) and titanium Ti6Al4V disks. The genetic stability was evaluated by sCGH following each differentiation step. A diagram of the experimental design is provided in Fig. 1.

Collagen CCC

The CCC sheets (Viscofan Bio Engineering) were equilibrated overnight in distilled water (200 μL per sheet) at 37 °C. The disks were transferred into 48-well plates pre-loaded with distilled water. After the removal of residual water, the culture plates containing the CCC sheets were dried overnight at RT under sterile conditions in a Laminar Air-Flow Cabinet. Before cell seeding, the dried CCC sheets were equilibrated with culture medium for 10 min at 37 °C. Due to the drying process, the collagen sheets firmly attached to the plastic well without the entrapment of air. Thus, cells could only adhere to the upper surface of the CCC. DPPSC were seeded at a density of 5×10^3 cells/cm^2 with osteogenic medium for 21 days. SAOS-2 cells were seeded and differentiated under the same conditions as a control.

Ti6Al4V disks

Ti6Al4V disks were obtained by cutting commercially available titanium alloy Ti6Al4V into 2.0-mm-thick disks with a 14 mm diameter and, subsequently, the surface was alumina-blasted and acid-etched to induce roughness, thus increasing the surface area (provided by MIS Implants Technologies Ltd.).

The osteogenic differentiation was performed on Ti6Al4V disks in 24-well plates at a density of 5×10^3 cells/cm^2 using osteogenic medium for 15 days.

For bone differentiation on titanium disks there are two models (see Fig 1). The first one, coded TI DPPSC,

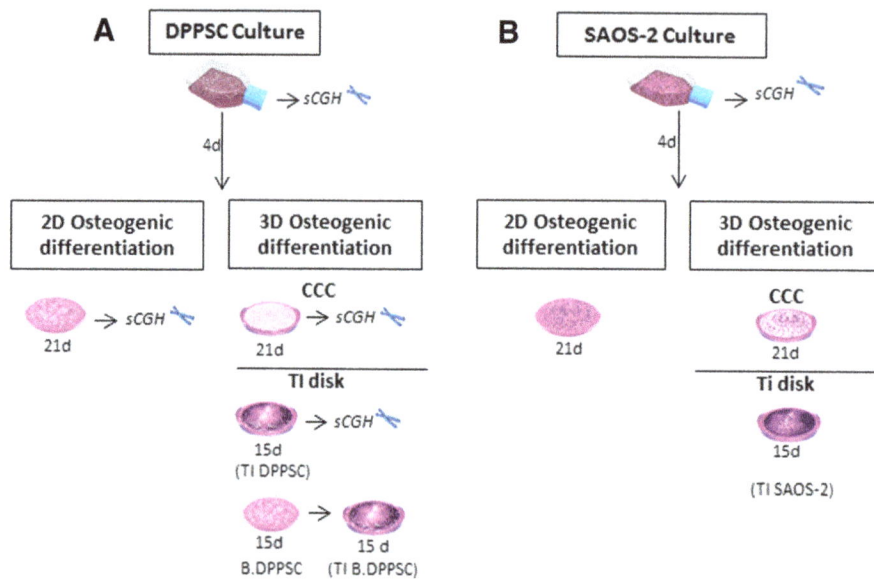

Fig. 1 Schematic diagram of the experimental design. **a** Osteogenic differentiation of DPPSC cultures. Undifferentiated DPPSC were cultivated during 15 passages, from 6 different donors. The genetic stability was checked at passages 1, 5, 10 and 15 by sCGH. 2D osteogenic differentiation: DPPSC from 3 of the 6 donors at P10 were differentiated during 21 days with osteogenic media in plastic culture plates. 3D osteogenic differentiation: DPPSC from 2 of the same donors (at passages 1, 5 and 10) used also in 2D differentiation were differentiated on biomaterials (CCC and Ti disks) during 21 and 15 days. DPPSC after 15 days of osteogenic 2D differentiation (B.DPPSC) were also seeded on titanium disks and maintained during 15 days more in osteogenic medium (TI B.DPPSC). sCGH was performed in all DPPSC differentiations. **b** SAOS-2 at P10 were used as a control cell line in all experiments. Genetic instability of SAOS-2 was evaluated before osteogenic induction experiments. *DPPSC, Dental pulp pluripotent stem cells; SAOS-2, human osteosarcoma cell line; B.DPPSC, Bone-like DPPSC, differentiated 15 days in 2D conditions; TI DPPSC, DPPSC differentiated on titanium disks; TI B. DPPSC, Bone-like DPPSC differentiated on titanium disks; CCC, Collagen I-based Cell Carrier; TI disks, Ti6Al4V disks; sCGH, short-chromosome genetic hybridization*

consist on undifferentiated DPPSC seeded directly on the disk surfaces. In the second model, coded as TI B.DPPSC, before being seeded on disks, DPPSC were differentiated for 15 days under 2D conditions, obtaining bone-like DPPSC (B.DPPSC). These osteogenic-like cells were then cultured on titanium disks and maintained during 15 days more in osteogenic medium.

Immunohistochemistry

For DPPSC analysis, the cells were fixed with Cytofix (BD Biosciences) for 1 h at room temperature and then incubated with 1% Tween-20 for 30 min at room temperature to increase cell permeability. The slides were sequentially incubated overnight at 4 °C with primary antibodies against SSEA-4 (1:50, BD Pharmingen) or OCT3/4 (1:50, BD Pharmingen) and then for 60 min at 37 °C with PE-coupled anti-rabbit IgG and FITC-coupled anti-mouse IgG (1:200, BD Pharmingen). For differentiated DPPSC at day 21, primary antibodies against osteocalcin (1:50, Millipore) and collagen IV (1:100, Abcam) were used. Alexa 568 anti-rabbit IgG (1:200, Life Technologies) and Alexa 488 anti-mouse IgG (1:200, Santa Cruz Biotechnology) were used as secondary antibodies, respectively. Between each step, the slides were washed with Perm/Wash 1X (1:200, BD

Biosciences). The cells were examined by confocal fluorescence microscopy.

Flow cytometry

To confirm the phenotype of the undifferentiated DPPSC, FACS analysis was performed. The following fluorochrome labelled monoclonal antibodies were used: CD105-FITC (R&D Systems), CD29-PE (R&D Systems), CD146-FITC (BD Pharmingen), CD45-PE (BD Pharmingen), NANOG-FITC and OCT3/4-FITC (R&D Systems). To analyze the control samples, different IgG isotypes coupled to PE and FITC fluorochromes (BD Pharmingen) were used. The cells were suspended in PBS with 2% FBS and were incubated for 45 min at 4 °C in the absence of light. Subsequently, the cells were washed twice with 2% FBS-PBS and centrifuged for 6 min at 1800 rpm, thereby removing any residual fluorochrome to avoid false positive results. The pellets were resuspended in volumes between 300 and 600 μl (depending on the number of cells) of PBS with 2% FBS. The flow cytometry measurements were made using a FACS cytometer (FACS Calibur, BD Biosciences) and analyzed with WinMDI 2.8 software. To detect and exclude non-specific unions and auto fluorescence, at least 5×10^5 cells were used for each sample.

Western blot analysis

Total protein was extracted from undifferentiated DPPSC at passages 5, 10 and 15 using Trizol Reagent (Life Technologies) according to the manufacturer's instructions. Protein quantification was performed using Bradford Reagent (Sigma). Aliquots of cell lysates at a concentration of 20 µg/µl were loaded on SDS-PAGE using 12% polyacrylamide gels and transferred onto nitrocellulose membranes. The membranes were then blocked with 1% (wt/vol) BSA in PBS containing 0.1% Tween-20. OCT3/4 and GAPDH primary antibodies (1:500, Abcam) were then incubated with the membranes, followed by washing and incubations with secondary antibodies (1:5000, Abcam). The primary antibodies used were anti-Osteocalcin (OC), anti-Osteopontin (OPN), anti-Collagen I (COL1) and anti-GAPDH as housekeeping (1:500, Abcam). The Western blot membrane was finally developed using Luminata Forte Western HRP substrate (Millipore).

Short-comparative genomic hybridization (sCGH)

The sCGH technique was performed as described in Rius M. et al. [23] in triplicates of 12-14 cells from each sample. Undifferentiated DPPSC at passages 1, 5, 10 and 15 from 6 different donors were analyzed. Differentiated DPPSC in 2D conditions (well plates) were analyzed at passages 1, 5 and 10 from 3 patients after 21 days of osteogenic differentiation. Differentiated DPPSC, from 2 donors at passage 10, cultured on biomaterials (titanium disks and collagen carrier) were evaluated after 21 days of osteogenic differentiation.

SAOS-2 cells at passage 10 were analyzed before differentiation under 2D and 3D conditions. A 47, XXY sample was used to perform the hybridization of the controls and samples. This procedure was performed by an external service (Universitat Autònoma de Barcelona).

Alkaline phosphatase (ALP) staining

ALP was determined in undifferentiated DPPSC (day 0) and after 3, 7, 15 and 21 days of DPPSC differentiation by ALP staining Kit (CosmoBio) following the manufacturer's instructions. Briefly, cells were fixed with 10% Formalin Neutral Buffer Solution, for 20 min at RT. After 3 washes with 1 ml of deionized water, 200 µl of Chromogenic Substrate were added to each well and incubated for 20 min at 37 °C. Finally, to stop the reaction, the stainings were washed again with 1 ml of deionized water and observed the obtained blue staining under an optical microscope.

Von Kossa Staining

Mineralization was determined in undifferentiated DPPSC (day 0) and after 3, 7, 15 and 21 days of DPPSC differentiation by Von Kossa Method for Calcium staining (Polysciences), following the manufacturer's protocol. Briefly, cells were fixed with 10% Formalin Neutral Buffer Solution, for 20 min at RT and stained with 0.5 ml of 3% Silver Nitrate Solution under UV light for 45 min. After 3 washes with 0.5 ml of deionized water, 0.5 ml of 5% sodium thiosulfate were added to each well for 2 min at RT, and washed again with 1 ml of deionized water 3 times. Finally, the samples were counterstained in Nuclear Fast Red for 5 min at RT and photographed with an optic microscope.

Alkaline phosphatase (ALP) activity

ALP activity was determined quantitatively in undifferentiated DPPSC (day 0) and after 7 and 15 days of differentiation in DPPSC, B. DPPSC and SAOS-2 cells on TI disks. An ALP Kit (BioSystems) was used in accordance with the manufacturer's instructions. In short, the ALP was measured by determining the velocity of the formation of 4-nitrophenol starting from 4-nitrophenol phosphate. In order to initiate the ALP enzyme reaction, 20 µl of each sample supernatant were added to 1 ml of working solution containing 4:1 stock substrate solution (4-nitrophenol phosphate) and buffer solution (1.2 M diethanoloamine, 0.6 mM $MgCl_2$). Finally, the absorbance of each sample was measured at 405 nm during 4 min. The increase of the absorbance was calculated with the average of absorbance per minute ($\Delta A/min$).

Scanning electron microscopy (SEM)

SEM analysis of the biomaterials was performed prior to differentiation and after DPPSC and SAOS-2 cells were differentiated on the biomaterials. The samples were fixed with 2.5% glutaraldehyde (Ted Pella Inc.) in 0.1 M Na-cacodylate buffer (EMS, Electron Microscopy Sciences, Hatfield, PA, USA) (pH 7.2) for 1 h on ice. After fixation, the samples were treated with 1% osmium tetroxide (OsO_4) for 1 h. The samples were then dehydrated in serial solutions of acetone (30–100%) with the scaffolds mounted on aluminum stubs. The samples were examined with a Zeiss 940 DSM scanning electron microscope.

RNA isolation, RT-PCR and qRT-PCR

Total RNA from undifferentiated cells, 2D differentiated cells (days 0, 7 and 21 days of differentiation) and differentiated cells on the biomaterials at the end of differentiation (day 21 in CCC and day 15 in Ti disks) was extracted by Trizol Reagent (Life Technologies) following the manufacturer's instructions. RNA aliquots of 2 µg were treated with DNase I (Life Technologies) and reverse-transcribed using a Transcription First Strand cDNA Synthesis Kit (Roche), following the manufacturer's instructions. PCR was performed using the primers listed in Table 1 for the amplification of OCT3/4, NANOG, SOX2, ALP, OC, COL1, RUNX2, VLA4, VCAM1, ITGα3, ITG and GAPDH

Table 1 Primer pairs used to assess osteoblast differentiation in the RT-PCR and qRT-PCR amplification

Gene	Accession number	Forward primer (5'-3')	Reverse primer (5'-3')	Product size (bp)	Use
OCT 3/4	NM_002701	GTGGAGAGCAACTCCGAT G	TGCAGAGCTTTGATGTCCTG	122	RT-PCR qRT-PCR
NANOG	NM_024865	CAGAAGGCCTCAGCACCTAC	ATTGTTCCAGGTCTGGTTGC	111	RT-PCR qRT-PCR
ALP	NM_000478	GGACATGCAGTACGAGCTGA	GTCAATTCTGCCTCCTTCCA	133	RT-PCR
ALP	NM_000478	CCGTGGCAACTCTATCTTTGG	GCCATACAGGATGGCAGTGA	79	qRT-PCR
COL1	NM_000088	ACTGGTGAGACCTGCGTGTA	CAGTCTGCTGGTCCATGTA	263	RT-PCR
COL1	NM_000088	CCCTGGAAAGAATGGAGATGAT	ACTGAAACC TCTGTGTCCCTTCA	139	qRT-PCR
OC	NM_199173	GTGCAGCCTTTGTGTCCAA	GCTCACACACCTCCCTCCT	129	RT-PCR
OC	NM_199173	AAGAGACCCAGGCGCTACC T	AAC TCGTCACAGTCCGGATTG	110	qRT-PCR
RUNX2	NM_001146038	TTACTGTCATGGCGGGTAAC	GGTTCCCGAGGTCCATCTA	220	RT-PCR
RUNX2	NM_001146038	AGCAAGGTTCAACGATCTGAGAT	TTTGTGAAGACGGTTATGGTCAA	81	qRT-PCR
VLA4	NM_000885	CGAACCGATGGCTCCTAGTG	CACGTCTGGCCGGGATT	115	RT-PCR
ITGα3	NM_002204	TCCGAGTCAATGTCCACAGA	GCTGGGCTACCCTATTCCTC	88	RT-PCR qRT-PCR
ITGαV	NM_002210	CCTTGCTGCTCTTGGAAC TC	ATTCTGTGGCTGTCGGAGAT	74	RT-PCR qRT-PCR
GAPDH	NM_002046	CTGGTAAAGTGGATATTGTTGCCAT	TGGAATCATATTGGAACATGTAAACC	81	RT-PCR qRT-PCR

OCT-3/4 octamer-binding transcription factor 4, *Nanog* Nanog homeobox, *ALP* alkaline phosphatase, *COL1* type I collagen, *OC* osteocalcin, *RUNX2* Runt-related protein 2, *VLA-4* Integrin alpha4beta1, *ITGα3* integrin alpha 3, *ITGαV* integrin, alpha V, *GAPDH* glyceraldehyde-3-phosphate dehydrogenase

cDNAs. The resulting amplicons were resolved by agarose gel electrophoresis. A human bone cDNA sample (Biocompare) and SAOS-2 cells were used as a positive control.

Quantitative RT-PCR was performed using a CFX96 thermocycler (Bio-Rad). A total of 50 ng of cDNA of TI DPPSC, TI B.DPPSC and TI SAOS-2 cells on Ti disks was used. cDNA samples were amplified using specific primers and SYBR Green Supermix (Bio-Rad Laboratories, Inc.). The expression levels of the genes of interest (OCT3/4, RUNX2, COL1, OC, VLA4, ITGα3 and ITGαV) were normalized against the housekeeping gene GAPDH. The relative expression levels were normalized to day 0 of the DPPSC cDNAs in 2D differentiation or with TI SAOS-2 cDNAs in titanium differentiation, which were assigned as 1. All analyses were performed using the $2^{-\Delta\Delta CT}$ method and 3 technical replicates.

Statistical analysis

Data from the osteogenic differentiation (qRT-PCR, calcium quantification, ALP activity) of DPPSC on 2D and titanium were analyzed by applying two-way analysis of variance or ANOVA for multiple factors. Statistical analysis was performed using the SPSS 21.0 software package. Confidence intervals were fixed at 95% ($P < 0.05$).

Results

Characterization of dental pulp pluripotent-like stem cells

DPPSC were isolated as a new subpopulation of DPSC with pluripotent-like characteristics. They were expanded

and characterized as previously described [22]. The results with their pluripotent-like characteristics during 15 culture passages comparing with DPMSC are shown in Additional file 1: Figure S1.

2D Osteogenic differentiation

Under 2D conditions, the osteogenic differentiation of DPPSC was investigated by examining the gene expression profile of pluripotent and osteogenic markers during 21 days of culture in osteogenic medium (Fig. 2a-b). Firstly, qRT-PCR analysis revealed a gradual decrease in the expression of the pluripotency markers OCT3/4 and NANOG during DPPSC differentiation (Fig. 2a). On the other hand, the transcript expression of the early osteogenic marker RUNX2 was shown to increase and peaked at day 7, showing notably lower expression at the end of the differentiation. Then, the OC expression was up-regulated reaching the highest levels at day 15 of differentiation (Fig. 2b).

Moreover, RT-PCR results showed that the expression levels of ALP and COL1 were pronouncedly increasing until week 3 and confirmed that RUNX2 expression was decreasing after first week of differentiation (Fig. 2c). Finally, in order to summarize the genetic pattern during the osteogenic differentiation of DPPSC, a schematic representation with the different stages was performed (Fig. 2d).

In addition, DPPSC morphology was examined by optical microscopy along the osteogenic differentiation

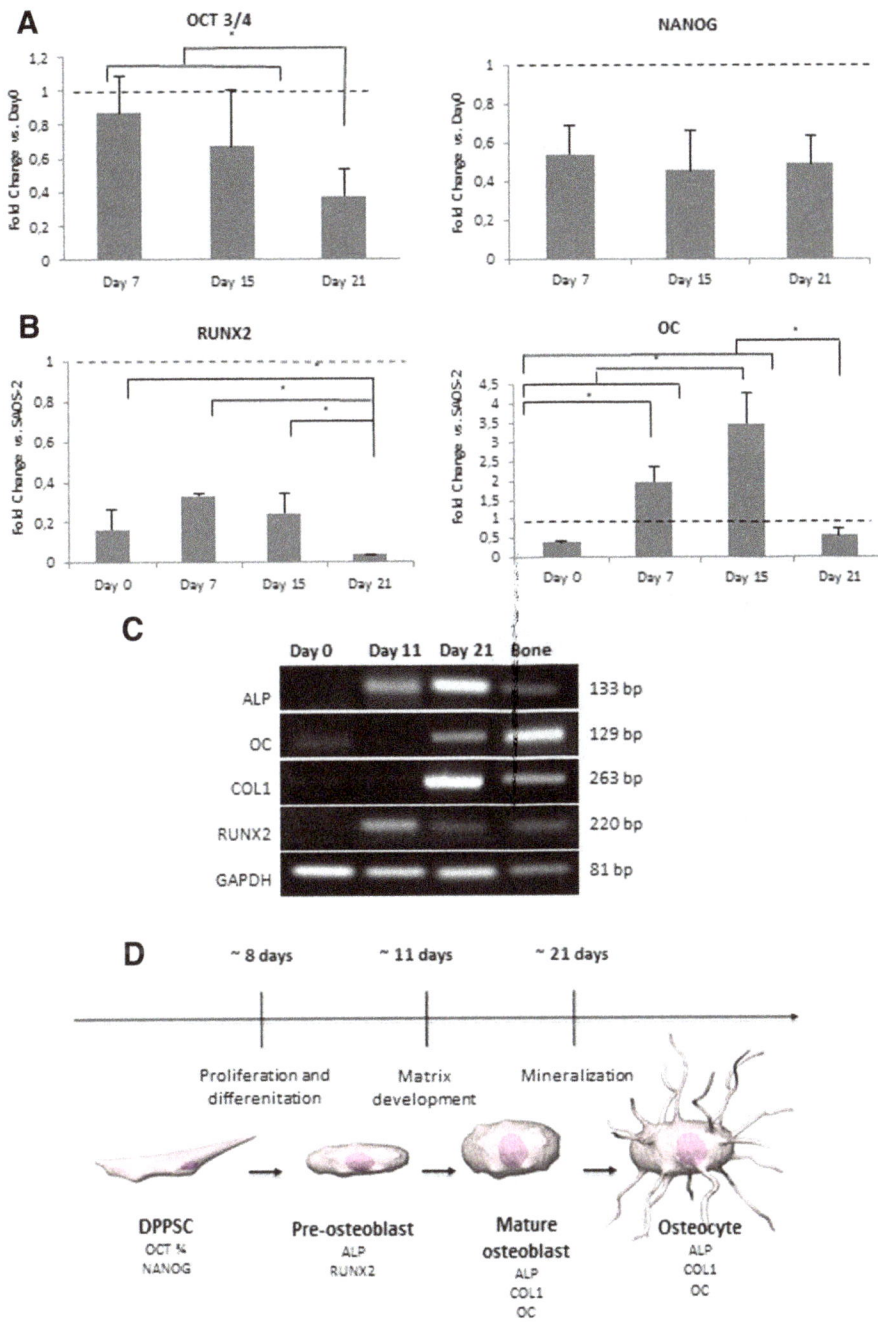

Fig. 2 Stages of DPPSC osteogenic differentiation during 21 days. **a** Relative expression of pluripotency (OCT3/4, NANOG) markers during the osteogenic differentiation of DPPSC versus DPPSC at Day 0 (dotted line). **b** Relative expression of osteogenic markers (RUNX2, OC) during the osteogenic differentiation of DPPSC versus SAOS-2 cells differentiated over 21 days in osteogenic medium (*dotted* line). **c** RT-PCR for osteogenic markers (ALP, OC, COL1, RUNX2) at days 0, 11 and 21 of differentiation. Bone cDNA was used as a positive control and GAPDH was used as a housekeeping. **d** Proposed scheme for osteogenic DPPSC culture with recognizable stages of differentiation. *$P < 0.05$

at different time points. At day 0, undifferentiated DPPSC were small-sized cells with rounded morphology, large nuclei and low cytoplasm content. By the second week of differentiation DPPSC adopted an elongated morphology and started to mineralize (Fig. 3a).

Furthermore, functional osteogenic activity was qualitatively detected at different time points. ALP activity enhancement was clearly observed in a time dependent manner over the differentiation process (Fig. 3b). Moreover, as a mineralization assay, von Kossa staining

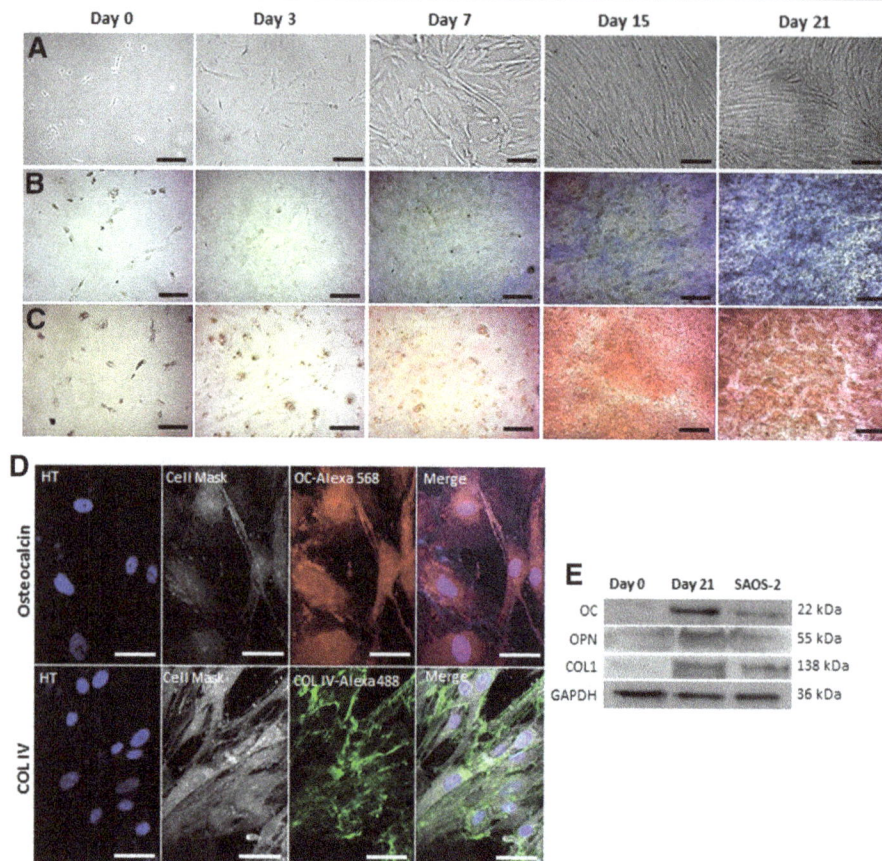

Fig. 3 Osteogenic capacity of DPPSC during 21 days. **a** Cell morphology of DPPSC observed with optic microscopy at days 0, 3, 7, 15 and 21 of osteogenic differentiation. **b** ALP activity observed with optic microscopy at days 0, 3, 7, 15 and 21 by an ALP staining (*blue*). **c** Images of mineralitzation at days 0, 3, 7, 15 and 21 stained by the von Kossa method showing mineralized bone (*brown*) and osteoid, supporting tissue and structures (*red*). **d**, **e** Osteogenic protein expression of DPPSC after 21 days. **d** Immunofluorescence analysis of DPPSC showing the expression of Osteogenic markers at day 21 of differentiation. Hoechst (HT) as a nucleus control. Scale bars: 200 μm (**a-c**), 50 μm (**d**). **e** Western Blot analysis of osteogenic markers at day 0 and 21 of differentiation. SAOS-2 cells were used as a positive control. GAPDH was used as a housekeeping control

showed the development of red regions which were rich in osteoid as well as, detected in brown, calcium phosphate depositions (Fig. 3c).

In order to corroborate the qRT-PCR results with the levels of protein expressed and hence, to evaluate the differentiation efficiency of DPPSC, the protein expression of the osteogenic markers was analyzed by immunofluorescence and Western blot analyses. Immunofluorescence analysis using specific anti- OC and Col IV antibodies showed the expression of the OC localized in the cytoplasm and the formation of the extracellular collagen matrix (Fig. 3d). In addition, higher protein expression of OC, OPN and COL 1 was detected at day 21 comparable to that observed in SAOS-2 cells (Fig. 3e).

Genetic stability

The genetic stability was checked in undifferentiated DPPSC and during their differentiation into bone-like tissue on culture plates and on biomaterials by sCGH. Undifferentiated and differentiated DPPSC exhibited a normal karyotype with no presence of any aneuploidy or any chromosome structural alteration. Therefore, results showed that DPPSC maintained the stability before and after the differentiation process (Fig. 4a-d).

Moreover, sCGH confirmed a normal karyotype of DPPSC cultured on CCC and Ti6Al4V disks after differentiation (Fig. 4e-f). In contrast, SAOS-2 cells showed genetic mutations in numerous chromosomes, probably due to their carcinogenic origin (Additional file 2: Figure S2).

Osteogenic differentiation on Biomaterials

Osteogenic differentiation on Collagen I-based Cell Carrier (CCC)

DPPSC were cultured on CCC and the differentiation was induced for 21 days (Fig. 5). SEM was utilized to visualize the scaffold/cells constructs and obtain a better understanding of the cell morphology. CCC scaffolds with the attached cells exhibited a dense microstructure. The cells were well dispersed and attached, covering the

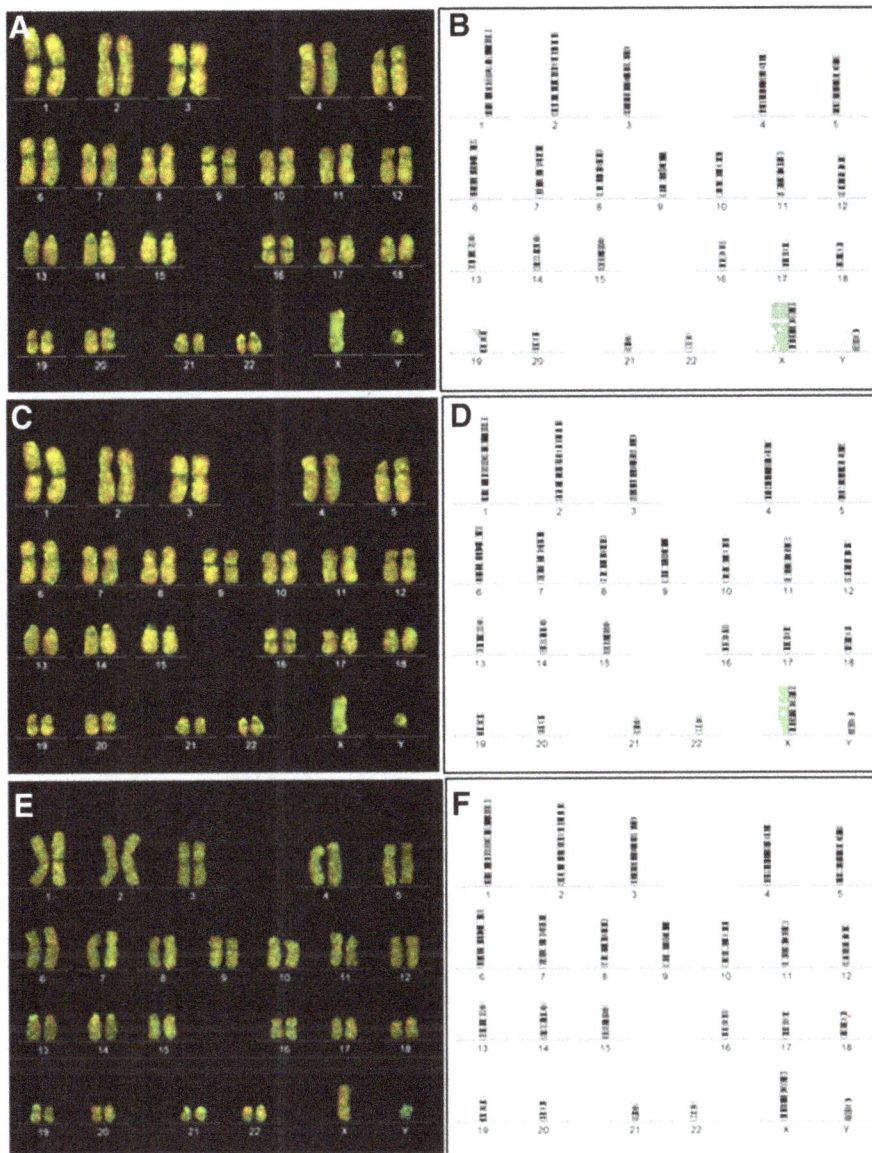

Fig. 4 Genetic stability of DPPSC, before and after osteogenic differentiation by sCGH analysis. **a**, **b** sCGH from undifferentiated DPPSC at passages 1, 5, 10 and 15 (N = 6, XX and XY donors); **a** Fluorochromes image; **b** Fixed limits summary. **c**, **d** sCGH at day 21 of 2D osteogenic differentiation from DPPSC at passages 1, 5 and 10 (N = 3, XX and XY donors). **c** Fluorochromes image; **d** Fixed limits summary. **e**, **f** CGH from DPPSC (P10) at day 21 of osteogenic differentiation on biomaterials: Collagen and Titanium carriers (N = 2, XX donors). **e** Fluorochromes image; **f** Fixed limits summary. 47, XXY control samples (labeled in *green*) and DPPSC samples (labeled in *red*) were mixed and co-hybridized onto 12 (46, XY) metaphases in triplicate. A gain in the X or Y chromosome dosage was due to sex differences

entire surface and penetrating inside the collagen scaffold (Fig. 5a-b).

Moreover, SEM images showed an extra cellular matrix formed by calcium phosphate depositions (Fig. 5c-d), also confirmed by an atomic microanalysis; with the presence of calcium and phosphorus ions, 0,43% and 3.16%, respectively (Fig. 5e). In addition, RT-PCR analysis for RNA extracted after the osteogenic induction, revealed an enhancement of the osteogenic markers ALP, OC and RUNX2 as well as adhesion markers, COL1, VCAM1 and

VLA4 (Fig. 5f) at similar levels to that of 2D- differentiated DPPSC or SAOS-2 cells. In addition, immunohistochemistry sections using specific OC and OPN antibodies showed the presence of these important proteins implied in the mineralization process (Fig. 5g-h).

Osteogenic differentiation on titanium alloy disks (TI disks)
DPPSC (TI DPPSC), 15 day bone-like DPPSC (TI B.DPPSC) and SAOS-2 cells (TI SAOS-2) were cultivated

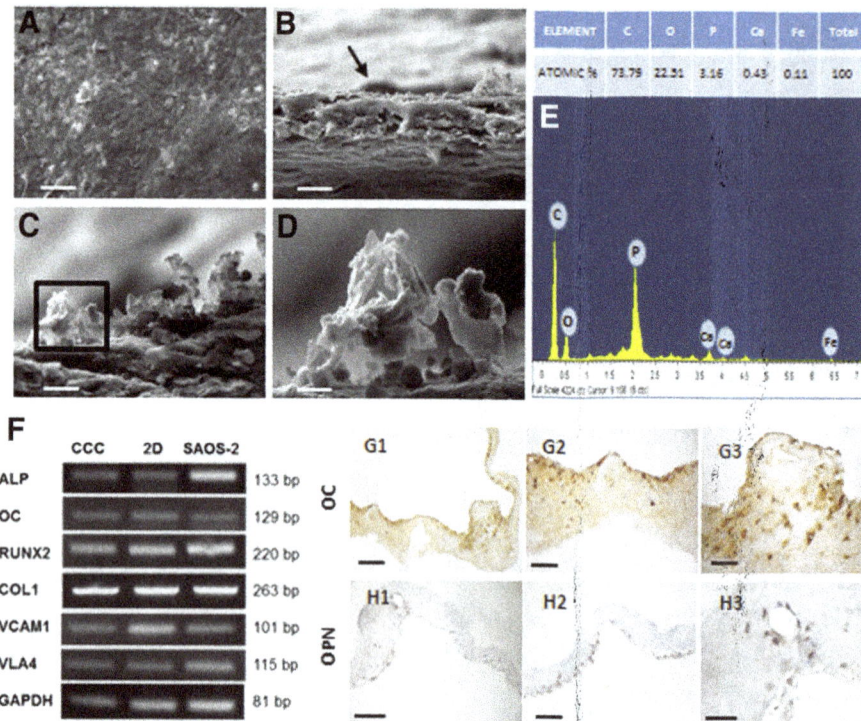

Fig. 5 Osteogenic differentiation on Collagen I-based Cell Carrier (CCC). **a, b** SEM images of differentiated DPPSC adhered on CCC surface (*black arrow*). **c** SEM image of differentiated DPPSC with hydroxyapatite deposition on CCC surface. **d** SEM image of hydroxyapatite deposition on CCC. Scale bars: 100 μm (**a**), 20 μm (**b**), 10 μm (**c**), 2 μm (**d**). **e** Microanalysis of the CCC surface with atomic concentrations. **f** RT-PCR gene expression analysis of differentiation markers (OC, ALP, COL1) and adhesion markers (VCAM1, VLA4) in DPPSC cultured on CCC, DPPSC cultured on 2D (plastic surface) and SAOS-2 cultured on CCC. GAPDH was used as a housekeeping. **g, h** Immunohistochemistry of differentiation markers (OC, OPN) in differentiated DPPSC on CCC. Scale bars: 1000 μm (**g1, h1**), 400 μm (**g2, h2**); 200 μm (**g3, h3**)

and differentiated on titanium alloy disks with osteogenic medium for 15 days.

SEM micrographs showed a high-density cell mass on the surface of the disk for all cell populations, indicating that the cells adhered and grew favorably (Fig. 6a-c). Moreover, TI B.DPPSC seemed to cover more surface than the other cell types (Fig. 6b2). The results of the ALP assay showed that the ALP activity increased significantly over time in TI DPPSC, TI B.DPPSC and TI SAOS-2, demonstrating that the cells acquired this functional activity during osteoblast differentiation (Fig. 6d). The behavior of TI DPPSC and TI B.DPPSC was similar; differences between cell types were only statistically significant when comparing TI SAOS-2 cells with the other cell types.

To further characterize the differentiation status of the cells, the expression levels of several osteogenic and adhesion markers were evaluated at the end of the osteogenic differentiation process on TI disks using SAOS-2 cells for normalization (Fig. 6e-g). Results showed that TI DPPSC and TI B.DPPSC showed less expression of the early osteogenic marker RUNX2 than SAOS-2 and more expression of the osteogenic advanced markers COL1 and OC, with the highest

expression in TI B.DPPSC (Fig. 6f). Furthermore, the analyses of the adhesion markers showed higher expression of the integrin genes in DPPSC than in SAOS-2 cells (Fig. 6g), confirming the high adhesion potential of DPPSC to the titanium surface.

Discussion

The third molar represents a very accessible organ, which is often extracted for dental reasons, and due to its late development, allows the presence of progenitor cells. Previously, we identified DPPSC as a new subpopulation of DPSC cultivated in a media containing LIF, EGF and PDGF to maintain their pluripotent state. In addition we showed that DPPSC are able to differentiate into adult tissues generating all three embryonic germ layers, i.e., endothelial cells, neurons, bone and hepatocyte-like cells [9].

DPMSC and DPPSC are obtained from the same dental pulp but cultured at different cell densities and different medium conditions. In this study, the phenotypical analysis showed high expression levels of CD29 and CD105 markers, and low expression levels of CD45, indicating that DPPSC share several similarities with DPMSC. Nevertheless, DPPSC also express pluripotency markers such as OCT3/4, NANOG, SOX2 and SSEA4,

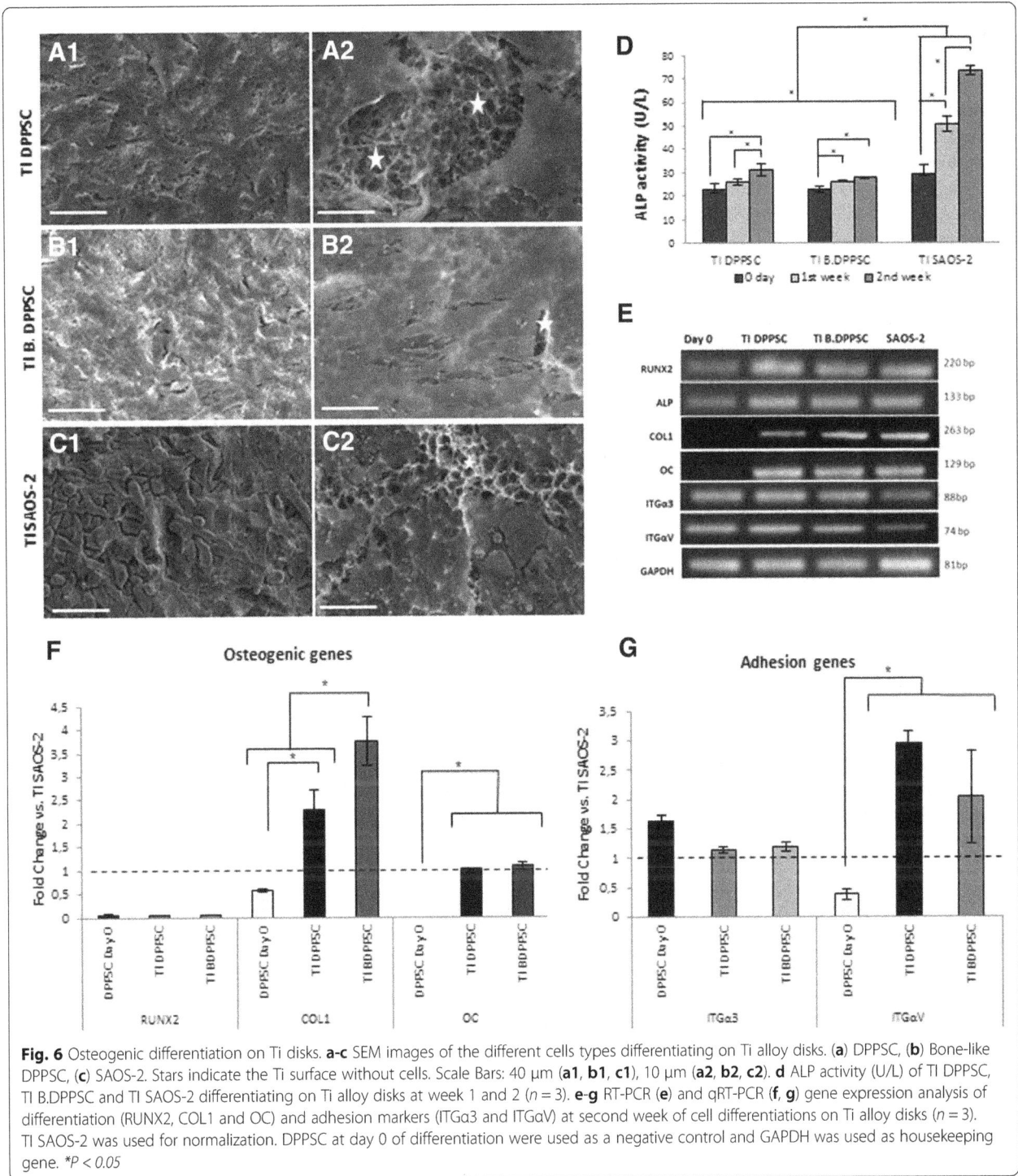

Fig. 6 Osteogenic differentiation on Ti disks. **a-c** SEM images of the different cells types differentiating on Ti alloy disks. (**a**) DPPSC, (**b**) Bone-like DPPSC, (**c**) SAOS-2. Stars indicate the Ti surface without cells. Scale Bars: 40 μm (**a1**, **b1**, **c1**), 10 μm (**a2**, **b2**, **c2**). **d** ALP activity (U/L) of TI DPPSC, TI B.DPPSC and TI SAOS-2 differentiating on Ti alloy disks at week 1 and 2 ($n = 3$). **e-g** RT-PCR (**e**) and qRT-PCR (**f**, **g**) gene expression analysis of differentiation (RUNX2, COL1 and OC) and adhesion markers (ITGα3 and ITGαV) at second week of cell differentiations on Ti alloy disks ($n = 3$). TI SAOS-2 was used for normalization. DPPSC at day 0 of differentiation were used as a negative control and GAPDH was used as housekeeping gene. *$P < 0.05$

which are indispensable for infinite stem cell division without affecting differentiation potential. On the other hand, we consider that pluripotent stem cells have higher value when testing the differentiation capacity of biomaterials, since these cells provide highly undifferentiated cells which need to be guided merely with the biomaterial. Hence, we consider that the use of osteoblastic-like cells

to determine the osteogenic capacity of biomaterials is meaningless compared to the use of pluripotent-like cells. For all these reasons, we elected DPPSC to perform the osteogenic differentiations on biomaterials.

To address the requirement to well-characterize the DPPSC differentiation process, one of the first aims of this study was to analyze the model system for osteogenesis of this new stem cell population and to establish the expression profile of bone related genes during their differentiation process.

In general, our results indicate that the osteogenic differentiation of DPPSC is in accordance with the human osteoblastic development, which can be divided into three chronologically stages: proliferation, matrix development and mineralization [24, 25]. Thus, RUNX2 is expressed during early osteoblast differentiation and is strictly required for the differentiation and appropriate functioning of osteoblasts [26]. Here, we found that RUNX2 was highly expressed during the first week of DPPSC differentiation, while the expression levels of pluripotency markers OCT3/4 and NANOG were reduced, indicating the end of the proliferation stage. At the second week, during the extracellular matrix maturation, RUNX2 expression levels were progressively down-regulated while differentiated cells began to express OC, in agreement with the reports that RUNX2 is an upstream gene of OC [25, 26]. Moreover, there was a remarkable peak of OC at day 15 that indicated an early mineralization of the differentiated DPPSC. On the other hand, by comparing the gene expression of differentiated DPPSC with differentiated SAOS-2, we can observe that SAOS-2 showed more expression of the initial osteogenic marker RUNX2 and similar levels of the advanced marker OC. It was probably due to the uncontrolled proliferation rate of SAOS-2, which constantly produces immature cells. SAOS-2 cells have been frequently used in applied biology since they are from human origin, they provide an unlimited number of cells and therefore, they are a fast and a cheap cell model to test osteogenesis in biomaterials. However, we consider that the use of pluripotent-like cells that can potentially differentiate into any lineage assesses better the role of the biomaterial in the osteogenic differentiation. Moreover, as DPPSC are in a previous stage of differentiation than SAOS-2 cells, this allows the analysis of the osteogenic differentiation since the beginning of the process.

Furthermore, RT-PCR analysis revealed that ALP was also expressed during the phase of matrix development, increasing from the first week until the end of the differentiation. This triggers the mineralization stage, commonly observed by the production of hydroxyapatite crystals [27].

At the same time, COLI was detected during the matrix development until the mineralization stage indicating that the maturation process of osteoblast-like cells produced abundant matrix proteins that were deposited as osteoid or non-mineralized bone matrix [28]. In addition, results showed very similar osteogenic expression levels between differentiated DPPSC and human bone cDNA, in accordance with a previous study of the osteogenic differentiation of DPPSC, where the bone-like tissue formed by DPPSCs in 3D were similar in complexity to human bone tissue [8].

In order to assess the functional activity of bone-like DPPSC, mineralization and ALP stainings were performed during DPPSC differentiation. By the third week, both stainings demonstrated an osteoid mineralization by the accumulation of calcium phosphate in the form of hydroxyapatite. In addition, protein analysis with high expression of late osteogenic markers, OC and OPN, confirmed the osteocyte-like phenotype of DPPSC. In summary, our results indicated that differentiated DPPSC show a behavior pattern similar to human primary osteoblasts [2, 29]. These results support the capacity of DPPSC to differentiate into bone-like tissue similar to the traditionally used osteoblastic progenitors.

After analyzing the expression profile of osteogenic genes during DPPSC differentiation, we examined the genetic stability of this pluripotent-like stem cell population before and after their differentiation process on biomaterials. The rationale was based on previous studies that used adult stem cells with genetic stability to assess the quality and osteogenic capacity of biomaterials [8, 30]. Thus, we analyzed the genetic stability of DPPSC and SAOS-2 by sCGH, a direct aneuploidy screening that allows the detection of chromosome imbalances generated by aberrant segregation and structural differences for fragments larger than 10–20 Mb [31]. Our results showed a normal chromosomal dosage during DPPSC culture expansion until passage 15. This was also evident at the end of the differentiation process, both in culture plates and on biomaterials. Nevertheless, we confirmed some genetic instability in SAOS-2 cells that have been probably related to the progression and genesis of osteosarcoma. This low stability of SAOS-2 could induce phenotype alterations, aberrations in mitotic processes or lack of growth inhibition affecting the results of biomaterial testing [5]. Furthermore, some reports demonstrate a direct correlation between culture density and the occurrence of DNA damage and genomic alterations during the culture of stem cells in vitro [32]. These effects are largely caused by the accumulation of lactic acid in the culture medium and the associated medium acidification [32, 33]. Here, the particular culture conditions of DPPSC (medium composition, low serum levels, low cell confluence before passaging and low cell culture density) could facilitate the preservation of the genomic stability, making DPPSC a more stable cell model for testing biomaterials in bone regeneration studies.

It is known that natural materials, metals and synthetic polymers scaffolds organize stem cells into complex spatial groupings which mimics native tissue [34]. In this study, we evaluated the capacity of DPPSC to grow and differentiate in a natural collagen scaffold (CCC). Natural biological and mechanical properties of native collagen provides a bio-mimetic environment for stem cells as well as a mechanical support, providing collagen based biomaterials as commonly used support for cell culture and tissue engineering [35]. Analysis of the scaffold surface and cell morphology by electron microscopy emphasized the high affinity of DPPSC to grow in CCC, the homogenous cell distribution and the high level of calcification at the final stage of the differentiation. The expression of osteogenic and adhesion markers in differentiated DPPSC in CCC was comparable to DPPSC differentiated on culture plates or SAOS-2 cultured in CCC. Moreover, an immunohistochemistry assay showed the expression of advanced osteogenic markers (OC, OPN), corroborating the presence of osteocyte-like cells and the complete differentiation and biocompatibility of DPPSC over CCC in vitro.

On the other hand, it has been demonstrated that metal scaffolds are also suitable for hard-tissue applications. The loosening of implants from bone tissues has been a problem in reconstructive surgery and joint replacement. Brånemark introduced the term "osseointegration" to describe this modality for stable fixation of titanium to bone tissue [36]. Some studies report that titanium disks favor the osseointegration, stimulating the functions of osteoblasts on the surface, such as the adhesion, proliferation or secretion of specific proteins composing the matrix [37]. The most common cell sources used for this propose, are immortalized cell lines or fibroblasts [38, 39]. However, these cell types show different characteristics from stem cells involved in bone regeneration in vivo. Currently, there are few studies regarding the osseointegration of titanium with stem cells [40]. Therefore, this study attempts to evaluate the osteogenic capacity of TI disks with an adult pluripotent-like stem cell population. For this purpose, two populations of DPPSC were used: DPPSC and B.DPPSC. B.DPPSC were DPPSC pre-differentiated for 15 days, when they reached the beginning of the mineralization stage, corresponding with the peak of OC expression. This previous differentiation of the cells on plates before transferring to titanium surfaces can improve the evaluation of biomaterials, reducing the costs (differentiated cells expand more rapidly and can be cultured at higher densities) and accelerating the process.

SEM images obtained of TI disks showed a complete coverage of the surface with all cell types, suggesting a major density, expansion and adhesion of TI B.DPPSC. Furthermore, at the second week of differentiation, ALP concentration increased in all cell types, indicating a mineralization stage with the development of a calcified matrix at 15 days of osteogenic induction. On the other hand, the high ALP activity levels in SAOS-2 could be explained by some studies which revealed that this property of SAOS-2 can differ considerably from primary osteoblasts behavior [2].

At the end of differentiation, TI SAOS-2 cells showed higher expression of initial markers ALP and RUNX2 and lower expression of advanced markers COL1 or OC. These results could indicate that this carcinogenic cell line presents cells in a more immature stage than DPPSC lines, probably due to their constant and uncontrolled proliferation rate. Moreover, we found that the expression of COL1, the most important protein in the non-mineralized matrix, was higher in TI B.DPPSC. Finally, TI DPPSC and TI B.DPPSC demonstrated also a higher expression of the adhesion markers than TI SAOS-2 cells, suggesting that DPPSC had a strong capacity to adhere on titanium surfaces.

Conclusions

In conclusion, these results support the use of DPPSC, a new pluripotent-like subpopulation of DPSC, as a good alternative model to evaluate the biocompatibility and the differentiation capacity of different types of biomaterials commonly used for bone regeneration studies. DPPSC showed high osteogenic and adhesion potential whilst seemed to maintain genetic stability during culture expansion and differentiation. However, further studies assessing their genetic stability by means of more accurate methods and their in vivo biocompatibility will be necessary before testing them in clinical applications.

Additional files

Additional file 1: Figure S1. Characterization of undifferentiated DPPSC. **a** Cell morphology of DPPSC from passage 10 observed with optic microscopy. DPPSC are characterized as small-sized cells with large nuclei and low cytoplasm content. **b** Immunofluorescence analysis of OCT3/4-FITC, SSEA4-PE, and Merge. Hoechst (HT) as a nucleus control. DPPSC were positive for these embryonic markers, and both were located in the nucleus. **c** FACS analysis of DPPSC. **c1** FACS analysis of membrane markers: CD105 (92,15%), CD29 (99,63%), CD146 (15,54%) and CD45 (0.04%). **c2** FACS analysis of pluripotency nuclear markers: OCT3/4 (76,72%) and NANOG (30,18%). **d** RT-PCR of OCT3/4, NANOG and SOX2 expresions in DPPSC and DPMSC. **e** Western Blot analysis of OCT3/4 in DPPSC and DPMSC at different time points (5, 10 and 15 passages). GAPDH as a housekeeping. (TIF 1031 kb)

Additional file 2: Figure S2. sCGH summary from SAOS-2 cells at passage 10. 47, XXY control samples (labelled in green) and SAOS-2 samples (labelled in red) were co-hybridized onto 46, XY metaphases. (TIF 234 kb)

Abbreviations

ALP: Alkaline phosphatase; B.DPPSC: Bone-like DPPSC 15 days differentiated; CCC: Collagen I-based Cell Carrier; DPMSC: Dental pulp mesenchymal stem cells; DPPSC: Dental pulp pluripotent-like stem cells; DPSC: Dental pulp stem cells; SAOS-2: Human osteosarcoma cell line; sCGH: Short-comparative genomic hybridization; TI B.DPPSC: B.DPPSC differentiating on TI alloy disks; TI DPPSC: DPPSC differentiating on Ti alloy disks; TI SAOS-2: SAOS-2 differentiating on Ti alloy disks

Acknowledgments

We would like to thank Joaquima Navarro and Javier Del Rey (Universitat Autònoma de Barcelona) for their expertise in cytogenetic analysis; to Reiner Tal and Samet Nachum (MIS implantology) for their excellent technical assistance; to Daniel Blanch for the images design; to Marina Urbano for the histological analysis support; to Eduard Ferrés Padró for the provision of study patients; to Román Pérez for fruitful discussions and to Lluís Giner for his administrative support.

Funding
This study was funded by the Universitat International de Catalunya (UIC) and the Agència de Gestió d'ajuts Universitaris i de Recerca, Generalitat de Catalunya (SGR 2014; No 1060). The authors R. Núñez-Toldrà, E. Martínez-Sarrà and C. Gil-Recio were funded by the predoctoral grant Junior Faculty awarded by the Obra Social La Caixa and the UIC. A. Al Madhoun was funded by Kuwait Foundation for the Advancement of Sciences (KFAS) (project number RA-2013-009).

Authors' contributions
RNT contributed to data acquisition, analysis, interpretation and drafted the manuscript; EMS contributed to data acquisition, analysis and interpretation. CGR contributed to data acquisition and analysis; MAC contributed to data acquisition and analysis; AAM contributed to analysis and critically revised the manuscript; SM contributed to interpretation and drafted the manuscript; MA contributed to conception, design and drafted the manuscript. All authors gave final approval and agreed to submit the work.

Competing interests
The authors declare no potential conflicts of interest with respect to the authorship and/or publication of this article.

Author details
[1]Regenerative Medicine Research Institute, Universitat Internacional de Catalunya, Barcelona, Spain. [2]Chair of Regenerative Implantology MIS-UIC, Barcelona, Spain. [3]Area of Pathology, Universitat Internacional de Catalunya, Barcelona, Spain. [4]Research Division, Dasman Diabetes Institute, Dasman, Kuwait. [5]Surgery and Oral Implantology Department, Universitat Internacional de Catalunya, Barcelona, Spain.

References
1. Frohlich M, Grayson WL, Wan LQ, Marolt D, Drobnic M, Vunjak-Novakovic G. Tissue engineered bone grafts: biological requirements, tissue culture and clinical relevance. Curr Stem Cell Res Ther. 2008;3(4):254–64.
2. Czekanska EM, Stoddart MJ, Ralphs JR, Richards RG, Hayes JS. A phenotypic comparison of osteoblast cell lines versus human primary osteoblasts for biomaterials testing. J Biomed Mater Res A. 2014;102(8):2636–43.
3. Amini AR, Laurencin CT, Nukavarapu SP. Bone tissue engineering: recent advances and challenges. Crit Rev Biomed Eng. 2012;40(5):363–408.
4. Rodan SB, Imai Y, Thiede MA, Wesolowski G, Thompson D, Bar-Shavit Z, et al. Characterization of a human osteosarcoma cell line (Saos-2) with osteoblastic properties. Cancer Res. 1987;47(18):4961–6.
5. Hausser HJ, Brenner RE. Phenotypic instability of Saos-2 cells in long-term culture. Biochem Biophys Res Commun. 2005;333(1):216–22.
6. Kassem M. Mesenchymal stem cells: biological characteristics and potential clinical applications. Cloning Stem Cells. 2004;6(4):369–74.
7. Wagner W, Bork S, Horn P, Krunic D, Walenda T, Diehlmann A, et al. Aging and replicative senescence have related effects on human stem and progenitor cells. PLoS One. 2009;4(6):e5846.
8. Atari M, Caballe-Serrano J, Gil-Recio C, Giner-Delgado C, Martinez-Sarra E, Garcia-Fernandez DA, et al. The enhancement of osteogenesis through the use of dental pulp pluripotent stem cells in 3D. Bone. 2012;50(4):930–41.
9. Atari M, Gil-Recio C, Fabregat M, Garcia-Fernandez D, Barajas M, Carrasco MA, et al. Dental pulp of the third molar: a new source of pluripotent-like stem cells. J Cell Sci. 2012;125(Pt 14):3343–56.
10. Perez RA, Choi S-J, Han C-M, Kim J-J, Shim H, Leong KW, et al. Biomaterials control of pluripotent stem cell fate for regenerative therapy. Prog Mater Sci. 2016;82:234–93.
11. Perez RA, Seo S-J, Won J-E, Lee E-J, Jang J-H, Knowles JC, et al. Therapeutically relevant aspects in bone repair and regeneration. Mater Today. 2015;18(10):573–89.

12. Chatakun P, Nunez-Toldra R, Diaz Lopez EJ, Gil-Recio C, Martinez-Sarra E, Hernandez-Alfaro F, et al. The effect of five proteins on stem cells used for osteoblast differentiation and proliferation: a current review of the literature. Cell Mol Life Sci. 2014;71(1):113–42.
13. Malafaya PB, Silva GA, Reis RL. Natural-origin polymers as carriers and scaffolds for biomolecules and cell delivery in tissue engineering applications. Adv Drug Deliv Rev. 2007;59(4-5):207–33.
14. Chevallay B, Herbage D. Collagen-based biomaterials as 3D scaffold for cell cultures: applications for tissue engineering and gene therapy. Med Biol Eng Comput. 2000;38(2):211–8.
15. Lee CH, Singla A, Lee Y. Biomedical applications of collagen. Int J Pharm. 2001;221(1-2):1–22.
16. Delgado LM, Bayon Y, Pandit A, Zeugolis DI. To cross-link or not to cross-link? Cross-linking associated foreign body response of collagen-based devices. Tissue Eng B Rev. 2015;21(3):298–313.
17. Lee W, Debasitis JC, Lee VK, Lee JH, Fischer K, Edminster K, et al. Multi-layered culture of human skin fibroblasts and keratinocytes through three-dimensional freeform fabrication. Biomaterials. 2009;30(8):1587–95.
18. Micol LA, Ananta M, Engelhardt EM, Mudera VC, Brown RA, Hubbell JA, et al. High-density collagen gel tubes as a matrix for primary human bladder smooth muscle cells. Biomaterials. 2011;32(6):1543–8.
19. Schmidt T, Stachon S, Mack A, Rohde M, Just L. Evaluation of a thin and mechanically stable collagen cell carrier. Tissue Eng Part C Methods. 2011;17(12):1161–70.
20. Kato H, Nakamura T, Nishiguchi S, Matsusue Y, Kobayashi M, Miyazaki T, et al. Bonding of alkali- and heat-treated tantalum implants to bone. J Biomed Mater Res. 2000;53(1):28–35.
21. Covani U, Giacomelli L, Krajewski A, Ravaglioli A, Spotorno L, Loria P, et al. Biomaterials for orthopedics: a roughness analysis by atomic force microscopy. J Biomed Mater Res A. 2007;82(3):723–30.
22. Atari M, Barajas M, Hernandez-Alfaro F, Gil C, Fabregat M, Ferres Padro E, et al. Isolation of pluripotent stem cells from human third molar dental pulp. Histol Histopathol. 2011;26(8):1057–70.
23. Rius M, Obradors A, Daina G, Cuzzi J, Marques L, Calderon G, et al. Reliability of short comparative genomic hybridization in fibroblasts and blastomeres for a comprehensive aneuploidy screening: first clinical application. Hum Reprod. 2010;25(7):1824–35.
24. Aubin JE, Liu F, Malaval L, Gupta AK. Osteoblast and chondroblast differentiation. Bone. 1995;17(2 Suppl):77S–83S.
25. Miron RJ, Zhang YF. Osteoinduction: a review of old concepts with new standards. J Dent Res. 2012;91(8):736–44.
26. Komori T. Regulation of bone development and extracellular matrix protein genes by RUNX2. Cell Tissue Res. 2010;339(1):189–95.
27. Wennberg C, Hessle L, Lundberg P, Mauro S, Narisawa S, Lerner UH, et al. Functional characterization of osteoblasts and osteoclasts from alkaline phosphatase knockout mice. J Bone Miner Res Off J Am Soc Bone Miner Res. 2000;15(10):1879–88.
28. Dallas SL, Bonewald LF. Dynamics of the transition from osteoblast to osteocyte. Ann N Y Acad Sci. 2010;1192:437–43.
29. Karner E, Backesjo CM, Cedervall J, Sugars RV, Ahrlund-Richter L, Wendel M. Dynamics of gene expression during bone matrix formation in osteogenic cultures derived from human embryonic stem cells in vitro. Biochim Biophys Acta. 2009;1790(2):110–8.
30. Mendonca G, Mendonca DB, Simoes LG, Araujo AL, Leite ER, Duarte WR, et al. The effects of implant surface nanoscale features on osteoblast-specific gene expression. Biomaterials. 2009;30(25):4053–62.
31. Griffin DK, Sanoudou D, Adamski E, McGiffert C, O'Brien P, Wienberg J, et al. Chromosome specific comparative genome hybridisation for determining the origin of intrachromosomal duplications. J Med Genet. 1998;35(1):37–41.
32. Jacobs K, Zambelli F, Mertzanidou A, Smolders I, Geens M, Nguyen HT, et al. Higher-Density Culture in Human Embryonic Stem Cells Results in DNA Damage and Genome Instability. Stem Cell Reports. 2016;6(3):330–41.
33. Chen X, Chen A, Woo TL, Choo AB, Reuveny S, Oh SK. Investigations into the metabolism of two-dimensional colony and suspended microcarrier cultures of human embryonic stem cells in serum-free media. Stem Cells Dev. 2010;19(11):1781–92.
34. Zhang J, Niu C, Ye L, Huang H, He X, Tong WG, et al. Identification of the haematopoietic stem cell niche and control of the niche size. Nature. 2003;425(6960):836–41.

35. Glowacki J, Mizuno S. Collagen scaffolds for tissue engineering. Biopolymers. 2008;89(5):338–44.

36. Branemark R, Branemark PI, Rydevik B, Myers RR. Osseointegration in skeletal reconstruction and rehabilitation: a review. J Rehabil Res Dev. 2001;38(2):175–81.

37. Barrilleaux B, Phinney DG, Prockop DJ, O'Connor KC. Review: ex vivo engineering of living tissues with adult stem cells. Tissue Eng. 2006;12(11):3007 19.

38. Nothdurft FP, Fontana D, Ruppenthal S, May A, Aktas C, Mehraein Y, et al. Differential Behavior of Fibroblasts and Epithelial Cells on Structured Implant Abutment Materials: A Comparison of Materials and Surface Topographies. Clin Implant Dent Relat Res. 2014;17(6):1237–1249.

39. Vandrovcova M, Jirka I, Novotna K, Lisa V, Frank O, Kolska Z, et al. Interaction of human osteoblast-like Saos-2 and MG-63 cells with thermally oxidized surfaces of a titanium-niobium alloy. PLoS One. 2014;9(6):e100475.

40. Hou Y, Cai K, Li J, Chen X, Lai M, Hu Y, et al. Effects of titanium nanoparticles on adhesion, migration, proliferation, and differentiation of mesenchymal stem cells. Int J Nanomedicine. 2013;8:3619–30.

5

External magnetic field promotes homing of magnetized stem cells following subcutaneous injection

Yu Meng[1†], Changzhen Shi[2†], Bo Hu[1], Jian Gong[3], Xing Zhong[3], Xueyin Lin[3], Xinju Zhang[4], Jun Liu[4], Cong Liu[4] and Hao Xu[3*]

Abstract

Background: Mesenchymal stem cells (MSCs) are multipotent stromal cells that have the ability to self-renew and migrate to sites of pathology. In vivo tracking of MSCs provides insights into both, the underlying mechanisms of MSC transformation and their potential as gene delivery vehicles. The aim of our study was to assess the ability of superparamagnetic iron oxide nanoparticles (SPIONs)-labeled Wharton's Jelly of the human umbilical cord-derived MSCs (WJ-MSCs) to carry the green fluorescent protein (GFP) gene to cutaneous injury sites in a murine model.

Methods: WJ-MSCs were isolated from a fresh umbilical cord and were genetically transformed to carry the GFP gene using lentiviral vectors with magnetically labeled SPIONs. The SPIONs/GFP-positive WJ-MSCs expressed multipotent cell markers and demonstrated the potential for osteogenic and adipogenic differentiation. Fifteen skin-injured mice were divided into three groups. Group I was treated with WJ-MSCs, group II with SPIONs/GFP-positive WJ-MSCs, and group III with SPIONs/GFP-positive WJ-MSCs exposed to an external magnetic field (EMF). Magnetic resonance imaging and optical molecular imaging were performed, and images were acquired 1, 2, and 7 days after cell injection.

Results: The results showed that GFP could be intensively detected around the wound in vivo 24 h after the cells were injected. Furthermore, we observed an accumulation of WJ-MSCs at the wound site, and EMF exposure increased the speed of cell transport. In conclusion, our study demonstrated that SPIONs/GFP function as cellular probes for monitoring in vivo migration and homing of WJ-MSCs. Moreover, exposure to an EMF can increase the transportation efficiency of SPIONs-labeled WJ-MSCs in vivo.

Conclusions: Our findings could lead to the development of a gene carrier system for the treatment of diseases.

Keywords: Mesenchymal stem cells (MSCs), Stem cells, UC, Magnetic resonance imaging (MRI), Rat

Background

There is significant potential for the use of mesenchymal stem cells (MSCs) in cell therapy [1]. However, their clinical application still faces various challenges, such as the fact that an efficient strategy for stem cell homing to target sites has not yet been identified. Several other factors limit the clinical application of stem cells, including the time and method of drug administration, cell concentration, the transmission medium, and cell homing [1, 2]. Homing of stem cells is achieved through direct local injection, local perfusion, and systematic administration. However, intravenous injection of stem cells results in the accumulation of a significant number of cells in the lungs and spleen, with a very low percentage of cells reaching the arterial system (about 5%), and an even lower percentage reaching the target organ (0.0005%) [2, 3]. Simple direct injections and local organ perfusions are limited to superficial organs or to organs directly connected to the main artery. In fact, although an intra-arterial injection ensures highest cell numbers for transplant, this method of stem cell transplantation

---

* Correspondence: xuhaodoc@163.com
†Equal contributors
3Department of Nuclear Medicine, the First Hospital Affiliated to Jinan University, No. 613 Huangpu West Road, Guangzhou 510630, China
Full list of author information is available at the end of the article

increases the death rate by 41%, with animals succumbing to arterial embolism [4]. Therefore, an important prerequisite for treatments is to transplant MSCs with a differentiation potential directly to the target area.

Compared with traditional preparations, magnetic targeted drug delivery systems, which have been studied for years [5–8], are characterized as methods for improving drug targeting, enhancing the curative effect of drugs, and decreasing toxic side effects. Superparamagnetic iron oxide nanoparticles (SPIONs) are an excellent transmission medium based on the magnetic targeted drug delivery system [9–14]. Recent studies have revealed that labeling stem cells with magnetic nanoparticles for magnetic resonance imaging (MRI)-mediated tracking of stem cells has evolved. An improved curative effect on common carotid artery injuries was observed using magnetized endothelial progenitor cells, obtained from in situ intra-arterial treatment of spinal cord injured animals, using a magnetic field to direct the stem cells [15]. Recently, studies have used anionic magnetic nanoparticles to load endothelial progenitor cells, and have successfully controlled cell movement in the vessel network using a magnetic field [16]. Although the results are exciting, most of these studies involved the use of a constant electromagnetic field or an internal magnetic field.

Studies involving noninvasive external magnetic fields (EMFs) with the use of permanent magnets are rare. The superparamagnetism of SPIONs can be utilized to bring about directional movement of magnetized stem cells under the influence of an EMF. In the present study, human umbilical cord MSCs were transfected with SPIONs and green fluorescent protein (GFP), and injected into the subcutaneous tissues of nude mice, specifically into partial cells at some cell intervals, following a skin injury. In summary, although some studies have shown that noninvasive EMFs can increase magnetized cell homing following an intra-arterial injection, intravenous injection, or intrathecal injection, no evidence has been provided showing the same effect following a subcutaneous injection. The current study tested the hypothesis that an EMF can promote homing and guide the magnetized stem cells to make rapid directional movements, following subcutaneous injection. MRI as well as fluorescence imaging was used to track the stem cells in vivo. Our findings demonstrated an improved method of injury treatment, using MSCs as drug or gene carriers, which can also be applied for directional drug treatment or gene therapy, both of which have a significant importance in the clinical setting.

Methods
Cell culture
The study was approved by the regional medical ethical review board (Jinan University and Shenzhen people's hospital). After obtaining a written informed consent, the human umbilical cord Wharton's Jelly-derived MSCs (WJ-MSCs) were isolated as described previously [15], from the umbilical cords of full-term newborns, delivered at the Shenzhen People's Hospital, Guangdong, China. Cells were cultured in low glucose-DMEM (Hyclone, Logan, Utah, USA) containing 2 ng/mL of basic fibroblast growth factor (bFGF) and 10% fetal bovine serum (FBS) (Gibco, Gran Island, NY, USA), and maintained at 37 °C in a humidified atmosphere of 5% CO_2. The medium was replaced every 3 days, and the human umbilical cord-derived MSCs (HUCMSCs) were collected by trypsin (0.25%, Invitrogen, USA) digestion. All experiments were performed using MSCs at 3–5 passages [17].

Flow cytometry analysis
Passage 3 WJ-MSCs were trypsinized, dissociated into a single cell suspension, and allowed to reach 60% confluency. Cells were then rinsed with phosphate buffered saline (PBS) and incubated with anti-human CD73-PE (BioLegend, 344,004), anti-human CD105-PE (BioLegend, 323,206), anti-human CD90-PE (BioLegend, 328,110), anti-human CD34-PE (BioLegend, 343,506), and anti-human CD45-FITC (BioLegend, 304,006) for 15 min at room temperature. After incubation, the cells were rinsed with PBS, read on a FACSCalibur (BD, USA) flow cytometer, and analyzed using the WinMDI 2.8 software. Mouse IgG1-PE (BioLegend, 400,114) and mouse IgG1-FITC (BioLegend, 400,107) were used as the isotype controls.

Osteogenic differentiation
At approximately 80% confluency, WJ-MSCs were rinsed with PBS and cultured in osteogenic differentiation medium (Gibco, Gran Island, NY, USA). The medium was changed twice a week, and after 3 weeks, the cells were washed with PBS and fixed in 4% paraformaldehyde for 30 min. The cells were then stained with 0.1% Alizarin Red S water solution for 30 min.

Adipogenic differentiation
At 80% confluency, WJ-MSCs were rinsed with PBS and cultured in adipogenic differentiation medium (Gibco, Gran Island, NY, USA). The medium was changed twice a week, and after 3 weeks, the cells were washed with PBS and fixed in 4% paraformaldehyde for 30 min. The cells were then stained with 0.3% Oil Red O solution for 30 min.

Lentiviral transfection of WJ-MSCs
A lentiviral vector (Fitgene, Guangzhou, China) expressing GFP with a cis-acting element, CMV-GFP-puro, was packaged and used to infect the WJ-MSCs, according to

the manufacturer's protocol. The transduced cells were grown in low glucose-DMEM containing 2 ng/mL bFGF, 10% FBS, and 400 µg/mL puromycin. Puromycin-resistant WJ-MSCs, overexpressing GFP, were obtained after 3 days of puromycin selection. Stable clones were identified by the expression of GFP protein [16]. The efficiency of the target cells was estimated by fluorescence microscopy using an inverted fluorescence microscope–the GFP positive cells in each field (10X) visible/all the cells in each field (10X).

Labeling of WJ-MSCs with SPIONs

Approximately 1×10^3 genetically modified WJ-MSCs were seeded into each well of a 24-well plate. After 12 h of incubation in MSC growth medium, the cells were magically labeled with SPIONs (25 µg/mL; SPIONs-MSCs) (Molday ION Rhodamine B™, BioPhysics Assay Laboratory, Inc., Worcester, MA, USA; CL-50Q02-6A-50) and complexed to poly-L-Lysine (0.75–1 µg/mL) (BioPhysics Assay Laboratory, Inc., CL-00-01). The cells were then collected, washed twice in PBS, counted, and resuspended at the appropriate cell density for in vivo analyses.

Cell vitality test

For Trypan blue staining, 200 µL of cells was aseptically transferred to a 1.5 mL clear Eppendorf tube, and incubated for 3 min at room temperature with an equal volume of 0.4% (w/v) Trypan blue solution prepared in 0.81% NaCl and 0.06% (w/v) dibasic potassium phosphate. Cells were counted using a dual-chamber hemocytometer and a light microscope. Viable and nonviable cells were recorded separately, and the means of the two cell counts were pooled for analysis.

Full-thickness skin defect model and cell transplantation

In total, 30 specific pathogen-free BALB/C nude mice were randomly divided into 3 groups: 10 mice in the control group, 10 mice in the group receiving SPIONs-MSCs, and 10 mice in the group receiving SPIONs-MSCs exposed to an EMF (SPIONs-MSCs + magnetic field). The full-thickness skin defect model was implemented in all experimental mice, and cell transplantation was performed at a 1.5 cm distance from the wound using a 1 mL syringe. The control group was injected with 200 µL 0.9% normal saline, the SPIONs-MSCs group was injected with 2×10^6 SPIONs and GFP double-labeled MSCs, and the SPIONs-MSCs + magnetic field group was injected with 2×10^6 SPIONs and GFP double-labeled MSCs exposed to an EMF generated by a permanent magnet for 6 h/day. The effects of different EMF exposure times on stem cells were analyzed by the MTT assay and western blotting, by determining the apoptosis markers (Additional file 1: Figure S1). The

full-thickness skin defect model was generated as follows: mice were anesthetized by an intraperitoneal injection of 10% chloral hydrate (0.3 mL/100 g); next, following disinfection, a 4-mm diameter circular skin defect was created to the depth of the deep fascia on the back of the mice, near the double hind limb. After completion of stem cell transplantation, a representative animal from each group was immediately subjected to MRI and in vivo fluorescence examination for baseline analysis. Three mice per group were subjected to MRI and fluorescence imaging at 24 h, 48 h, and 7 days post-cell transplantation, and the healed skin lesions were removed for pathological examination on day 7. Healed wounds were identified by the presence of the following: wound closure, epithelium, proper activity intensity, no wound dehiscence, no ulceration, an appropriate time lapse after wound generation, the ability to tolerate a certain tension and pressure at the wound site, gradual fading of skin color at the wound site, and similarity of the wound skin color to that of the surrounding healthy skin to maintain the skin barrier integrity. All experimental procedures were approved by the Animal Experimentation Ethics Committee of First Hospital Affiliated to Jinan University, Guangzhou, China.

In vivo injection of magnetic WJ-MSCs

Mice were anesthetized with isoflurane (4% induction, 1.5% maintenance). The wounds were exposed to a magnetic field of 0.5 T using a small permanent neodymium (FeNdB) magnet (8 × 2 mm) for 6 h/day (Additional file 2: Figure S2). Subsequently, 2×10^6 WJ-MSCs, which were previously magnetized using 25 µg/mL of SPIONs, were hypodermically injected (hypodermis) with 150 µL of PBS. Control animals received an identical cell infusion without the magnet implantation. Magnets were placed for 6 h daily.

In vivo MRI

The animals used were BALB/C (nu/nu) nude mice ($n = 3$). One day, two days, or seven days after transplantation, the mice were anesthetized using chloral hydrate and underwent in vivo MRI (gradient echo scan), using a 3 T MRI scanner (Discovery MR750, GE Healthcare, USA). For image acquisition, a quadrature birdcage volume coil of 7 cm inner diameter was used. Axial images were taken with the following parameters: field of view = 4 × 4 cm^2, matrix size = 256 × 256, slice thickness = 1 mm, TE = 6 ms, and TR = 700 ms. Following completion of the scan, the raw data and images were processed using the built-in professional software Discovery MR750, with each injected area defined as a unit, and using each of the 4 selected regions of interest (ROI) for measurements. MRI parameters included injection of cell volume, displacement, carrier-to-noise ratio (CNR), and signal-to-noise ratio (SNR).

Fluorescence stereomicroscopy
To assess the distribution of SPIONS-MSCs in vivo, anesthetized rats were imaged for GFP fluorescence using a whole-body imaging system (IVIS Lumina II, Caliper, France). Filters of 480 nm (±10 nm) and 505 nm (±5 nm) represented the excitation and emission signals, respectively. High-resolution images were captured directly on a computer and analyzed using Living Image software (Xenogen Corporation, Almeda, California, USA). Results were expressed as number of photons/s/ROI.

Prussian blue staining and tissue specimens
Prussian blue staining was performed to identify the SPIONs-MSCs. Cells were incubated for 30 min with 2% potassium ferrocyanide in 6% hydrochloric acid, and then counterstained with nuclear fast red for 30 s. A blue color indicated the presence of iron within the cells, thereby corresponding to the SPIONs-MSCs. Similarly, in skin tissue sections, SPIONs appeared as blue precipitates in the cytoplasm and pink in the nucleus. Tissue specimens were frozen in optimal cutting temperature (OCT) compound (Sakura Finetek Inc., Torrance, California, USA) in liquid nitrogen, and 10-μm sections were prepared using a cryostat microtome (CM1850; Leica Microsystems GmbH).

Statistical analysis
The statistical significance of intergroup differences was assessed using the Student's t-test or ANOVA followed by Tukey's post hoc test. A P value <0.05 was considered statistically significant at the 95% confidence level. All values in the bar and line graphs are expressed as mean ± standard deviation (SD). The number of independent experiments analyzed has been stated in each figure legend.

Results
Lentivirus infection and SPIONs labeling
In our study, GFP was used for the in vivo tracking of WJ-MSCs. GFP was incorporated into a lentiviral vector containing independent puromycin expression frames. WJ-MSCs were isolated from fresh umbilical cords and cultured in MSC medium for several passages. Lentivirus-infected WJ-MSCs were selected in MSC medium with puromycin for 3 days. Stable clones were GFP positive (>99%), as detected by fluorescence microscopy (Fig. 1a and b). GFP expression was observed under a fluorescence microscope. Using an inverted fluorescence microscope, we observed the HUCMSCs for a green fluorescence signal at 12 h post-transfection; however, the signal was weak and expressed only by a few cells. The number of GFP-positive cells increased constantly 24 h post-transfection, with 4-10 GFP-positive cells in each visual field (10X) at 48 h, and more than 10 GFP-positive cells in each visual field (10X) at 72 h. The GFP transfection efficiency with lentivirus infection was over 99%. The GFP-positive WJ-

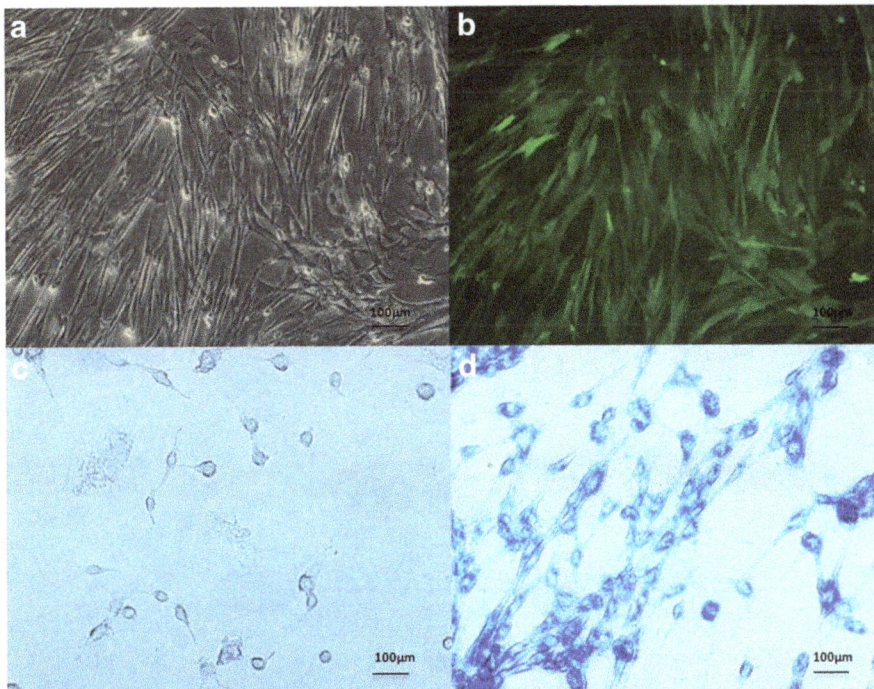

Fig. 1 WJ-MSCs labeled with GFP/SPIONs. GFP-positive cells under fluorescence microscope (**a**) and (**b**). Cells labeled with SPIONs (**c**) and (**d**)

MSCs were then transfected with SPIONs, and the transfection efficiency was evaluated by Prussian blue staining. Results demonstrated that more than 80% of the cells were labeled with SPIONs (Fig. 1c and d).

Immunophenotype, differentiation potential, and vitality of WJ-MSCs

The immunophenotype of passage 3 WJ-MSCs, which represent typical fibroblastic cells, and GFP/SPIONs-positive WJ-MSCs was evaluated by flow cytometry. The results showed that all cells expressed CD73, CD105, and CD90 (>95%), but did not express CD34 or CD45 (<2%) (Fig. 2). Furthermore, both untransfected WJ-MSCs and GFP/SPIONs-positive WJ-MSCs were evaluated for their osteogenic and adipogenic differentiation potential. After a 3-week induction under osteogenic conditions, these cells were stained with 0.1% Alizarin Red S water solution. Results showed that majority of the WJ-MSCs were alkaline phosphatase-positive, indicating their osteogenic differentiation potential (Fig. 3a

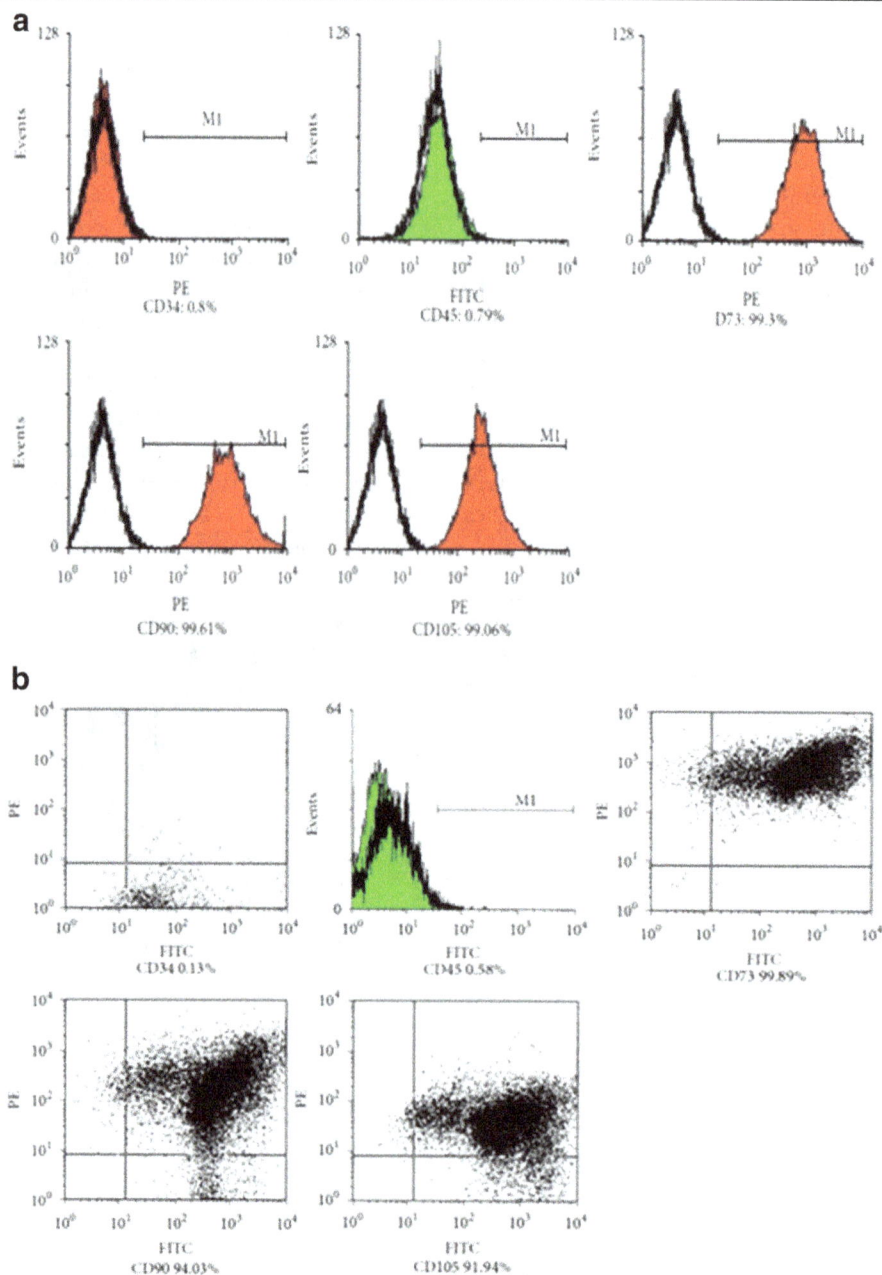

Fig. 2 Immunophenotype of WJ-MSCs. Untransfected WJMSCs (**a**) and GFP/SPION-positive WJMSCs (**b**). Flow cytometry data showing negative expression of CD34, CD45 and positive expression of CD73, CD90 and CD105 in the WJMSCs

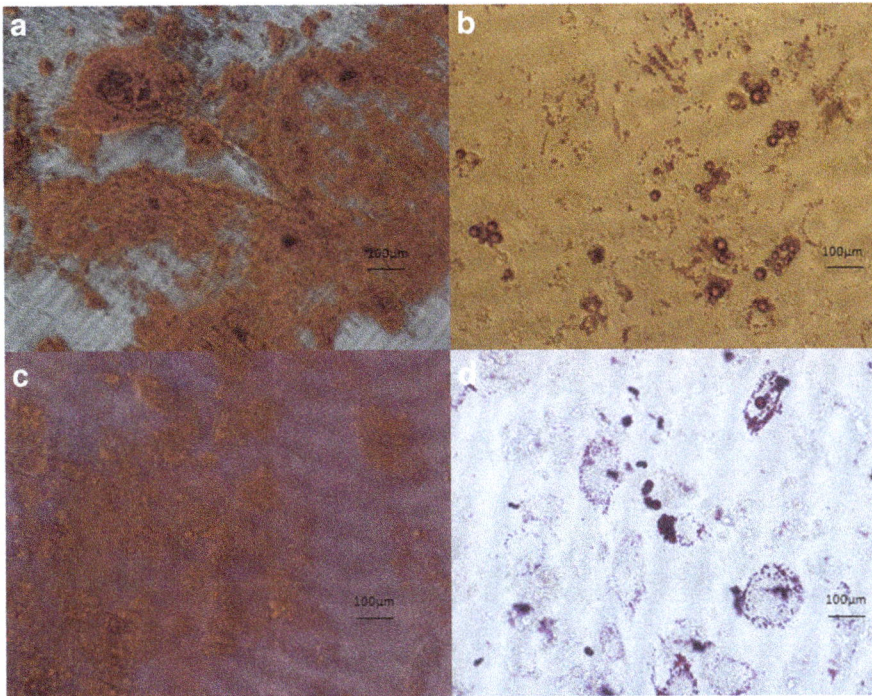

Fig. 3 Differentiation of WJMSCs. Osteogenic differentiation analysis of untransfected WJMSCs (**a**) and GFP/SPION-positive WJMSCs (**c**). Adipogenic differentiation analysis of untransfected WJMSCs (**b**) and GFP/SPION-positive WJMSCs (**d**)

and c). To assess their adipogenic differentiation potential, another culture plate of passage 3 WJ-MSCs was incubated with adipogenic differentiation medium for 3 weeks and then stained with 0.3% Oil Red O. We observed that majority of the cells contained numerous Oil Red O-positive lipid droplets, indicating that WJ-MSCs underwent adipogenic differentiation. (Fig. 3b and d). Growth of GFP/SPIONs-positive WJ-MSCs was seen in the two multiplication cycles; the first multiplication cycle started on day 3 and 4, and the second one in the first 4-7 days. Compared with the control group (HUCMSCs), there was no significant difference at all time points ($t = 2.05$, $p > 0.05$) (Additional file 3: Figure S3).

In vivo cell tracking using MRI

GFP/SPIONs-positive WJ-MSCs were injected around the cutaneous wound areas in injured mice. MRI was performed at 0, 24, and 48 h and 7 days post-transplantation. As expected, GFP/SPIONs-positive WJ-MSCs were successfully directed to the subcutaneous areas of the skin under the influence of the magnetic gradient created by the implanted magnet (Fig. 4). The effect of the EMF on stem cell targeting was significant compared to WJ-MSCs without exposition to magnetic field. The 24-h MR images confirmed that more than 80% of SPIONs/GFP-labeled WJ-MSCs (low signal distribution) reached the trauma center within the first 24 h. However, only low signals were detected around

the wounds that were not exposed to the EMF. On day 7, post-cell transplantation, MRI results demonstrated a hypointense signal distribution in the wound center of mice regardless of the magnetic field interference. However, the MRI parameters were not significantly different between the SPIONS-MSCs and SPIONs-MSCs + magnetic field groups. Parameters such as area, SNR, CNR, and displacement can be used as indicators of the concentration and movement speed of cell clusters. In the SPIONs-MSCs + magnetic field group, we observed a significantly reduced area (Fig. 5a) and an increased SNR, CNR, and displacement 24 h, 48 h, and 7 days post-cell transplantation, particularly in the first 24 h (Fig. 5b). The CNR was significantly different between the two groups, especially at the 24 h time point (Fig. 5c). The highest value of SNR was observed at the 7 day time point in the two groups (Fig. 5d).

Moreover, we observed that the wound recovery rate was enhanced in the SPIONs/GFP-MSCs + magnetic field group as compared with the SPIONs/GFP-MSCs group and the control group. Qualitative skin analysis via Prussian blue staining and fluorescence imaging was performed to validate that the hypointense signals detected by MRI indeed corresponded to the presence of MSCs at the target site. The presence of blue cells (Prussian blue) was confirmed in matching skin injury areas, indicating the presence of MSCs in the skin tissue (Fig. 6). Both groups of SPIONs-MSCs, with or without

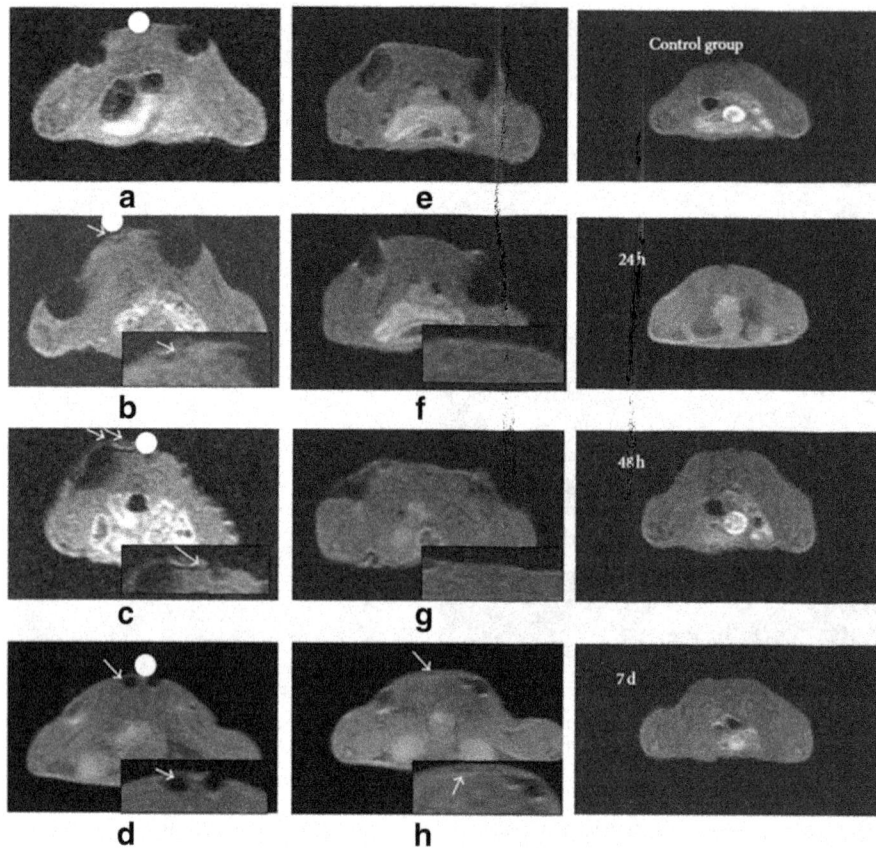

Fig. 4 T2-weighted MR images and in vivo cell tracking. Magnetized cells can be detected as hypointense signals (dots): e is FeNdB magnets (8 mm × 2 mm) with a magnetic field of 0.5 T. **a–d** the SPIONs-MSCs + magnetic field group, MR images were taken at 0 h (**a**), 24 h (**b**), 48 h (**c**), and 7 d (**d**) after cell transplantation, the MRI image in (**b**) confirms that labeled stem cells (low signal distribution) entered into the trauma center within the first 24 h. **e–h** SPIONs-MSCs group without magnetic field, MR images were taken at 0 h (**e**), 24 h (**f**), 48 h (**g**), and 7 d (**h**) after cell transplantation, the 24-h and 48-h MRI images in this group do not demonstrate a low signal area, corresponding to the skin surrounding the wound

exposure to a magnetic influence, were observed in the skin injury areas of mice 7 days after cell transplantation; however, far fewer positive cells were observed in the SPIONs-MSCs group that did not receive a magnetic implantation. Magnetic implantation, along with injection of SPIONs/GFP-MSCs was found to be safe, as all animals survived and no major signs of tissue injury were observed in vivo by MRI, or ex vivo in the skin tissue.

In vivo optical molecular imaging

To evaluate the role of the magnetic gradient created by the EMF on the activity of SPIONs/GFP-MSCs, we used in vivo optical molecular imaging as a tracer technique to observe the living cells, which express the fluorescent protein. Fluorescence imaging at 0, 24, and 48 h demonstrated a trend of targeted SPIONs-MSCs movement under the EMF, which was consistent with MRI results (Fig. 7). However, the optical molecular imaging technique was not as sensitive as MRI.

Discussion

Our study demonstrated an EMF-targeted approach that promotes the directional movement of SPION-labeled stem cells, enhancing their ability to repair damaged skin tissues. SPION-labeled human MSCs exhibit excellent paramagnetism, and can accurately target lesion locations under the effect of an EMF, thereby increasing stem cell concentrations at the target site. We confirmed that MSCs doubly labeled with SPIONs and GFP reporter gene demonstrate better dry phenotypes in in vivo experiments. We also demonstrated that SPIONs and GFP labeling does not affect MSC proliferation or vitality, as no significant difference in cell viability was observed before and after labeling. The osteogenic and adipogenic differentiation potential is not affected when the amount of SPIONs-labeled MSCs is 90% [18–21]. In the present study, SPIONs/GFP-labeled MSCs displayed bona fide stem cell features in vitro, similar to that in other studies [22–24]. Weakening of the magnetic flux by exposure to an electromagnetic field, with strength

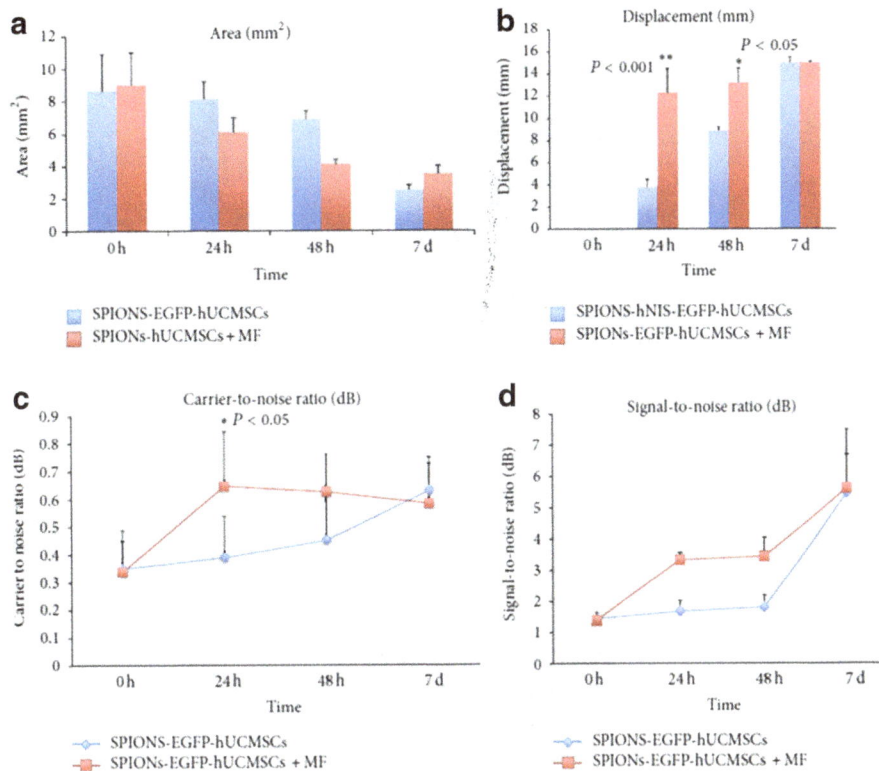

Fig. 5 Parameters of MRI image analysis. **a** Reduction in the cluster area of labeled MSCs with time. **b** 24-h (1 d) and 48-h (2 d) displacement of SPIONs-MSCs increased significantly under magnetic field (*P < 0.05, **P < 0.001). **c** CNR was the highest at 24 h, and an MRI shows obvious increase in the SPIONs-MSCs + magnetic field group. **d** SNR increased in the SPIONs-MSCs + magnetic field group by 24 h, and was high in both groups by 7 d

less than 0.1 mT, can promote human umbilical vein endothelial cell proliferation [25]. Exposure of bone marrow-derived MSCs to a 1 mT magnetic field for 1 h per day has been shown to promote cell proliferation and differentiation at an early stage [26–28]. The characteristics of a permanent magnet are far more complex than that of an electromagnetic field, and the force is distance-dependent; therefore, magnetic field properties were not analyzed in this study. There are very few stem cell studies involving the generation of outside target fields by use of a permanent magnet; however, some studies have successfully applied this method. We put an 8 mm × 2 mm permanent magnet with a surface residual magnetism of 0.5 T and a coercive force of 900 KA/m on the surface of the wound, 1.5 cm away from the transplanted cells. The magnetic targeting effect was evident from our results. Since the dead labeled-stem cells also release iron-containing nanoparticles, which can be taken up by the surrounding unlabeled stem cells, we utilized MRI and in vivo fluorescence imaging for synergistic monitoring to avoid any false positive results. However, MRI was more sensitive than in vivo fluorescence imaging. MRI demonstrates high spatial resolution; therefore, a small number of cells can be studied and quantified in a very simple manner using this technique. In the present study, in vivo fluorescence imaging was unable to track cells that had been transplanted for 48 h; however, MRI maintained an excellent tracer capacity until the end of the experiment (up to 7 days). In the case of a 6-h magnetic field exposure each day, both tracer techniques demonstrated that labeled stem cells possess a quick and clear central tendency for magnetic fields. MRI was also very sensitive in demonstrating that a considerable number of stem cells entered the epidermal trauma center within 24 h, and revealed that the maximum displacement of cells extending to the center of the magnetic field was 1.5 cm. Furthermore, using MRI, we also determined that the change in maximum displacement after 48 h was smaller than that in the first 24 h, with a maximum displacement of 0.3 cm. The MRI technique also demonstrated increased cell proliferation in the wound area on day 7 post-cell transplantation. A directed stem cell homing to trauma centers was observed in mice exposed to an EMF; however, other parameters such as displacement, area, and SNR were significantly reduced as compared to in mice that were not exposed to the EMF.

Mice in the EMF group displayed a distinct advantage in the overall wound healing time. The fact that GFP is expressed only in surviving stem cells was used to

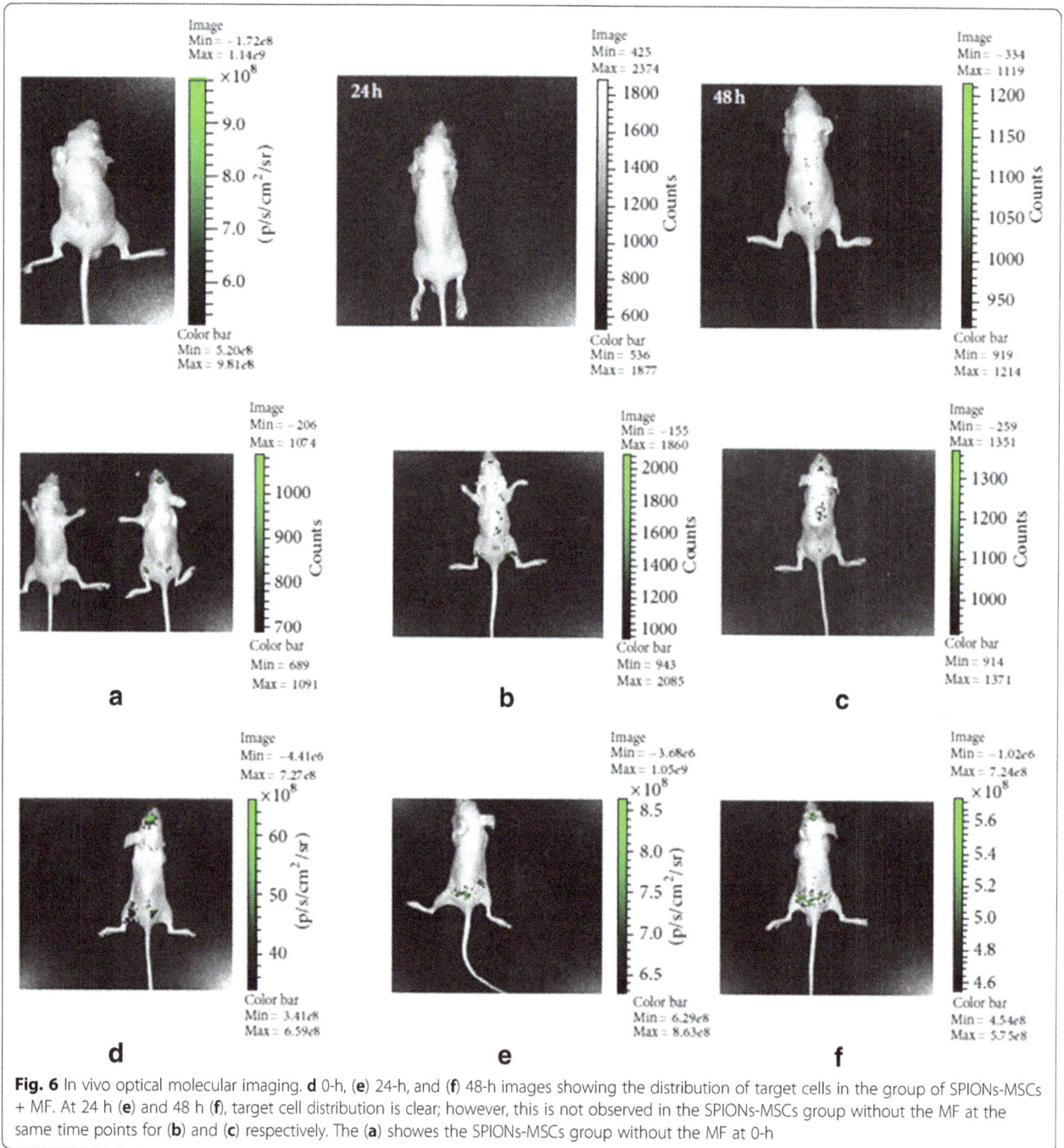

Fig. 6 In vivo optical molecular imaging. **d** 0-h, (**e**) 24-h, and (**f**) 48-h images showing the distribution of target cells in the group of SPIONs-MSCs + MF. At 24 h (**e**) and 48 h (**f**), target cell distribution is clear; however, this is not observed in the SPIONs-MSCs group without the MF at the same time points for (**b**) and (**c**) respectively. The (**a**) showes the SPIONs-MSCs group without the MF at 0-h

distinguish the false positive cells, and reflect the true state of surviving stem cells. Its advantage was pronounced in pathological tissue sections, and was confirmed by Prussian blue staining. Fluorescence imaging of pathological tissue sections also demonstrated that the aggregation ability of the targeted stem cells under the effect of an EMF was superior to that of control and non-magnetic field stem cell groups.

MRI tracking of SPION-labeled stem cells has so far proven to be an effective technique, resulting in limited concentration-dependent cellular toxicity. A previous study reported a labeling efficiency of 99%, at concentrations of 25-50 mg Fe/L, without any adverse effects on cell viability, growth, differentiation, and other biological activities [29]. In another study, culture mediums with a SPION concentration of 11.2, 22.4, and 44.8 mg Fe/L led to no changes in stem cell viability and proliferation [30]. Proliferation of liver stem cells has been shown to be inhibited by SPIONs at concentrations higher than 100 mg Fe/L [31]. It has therefore been recommended

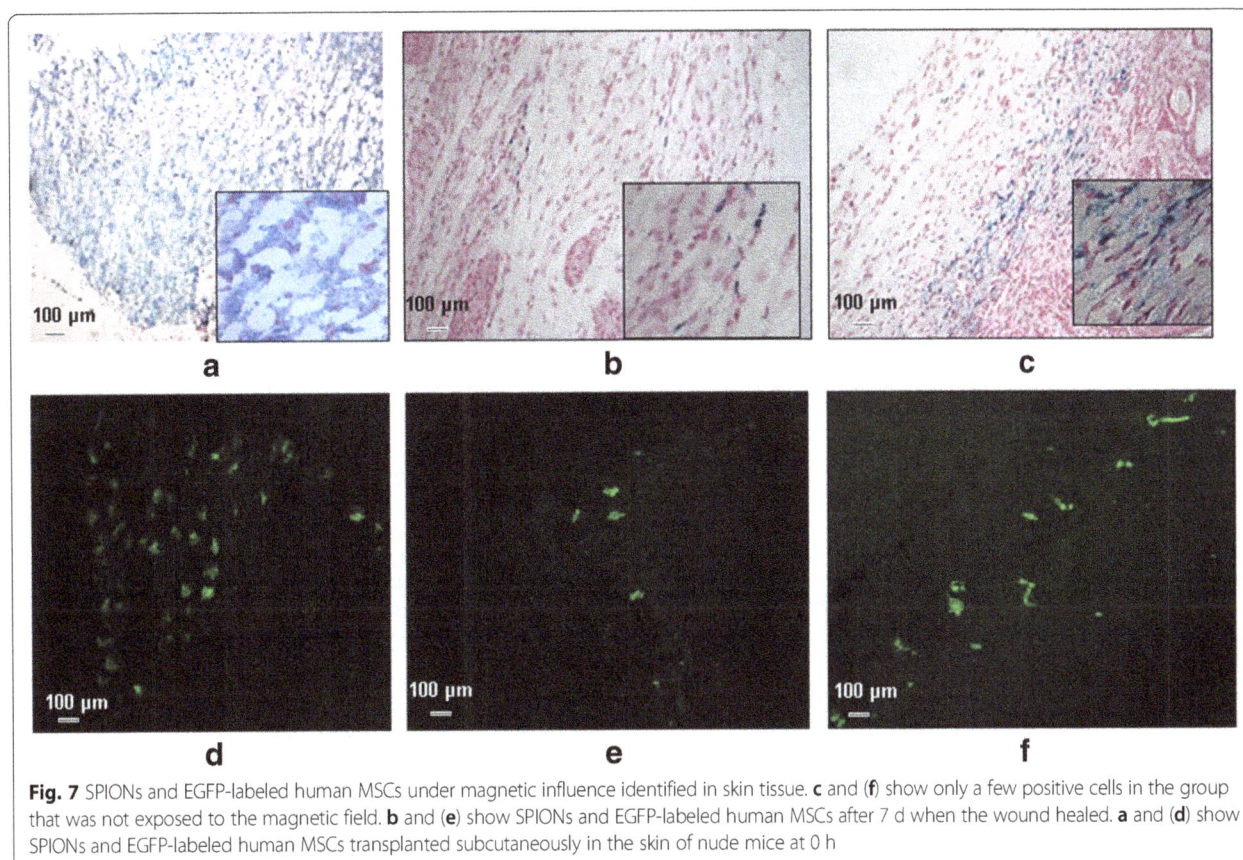

Fig. 7 SPIONs and EGFP-labeled human MSCs under magnetic influence identified in skin tissue. **c** and (**f**) show only a few positive cells in the group that was not exposed to the magnetic field. **b** and (**e**) show SPIONs and EGFP-labeled human MSCs after 7 d when the wound healed. **a** and (**d**) show SPIONs and EGFP-labeled human MSCs transplanted subcutaneously in the skin of nude mice at 0 h

that a SPION-labeling concentration of 25 mg Fe/L be used [32], which minimally affects the physiological activity of stem cells. In the present study, we used the same SPION-labeling concentration, and observed no changes in cellular morphology or activity.

Limitations

Our study is limited by the use of relatively few animals, a short duration of the disease, and lack of another control group with only a magnet implant without MSCs injection, to test the effect of magnetic field exposure on skin wound healing. Nevertheless, no histology but the wound area determines whether the skin defect is healed [33]. Moreover, the differentiation potential of SPIONs and GFP-labeled MSCs could have been affected by the iron particles. Further studies are required to explore whether exposure to a magnetic field has the same effect on MSCs from a larger distance and in diseases that are more complex.

Conclusions

In summary, SPION-labeled stem cells are excellent and safe magnetization and tracer agents, and together with exposure to an EMF generated by permanent magnets, they can be used as a new method of magnetic guidance of targeted stem cells. This method was shown to be safe and effective by MRI and fluorescence analysis of tissue sections. The findings of this study provide a platform for the development of stem cell targeted therapies, and can be further applied for drug and gene targeted therapies.

Additional files

Additional file 1: Figure S1. The Effect of different external magnetic field exposure time on stem cells was tested by MTT (A), western-blot testing apoptosis marker (B, C). (a) MTT testing showed when the SPION/GFP positive MSCs exposure 6 h/d under magnetic field the cell viability increased, and (b, c) apoptosis marker were low. (TIFF 134 kb)

Additional file 2: Figure S2. Mice were anesthetized with isoflurane (4% induction, 1.5% maintenance), and a small permanent neodymium (FeNdB) magnet (8 × 2 mm) with a magnetic field of 0.5 T was put on the wound of mice for 6 h/day. (TIFF 1394 kb)

Additional file 3: Figure S3. The distributions of cell growth. (TIFF 589 kb)

Abbreviations

bFGF: Basic fibroblast growth factor; CNR: Carrier-to-noise ratio; EMF: External magnetic field; FBS: Fetal bovine serum; GFP: Green fluorescent protein;; HUCMSCs: Human umbilical cord-derived MSCs; MRI: Magnetic resonance imaging; MSCs: Mesenchymal stem cells; SD: Standard deviation; SNR: Signal-to-noise ratio; SPIONs: Superparamagnetic iron oxide nanoparticles; WJ-MSCs: Wharton's Jelly of the human umbilical cord-derived MSCs

Acknowledgments

This study was supported by grants from the Science and Technology Planning Project of Guangdong Province, China (#2016A040403054). And Important Guangdong Province Science & Technology Specific Projects (#2003A3080501).

Authors' contributions

HX designed the study and the experiments. YM, CS, BH, JG and XZ performed data collection; XZ, JL, CL and XL analyzed the date. YM and CS wrote the manuscript. HX supervised the writing of the whole paper. All authors read and approved the final manuscript.

Competing interests

The authors declare that they have no competing interests.

Author details

[1]Department of Nephrology, the First Hospital Affiliated to Jinan University, No. 613 Huangpu West Road, Guangzhou 510630, China. [2]Department of Radiology, the First Hospital Affiliated to Jinan University, No. 613 Huangpu West Road, Guangzhou 510630, China. [3]Department of Nuclear Medicine, the First Hospital Affiliated to Jinan University, No. 613 Huangpu West Road, Guangzhou 510630, China. [4]Shenzhen Engineering Laboratory for Genomics-Assisted Animal Breeding, BGI-Shenzhen, Shenzhen 518083, China.

References

1. Gao Z, Zhang L, Hu J, Sun Y. Mesenchymal stem cells: a potential targeted-delivery vehicle for anti-cancer drug, loaded nanoparticles. Nanomedicine. 2013;9(2):174–84.
2. Schrepfer S, Deuse T, Reichenspurner H, Fischbein MP, Robbins RC, Pelletier MP. Stem cell transplantation: the lung barrier. Transplant Proc. 2007;39(3):573–6.
3. Fischer UM, Harting MT, Jimenez F, et al. Pulmonary passage is a major obstacle for intravenous stem cell delivery: the pulmonary first-pass effect. Stem Cells Dev. 2009;18(5):683–92.
4. Li L, Jiang Q, Ding G, et al. Effects of administration route on migration and distribution of neural progenitor cells transplanted into rats with focal cerebral ischemia, an MRI study. J Cereb Blood Flow Metab. 2010;30(3):653–62.
5. Jordan A, Scholz R, Maier-Hauff K, et al. Presentation of a new magnetic field therapy system for the treatment of human solid tumors with magnetic fluid hyperthermia. J Magn Magn Mater. 2001;225(1-2):118–26.
6. Jun YW, Lee JH, Cheon J. Chemical design of nanoparticle probes for high-performance magnetic resonance imaging. Angew Chem Int Ed Engl. 2008;47(28):5122–35.
7. Song HT, Choi JS, Huh YM, et al. Surface modulation of magnetic nanocrystals in the development of highly efficient magnetic resonance probes for intracellular labeling. J Am Chem Soc. 2005;127(28):9992–3.
8. Lewin M, Carlesso N, Tung CH, et al. Tat peptide-derivatized magnetic nanoparticles allow in vivo tracking and recovery of progenitor cells. Nat Biotechnol. 2000;18(4):410–4.
9. Na HB, Song IC, Hyeon T. Inorganic nanoparticles for MRI contrast agents. Adv Mater. 2009;21(21):2133–48.
10. Jordan A, Scholz R, Wust P, Fähling H, Felix R. Magnetic fluid hyperthermia (MFH): cancer treatment with AC magnetic field induced excitation of biocompatible superparamagnetic nanoparticles. J Magn Magn Mater. 1999;201(1-3):413–9.
11. Sonvico F, Mornet S, Vasseur S, et al. Folate-conjugated iron oxide nanoparticles for solid tumor targeting as potential specific magnetic hyperthermia mediators: synthesis, hysicochemical characterization, and in vitro experiments. Bioconjug Chem. 2005;16(5):1181–8.
12. Chan DCF, Kirpotin DB, Bunn PA Jr. Synthesis and evaluation of colloidalmagnetic iron oxides for the site-specific radiofrequency-induced hyperthermia of cancer. J Magn Magn Mater. 1993;122(1-3):374–8.
13. Hou CH, Hou SM, Hsueh YS, Lin J, Wu HC, Lin FH. The in vivo performance of biomagnetic hydroxyapatite nanoparticles in cancer hyperthermia therapy. Biomaterials. 2009;30(23-24):3956–60.
14. Manca MF, Zwart I, Beo J, et al. Characterization of mesenchymal stromal cells derived from full-term umbilical cord blood. Cytotherapy. 2008;10(1):54–68.
15. Yin JL, Shackel NA, Zekry A, et al. Real-time reverse transcriptase-polymerase chain reaction (RT-PCR) for measurement of cytokine and growth factor mRNA expression with fluorogenic probes or SYBR green I. Immunol Cell Biol. 2001;79(3):213–21.
16. Kyrtatos PG, Lehtolainen P, Junemann-Ramirez M, et al. Magnetic tagging increases delivery of circulating progenitors in vascular injury. JACC Cardiovasc Interv. 2009;2(8):794–802.
17. Hua Z, Liu L, Shen J, et al. Mesenchymal stem cells reversed morphine tolerance and Opioid-induced Hyperalgesia. Sci Rep. 2016;6:32096.
18. Kostura L, Kraitchman DL, Mackay AM, Pittenger MF, Bulte JW. Feridex labeling of mesenchymal stem cells inhibits chondrogenesis but not adipogenesis or osteogenesis. NMR Biomed. 2004;17(7):513–7.
19. Bulte JW, Kraitchman DL, Mackay AM, Pittenger MF. Chondrogenic differentiation of mesenchymal stem cells is inhibited after magnetic labeling with ferumoxides. Blood. 2004;104(10):3410–2.
20. Soenen SJ, Himmelreich U, Nuytten N, De Cuyper M. Cytotoxic effects of iron oxide nanoparticles and implications for safety in cell labeling. Biomaterials. 2011;32(1):195–205.
21. Balakumaran A, Pawelczyk E, Ren J, et al. Superparamagnetic iron oxide nanoparticles labeling of bone marrow stromal (Mesenchymal) cells does not affect their 'sternness'. PLoS One. 2010;5(7):e11462.
22. Arbab AS, Bashaw LA, Miller BR, et al. Characterization of biophysical and metabolic properties of cells labeled with superparamagnetic iron oxide nanoparticles and transfection agent for cellular MR imaging. Radiology. 2003;229(3):838–46.
23. Chen R, Yu H, Jia ZY, Yao QL, Teng GJ. Efficient nano iron particle-labeling and noninvasive MR imaging of mouse bone marrow-derived endothelial progenitor cells. Int J Nanomedicine. 2011;6:511–9.
24. Yang JX, Tang WL, Wang XX. Superparamagnetic iron oxide nanoparticles may affect endothelial progenitor cell migration ability and adhesion capacity. Cytotherapy. 2010;12(2):251–9.
25. Smith CA, de la Fuente J, Pelaz B, Furlani EP, Mullin M, Berry CC. The effect of static magnetic fields and tat peptides on cellular and nuclear uptake of magnetic nanoparticles. Biomaterials. 2010;31(15):4392–400.
26. Wiskirchen J, Groenewaeller EF, Kehlbach R, et al. Longterm effects of repetitive exposure to a static magnetic field (1.5T) on proliferation of human fetal lung fibroblasts. Magn Reson Med. 1999;41(3):464–8.
27. Nakahara T, Yaguchi H, Yoshida M, Miyakoshi J. Effects of exposure of CHO-K1 cells to a 10-T static magnetic field. Radiology. 2002;224(3):817–22.
28. Wiskirchen J, Grönewäller EF, Heinzelmann F, et al. Human fetal lung fibroblasts: in vitro study of repetitive magnetic field exposure at 0.2, 1.0, and 1.5 T. Radiology. 2000;215(3):858–62.
29. Ko IK, Song HT, Cho EJ, Lee ES, Huh YM, Suh JS. In vivo MR imaging of tissue-engineered human mesenchymal stem cells transplanted to mouse: a preliminary study. Ann Biomed Eng. 2007;35(1):101–8.
30. Himes N, Min JY, Lee R, et al. In vivo MRI of embryonic stem cells in a mousemodel ofmyocardial infarction. Magn Reson Med. 2004;52(5):1214–9.
31. Bos C, Delmas Y, Desmoulière A, et al. In vivo MR imaging of intravascularly injected magnetically labeled mesenchymal stem cells in rat kidney and liver. Radiology. 2004;233(3):781–9.
32. Dai GH, Xiu JG, Zhou ZJ, et al. Effect of superparamagnetic iron oxide labeling on neural stem cell survival and proliferation. Nan Fang Yi Ke Da Xue Xue Bao. 2007;27(1):49–55.
33. Lee H, Hong Y, Kwon SH, Park J, Park J. Anti-aging effects of Piper Cambodianum P. Fourn. Extract on normal human dermal fibroblast cells and a wound-healing model in mice. Clin Interv Aging. 2016;11:1017–26.

6

Cladosporol A triggers apoptosis sensitivity by ROS-mediated autophagic flux in human breast cancer cells

Mytre Koul[1,4], Ashok Kumar[1,4], Ramesh Deshidi[2,4], Vishal Sharma[3,4], Rachna D. Singh[5], Jasvinder Singh[1,4], Parduman Raj Sharma[1,4], Bhahwal Ali Shah[2,4*], Sundeep Jaglan[3,4] and Shashank Singh[1,4*]

Abstract

Background: Endophytes have proven to be an invaluable resource of chemically diverse secondary metabolites that act as excellent lead compounds for anticancer drug discovery. Here we report the promising cytotoxic effects of Cladosporol A (HPLC purified >98%) isolated from endophytic fungus *Cladosporium cladosporioides* collected from *Datura innoxia*. Cladosporol A was subjected to in vitro cytotoxicity assay against NCI60 panel of human cancer cells using MTT assay. We further investigated the molecular mechanism(s) of Cladosporol A induced cell death in human breast (MCF-7) cancer cells. Mechanistically early events of cell death were studied using DAPI, Annexin V-FITC staining assay. Furthermore, immunofluorescence studies were carried to see the involvement of intrinsic pathway leading to mitochondrial dysfunction, cytochrome c release, Bax/Bcl-2 regulation and flowcytometrically measured membrane potential loss of mitochondria in human breast (MCF-7) cancer cells after Cladosporol A treatment. The interplay between apoptosis and autophagy was studied by microtubule dynamics, expression of pro-apoptotic protein p21 and autophagic markers monodansylcadaverine staining and LC3b expression.

Results: Among NCI60 human cancer cell line panel Cladosporol A showed least IC_{50} value against human breast (MCF-7) cancer cells. The early events of apoptosis were characterized by phosphatidylserine exposure. It disrupts microtubule dynamics and also induces expression of pro-apoptotic protein p21. Moreover treatment of Cladosporol A significantly induced MMP loss, release of cytochrome c, Bcl-2 down regulation, Bax upregulation as well as increased monodansylcadaverine (MDC) staining and leads to LC3-I to LC3-II conversion.

Conclusion: Our experimental data suggests that Cladosporol A depolymerize microtubules, sensitize programmed cell death via ROS mediated autophagic flux leading to mitophagic cell death.

Keywords: Apoptosis, Breast cancer, *Cladosporium cladosporioides*, Cladosporol a, Endophytes, Reactive oxygen species

Background

Natural products from microbial source produce a vast wealth of specialized metabolites with wide range of structural diversity and biological activities. Secondary metabolites from fungal endophytes and their synthetic analogues have been widely used in pharmaceutical industry. In the past two decades variety of promising bioactive compounds have been successfully discovered from fungal endophytes having insecticidal, antimicrobial and anticancer activities [1]. It has been estimated that, in last 50 years about 47% of the anticancer drugs have been obtained either as natural products or synthetic compounds derived from them [2]. Especially fungal endophytes have produced a plethora of bioactive compounds having significant cytotoxic potential e.g. Paclitaxel, Podophyllotoxin, Vinblastine, Vincristine, Camptothecine, Mycophenolic acid, Emodin, Wortmannin, Dicatenarin etc. [3–5]. Among various fungal endophytes, *Cladosporium* species are also well known to produce variety of secondary metabolites which find promising application in pharmaceutical industry [6]. In regard to their

* Correspondence: bashah@iiim.ac.in; sksingh@iiim.ac.in
[2]Natural Product Chemistry, CSIR-Indian Institute of Integrative Medicine, Jammu, India
[1]Cancer Pharmacology Division, CSIR-Indian Institute of Integrative Medicine, Jammu, India
Full list of author information is available at the end of the article

anticancer potential, Earlier in 2007 Wang et al. isolated endophytic *Cladosporium Sp* from *Quercus variabilis.* They further purified and characterized a unique fungal metabolite Brefeldin A from crude extract of *Cladosporium sp.* [7]. Brefeldin A exhibited significant antitumor activity, its treatment causes apoptosis in several human cancer cell lines e.g. human glioblastoma (SA4, SA146, U87MG) and colon cancer cells (HCT-116) via arresting $G_0/G1$ phase of cell cycle independent of Bcl-2/ Bax Mcl-1 and p53 [5]. In 2009 Zang and collegues, isolated Taxol from *Cladosporium cladosporioides* [8]. In the present study we isolated an endophytic fungus from *Datura innoxia* a well-known Indian annual medicinal plant. It belongs to the family Solanaceae [9]. *Datura innoxia* has been widely used as a traditional medicine in ayurveda since long times due to its immense medicinal properties, as all parts of the plants i.e. flowers, leaves, seed, root have appropriate medicinal applications. Its medicinal properties are due to the presence of about more than 30 alkaloids including atropine, hyoscyamine, scopolamine, withanolides (lactones) and other tropanes as well [10]. The methanolic leaf extract of *Datura innoxia* has shown to induce apoptosis in human colon adenocarcinoma (HCT 15) and larynx (Hep-2) cancer cell lines via inhibiting the expression of antiapoptotic Bcl-2 protein [11]. In view of its (*Datura innoxia*), promising cyotoxic effects, we isolated an endophytic fungus *Cladosporium cladosporioides* from it. We further isolated, purified and characterized a secondary metabolite Cladosporol A from endophytic *Cladosporium cladosporioides* and investigated the cyotoxic effects of Cladosporol A treatment against various human cancer cell lines. It exhibited promising cytotoxic effect against human breast (MCF-7) cancer cell line having minimum IC_{50} 8.7 μM. We next, ascertained mechanistically the cell death caused by Cladosporol A against breast cancer (MCF-7) cells. Breast cancer represents the second leading cancer in women worldwide. It is molecularly and clinically heterogeneous disease representing about 25% of all cancers in women and 12% of all new cancer cases [12]. It usually occurs in the breast tissue; starting in the lobules or ducts. The two major routes of cell death i.e. apoptosis and autophagy are highly controlled and dynamic processess that are used to remove damaged and defective cells. Upregulation of mitochondrial apoptosis pathway in response to antitumor agents is considered a signature of intrinsic apoptosis pathway in tumor cell lines. Apoptotic signals that trigger activation of mitochondrial pathway will result in MMP loss and cytochrome c release in mitochondrial inter- membrane space [4]. Autophagy, is a complex process which involves sequestration of intracellular organelles and cytoplasmatic portions into vacuoles called autophagosomes which further fuse with lysosomes to generate autophagolysosomes and mature lysosomes, where the whole material is degraded ultimately leading to cell death [13]. In addition, redox status of the cell i.e.

reactive oxygen species (ROS) generation is a determining factor in regulating cell death pathways [14]. Here we first time report the involvement of ROS generation as major features of the apoptotic cell death caused by Cladosporol A in human breast (MCF-7) cancer cell line. Cladosporol A treatment induces membrane potential loss of mitochondria, cytochrome c release, Bax upregulation and Bcl-2 down regulation, thereby inducing mitochondrial activation mediated apoptosis. Cladosporol A also inhibited the assembiling of microtubules and induction of p21 a pro-apoptotic protein. Furthermore, Cladosporol A treatment also induced mild autophagic flux in human breast (MCF-7) cell line. Collectively the data, suggest that Cladosporol A, a microtubule de-polymerizer triggers mitochondrial cell death machinery and could be used as potential chemotherapeutic agent against human breast cancer.

Results

Identification, characterization and phylogenetic analysis of endophytic fungus (MRCJ-314) revealed it as *Cladosporium cladosporioides*

The morphological characteristics of isolated endophytic fungus from *Datura innoxia* MRCJ-314 (DIE-10) supports that it belongs to genus *Cladosporium* [15]. Morphologically, in obverse view on PDA (potato dextrose agar plate), MRCJ-314 (DIE-10) showed dark olive green growth, velvety and on reverse view it seems olivaceous black (Fig. 1).

To designate MRCJ-314 (DIE-10) taxonomically up to species level, molecular technique was employed. The NCBI GenBank search for DNA sequence similarity showed that ITS1–5.8S–ITS2 sequence of DIE-10 has 99% homology with *Cladosporium cladosporioides* (GenBank Accession No. EU497597). Sequences of the maximum identity greater than 90% were retrieved, aligned with the sequence of strain MRCJ-314 (DIE-10), using clustal W module of MEGA6 software further subjected to neighbor-joining (NJ) analysis to obtain the phenogram (Fig. 2). The ITS sequence of strain (MRCJ-314) DIE-10 has highest nucleotide similarities with

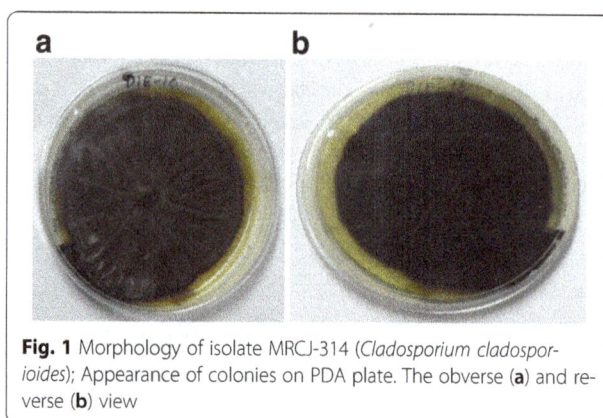

Fig. 1 Morphology of isolate MRCJ-314 (*Cladosporium cladosporioides*); Appearance of colonies on PDA plate. The obverse (**a**) and reverse (**b**) view

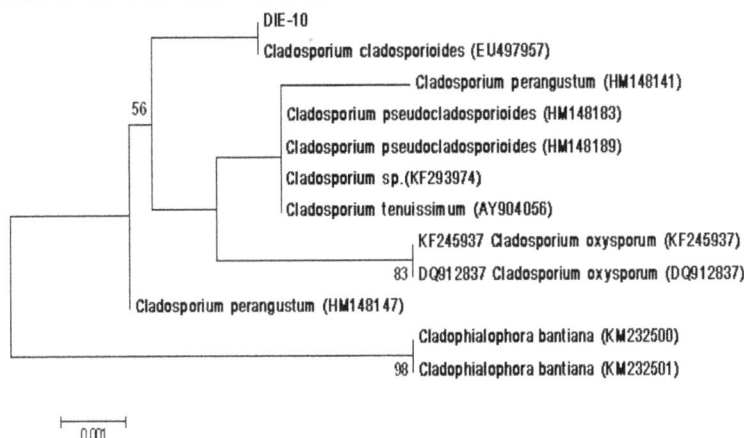

Fig. 2 Neighbor-Joining tree of fungal endophyte MRCJ-314 (*Cladosporium cladosporioides*) based on ITS1–5.8S–ITS2 rDNA sequences. Confidence values above 50% obtained from a 1000-replicate bootstrap analysis are shown at the branch nodes

Cladosporium cladosporioides (EU497597), formed a clade with 100% bootstrap support indicating MRCJ-314 (DIE-10) as *C. cladosporioides.*

Cladosporol A treatment effectively inhibited proliferation of human breast cancer (MCF-7) cell as well as their colony formation capacity

Antiproliferative effect of Cladosporol A (Fig. 3) was demonstrated against human cancer cell line, panel via MTT assay. Cladosporol A treatment significantly inhibited cell proliferation of human breast cancer (MCF-7) cells in concentration-dependent manner with least IC_{50} 8.7 μM after 48 h (Table 1). Furthermore, on determining the ability of human breast (MCF-7) cancer cells to form colonies after Cladosporol A treatment revealed that it causes reduction in both the number as well as size of MCF-7 colonies as compared to untreated cells thereby suppressing ability of MCF-7 cells to form colonies in a concentration dependent manner (Fig. 4).

Cladosporol A treatment causes G1 phase arrest, depolymerizes microtubules and increases the expression of protein p21 of human breast cancer (MCF-7) cells

The cell cycle distribution of Cladosporol A treated human breast cancer (MCF-7) cells was observed via flow-cytometry. After 24 h treatment with different concentrations of Cladosporol A, it causes G0/G1 phase arrest at higher concentration and accumulation of some population of cells in the G2/M phase at lower concentration (Fig. 5a). The percentage of G0 phase increased up to a high level at 20 μM (45%). To mechanistically investigate the basis of growth inhibitory effects of Cladosporol A, we next elucidated its effect on the expression of key proteins involved in cell proliferation by immunofluorescence microscopic studies.

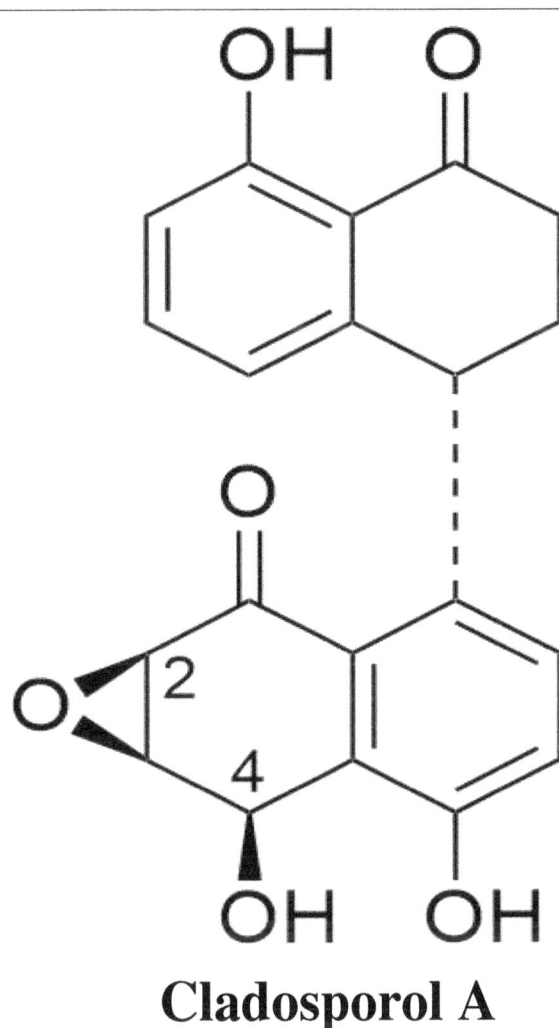

Cladosporol A

Fig. 3 Structure of Cladosporol A

Table 1 IC_{50} values of Cladosporol A against NC1 60 panel of human cancer cell lines of different tissue origin. Data are expressed as the mean ± SD of three similar experiments

$IC_{50}(\mu M)$

S.No	MCF-7 (Breast cancer)	A549 (Lung cancer)	HCT-116 (Colon cancer)	PC-3 (Prostate cancer)	OVCAR-3 (Ovarian Cancer)
Cladosporol A	8.7 ± 0.205	11.7 ± 0.505	12 ± 0.505	15.6 ± 0.360	10.3 ± 0.556
Paclitaxel	0.0.25 ± 0.001	0.75 ± 0.05	8.5 ± 0.620	0.48 ± 0.003	2.6 ± 0.120
Doxorubicin	0.1 ± 0.006	0.087 ± 0.009	0.096 ± 0.005	0.05 ± 0.006	0.13 ± 0.008

We investigated the effect of Cladosporol A on microtubules by immunofluorescence microscopy. Our data elucidated a concentration-dependent reduction in microtubule density after 24 h Cladosporol A treatment whereas control cells showed a typical array of radial, interphase microtubules (Fig. 5b). Thus indicating that Cladosporol A depolymerizes microtubules in MCF-7 cells. Further, we analysed the effect of Cladosporol A treatment on expression of cyclin-dependent kinase inhibitor p21 via immunofluorescence microscopic studies as well as western blot analysis. Our data suggested a concentration-dependent increase in expression of cyclin-dependent kinase inhibitor p21 (Fig. 5c and d).

Cladosporol A induced significant chromatin condensation and triggers apoptosis in human breast cancer (MCF-7) cells

Apoptosis induction in human breast cancer (MCF-7) cells via Cladosporol A treatment was determined microscopically by analysis of DAPI stained cells. Cladosporol A significantly induced condensation of chromatin and its fragmentation within the nucleus in concentration dependent manner in MCF-7 cells after treatment of 24 h. Promising chromatin condensation was also observed in paclitaxel treated cells (Fig. 6a). Further, a specific marker which detects the phosphatidylserine externalization during apoptosis, a phosphatidylserine-binding protein (annexin V) was utilized to assess the

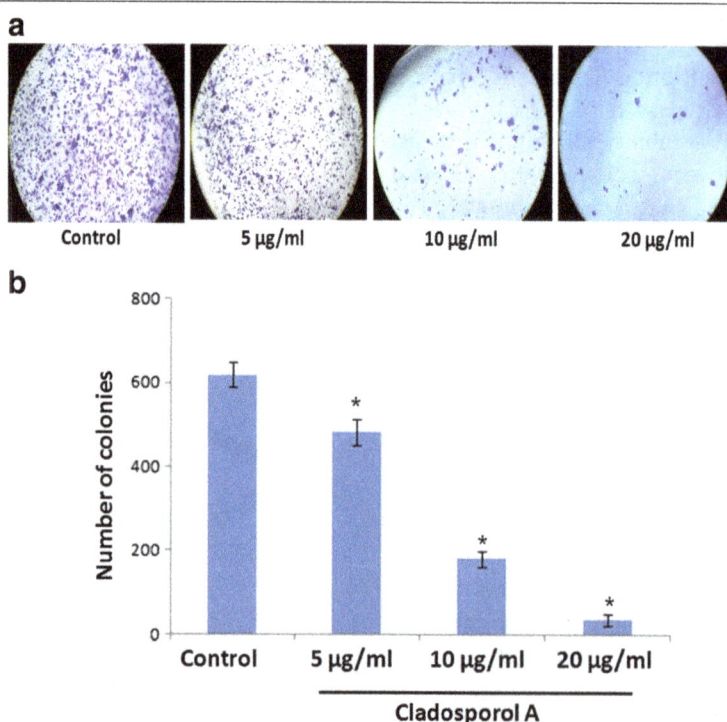

Fig. 4 Clonogenic ability was assessed in human breast cancer (MCF-7) cells, after Cladosporol A treatment. The representative images are shown. **a** Reduction in colony forming capability was determined in MCF-7 cells. Cells (1×10^3/ml/well) were seeded in six well plates and treated with different concentrations of Cladosporol A (5, 10, 20 μM). The number of crystal violet stained colony were counted randomly after seven days, quantified and photographed. **b** Data are presented as mean ± S.D., statistical analysis was done with **$p < 0.01$ and ***$p < 0.001$

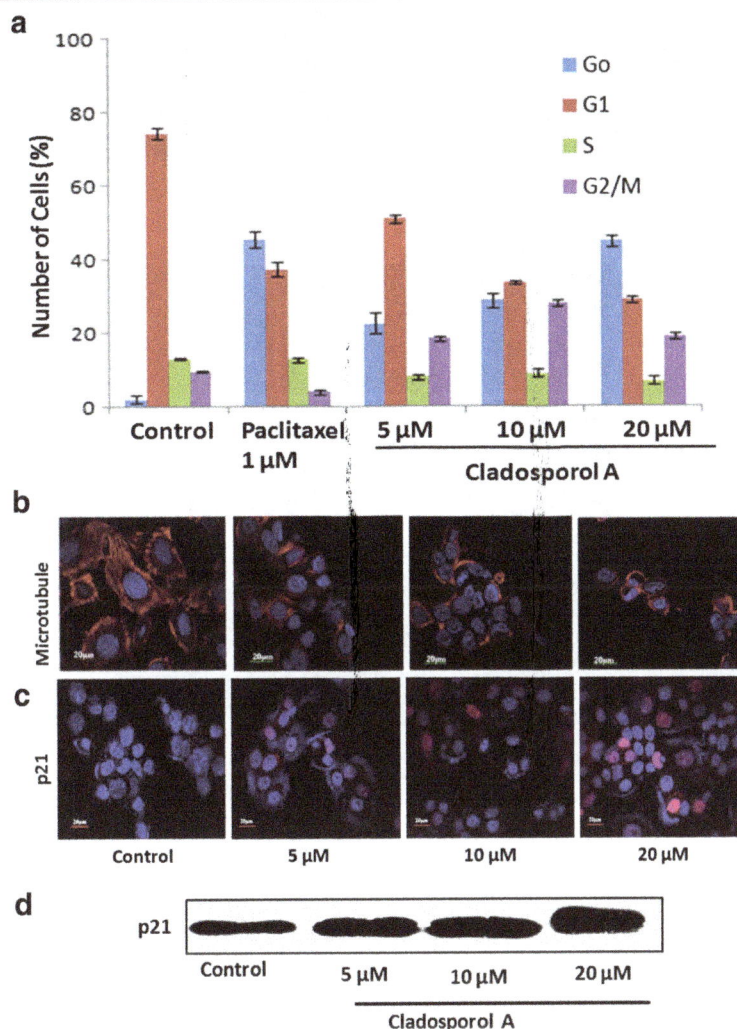

Fig. 5 a Cell cycle distribution in human breast cancer (MCF-7) cells treated with 5, 10, 20 µM of Cladosporol A. Data are expressed as the mean ± SD of three similar experiments. **b** Cladosporol A disrupts the microtubules of human breast cancer (MCF-7) cells. Cells were treated with indicated concentrations of Cladosporol A for 24 h. Immunocytochemical staining was conducted using anti α-tubulin antibody and Alexa Flour-555-labelled secondary antibody and nuclei were stained with DAPI (**c**). Analysis of p21 protein expression of human breast cancer (MCF-7) cells after 24 h of Cladosporol A (5, 10, 20 µM) treatment by immunofluorescence microscopy (**d**). Western blot indicating the increase in p21 protein expression after Cladosporol A treatment in concentration dependent manner

extent of apoptosis of human breast cancer (MCF-7) cells after Cladosporol A treatment. The results of Annexin V-FITC and PI dual staining suggested that both the two dyes did not stained the normal cells, whereas Annexin V-FITC stained only the early apoptotic cells and both Annexin V-FITC and PI stained the late apoptotic cells. As it is evident from Fig. 6b, maximum of the Cladosporol A and paclitaxel treated cells were Annexin V-FITC (green) stained only indicating their early apoptotic stage. Additionally, some percentage of the cells, treated with Cladosporol A and paclitaxel, were stained by both dyes indicating that these cells were in late apoptotic stage. In Fig. 6c, at lower concentration (10 µM) Cladosporol A induces 16% early stage apoptosis and

6% of late stage apoptosis. At 20 µM concentration Cladosporol A induces 25% early stage apoptosis and 9% late apoptotic stage.

Cladosporol A induces ROS-mediated cell death

Subsequent studies were conducted to assess levels of reactive oxygen species (ROS) following treatment with Cladosporol A measured by the fluorescent indicator ',7'-dichlorofluorescein diacetate. After loading human breast cancer (MCF-7) cells with 2',7'-dichlorofluorescein diacetate a ROS-sensitive fluorescent probe, intracellular ROS levels were measured by flow cytometry and fluorescence microscope (Fig. 7a & b). ROS generation was induced in a significant manner by

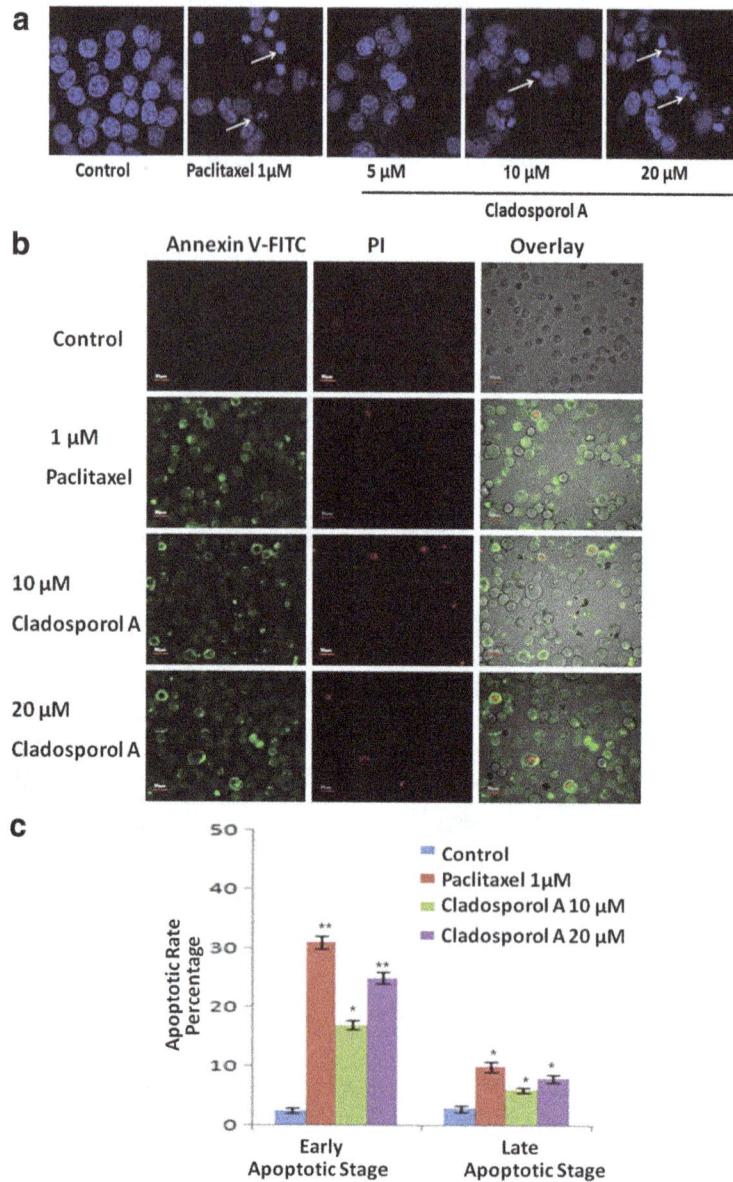

Fig. 6 a Nuclear morphology analysis of human breast cancer (MCF-7) cells (2×10^5/ml/well) using DAPI after treatment with different concentrations of Cladosporol A (5, 10, 20 μM) for 24 h and examined using fluorescence microscopy (40X). Paclitaxel (1 μM) was used as positive control. With increase in concentration of Cladosporol A there is significant increase in nuclear condensation and formation of apoptotic bodies (**b**). The effect of Cladosporol A on the exposure of phosphatidylserine (PS) in human breast cancer (MCF-7) cells after 24 h treatment was analysed. Phosphatidylserine exposure was assessed by the annexin-V/ propidium iodide assay, as described in methodology and analyzed by confocal microscopy (**c**). The percentage of cells in early and late stages of apoptosis obtained by analysis of the cell images (mean ± SD, 3 experiments), p*< 0.05 and **p < 0.01

Cladosporol A treatment, as indicated by the increase in fluorescence intensity. In comparison with untreated cells, nearly 45% ROS was generated in the cells treated with 20 μM Cladosporol A after 24 h treatment. Furthermore, to determine the involvement of ROS generation in cell death induced by Cladosporol A, ROS accumulation as well as cell viability was studied in the presence or absence of NAC (N-acetyl-cysteine) (ROS scavenger) was performed. In the presence of NAC, Cladosporol A treatment was unable to elevate ROS levels (Fig. 7a & b) and failed to induce cell death significantly as well (Fig. 7c), indicating that cell death was ROS-mediated.

Cladosporol A induces loss in mitochondrial membrane potential (ΔΨm) and causes cytochrome c release in human breast cancer (MCF-7) cells

Flow cytometry and laser scanning confocal microscope (LSCM) were used to measure of mitochondrial

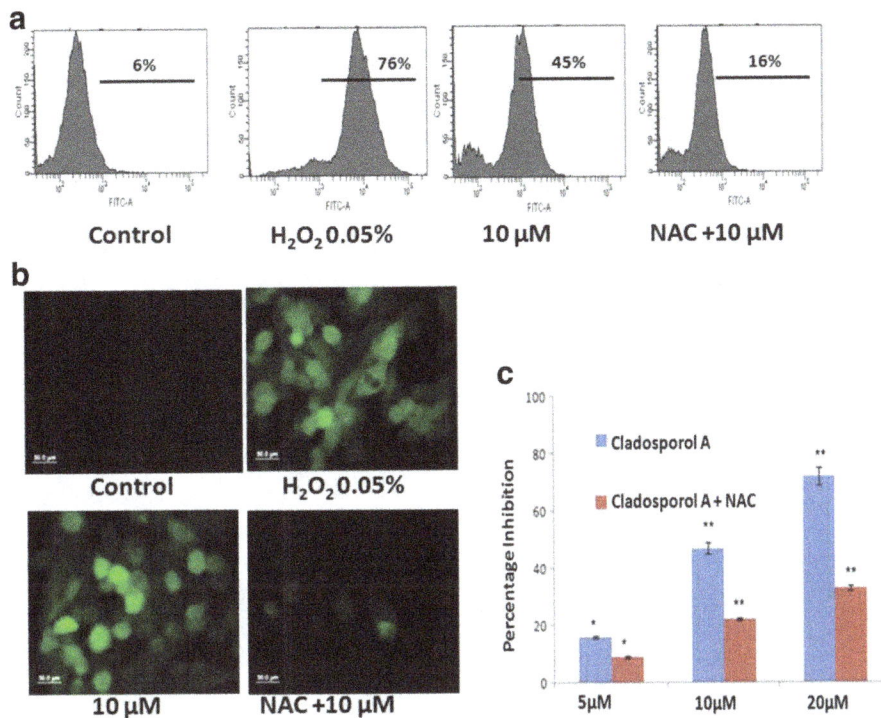

Fig. 7 a Intracellular ROS level was measured by flow cytometry analysis using DCFH- DA after 24 h. Human breast cancer (MCF-7) cells were treated with indicated concentrations of Cladosporol A and 0.05% H_2O_2 and incubated with 5 μM DCFH-DA. **b** Detection of ROS by fluorescence microscope. After incubation with DCFH-DA and 0.05% H_2O_2, MCF-7 cells (2×10^5/ml/well) were washed and examined by fluorescence microscope (40X). **c** MCF-7 cell growth inhibition in the presence of ROS scavenger (NAC) was also determined. The cells were treated with indicated concentrations of Cladosporol A in the presence or absence of NAC (150 μM)

membrane potential (ΔΨm) loss (Fig. 8 a & b) with Rh123 staining. Human breast cancer (MCF-7) cells after Cladosporol A treatment for 24 h induced a significant mitochondrial membrane potential loss in human breast cancer (MCF-7) cells in a concentration dependent manner. Doxorubicin (positive control) treated human breast cancer (MCF-7) cells also showed a significant reduction in mitochondrial membrane potential. These data suggested that Cladosporol A induces apoptosis via MMP (ΔΨm) loss. In Fig. 8 C Flow cytometric analysis showed that Cladosporol A at 20 μM concentration induces 27% of MMP loss while at 10 μM concentration causes 12% of the MMP loss. Colocalization between mitochondria and cytochrome c was studied by laser scanning confocal microscopy. MitoTracker red dye was used to label functional mitochondria. Localization of cytochrome c was ascertained by immunofluorescence using a specific antibody of cytochrome c. In control cells, cytochrome c was colocalized with mitochondria (Fig. 9). In contrast, Cladosporol A and doxorubicin treatments induced a significant reduction in mitochondria and release of cytochrome c to the cytosol respectively.

Cladosporol A treatment induces upregulation of Bax and downregulation of Bcl-2 expression in human breast cancer (MCF-7) cells

Pro-apoptotic protein Bax and anti-apoptotic Bcl-2 plays a very significant role in mitochondrial mediated apoptotic pathway. The expression levels of Bax and Bcl-2 in human breast cancer (MCF-7) cells after Cladosporol A treatement for 24 h were evaluated, using immunofluorescence microscopic studies and western blot analysis. Results of immunofluorescence microscopy and western blot analysis suggested that the fluorescence intensity of Bcl-2 reduced whereas in case of Bax increased after Cladosporol A treatment in concentration dependent manner (Fig. 10 a and c). Thus, apoptosis induction via Cladosporol A, was accompanied by upregulation of Bax and downregulation of Bcl-2 in a concentration dependent manner.

Cladosporol A treatment also induces autophagic cell death in human breast cancer (MCF-7) cells

Changes in the fluorescence intensity of MDC (monodansylcadaverine staining), a specific marker for autophagic vacuoles was used to detect the autophagic activity of Cladosporol A (Fig. 11a). Compared to the untreated

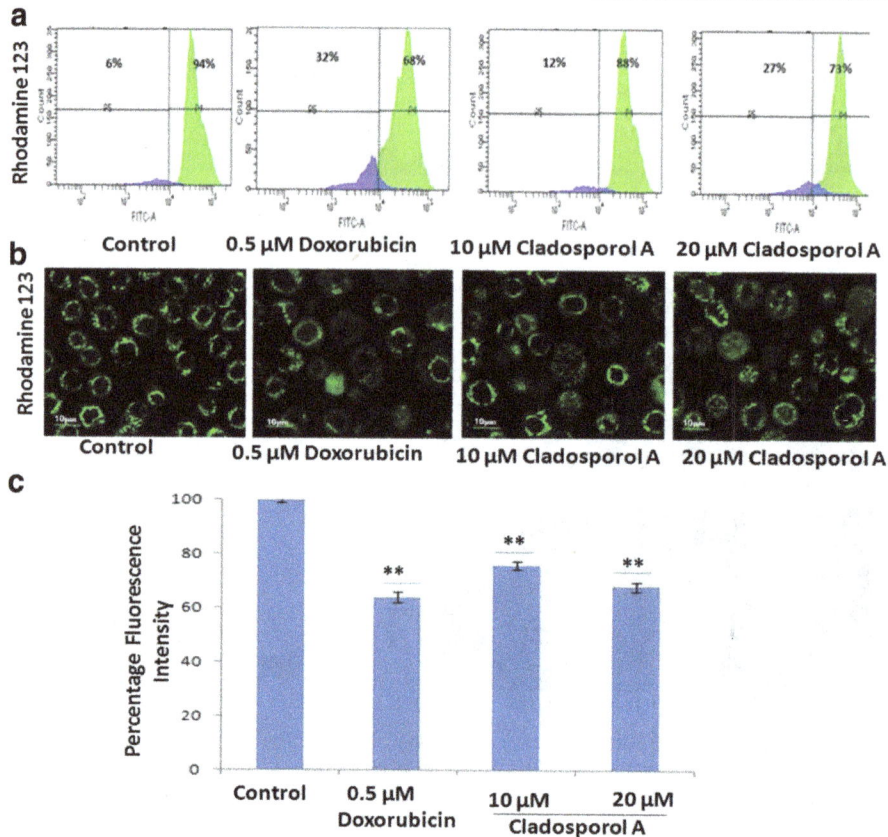

Fig. 8 Cladosporol A induces loss of mitochondrial transmembrane potential. Human breast cancer (MCF-7) cells were incubated with different concentrations of Cladosporol A for 24 h. Thereafter, cells were stained with Rh-123 (1 µM) for 20 min and analyzed by (**a**) Flow cytometer (**b**). Confocal microscopy (**c**). Histogram showing the effect of Cladosporol A on $\Delta\psi m$ measured with laser scanning confocal microscope by staining with Rhodamine 123. Data are presented as mean ± S.D., statistical analysis was done with **$p < 0.01$

control cells, in Cladosporol A treated human breast (MCF-7) cancer cells, the number of autophagic vacuoles stained by MDC were much higher. Furthermore, to mechanistically confirm the Cladosporol A induced autophagy induction, a set of autophagy-related factors including LC3-I and LC3-II in the human breast (MCF-7) cancer cells were investigated by immunofluorecence microscopy and western blot analysis after treatment with

Cladosporol A at various concentrations for 24 h (Fig. 11 b and c). Cladosporol A treatment significantly induced conversion of LC3-I to LC3-II.

Discussion

Natural products represent an unsurpassed source of bioactive compounds and constitute a relevant resource for the pharmaceutical industry. Endophytes represent a

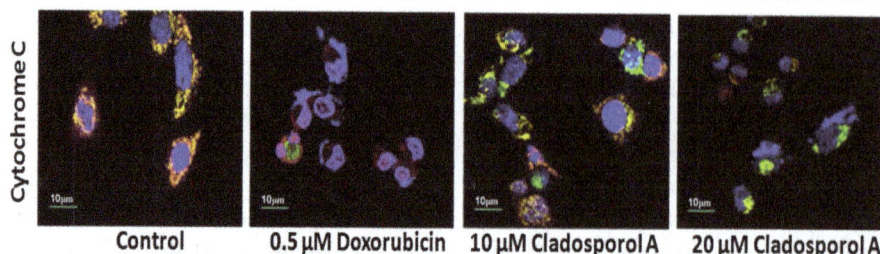

Fig. 9 Colocalization of cytochrome c and mitochondria was determined by confocal microscopy. Human breast cancer (MCF-7) cells were immunostained for cytochrome c release (green) and the mitochondria of the cells were stained with MitoTracker (red). MCF-7 cells were treated with different concentrations of Cladosporol A and doxorubicin (500 nM) for 24 h and stained with anti-cytochrome c antibody and Alexa Fluor 488-labeled secondary antibody

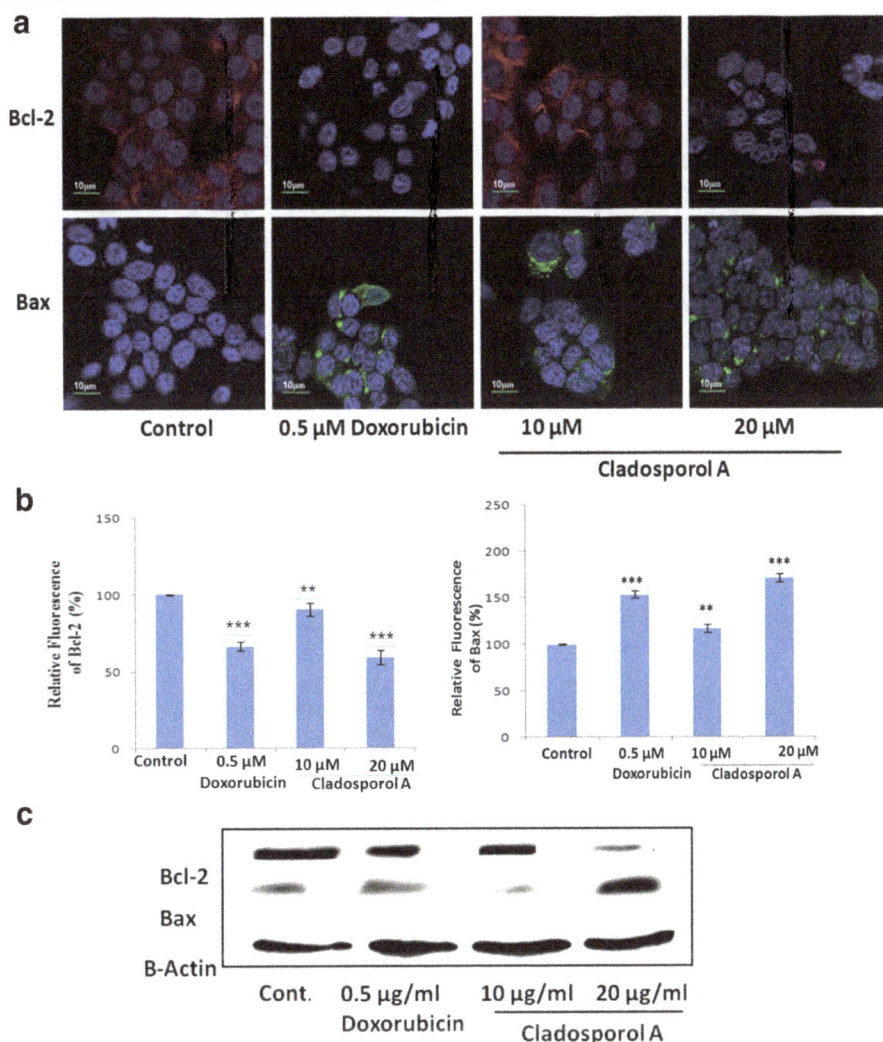

Fig. 10 Effects of Cladosporol A and doxorubicin treatment on the expression of apoptosis related proteins. (**a**) Representative images of immunofluorescence analysis of effects of the expression of; antiapoptotic Bcl-2 and pro-apoptotic Bax by confocal microscopy (using 40× oil immersion lens) in human breast cancer (MCF-7) cells treated with Cladosporol A for 24 h (**b**). Statistical analysis of the expression of Bcl-2 and Bax. The relative fluorescence intensity was compared to the control group. Data are mean ± S.D. of three similar experiments; statistical analysis was done with **$p < 0.01$ and ***$p < 0.001$. **c** Western blot indicating the increase in bax and decrease bcl-2 protein expression after Cladosporol A treatment in concentration dependent manner

potential source for isolation of variety of lead compounds for the development of drugs for various malignancies. Breast cancer is the frequently diagnosed malignancy leading to cancer related deaths worldwide. Various gene mutations taking place in luminal or basal progenitor cell population giving rise to phenotypically heterogeneous cell population is responsible for breast cancer [16]. There is a need to develop novel strategies for achieving additional means to control development and progression of breast cancer. In the search for novel anticancer agents against breast cancer with effective chemotherapeutic spectrum various research groups have been motivated to isolate and characterize natural compounds from fungal endophytes that will increase the number of these

therapeutic tools to inhibit breast cancer cells proliferation [17]. In the present study, Cladosporol A, a potent, natural compound was isolated and purified from endophytic fungus *Cladosporium cladosporioides* isolated from *Datura innoxia* plant. *Cladosporium cladosporioides* is main producer of antifungal metabolites i.e. cladosporin, 5 hydroxyasperentin and isocladosporin. These molecules (metabolites) have shown promising activity in the treatment and control of various plant-infected diseases [18]. We determined the antiproliferative activity of the Cladosporol A, a potent, natural compound isolated against NCI60 human cancer cell lines. It has shown least IC_{50} value of 8.7 μM against human breast (MCF-7) cancer cells (Table 1). We next ascertained the effect of

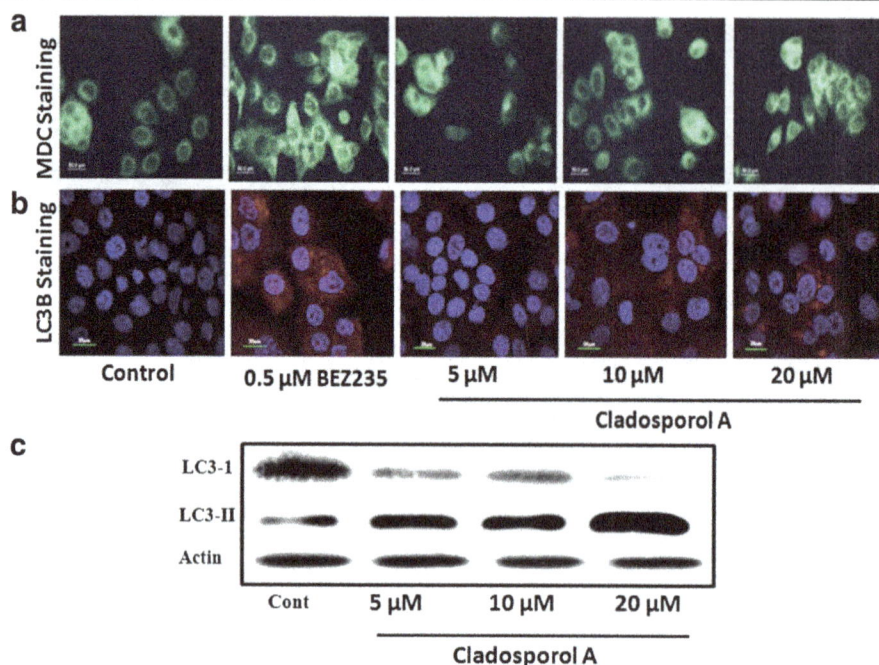

Fig. 11 Cladosporol A induces autophagy in human breast cancer (MCF-7 cells) (**a**). The autophagic vacuoles were observed under fluorescence microscope (40×) with MDC staining. The treatment of Cladosporol A and BEZ235 (positive control group) induced concentration-dependent formation of autophagic vacuoles in MCF-7 cells after 24 h. **b** Detection of autophagy with LC3b antibody using confocal microscopy. Immunocytochemical staining was conducted using anti-LC3b antibody and Alexa Flour-555-labelled secondary antibody. Nuclei were stained with DAPI. **c** Detection of autophagy by western blot analysis after Cladosporol A treatment in concentration dependent manner

Cladosporol A treatment on proliferation, growth and clonogenic ability of human breast (MCF-7) cancer cells. Cladosporol A treatment produced a concentration-dependent inhibitory effect on the ability of human breast (MCF-7) cancer cells to reproduce and form large colonies (Fig. 4 a & b). We therefore used human breast (MCF-7) cancer cell line as a model cell line for further mechanistic studies. We examined the mechanism of changes in cell cycle distribution caused by Cladosporol A in MCF-7 cells. It causes G0/G1 phase arrest and also induces accumulation of some cells in the G2/M phase (Fig. 5a). Microtubules play an important role in different phases of the cell cycle. Their differential dynamic behaviour has qualified them as targets for treating several diseases including cancer. Several antimicrotubule agents have been evaluated for their promising uses in cancer chemotherapy e.g. paclitaxel, vinblastine, griseofulvin etc. These agents have shown to induce apoptosis via targeting microtubule assembly dynamics [19]. Immunofluorescence confocal microscopic studies on MCF-7 cells after Cladosporol A treatment revealed that it induces disruption of microtubules by inhibiting microtubule polymerization in concentration-dependent manner (Fig.5b). Further, we systematically analyzed the involvement of p21 in Cladosporol A mediated cell death. Our immunofluorescence studies as well as western blot analysis revealed increase

in p21 protein expression in human breast (MCF-7) cancer cells after Cladosporol A treatment in a concentration dependent manner (Fig. 5c and d). Further the induction of apoptosis in human breast cancer (MCF-7) cells after Cladosporol A treatment was determined microscopically by analysis of DAPI stained cells. Cladosporol A treatment induced promising condensation of chromatin and its fragmentation within the nucleus in concentration dependent manner after 24 h treatment (Fig. 6a). Next, Annexin V-FITC/PI dual staining was performed to further determine the molecular mechanism of Cladosporol A induced cell death (apoptosis) in human breast (MCF-7) cancer cells by laser scanning confocal microscopy. The data indicated that following Cladosporol A treatment annexin V-FITC positive cells were seen in a significant amount whereas untreated cells were negative for annexin V-FITC and PI staining (Fig. 6b). ROS plays a vital role in various cellular biological activities including the processes of tumor metastasis and progression. In comparison to normal cells, cancer cells usually possess elevated ROS levels as well as antioxidant activities in an uncontrolled manner [20, 21]. Earlier it has been elucidated that targeting ROS is a critically significant cancer therapeutic strategy. This has promisingly contributed to the potent anticancer effects of paclitaxel thereby improving its clinical use. Thus, anticancer agents which induce the production of ROS are useful for eliminating the bulk of cancer cells [22]. We determined ROS generation induced by Cladosporol A in human breast cancer (MCF-7) cells. Cladosporol A induced prominent ROS generation in concentration dependent manner (Fig. 7a&b). To determine the role of ROS in inducing apoptosis, we tested the effect of NAC on Cladosporol A treated human breast cancer (MCF-7) cells and found that ROS generation was almost completely inhibited by NAC pretreatment of cells treated with Cladosporol A for 24 h (Fig. 7c). Mitochondria play a pivotal role as energetic centres and contain various pro-apoptotic molecules that further activate cytosolic proteins to initiate apoptosis. The mitochondrial dysfunction and changes in the $\Delta\Psi m$ are considered an early event in apoptosis [23, 24]. In the cell major source of oxidative stress is mitochondria-mediated ROS generation [4, 25]. Flow cytometry and Laser scanning confocal microscopy (LCMS) results suggested that human breast (MCF-7) cancer cells treated with different concentrations of Cladosporol A (Fig. 8 a, b,c) considerable caused significant mitochondrial membrane potential loss. To further mechanistically investigate the apoptotic pathway, cytochrome c release induced by Cladosporol A treatment was studied. The results of our immunofluorescence data revealed, cytochrome c to be highly colocalized with mitochondria, in untreated control cells. In contrast, treatment of human breast (MCF-7) cancer cells with Cladosporol A and positive control (doxorubicin) caused diffused cytoplasmic distribution of cytochrome c from the transition pores of the mitochondria into the cytosol (Fig. 9). The best characterized protein family that play an important role in apoptotic cell death regulation are Bcl-2 family proteins. The ratio of pro- to anti-apoptotic proteins, of the Bcl-2 family proteins critically mediate mitochondrial cell death pathway [26]. Disruption of $\Delta\Psi m$ and cytochrome c excretion into the cytosol is promoted by Bax, the pro-apoptotic protein on the other hand mitochondrial integrity is preserved by Bcl-2, an anti-apoptotic thereby preventing the apoptosis [27]. Bax and Bcl-2 balance is thus, very critical in apoptotic pathway mediated by mitochondrial dysfunction [28]. Our results reflected the involvement of Bcl-2 family proteins in Cladosporol A mediated cell death. Treatment of human breast cancer (MCF-7) cells with Cladosporol A decreased $\Delta\Psi m$, upregulated Bax expression and downregulated Bcl-2 levels in a concentration-dependent manner, thereby increasing the ratio of Bax/Bcl-2 protein levels (Fig. 10). Thus the study clearly revealed the involvement of mitochondrial mediated pathway in Cladosporol A induced cell apoptosis. Several anticancer agents are known to induce tumor growth inhibition thereby causing mitophagic cell death [29]. Furthermore, another target for cancer treatment is autophagy. It has provided new opportunity for discovery and development of novel cancer therapeutics. To elucidate the role of

Cladosporol A in mediating autophagic cell death autophagolysosomes formation induced by it was examined by staining human breast cancer (MCF-7) cells by MDC (monodansylcadaverine), a tracer for autophagic vacuoles [30]. Cladosporol A treatment significantly increased MDC fluorescence in a concentration-dependent manner (Fig. 11a). Furthermore, the microtubule associated protein light chain 3 (LC3) is the another signature marker of autophagosomes. A key step in autophagy is the cleavage of the 18 kDa full length LC3, known as LC3-I, to a 16 kDa form, known as LC3-II, resulting in its recruitment to double-layered membrane of autophagosomes [31]. Our immunofluorescence confocal microscopic studies revealed that human breast (MCF-7) cancer cells treated with Cladosporol A significantly showed concentration-dependent increase in conversion of LC3-I to LC3-II (Fig. 11b and c).

Conclusions

Present data indicated that Cladosporol A isolated from endophytic *Cladosporium cladosporioides* from *Datura innoxia* mechanistically showed marked growth inhibition against panel of NCI60 human cancer cell lines especially against human breast (MCF-7) cancer cells. It triggers ROS-mediated mitophagic cell death inducing loss in MMP, releasing cytochrome c, in turn up regulating the expression of Bax and down regulating the level of Bcl-2 proteins sensitizing apoptosis via autophagic flux due to increased conversion of LC3-I to LC3-II. It also causes microtubule depolymerisation and increase in p21 protein expression. Our study provides a molecular basis for its development as novel anticancer lead for human breast cancer.

Methods
Chemicals and antibodies
PDA (Potato Dextrose Agar), lactophenol-cotton-blue, 1% sodium hypochlorite, Glycerol 15% was obtained from Difco. From Qiagen Purification kit for DNA was obtained, DNA MiniPrep™ kit (Sigma), HiPurA™ PCR product purification kit (HiMedia Laboratories), Penicillin G, dimethyl sulfoxide (DMSO), streptomycin, trypsin-EDTA, 3-(4,5-dimethylthiazole-2-yl)-2,5-diphenyltetrazolium bromide (MTT), Propidium iodide (PI), $2',7'$-dichlorofluorescein diacetate, DNase-free RNase, paraformaldehyde, annexin V-FITC Kit, crystal violet, N-acetyl cysteine, doxorubicin and paclitaxel were purchased from Sigma Chemicals Co. (St. Louis, MO). DAPI mounting medium was purchased from Cell signaling technology (CST). Fetal bovine serum (10%) was purchased from Gibco. Antibodies were purchased from different commercial sources: Anti cytochrome c, anti p21, anti bax and anti bcl-2 were procured from Santa Cruz biotechnology incorporation and anti α-

tubulin was procured from Sigma Chemical, St. Louis, MO. Secondary antibodies were obtained from CST. Other reagents were of analytical grade and were purchased from local sources.

Cell lines, cell culture, growth conditions and treatment

Colon (HCT-116) cancer cells, ovarian (OVCAR-3) cancer cells, lung (A549) cancer cells, breast (MCF-7) cancer cells and prostate (PC3) cancer cells, all were procured from NCI (National Cancer Institute) and were cultured in RPMI 1640/ McCoy's supplementation of medium was done with penicillin (100 units/ml), streptomycin (100 µg/ml), 10% FBS (fetal bovine serum), sodium pyruvate (550 mg/ml), L-glutamine (0.3 mg/ml) and $NaHCO_3$ (2 mg/ml). Cells were cultured at 37° C in CO_2 incubator (Heraeus, GmbH, Germany) with 95% humidity and 5.0% CO_2. In DMSO Cladosporol A was dissolved and cells were treated with it, while only vehicle was added to untreated control cultures. The concentration of DMSO added to the cultures was maintained at <1%.

Isolation of endophytic fungi

Root sample of *Datura innoxia* (Family: Solanaceae) was collected in January 2013, from healthy and symptomless mature plant, located in Jammu (32.72°N, 74.77°E), India. Sample was taken in sterilized polythene bag kept in icebox and processed within 24 h of collection. Sample was washed with the running tap water followed by thrice washing with autoclaved distilled water and air dried. Isolation of endophytic fungi was done by the procedure as described in our previous work [4]. The mycelium was picked and transferred to potato dextrose agar media (PDA), incubated at 28 °C and pure culture was obtained. The pure culture of the endophytic fungus DIE-10 was deposited to the Col Sir R. N. Chopra, Microbial Resource Center Jammu (MRCJ), India under accession number MRCJ-314.

Identification, characterization and phylogenetic analysis of endophytic fungus MRCJ-314

ZR Fungal/Bacterial DNA MiniPrep™ kit was used to obtain the pure genomic DNA. Spectrophotometrically (NanoDrop 2000) analyzed DNA was used for the PCR amplification of ITS1–5.8S–ITS2 regions. PCR was performed by universal primers ITS5: 5'-GGAAGTAA AAGTCGTAACAAGG-3' and ITS4: 5'-TCCTCCGCTT ATTGATATGC-3' [32]. The thermal cycling program used was as per our previous study [4]. The sequences obtained were used as query sequence for similarity search by using BLAST algorithm against the database maintained at NCBI. The ITS region sequence of (DIE-10) MRCJ-314 isolate was aligned with the most similar reference sequences of the taxa by using the clustal W module of MEGA6 software [33]. A phylogenetic tree was constructed subsequently analyzed for evolutionary distances by the neighbor joining method. The robustness of clades was determined by analysis of bootstrap with 1000 replications. The contiguous rDNA sequences of the representative isolate have been deposited in GenBank under the accession number KX553964.

Fermentation and isolation of Cladosporol A

The fungus was fermented in potato dextrose broth (PDB) medium for 7 days, pH 5.5 at 28 °C in dark with constant shaking in six 1 L flasks containing 300 ml broth. The fermentation broth was then extracted with dichloromethane following the National Cancer Institute's protocol [34]. The resulting extract was then concentrated in vacuum and subjected to column chromatography on silica gel using hexane-ethyl acetate with increasing polarity leading to the isolation of cladosporol A (Fig 2). The ^1H and ^{13}C NMR data of cladosporol A was found to be in identical range to that of reported in literature [35]. The purity of cladosporol A was established by HPLC: tR = 34.4 min (>99% purity) (please see Additional file 1: Figure S1 for spectral graphs).

Analytical data of Cladosporol A

White Solid, ^1H NMR (400 MHz, $CDCl_3$) δ 12.59 (d, $J = 6.7$ Hz, 1H), 8.62 (s, 1H), 7.33–7.20 (m, 2H), 7.01 (t, $J = 15.8$ Hz, 1H), 6.95 (d, $J = 8.6$ Hz, 1H), 6.80 (d, $J = 8.3$ Hz, 1H), 6.23 (d, $J = 7.6$ Hz, 1H), 5.43 (d, $J = 8.7$ Hz, 1H), 4.88 (dd, $J = 8.2, 4.8$ Hz, 1H), 4.10–4.04 (m, 1H), 3.87 (d, $J = 4.5$ Hz, 1H), 3.53 (t, $J = 14.0$ Hz, 1H), 2.76 (t, $J = 6.6$ Hz, 2H), 2.51 (dt, $J = 18.7, 5.5$ Hz, 1H), 2.21 (dt, $J = 22.4, 8.2$ Hz, 1H); ^{13}C NMR (125 MHz, $CDCl_3$) δ 205.3, 194.9, 162.7, 137.3, 136.4, 132.2, 122.5, 120.0, 115.8, 67.5, 56.1, 55.2, 40.1, 36.8, 31.0. HRMS (ESI$^+$) calculated for $C_{20}H_{16}O_6$ $[M ± H]^+$: 353.1020, found: 353.1015.

Cell viability (MTT) assay

MTT dye was used for performing cell proliferation assay. Various human cancer cell lines at a density of $8 × 10^4$ cells/well were seeded in 96-well culture plates. Cells were incubated with various concentrations of Cladosporol A after 24 h and untreated cells served as control. After 48 h of incubation of the test material, MTT dye, 3[4,5-dimethylthiazol- 2-yl]-2,5-diphenyl-tetrazolium bromide, was then added before 4 h of the experiment termination (2.5 mg/mL). The resulting MTT formazan crystals obtained were dissolved in DMSO (150 ml) and at 570 nm (reference wavelength 620 nm), OD was measured. Further, by comparing the absorbance of treated and untreated cells, percentage viability (cell growth) was calculated [36].

DNA content and cell cycle phase distribution

Human breast (MCF-7) cancer cell line at a seeding density of $2 × 10^5$ cells/well, were seeded in six-well plates and treated with 5, 10 and 20 µM concentrations of Cladosporol A and 1000 nM concentration of paclitaxel. Treated cells were harvested after 24 h, PBS washed and fixed overnight at −20 °C in 70% cold ethanol. Later on, cells were again given PBS washing and further treated for RNase digestion (0.1 mg/ml) at 37 °C for 90 min. PI (50 µg/ml) was used to stain the cells and with the help of flow-cytometry (BD FACS Aria) further analysis of the relative DNA content was done [37].

Nuclear morphology analysis

The presence of apoptotic cells were determined by staining MCF-7 (human breast cancer cells) with DAPI. MCF-7 cells were seeded in 60 mm culture dishes at a seeding density of 2×10^5 cells/well and incubated for 24 h. After that cells were treated with various concentrations of Cladosporol A (5, 10, 20 µM) and again incubated for 24 h. The media was decanted, collected and PBS wash was given to the cells. Trypsinization was carried out to detach the adherent cells. In the analysis collectively, floating and poorly attached cells were included. Cells were centrifuged and air-dried smears of MCF-7 cells were prepared. Later on in methanol at −20 °C, air-dried smears of MCF-7 cells were fixed for 20 min and DAPI (1 µg/ml in PBS) stained for 20 min at room temperature in the dark. Glycerol-PBS (1:1) was used to mount the slides and prepared slides were observed via inverted fluorescence microscope (Olympus, 1X81) [4, 38].

Apoptosis detection by Annexin V-FITC and PI dual staining

To investigate the apoptosis-inducing effect of Cladosporol A, we analyzed the percentage of early and late apoptotic cells by Annexin V-FITC and propidium iodide (PI) dual staining. Human breast (MCF-7) cancer cells at a seeding density of 2×10^5 cells/well, were seeded in six-well plates. After treatment with different concentrations of Cladosporol A (10 and 20 µM) and paclitaxel (1 µM) for 24 h, harvested cells were PBS washed two times and suspended in binding buffer. After that, Annexin V/FITC and PI was used to stain the cells in dark for 15 min and cells were further observed by LSCM (laser scanning confocal microscope) [4, 39].

Determination of intracellular ROS

To identify the role of ROS in mediating Cladosporol A anti-cancer effects, intracellular ROS levels were monitored using 2′,7′-dichlorofluorescein diacetate (DCFH-DA). Human breast cancer (MCF-7) cells at a seeding density of $2X10^5$ cells/well were seeded in a six well plate and incubated overnight. After that, cells were treated with Cladosporol A for 24 h and H_2O_2 was used as a positive control. The cells were then treated with DCFH-DA (5 µM) at 37 °C for 30 min. Further, flow cytometry and

fluorescence microscopy was used to measure the fluorescence intensity of the treated cells [4, 25].

Loss of mitochondrial membrane potential (MMP)

Alterations in MMP (Δψm) were determined, by using the fluorescent dye rhodamine 123 (Rh-123) staining method and were studied using confocal microscopy. Human breast cancer (MCF-7) cells at a seeding density of 2×10^5 cells/well were seeded in 6-well plates and treated with different concentrations of Cladosporol A (10 and 20 µM) and doxorubicin (0.5 µM /mL) for 24 h. Later on, trypsinization was carried to detach the cells and resulting cell pellets were given two times PBS washing. 2 ml fresh medium containing Rh123 (Rhodamine 123) (1.0 µM) was then added to the cell pellets and were incubated at 37 °C for 20 min with gentle shaking. The cells were centrifuged and resulting cell pellets were washed again with PBS two times, then observed using a laser scanning confocal microscope (Olympus Fluoview FV1000) [4, 40].

Autophagy analysis by monodansylcadaverine staining

Development of autophagic vacuoles determines the level of autophagy. Monodansylcadaverine (MDC) has been regarded as a specific marker for the detection of autophagic vacuoles. By treating the cells with MDC (0.05 mM) in PBS at 37 °C for 1 h, autophagic vacuoles were labelled. After completion of incubation, Cladosporol A treated cells were washed thrice with PBS and immediately observed by fluorescence microscope using 40× lens [41].

Detection of p21, α-tubulin, cytochrome c, Bcl-2, Bax and LC3b by immunofluorescence confocal microscopy

Human breast cancer (MCF-7) (2×10^5 cells/well) cells, after treatment with different concentrations of Cladosporol A (10 and 20 µM) and doxorubicin (0.5 µM /ml) were processed for immunofluorescence microscopic studies. Briefly the cells were grown on cover slips, 4% paraformaldehyde was used to fix the cells for 10 min at room temperature and then cells were permeabilized using 0.5% Triton-X in PBS for 5 min. 10% goat serum was used to block the cells for 20 min at room temperature. The expression of proteins such as p21, α-tubulin, cytochrome c, Bcl-2, Bax and LC3b was observed by treating the cells for 1 h at room temperature with their specific primary antibodies. Cells were then incubated for 1 h at room temperature with respective conjugated secondary antibodies, alexa fluor 488 and 555. Further, the cells were given PBS washing thrice and were further stained with DAPI (1 µg/ml in PBS). Then over glass slides, cover slips were mounted and cells were observed and imaged via laser scanning confocal microscope (LSCM) [4, 42, 43].

In vitro clonogenic assay

To test the effect of Cladosporol A treatment on the colony formation ability of MCF-7 cells, clonogenic assay was performed. MCF-7 cells were seeded at 1×10^3 cells/ml/well in 6-well plates. The culture medium was changed after 24 h, and fresh medium was added, and Cladosporol A (5, 10, 20 µM) treatment was given to cells for 7 days in a 37 °C incubator in 5% CO_2. Thereafter, 4% paraformaldehyde was used to fix the obtained colonies which were further stained with crystal violet solution (0.5%). Thereafter, from the plates colonies were observed, counted and photographed [4, 11, 44].

Western blot analysis

Cladosporol A treated MCF-7 cells and untreated control cells were centrifuged at 4 °C at 400 g. The resulting cell pellets were given PBS washing and whole-cell lysate was prepared. The cell pellets were further lysed with RIPA buffer. Thereafter, for SDS-PAGE, equal amount of protein (60 µg) was loaded into each well. Further, primary antibodies; p21, Bcl-2 and Bax, LC3b were used to incubate the resulting blots and chemiluminiscence was captured on hyper film after incubating the blots in ECL plus solution [37].

Statistical analysis

Three independent experiment results were expressed as mean ± SD. Statistical analysis was performed using an unpaired t test, $^*p < 0.05$, $^{**}p < 0.01$, and $^{***}p < 0.001$.

Abbreviations

DAPI: 4,6-diamidino-2-phenylindole; DMSO: Dimethyl sulfoxide; FBS: Fetal bovine serum; MTT: 3-(4,5-dimethylthiazole-2- yl)-2,5-diphenyltetrazolium bromide; NAC: N-acetyl-cysteine; PDA: Potato dextrose agar; PI: Propidium iodide; ROS: Reactive oxygen species

Acknowledgements

We thank our Director Dr. Ram A. Vishwakarma for encouraging us to complete this work. Mytre Koul is thankful to Department of Science and Technology, Government of India for INSPIRE fellowship. Mytre Koul is grateful to Dr. B. K Koul and Rajni Koul for providing encouragement and support for work. We also thankful to CSIR for providing financial support from supra institutional projects BSC0205 and BSC0108.

Authors' contributions

MK: involved in the conception of the research idea, laboratory work, interpretation of the results, preparation of manuscript, AK: laboratory work, performed data analysis, RD: isolated the pure compound, VS: collected plant samples and characterized the endophytic fungus, RDS, JS, PRS,BAS and SJ: drafting the manuscript and editing, SS: supervised the study and contributed in the designing of the study, proposal write-up, data analysis, result interpretation, manuscript preparation and editing. All authors read and approved the final manuscript.

Competing interests

The authors declare that they have no competing interests.

Author details

[1]Cancer Pharmacology Division, CSIR-Indian Institute of Integrative Medicine, Jammu, India. [2]Natural Product Chemistry, CSIR-Indian Institute of Integrative Medicine, Jammu, India. [3]Microbial Biotechnology Division, CSIR-Indian Institute of Integrative Medicine, Jammu, India. [4]Academy of Scientific & Innovative Research (AcSIR), CSIR, New Delhi, India. [5]Department of Conservative Dentistry & Endodontics, Indira Gandhi Govt. Dental College and Hospital, Jammu, India.

References

1. Zhang HW, Song YC, Tan RX. Biology and chemistry of endophytes. Nat Prod Rep. 2006;23:753–71.
2. Newman DJ, Cragg GM. Natural products as sources of new drugs over the 30 years from 1981 to 2010. J Nat Prod. 2012;75:311–35.
3. Zhao J, Zhou L, Wang J, Shan T, Zhong L, Liu X, et al. Endophytic fungi for producing bioactive compounds originally from their host plants. Current Research, Technology and education Topics in applied Microbiology and Biotechnology. 2010;1:567–76.
4. Koul M, Meena S, Kumar A, Sharma PR, Singamaneni V, Hassan SRU, et al. Secondary metabolites from endophytic fungus Penicillium pinophilum induce ROS-mediated apoptosis through mitochondrial pathway in pancreatic cancer cells. Planta Med. 2016;82:344–55.
5. Koul M, Singh S. Penicillium spp.: prolific producer for harnessing cytotoxic secondary metabolites. Anti-Cancer Drugs. 2017;28:11–30. doi:10.1097/CAD. 0000000000000423.
6. Gallo ML, Seldes AM, Cabrera GM. Antibiotic long-chain and a, b-unsaturated aldehydes from the culture of the marine fungus Cladosporium sp. Biochem Syst Ecol. 2004;32:545–51.
7. Wang FW, Jiao RH, Cheng AB, Tang SH, Song YC. Antimicrobial potentials of endophytic fungi residing in Quercus variabilis and Brefeldin a obtained from Cladosporium sp. World J Microb Biot. 2007;23:79–83.
8. Zhang P, Zhou P, Yu L. An endophytic taxol-producing fungus from Taxus media, Cladosporium cladosporioides MD2. Curr Microbiol. 2009;59:227–32.
9. Priya KS, Gnanamani A, Radhakrishnan N, Babu M. Healing potential of Datura Alba on burn wounds in albino rats. J Ethnopharmacol. 2002;83:193–9.
10. Maheshwari NO, Khan A, Chopade BA. Rediscovering the medicinal properties of Datura sp.: a review. J Med Plants Res. 2013;7:2885–97.
11. Arulvasu C, Babu G, Manikandan R, Srinivasan P, Sellamuthu S, Prabhu D, et al. Anti-cancer effect of Datura innoxia P.Mill. Leaf extract in vitro through induction of apoptosis in human Colon Adenocarcinoma and larynx cancer cell lines. J Pharm Res. 2010;3:1485–8.
12. Ferlay J, Soerjomataram I, Parkin EM. GLOBOCAN cancer incidence and mortality worldwide: IARC CancerBase no. 11: 2012 http://globocan.iarc.fr.
13. Baehrecke EH. Autophagy dual roles in life and death? Nat Rev Mol Cell Biol. 2005;6:505–10.
14. Kamata H, Honda S, Maeda S, Chang L, Hirata H, Karin M. Reactive oxygen species promote TNF alpha-induced death and sustained JNK activation by inhibiting MAP kinase phosphatises. Cell. 2005;120:649–61.
15. Rafal O, Agnieszka L, Wojciech P, Anna M, Paulina M. Characteristics and taxonomy of Cladosporium Fungi. Mikologia. 2012;19:80–5.
16. Skibinski A, Kuperwasser C. The origin of breast tumor heterogeneity. Oncogene. 2015;34:5309–16.
17. Buommino E, Boccellino M, Filippis AD, Petrazzuolo M, Cozza V, Nicoletti R, et al. 3-O-methylfunicone produced by penicillium pinophilum affects cell motility of breast cancer cells, downregulating αvβ5 integrin and inhibiting metalloproteinase-9 secretion. Mol Carcinog. 2007;46:930–40.
18. Wang X, Radwan MM, Taráwneh AH, Gao J, Wedge DE, Rosa LH, et al. Antifungal activity against plant pathogens of metabolites from the Endophytic fungus Cladosporium cladosporioides. J Agric Food Chem. 2013; 61:4551–5.
19. Singh P, Rathinasamy K, Mohan R, Panda D. Microtubule assembly dynamics: an attractive target for anticancer drugs. IUBMB Life. 2008;60:368–75.
20. Szatrowski TP, Nathan CF. Production of large amounts of hydrogen peroxide by human tumor cells. Cancer Res. 1991;51:794–8.
21. Gorrini C, Harris IS, Mak TW. Modulation of oxidative stress as an anticancer strategy. Nat Rev Drug Discov. 2013;12:931–47.
22. Alexandre J, Hu Y, Lu W, Pelicano H, Huang P. Novel action of paclitaxel against cancer cells: bystander effect mediated by reactive oxygen species. Cancer Res. 2007;67:3512–7.

23. Hellebrand EE, Varbiro G. Development of mitochondrial permeability transition inhibitory agents: a novel drug target. Drug Discov Ther. 2010;4:54–61.

24. Tomasello F, Messina A, Lartigue L, Schembri L, Medina C, Reina S, et al. Outer membrane VDAC1 controls permeability transition of the inner mitochondrial membrane in cellulo during stress-induced apoptosis. Cell Res. 2009;19:1363–76.

25. Bai J, Lei Y, An G, He L. Down-regulation of deacetylase HDAC6 inhibits the melanoma cell line A375.S2 growth through ROS-dependent mitochondrial pathway. PLoS One. 2015;10:1371–82.

26. Ling YH, Lin R, Perez-Soler R. Erlotinib induces mitochondrial-mediated apoptosis in human H3255 non-small-cell lung cancer cells with epidermal growth factor receptorL858R mutation through mitochondrial oxidative phosphorylation-dependent activation of BAX and BAK. Mol Pharmacol. 2008;74:793–06.

27. Kirkin V, Joos S, Zörnig M. The role of Bcl-2 family members in tumorigenesis. Biochim Biophys Acta. 2004;1644:229–49.

28. Yip KW, Reed JC. Bcl-2 family proteins and cancer. Oncogene. 2008;27:6398–06.

29. Gui YX, Fan XN, Wang HM, Wang G, Chen SD. Glyphosate induced cell death through apoptotic and autophagic mechanisms. Neurotoxicol Teratol. 2012;34:344–9.

30. Biederbick A, Kern HF, Elsässer HP. Monodansylcadaverine (MDC) is a specific in vivo marker for autophagic vacuoles. Eur J Cell Biol. 1995;66:3–14.

31. Mizushima N, Yoshimori T. How to interpret LC3 immuno-blotting. Autophagy. 2007;3:542–5.

32. White TJ, Bruns T, Lee S, Taylor J. Amplification and direct sequencing of fungal ribosomal RNA genes for phylogenetics. In: Innis MA, Gelfand DH, Sninsky JJ, White TJ, editors. PCR protocols: a guide to methods and application. San Diego: Academic Press; 1990. p. 315–22.

33. Tamura K, Stecher G, Peterson D, Filipski A, Kumar S. MEGA6: molecular evolutionary genetics analysis version 6.0. Mol Biol Evol. 2013;30:2725–9.

34. McCloud TG. High throughput extraction of plant, marine and fungal specimens for preservation of biologically active molecules. Molecules. 2010; 15:4526–63.

35. Sakagami Y, Sano A, Hara O, Mikawa T, Marumo S. Cladosporol, b-1,3-glucan biosynthesis inhibitor, isolated by the fungus Cladosporium cladosporioides. Tetrahedron Lett. 1995;36:1469–72.

36. Mosmann T. Rapid colorimetric assay for cellular growth and survival application to proliferation and cytotoxicity assays. J Immunol Meth. 1983; 65:55–3.

37. Zurlo D, Leone C, Assante G, Salzano S, Renzone G, Scaloni A, et al. Cladosporol a stimulates G1-phase arrest of the cell cycle by up-regulation of p21waf1/cip1 expression in human Colon carcinoma HT-29 cells. Mol Carcinog. 2013;52:1–17.

38. Rello S, Stockert JC, Moreno V, Gámez A, Pacheco M, Juarranz A, et al. Morphological criteria to distinguish cell death induced by apoptosis and necrotic treatments. Apoptosis. 2005;10:201–8.

39. Chen T, Pengetnze Y, Taylor C. Src inhibition enhances paclitaxel cytotoxicity in ovarian cancer cells by caspase-9-independent activation of caspase-3. Mol Cancer Ther. 2005;2:217–24.

40. Dai J, Wang J, Li F, Ji Z, Ren T, Tang W, et al. Scutellaria barbate extract induces apoptosis of hepatoma H22 cells via the mitochondrial pathway involving caspase-3. World J Gastroenterol. 2008;14:7321–8.

41. Munafo DB, Colombo MI. A novel assay to study autophagy: regulation of autophagosome vacuole size by amino acid deprivation. J Cell Sci. 2001;114: 3619–29.

42. Seervi M, Joseph J, Sobhan PK, Bhavya BC, Santhoshkuma TR. Essential requirement of cytochrome c release for caspase activation by procaspase-activating compound defined by cellular models. Cell Death Dis. 2011;8(2): e207. doi:10.1038/cddis.2011.90.

43. Kumar A, Singh B, Mahajan G, Sharma PR, Bharate SB, Mintoo MJ, Mondhe DM.A novel colchicine based microtubule inhibitor exhibits potent antitumor activity by inducing G2/M arrest, endoplasmic reticular stress and mitochondrial mediated apoptosis in MIA PaCa-2 pancreatic cancer cells. Tumor Biol. 2016;DOI 10.1007/s13277-016-5160-5.

44. Munshi A, Hobbs M, Meyn RE. Clonogenic cell survival assay. Methods Mol Med. 2005;110:21–8.

7

The M-phase specific hyperphosphorylation of Staufen2 involved the cyclin-dependent kinase CDK1

Rémy Beaujois[2], Elizabeth Ottoni[2], Xin Zhang[2], Christina Gagnon[2], Sami HSine[2], Stéphanie Mollet[2], Wildriss Viranaicken[1] and Luc DesGroseillers[2*]

Abstract

Background: Staufen2 (STAU2) is an RNA-binding protein involved in the post-transcriptional regulation of gene expression. This protein was shown to be required for organ formation and cell differentiation. Although STAU2 functions have been reported in neuronal cells, its role in dividing cells remains deeply uncharacterized. Especially, its regulation during the cell cycle is completely unknown.

Results: In this study, we showed that STAU2 isoforms display a mitosis-specific slow migration pattern on SDS-gels in all tested transformed and untransformed cell lines. Deeper analyses in hTert-RPE1 and HeLa cells further indicated that the slow migration pattern of STAU2 isoforms is due to phosphorylation. Time course studies showed that STAU2 phosphorylation occurs before prometaphase and terminates as cells exit mitosis. Interestingly, STAU2 isoforms were phosphorylated on several amino acid residues in the C-terminal half via the cyclin-dependent kinase 1 (Cdk1), an enzyme known to play crucial roles during mitosis. Introduction of phospho-mimetic or phospho-null mutations in STAU2 did not impair its RNA-binding capacity, its stability, its interaction with protein co-factors or its sub-cellular localization, suggesting that STAU2 phosphorylation in mitosis does not regulate these functions. Similarly, STAU2 phosphorylation is not likely to be crucial for cell cycle progression since expression of phosphorylation mutants in hTert-RPE1 cells did not impair cell proliferation.

Conclusions: Altogether, these results indicate that STAU2 isoforms are phosphorylated during mitosis and that the phosphorylation process involves Cdk1. The meaning of this post-translational modification is still elusive.

Keywords: Staufen2, Cell cycle, Mitosis, Cyclin-dependent kinase, Phosphorylation, RNA-binding protein

Background

Cell cycle can be defined as a succession of events that allow cell to replicate DNA and segregate chromosomes into two daughter cells. Proper proceeding through each step of the cell cycle depends on the expression and activity of many critical proteins such as proto-oncogenes, tumour suppressors and other regulators [1]. When misregulated by mutations, these effectors cause cell proliferative disorders, genomic instability and cell injuries leading to tumour emergences [2]. Among the crucial regulators of the cell cycle are the evolutionarily conserved cyclin-dependant kinases (Cdk). Cell-cycle dependent activation of members of this important family of serine/threonine kinases is mediated by their association with sequentially expressed cyclin partners [3–8]. The control of the checkpoint pathways is granted by the balance between protein synthesis and degradation and by post-translational modifications of its effectors.

The most fundamental and dominant mechanism of post-translational modifications in eukaryotes involves site-specific protein phosphorylation. The reversible transfer of a γ-phosphate to threonine, serine or tyrosine residues is mediated by a super-family of protein kinases that causes conformational changes to the proteins. As a result, the recognition site or the binding properties of target proteins and/or the activity of modified enzymes

* Correspondence: luc.desgroseillers@umontreal.ca

[2]Département de biochimie et médecine moléculaire, Faculté de médecine, Université de Montréal, 2900 Edouard Montpetit, Montréal, QC H3T 1J4, Canada

Full list of author information is available at the end of the article

are altered [9–12]. Importantly, 518 protein kinases have been identified in the human kinome [13]. They are pivotal regulators of cell signalling cascades and networks and represent important putative targets for the development of inhibitors with potential therapeutic application [14–17]. Based on large-scale proteomics analysis performed on the HeLa cell line, a peak of protein phosphorylation appears at the onset of mitosis whereas a drastic reduction of phosphopeptides is observed in late mitosis (anaphase, telophase) [18–20]. Such modulation in protein regulation profoundly alters the behaviour of a significant proportion of proteins. Cdk1-cyclin B is one of the most active and critical complexes during mitosis. It orchestrates the G_2/M transition and phosphorylates a plethora of M-phase regulators. In anaphase, cyclin B is released from the complex and degraded by the anaphase-promoting complex/cyclosome (APC/c), leading to inactivation of Cdk1 and mitosis exit [21]. It is admitted that phosphorylation of proteins by Cdks coordinates cell division as well as essential cellular processes such as transcription [22], mRNA splicing [23, 24] and translation [25–28]. However, it is not yet understood if Cdks regulate post-transcriptional mechanisms involved in coordinating the expression of mRNAs coding for cell cycle regulators.

STAU2 is a double-stranded RNA-binding protein that is expressed as 4 protein isoforms (52, 56, 59 and 62 kDa) generated by differential splicing of the *STAU2* gene [29, 30]. In mammals, the *STAU2* gene is highly expressed in brain and heart [29] and ubiquitously expressed in all tested cell lines. STAU2 is a component of ribonucleoprotein complexes [29, 31, 32] involved in microtubule-dependent mRNA transport in many species [29, 30, 33–41]. Interestingly, chemical induction of long term depression in hippocampal neurons causes a reduction in the amount of Stau2 in dendrites allowing the release of Stau2-bound mRNAs and their translation on polysomes [40]. Therefore, STAU2 can sequester sub-populations of mRNAs and allow their release and local translation according to cell needs. In addition to transport, STAU2 was shown to increase the translation of reporter proteins [42] or decay of mRNA [43]. In a high throughput experiment, STAU2 was also found to be required for differential splicing [44]. Using a genome-wide approach, we found that STAU2-bound mRNAs code for proteins involved in catabolic process, post-translational protein modifications, RNA metabolism, splicing, intracellular transport, and translation [45, 46]. Accordingly, STAU2 was linked to multiple cell processes. Stau2 down-regulation in neurons impairs mRNA transport, causes dendritic spines defects and prevents hippocampal long term depression [30, 34, 40]. In addition, Stau2 induces neural stem cell differentiation [47, 48]. Similarly, stau2 is required for survival and migration of primordial germ cells [37] in zebrafish, while it is involved in anterior endodermal organ formation in *Xenopus* [49]. In chicken, STAU2 down-regulation reduced cell proliferation with no evidence of cell death or apoptosis [50]. We recently showed that STAU2 down-regulation increases DNA damage in human cells and promotes apoptosis when cells are challenged with DNA-damaging agents [51]. However, not much is known about STAU2 regulation, although phosphorylation may account for the control of at least some of its functions. Indeed, in Xenopus oocytes, stau2 was shown to be transiently phosphorylated by the mapk pathway during meiotic maturation, a time period that coincides with the release of anchored RNAs from their localization at the vegetal cortex [33]. In rat neurons, the activity-stimulated transport of Stau2-containing complexes in dendrites of neurons is dependent on Mapk activity [35]. Stau2 contains a docking site for Erk1/2 in the RNA-binding domain interregion and this site is required for proper transport of Stau2-containing complexes [36].

Here, we report that STAU2 is hyperphosphorylated during mitosis and that CDK1 participates in the process. Several phosphorylated amino acids residues were localized as clusters in the C-terminal region of STAU2. Taking together, our results highlight for the first time the fact that the RNA-binding protein STAU2 is finely regulated in a cell-stage-dependant manner.

Methods
Plasmids and cloning strategies
The human STAU2[59] coding sequence was generated by PCR amplification of a commercial clone (ATCC) using sense (ATAAGATATCGCCACCATGCTTCAAATAAAT-CAGATGTTC) and antisense (ATAAGATATCTTAT-CAGCGGCCGCCGACGGCCGAGTTTGATTTC) oligonucleotides. The PCR product was then cloned in the retroviral pMSCV puromycin vector after EcoRV digestion and blunt ligation. Subsequently, a C-terminal FLAG₃ tag was inserted at the Not1 site using complementary sense (5'TCGAGATGGGCGGCCGCGACTACAAAGACCATG ACGGTGATTATAAAGATCATGACATCGACTACAAG-GATGACGATGACAAGTGATAAGCGGCCGCG3') and antisense (5'ATTTCGCGGCCGCTTATCACTTGTCATC GTCATCCTTGTAGTCGATGTCATGATCTTTATAATC ACCGTCATGGTCTTTGTAGTCGCGGCCGCCCATC3') oligonucleotides. The same strategy was used to generate STAU2[52]-FLAG₃: PCR-amplification from STAU2[59] with sense (5'TTAAGATATCTCAAGCGGCCGCCTACCTGA AAGCCTTGAATCCTTGC3') and anti-sense (5'TTAA-GATATCTCAAGCGGCCGCCTACCTGAAAGCCTTGA ATCCTTGC3') oligonucleotides, cloning into the pMSCV vector and addition of FLAG₃ tag at the NotI site. Similarly, STAU2[N-ter]-FLAG₃ was generated from STAU2[52] with sense (5'AATTGATATCATGCTTCAAATAAATCAGATG TTCTCAGTGCAG3') and antisense (5'TTAAGATAT

CTCATGCGGCCGCCATTAGTGGATGCTTTATAACC AAGTTG3') oligonucleotides. STAU2$^{52C\text{-ter}}$-YFP and STAU2$^{59C\text{-ter}}$-mCherry were PCR amplified from STAU2^{52}-YFP and STAU2^{59}-mCherry, respectively, using sense (5'AATTGATATCATGTTACAACTTGGTTATAAAG-CATCCACTAAT3') and antisense (5'AATTGATAT-CAGCGGCCGCTTATCACTTGTACAGCTCGTCCAT GCCG3'). oligonucleotides.

To construct the P(7) phospho-mutants, DNA fragments of 280 bp containing the 7 mutated residues (T^{373}, T^{376}, S^{384}, S^{394}, S^{408}, S^{423} and S^{426}) were in vitro synthesized (Life Technologies) and cloned in the STAU2^{52}- and STAU2^{59}-FLAG pMSCV puromycin vector using AanI and MunI restriction enzymes. Threonine and Serine residues were simultaneously substituted for aspartic acid or alanine amino acids to generate the P(7) phospho-mimetic or phospho-null mutants, respectively.

PCR-mediated site specific mutagenesis was used to convert S454, T456, S460 into S454A, T456A, S460A and S454D, T456D, S460D, respectively. The resulting PCR fragments were digested with MunI and DraIII and cloned into pMSCV-STAU2^{52}-FLAG$_3$ and pMSCV-STAU2^{59}-FLAG$_3$ to create P(3)- and P(3) + mutants, and into P(7)- and P(7) + vectors to generate P(10)- and P(10) + mutants.

Cell culture and synchronization

Human cell lines (American Type Culture Collection) were cultured in Dulbecco modified Eagle's medium supplemented with 10% (v/v) fetal bovine serum (Wisent), 100 µg/ml streptomycin and 100 units/ml penicillin (Wisent); here referred to be a complete DMEM. Cells were cultured at 37 °C under a 5% CO_2 atmosphere. For synchronization in prometaphase, hTert-RPE1 and HeLa cell lines were treated with nocodazole (200 and 60 ng/ml, respectively), for 15 to 18 h. Round shaped mitotic cells were recovered and enriched by gentle shaking of the culture dish (shake-off), then directly collected (hours after release = 0) or re-plated in drug free complete DMEM for kinetics. Cells were synchronized at the G$_1$/S border using a double thymidine-block (DTB) protocol [52]. Briefly, hTert-RPE1, HeLa and HCT116 cells were treated with 5, 2,5 and 2 mM thymidine for 16 h and released for 8 h in fresh medium before the second thymidine block was performed for another 16 h. Cells were then washed three times in phosphate buffered saline (PBS) and either collected (hour after treatment = 0) or released in fresh medium for time course. Cells were also synchronized at the G$_2$/M phase border using 10 µM of the CDK1 inhibitor, RO-3306 (Enzo Life Sciences) for 18 h or at the metaphase/anaphase transition by paclitaxel (Taxol - Sigma). Cells were then washed three times in PBS and either collected (hours after treatment = 0) or released in fresh medium for different time periods.

DNA transfection and viral transduction

Closed circular DNA plasmids purification was performed through CsCl-Ethidium bromide gradients protocol to obtain high DNA concentration and maximum purity. For transient expression, cells were transfected with Lipofectamine 2000 according to the manufacturer's instructions (Invitrogen). Alternatively, plasmids were transfected in ecotropic Phoenix cells, the virus-containing cell supernatant was collected two days after transfection and used to infect cells as previously described [51].

Colony forming assays

Cells were selected with puromycin as described above. For colony formation assays, 5000 cells in 6-wells plates were allowed to grow for 8 days. Cells were then washed two times with PBS, fixed 10 min directly in the dish using 1% (v/v) glutarhaldehyde in PBS, washed two times with PBS and stained with 0,1% (w/v) crystal violet in PBS for 45 min. After extensive washes in water, plates were dried, scanned and decoloured in a solution containing 10% (v/v) acid acetic in distillate water. Colony formation was determined by measuring absorbance at 590 nm. Even if hTert-RPE1 did not form colony, this method was suitable to quantify cell density following control condition based data normalization.

Sample preparation and gel electrophoresis

Total-cell extracts were prepared and boiled 10 min in lysis buffer (250 mM Tris-Cl pH 6.8, 5% (w/v) SDS and 40% (v/v) Glycerol). Before completing extracts with Laemmli sample buffer, protein concentrations were measured using BCA assays (ThermoScientific) and adjusted equally. Extracts were finally heated at 100 °C for 10 min to denature proteins. For separation, 10 to 20 µg of protein samples were submitted to an 8% or 10% SDS-PAGE prepared from 29:1 acrylamide/bisacrylamide mix. To characterize STAU2, Flag and Rsk-1 gel shit patterns, samples were loaded onto an 8% modified SDS-PAGE prepared from 16:0,215 acrylamide/bisacrylamide mix. Following gel migration, proteins were transferred on nitrocellulose membrane and stained with ponceau red (5 g/L Ponceau S (BioShop) to control transfer and loading.

Western blot analysis and antibodies

Nitrocellulose membranes were decoloured in distillate water for 10 min, saturated by 5% (w/v) skim milk in PBS-Tween20 (0,2% (v/v)), incubated in primary antibody (see below for antibody preparation) for 2 h or overnight, washed three times for 10 min in PBS-Tween20 (0,2% (v/v)), incubated in secondary antibody for 1 h, washed three times for 15 min in PBS-Tween20 (0,2% (v/v)), and finally subjected to HRP chemiluminescence reaction. Western blotting data were collected onto X-ray films (Fujifilm).

Primary antibodies were prepared with 1% (*w/v*) skim milk in PBS-Tween20 (0,2%). 0,1‰ (*w/v*) sodium azide was also added to prevent antibody contamination and allow its long-term conservation. Membranes were incubated 2 h at room temperature with monoclonal anti-cyclin A1 (Sigma) or anti-β-actin (Sigma), or with polyclonal anti-RSK1 (Santa Cruz) antibody. Overnight incubation was required for monoclonal anti-FLAG (Sigma), anti-cyclin B1 (Santa Cruz) and anti-MPM2 (Abcam) antibodies, as well for anti-STAU2 (Sigma) antibody. Secondary antibodies were prepared extemporaneously with 2,5% (*w/v*) skim milk in PBS-Tween20 (0,2%). Membranes were incubated at room temperature for 1 h with polyclonal goat anti-mouse (Dako) or anti-rabbit (Dako) HRP-conjugated secondary antibodies.

Other antibodies targeted nucleolin (Abcam), STAU1 [53], GFP (Millipore), RFP (Medical & Biological Laboratories LTD), histone H3 (Abcam), pS28-histone H3 (Millipore), pS10-histone H3 (Millipore), ubiquitin (Millipore), pCDK1/MAPK-substrates (Cell Signaling), pS22-lamina A/C (Cell Signaling), pY15-CDK1 (ThermoScientific), aurora A (Abcam), pT288/T232/T198-aurora A/B/C (Cell Signaling), aurora B (Cell Signaling), pT102/Y204-MAPK/ERK1/2 (Cell Signaling), MAPK/ERK1/2 (Cell Signaling).

In vitro dephosphorylation assay and phosphoprotein purification

Asynchronous or mitotic cells were prepared in lysis buffer (50 mM Tris-Cl pH 7.5, 150 mM NaCl, 10 mM $MgCl_2$, 1% (*v/v*) triton X-100 and Complete EDTA-free protease inhibitor cocktail (Roche)) and lysates were cleared by 15,000 *g* centrifugation for 15 min at 4 °C. 100 to 200 μg of proteins were then incubated in a 100 μl final volume including either 3 U/μl Calf Intestinal Alkaline Phosphatase (CIP, 37 °C for 3 h, New England BioLabs) with 1× buffer, or 8 U/μl Lambda Phosphatase (λPP, 30 °C for 3 h, New England BioLabs) with 1× buffer and 1 mM $MnCl_2$. For control conditions, the same quantity of each phosphatases was inactivated according to the manufacturer instructions, and then added with proteins, 50 mM EDTA, 50 mM NaF and 50 mM Na_3VO_4 to the 100 μl final volume sample mix. In vitro assayed extracts were completed with Laemmli sample buffer and heated at 100 °C for 10 min to denature proteins before western blotting analysis.

Phosphorylated proteins were also purified by affinity chromatography. For this purpose, lysates from asynchronous or mitotic extracts were submitted to a phosphoprotein purification kit (Qiagen) following the manufacturer instructions. Total cell extracts, flow through and eluates were analysed by western blotting to control and characterize phosphoprotein enrichment.

Flow cytometry analysis

Cell cycle distribution was determined by Fluorescence Activated Cell Sorting (FACS). Fixed overnight in 70% (*v/v*) ethanol at –20 °C, cells were washed in PBS and resuspended in PBS containing 40 μg/ml propidium iodide and 100 μg/ml RNase A (Sigma). Cells were incubated for 30 min at 37 °C and data was acquired using a BD LSRII apparatus. For each experiment, 10^4 cells were analyzed.

Phosphorylation inhibition in vivo

Cells were incubated with nocodazole for 16 h. Cells were then incubated with nocodazole, specific kinase inhibitors and the proteasome inhibitor MG132 (Sigma, 20 uM) for an additional 4 h. Cells were collected and extracts were analyzed by western blotting. Cells were treated with either DMSO (Sigma, 0.1 to 1‰), Purvalanol A (20 uM, Sigma-Aldrich), Roscovitine (100 uM, Millipore), Flavopiridol (2 uM, Cedarlane), RO-3306 (10 uM, Enzo Life Sciences), Alisertib (100 nM, Cedarlane), Barasertib (75 nM, Cedarlane), BI2536 (50 nM, Cedarlane), U0126 (20 uM, Millipore), AZD6244 (40 uM, Symansis), BI-D1870 (20 uM, Cedarlane), SB203580 (20 uM, Sigma-Aldrich), and SP600135 (20 uM, Sigma-Aldrich).

Results

Post-translational modification of STAU2 in mitosis

To determine if specific regulation of human STAU2 occurs during cell division, expression of endogenous STAU2 isoforms was monitored through the cell cycle. Untransformed hTert-RPE1 cells were first arrested at the G_1/S transition by a double thymidine block (DTB) and then released in fresh medium to allow cell progression through the S and G_2 phases (Fig 1a). In parallel, nocodazole was used to arrest cells in prometaphase (Figs 1a and b). Mitotic cells were recovered by a gentle shake-off, replated and released from the block by fresh medium to reach late mitosis and the G_1 phase. Cells were then harvested at different time points after release and analyzed by western blotting (Fig 1) and FACS (Additional file 1). G_1/S-arrested DTB-treated cells successfully reached the S and G_2-phases about 3 h and 8 h post-release, respectively. Protein analysis indicated that STAU2 expression is stable during these phases of the cell cycle. In contrast, prometaphase-arrested nocodazole-treated cells showed a strong shift in STAU2 isoforms migration that disappeared 2 h post-release (Fig 1a). A time course experiment indicated that the slow migration bands returned to a fast migration pattern 90 min post-release (Fig 1b). This corresponds to cell entry into the G_1-phase (Additional file 1B). Indeed, at these time points, the mitosis markers MPM2 and cyclin B1 were no longer visible (Fig 1b). Similar results were obtained in the

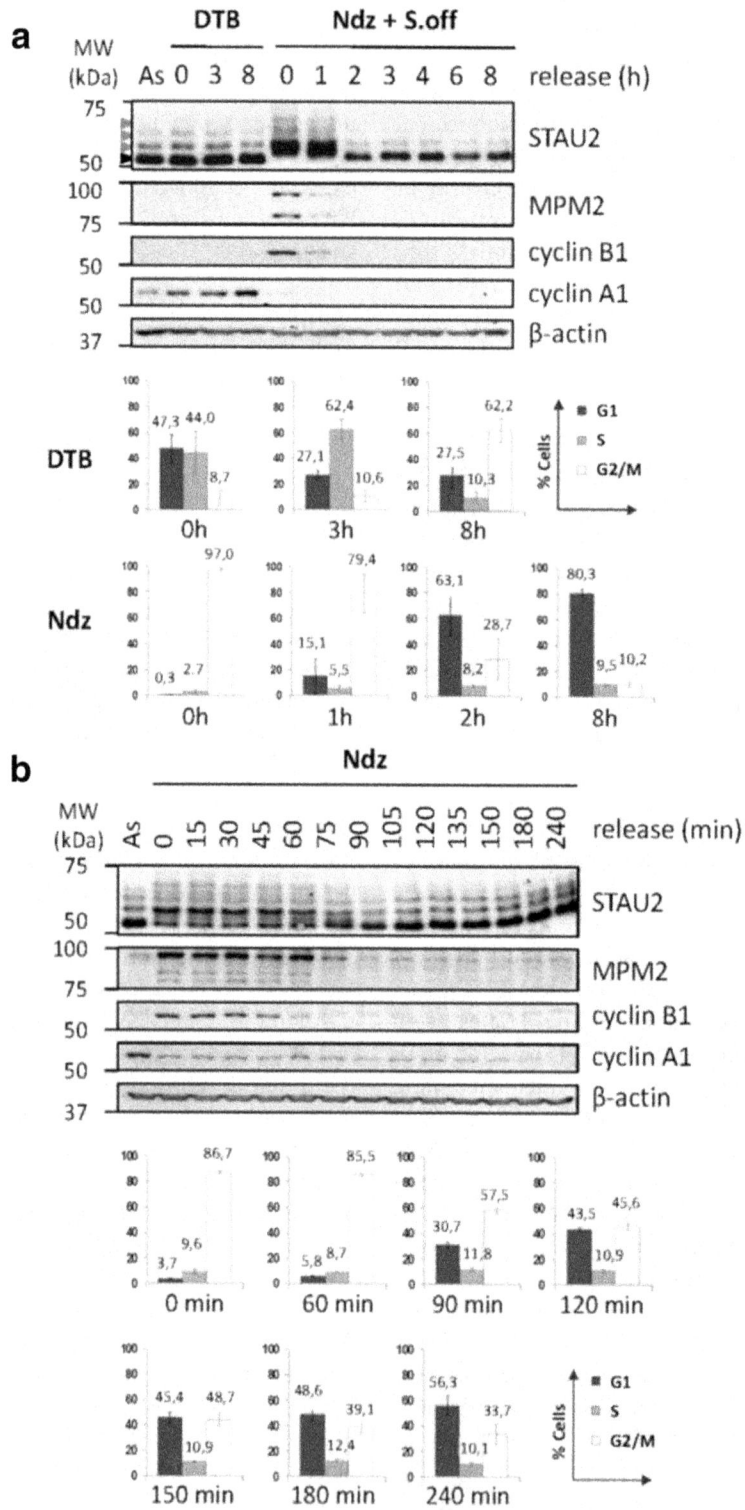

Fig. 1 (See legend on next page.)

(See figure on previous page.)
Fig. 1 STAU2 is differentially regulated through the cell cycle in hTert-RPE1 cells. **a** hTert-RPE1 cells were synchronized by a double thymidine block (DTB) or a nocodazole arrest (Ndz) followed by shake off (S.off). Cells were then released in a fresh medium for different time periods as indicated (Rel (h)). Asynchronous (As) cells were collected as controls. Protein extracts from synchronized cells were analyzed by SDS-PAGE and western blotting to investigate STAU2 pattern migration and expression of cell cycle markers (MPM2 and cyclins). β-actin was used as loading control. As control of synchronization, the percentage of cell population in the G_1, S or G_2/M phases was determined by FACS analysis ($n = 3$)(see also Additional file 1). **b** STAU2 migration dynamics was examined in hTert-RPE1 cells. Cells were blocked in prometaphase with nocodazole (Ndz) without shake-off, released in fresh medium and harvested every 15 min (Rel (min)). Extracts from untreated asynchronous (As) and nocodazole-treated cells were analyzed by western blotting. The percentage of cell population within the G_1, S or G_2/M phases was determined by FACS ($n = 3$). In both (**a**) and (**b**), Western blots are representatives of three independently performed experiments that showed similar profiles. Error bars represent the standard deviation

transformed HeLa cell line (Additional files 2, 3): stable STAU2 expression during the S and G_2 phases of the cell cycle and a slow migration pattern of STAU2 isoforms during mitosis. Immunodetection of cell cycle markers and FACS analyses confirmed the phases of the cell cycle.

To confirm that the slow migration of STAU2 isoforms during mitosis was not a non-specific effect of nocodazole treatment, we synchronized M-phase cells by three additional approaches and compared STAU2 isoform migration to that obtained in the presence of nocodazole. First, paclitaxel (taxol) was used to block hTert-RPE1 and HeLa cells in mitosis, as confirmed by western blotting (Fig 2a) and FACS analysis (Fig 2b). This drug caused a delay in STAU2 migration as did nocodazole (Fig 2a). Then, HeLa cells were arrested at the G_1/S transition by a DTB and release for 9 h to reach mitosis. In parallel, cells were arrested in late G_2 by the CDK1 inhibitor [54] RO-3306 and released for 1 h until appearance of sufficient amounts of round shaped mitotic cells. In both cases, mitotic cells were enriched by gentle shake-off and cell extracts were prepared for western blotting and FACS (Figs 2c and d, respectively). Following DTB and RO-3306 releases, cells reached mitosis inconsistently. Mitotic markers were weakly expressed and mitotic synchronization was less efficient, as compared to nocodazole-treated cells. Nevertheless, both treatments induced a comparable slow pattern of migration (Fig 2c). These results confirmed that the slow migration profile of STAU2 is mitosis-specific and suggest that STAU2 is post-translationally modified during mitosis.

The slow migrating pattern of STAU2 is observed in all tested cell lines

Our data indicate that STAU2 is modified during mitosis in both untransformed and transformed cell lines derived from two different tissues, normal retina and tumor from cervix, respectively. To determine how universal is this property, eight additional cancer cell lines derived from different organs were synchronized with nocodazole or paclitaxel (taxol) as above. Cell extracts were analyzed by western blotting (Fig 3) and FACS (Additional file 4). Interestingly, the slow

migrating bands of STAU2 isoforms were observed in all tested cell lines with both drugs indicating that STAU2 post-translational modification is ubiquitous and independent of cell origin. Efficient mitotic synchronization was confirmed in all cell lines by cell cycle markers MPM2 and cyclin B1 and by FACS.

STAU2 is hyper-phosphorylated in M-phase

Using RNA interference, we first proved that the bands observed during interphase and mitosis corresponded to STAU2 isoforms (Fig 4a). HeLa and hTert-RPE1 cells were infected with control shRNA or shRNA against STAU2, incubated in nocodazole-complemented medium and shaken-off to collect prometaphase arrested cells. Asynchronous cells were used as controls. Cell extracts were analyzed by SDS-PAGE (Fig 4a). In both cell lines, bands recognized by the STAU2 antibody disappeared as a consequence of infection with shRNA against STAU2, authenticating the shifted bands as STAU2 proteins. STAU1 was used as control to confirm the specificity of the RNA interference. Its partial degradation in mitosis is consistent with our previous data [55]. Then, human STAU2^{52}-FLAG$_3$ and STAU2^{59}-FLAG$_3$ (Fig 4b) were expressed in hTert-RPE1 cells and their migration in extracts of untreated and nocodazole-treated cells was observed by western blotting using anti-STAU2 and anti-FLAG antibodies (Fig 4c). As expected, both overexpressed isoforms appeared as slow migrating bands in M-phase-enriched cell extracts.

Differential migration pattern of proteins is usually caused by protein post-translational modifications such as phosphorylation. Interestingly, Xstaufen was previously shown to be phosphorylated during meiotic maturation in Xenopus oocyte [33]. Therefore, two different approaches were used to determine if STAU2 is phosphorylated during mitosis. First, untreated and nocodazole-treated cell extracts were loaded on phospho-specific columns. Column-trapped phosphoproteins were eluted and analyzed by western blotting (Fig 4d). Proteins in the flow-through were also analyzed. A significant fraction of STAU2 was found in the column eluate of nocodazole-treated cells and showed as expected a slow migration pattern. STAU2 fast-migrating bands were found in the flow-through. In untreated cells, a significant amount of fast-migrating bands was found in the column eluate

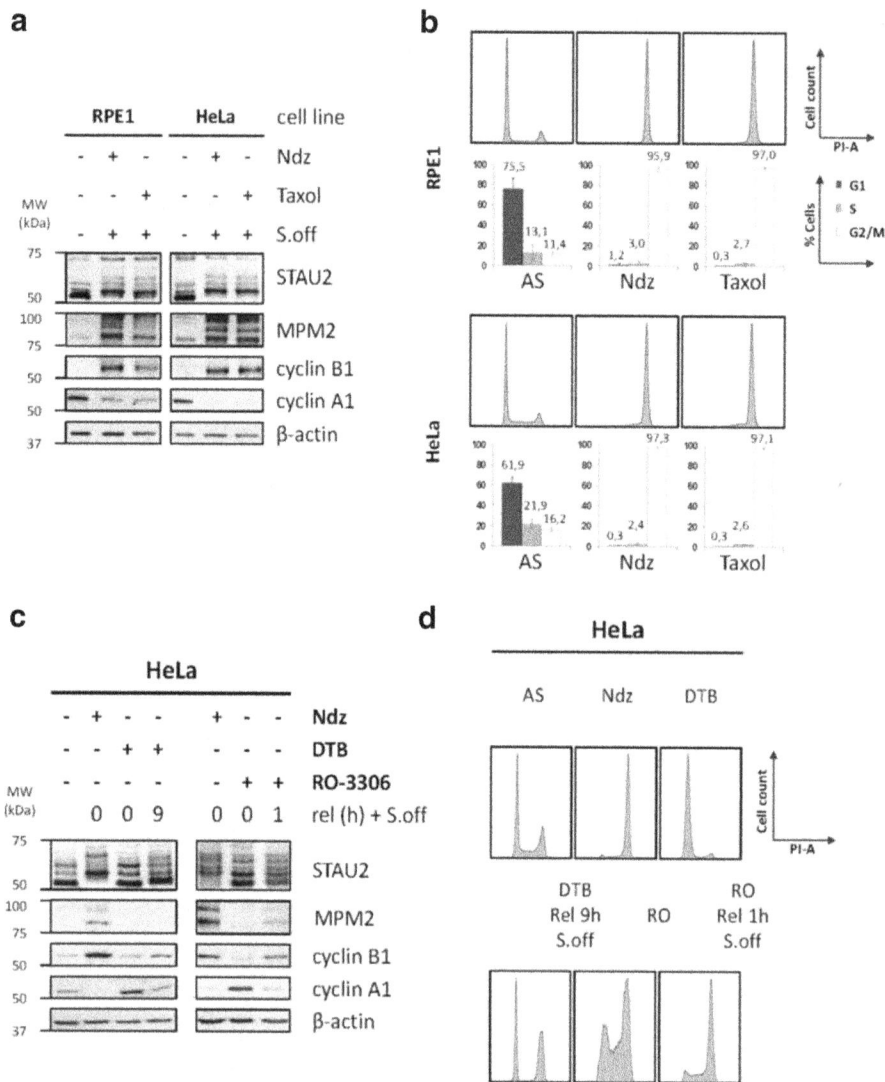

Fig. 2 Post-translational modification of STAU2 in mitosis. **a** hTert-RPE1 and HeLa cells were synchronized in mitosis with either nocodazole (Ndz) or paclitaxel (Taxol) and enriched by shake off (S.off). Protein extracts from synchronized cells were analyzed by SDS-PAGE and western blotting to investigate STAU2 migration pattern and expression of cell cycle markers (MPM2 and cyclins). β-actin was used as loading control. **b** As control of synchronization, the percentage of cell population in the G_1, S or G_2/M phases was determined by FACS analysis ($n = 3$). **c** HeLa cells were synchronized either in prometaphase by nocodazole (Ndz), in G_1/S transition by double-thymidine block (DTB) or in late G_2 by the CDK1 inhibitor RO-3306 (RO-3306) and released from the block for the indicated time periods to reach mitosis. Protein extracts from synchronized cells were analyzed by SDS-PAGE and western blotting to investigate STAU2 migration pattern and expression of mitotic markers (MPM2 and cyclins). β-actin was used as loading control. **d** As control of synchronization, the percentage of cell population in the G_1, S or G_2/M phases was determined by FACS analysis ($n = 3$). Error bars represent the standard deviation

indicating a basal phosphorylation pattern of STAU2 isoforms in interphase. RSK1, nucleolin and β-actin were used as controls for phosphorylated and non-phosphorylated proteins, respectively. Then, hTert-RPE1 and HeLa cell extracts from nocodazole-synchronized cells were treated with λ protein phosphatase or calf intestinal alkaline phosphatase (CIP) in vitro prior to analysis by SDS-PAGE (Fig 4e). While STAU2 isoforms in the mitotic cell extracts showed the slow migration pattern, treatment with λ

phosphatase completely abolished the migration shift and treatment with CIP had a partial effect (Fig 4e). Dephosphorylation of STAU2 isoforms was specific since heat-inactivation of phosphatases prior to incubation restored the slow migration of the bands. The phosphorylation pattern of RSK1 was used as control.

STAU2 is phosphorylated in the C-terminal region

To localize the sites of phosphorylation, we generated three truncated mutants: human STAU2$^{\text{N-ter}}$-FLAG$_3$

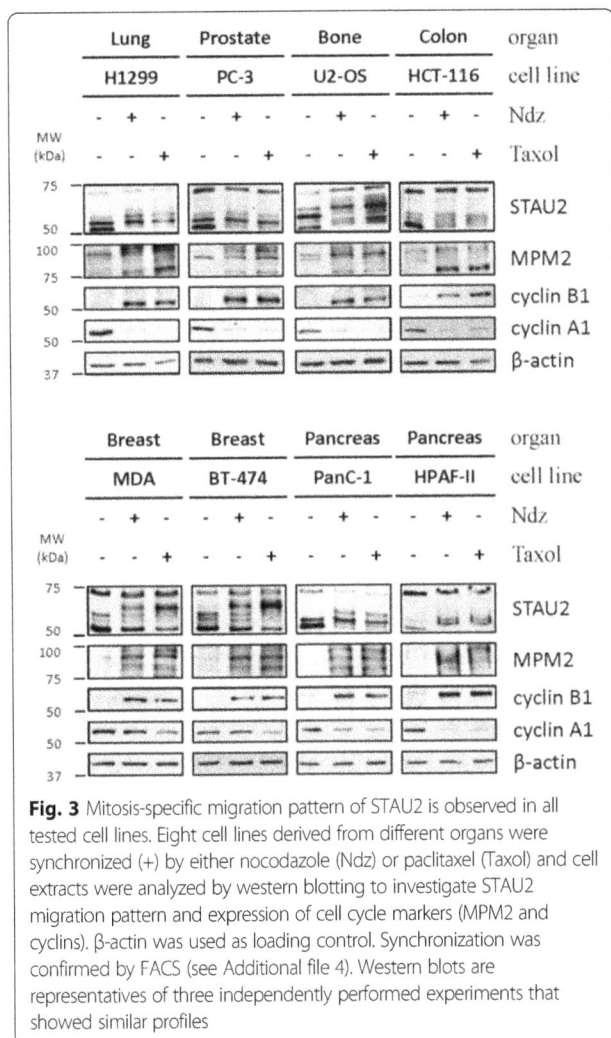

Fig. 3 Mitosis-specific migration pattern of STAU2 is observed in all tested cell lines. Eight cell lines derived from different organs were synchronized (+) by either nocodazole (Ndz) or paclitaxel (Taxol) and cell extracts were analyzed by western blotting to investigate STAU2 migration pattern and expression of cell cycle markers (MPM2 and cyclins). β-actin was used as loading control. Synchronization was confirmed by FACS (see Additional file 4). Western blots are representatives of three independently performed experiments that showed similar profiles

coding for the four N-terminal dsRBDs, human STAU2$^{52\text{-C-ter}}$-YFP coding for the STAU2^{52} tubulin-binding domain (TBD) and short C-terminal region and human STAU2$^{59\text{-C-ter}}$-mCherry coding for the STAU2^{59} TBD and long C-terminal region (Fig 5a). hTert-RPE1 cells were transfected with plasmids coding for the deletion mutants, synchronized or not in mitosis with nocodazole and analyzed by western blotting (Fig 5b). While the endogenous STAU2 proteins showed a typical gel shift pattern in mitosis, no difference in the migration of STAU2$^{N\text{-ter}}$-FLAG$_3$ was observed between asynchronous and mitotic cell extracts. In contrast, slow migrating bands appeared in mitotic cell extracts following expression of STAU2$^{52\text{-}}$$^{C\text{-ter}}$-YFP and STAU2$^{59\text{-C-ter}}$-mCherry. These results indicate that a region in the C-terminal half of the protein that is common to both STAU2^{52} and STAU2^{59} contains phosphorylation site(s) responsible for their slow migration pattern in mitosis.

Multiple residues are phosphorylated in the STAU2 C-terminal region during M-phase

Large-scale analysis of phospho-proteomes identified several mitosis-specific STAU2 phospho-residues [56]. Interestingly, seven of them were located within TBD and three in the short C-terminal region (Fig 6a). We thus muted as clusters the seven and the three amino acid residues found in TBD and C-terminal domain, respectively, as well as all ten residues. Residues were substituted for either aspartic acid or alanine to respectively generate phospho-mimetic (P(7)+; P(3)+; P(10)+) or phospho-null (P(7)-; P(3)-; P(10)-) STAU2^{52}-FLAG$_3$ and STAU2^{59}-FLAG$_3$ mutants (Fig 6a).

hTert-RPE1 cells were infected with viruses expressing either wild type (WT) STAU2^{52}-FLAG$_3$, the phospho-mimetic or the phospho-null mutant. Their migration patterns were analyzed by western blotting in extracts from both mitotic and unsynchronized cells (Fig 6b). STAU2^{52}-FLAG$_3$ was found as two slow migrating bands when extracted from mitotic cells as compared to unsynchronized cells, indicating the presence of two differentially phosphorylated populations of STAU2 in mitosis. Essentially the same migration pattern was observed with the WT, P(3) + and P(3)- mutants suggesting that residues in the P3 cluster did not contribute much to STAU2 phosphorylation pattern in mitosis. In contrast, a single slow migrating band was observed with the phospho-mimetic P(7) + mutants in both asynchronous and mitotic cells, suggesting that at least some residues are phosphorylated in this cluster. Interestingly, whereas the phospho-null P(7)- mutant exhibited the fast migrating pattern in asynchronous cells, as expected, it however showed a slight shift in mitosis, suggesting that other amino acids are also phosphorylated in mitosis. Since a similar shift was not detected with the P(10)- mutant, it is suggests that site(s) in the P3 cluster may accounted for the difference. Altogether, these results mapped the major mitotic phosphorylation residues within the P7 cluster and suggested that residues in the P3 cluster may also contribute to STAU2 phosphorylation. A slightly different pattern of phosphorylation was observed for STAU2^{59}-FLAG$_3$ (Fig 6c). First, STAU2^{59}-FLAG$_3$ showed a single phosphorylated band in mitosis. Second, both the P(3) and P(7) mutants showed altered migration patterns in asynchronous and mitotic cells as compared to the wild type protein, and both phospho-null P(3)- and P(7)- proteins slightly shifted in mitotic cells as compared to asynchronous cells. Nevertheless, as observed above for STAU2^{52}-FLAG$_3$ P(10) mutants, STAU2^{59}-FLAG$_3$ P(10) mutants did not shift during mitosis indicating that the major phosphorylation sites are located within the P3 and P7 clusters. These results suggest a more important role for residues in the P3 cluster for STAU2^{59}-FLAG$_3$ than for STAU2^{52}-FLAG$_3$.

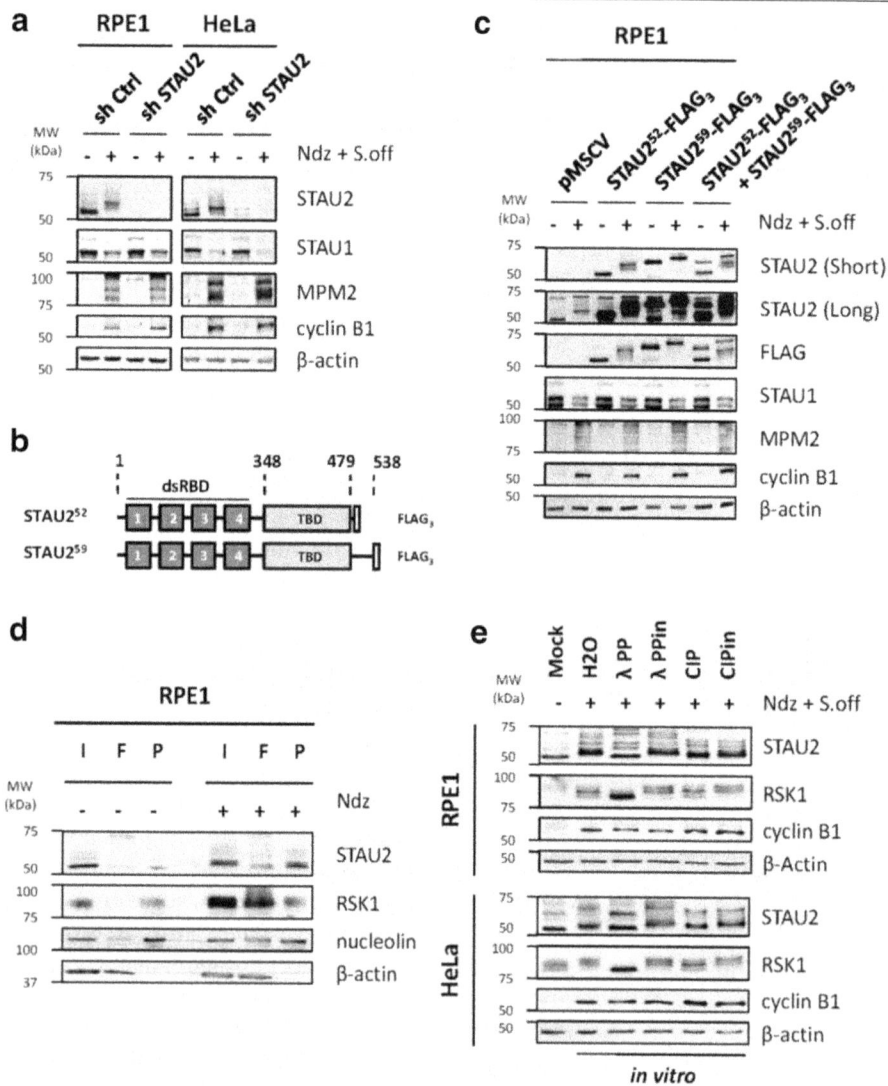

Fig. 4 STAU2 is hyperphosphorylated during the M-phase. **a** shRNA control (shCtrl) or against STAU2 (shSTAU2) were cloned in a retrovirus vector to infect hTert-RPE1 and HeLa cells. Cells were synchronized in prometaphase with nocodazole and mitotic cells were enriched by gentle shake off (Ndz + S.off). Protein extracts from asynchronous (−) and synchronized cells (+) were analyzed by western blotting to detect STAU2 migration and cell cycle markers (MPM2 and cyclins). β-actin was used as loading control. **b** Schematic representation of STAU2 expression vectors, STAU2^{52}-FLAG$_3$ and STAU2^{59}-FLAG$_3$. Dark gray boxes, double-stranded RNA-binding domains (dsRBD); light gray boxes, tubulin-binding domain (TBD); white boxes, FLAG$_3$. **c** hTert-RPE1 cells infected with viruses expressing the empty pMSCV vector (pMSCV), STAU2^{52}-FLAG$_3$, STAU2^{59}-FLAG$_3$ or both were synchronized (+) by nocodazole and shake off (Ndz + S.off). Migration of STAU2 proteins was detected by SDS-PAGE and western blotting. Both endogenous and overexpressed STAU2^{59}-FLAG$_3$ were analyzed with anti-STAU2 antibody, while anti-FLAG antibody was used to specifically recognize FLAG$_3$-tagged STAU2 isoforms. Mitotic marker accumulation was assessed with anti-MPM2 and anti-cyclin B1 antibodies. Loading was normalized with β-actin antibody. **d** Asynchronous (−) and nocodazole-treated (Ndz)(+) hTert-RPE1 cells were lysed and protein extracts were subjected to separation on phospho-columns. Input from total extracts (I), flow through (F) and phospho-eluates (P) were analyzed by western blotting using anti-STAU2. Anti-RSK1, anti-nucleolin and anti-β-actin antibodies were used as controls for phosphorylated and unphosphorylated proteins, respectively. **e** Protein extracts from nocodazole-treated (+) hTert-RPE1 and HeLa were incubated in vitro with either water (H2O), Lambda Phosphatase (λPP), inactivated Lambda Phosphatase (λPPin), Calf Intestinal Alkaline Phosphatase (CIP) or inactivated Calf Intestinal Alkaline Phosphatase (CIPin). STAU2 phosphorylation status was analyzed by SDS-PAGE and western blotting. Untreated cells (Mock) were used as control for dephosphorylated STAU2. All western blots are representatives of three independently performed experiments that showed similar profiles

STAU2 phosphorylation depends on CDK1

To further explore the STAU2 phosphorylation pathway in mitosis, hTert-RPE1 cells were synchronized in prometaphase with nocodazole and exposed for 4 h to kinase inhibitors in the presence of the proteasome inhibitor MG132 to prevent cyclin B degradation and subsequent mitotic exit. Several drugs were tested that specifically targeted CDK1 (Purvalanol A, Roscovitine, Flavopiridol, RO-3306), aurora-A (Alisertib), aurora-B (Barasertib), and polo-like kinase 1 (PLK1)(BI2536). Cell

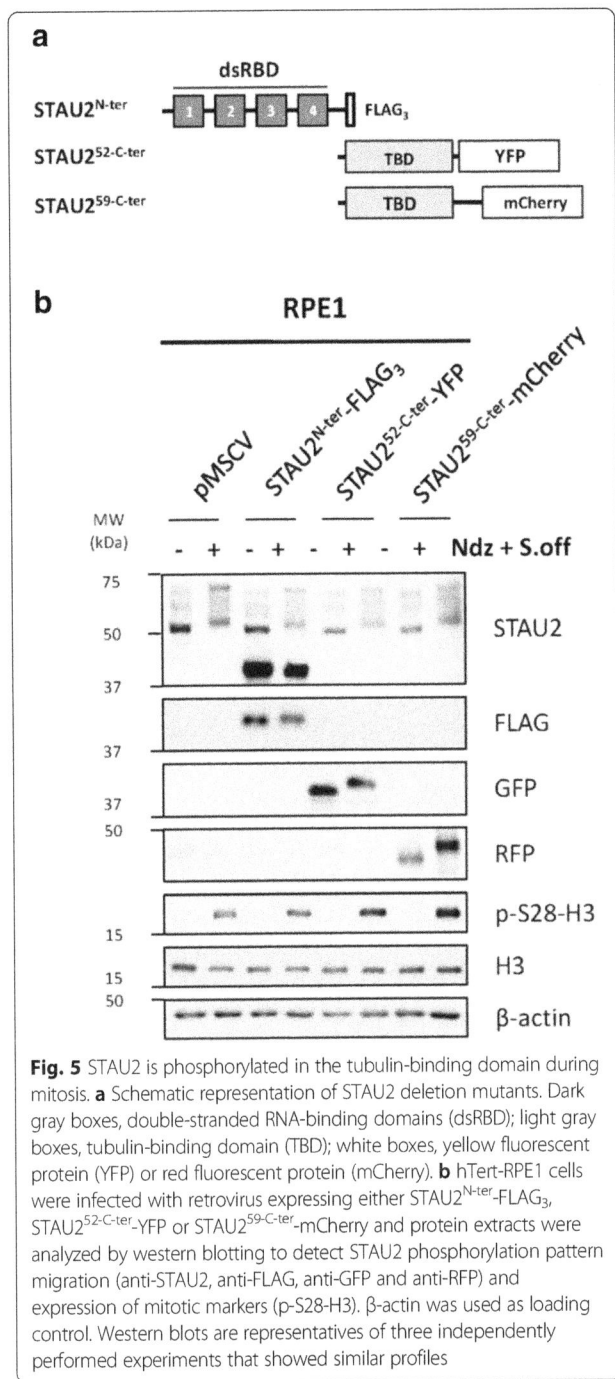

Fig. 5 STAU2 is phosphorylated in the tubulin-binding domain during mitosis. **a** Schematic representation of STAU2 deletion mutants. Dark gray boxes, double-stranded RNA-binding domains (dsRBD); light gray boxes, tubulin-binding domain (TBD); white boxes, yellow fluorescent protein (YFP) or red fluorescent protein (mCherry). **b** hTert-RPE1 cells were infected with retrovirus expressing either STAU2^{N-ter}-FLAG$_3$, STAU2^{52-C-ter}-YFP or STAU2^{59-C-ter}-mCherry and protein extracts were analyzed by western blotting to detect STAU2 phosphorylation pattern migration (anti-STAU2, anti-FLAG, anti-GFP and anti-RFP) and expression of mitotic markers (p-S28-H3). β-actin was used as loading control. Western blots are representatives of three independently performed experiments that showed similar profiles

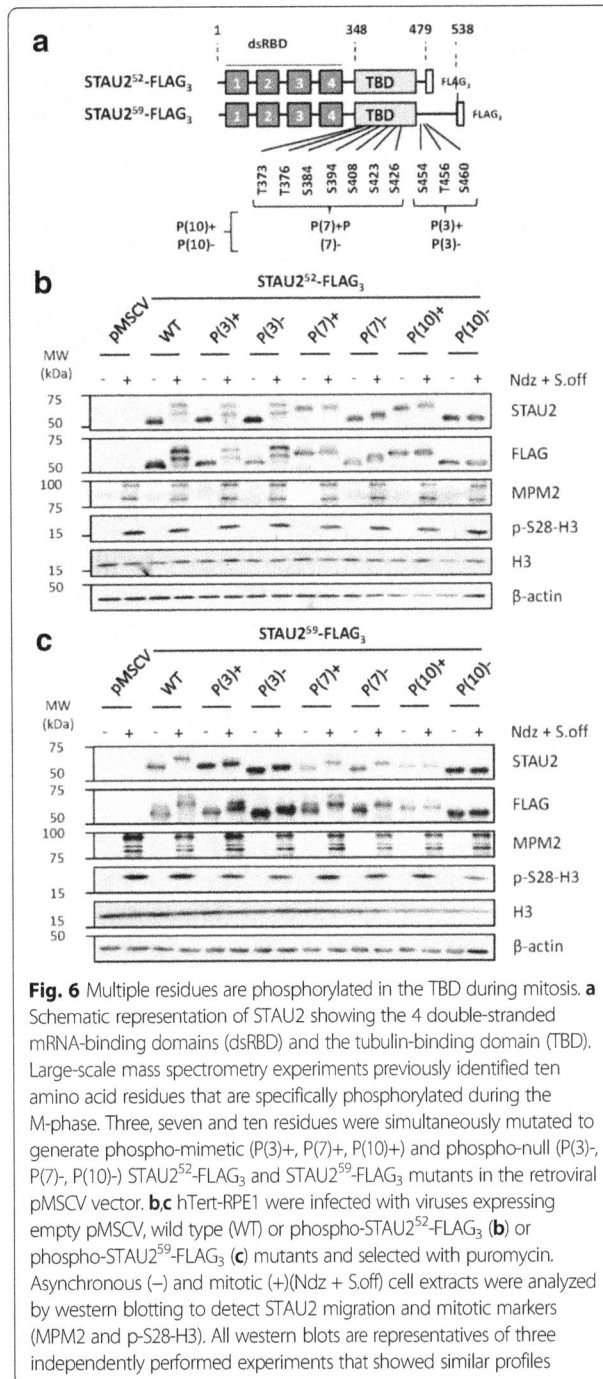

Fig. 6 Multiple residues are phosphorylated in the TBD during mitosis. **a** Schematic representation of STAU2 showing the 4 double-stranded mRNA-binding domains (dsRBD) and the tubulin-binding domain (TBD). Large-scale mass spectrometry experiments previously identified ten amino acid residues that are specifically phosphorylated during the M-phase. Three, seven and ten residues were simultaneously mutated to generate phospho-mimetic (P(3)+, P(7)+, P(10)+) and phospho-null (P(3)-, P(7)-, P(10)-) STAU2^{52}-FLAG$_3$ and STAU2^{59}-FLAG$_3$ mutants in the retroviral pMSCV vector. **b,c** hTert-RPE1 were infected with viruses expressing empty pMSCV, wild type (WT) or phospho-STAU2^{52}-FLAG$_3$ (**b**) or phospho-STAU2^{59}-FLAG$_3$ (**c**) mutants and selected with puromycin. Asynchronous (−) and mitotic (+)(Ndz + S.off) cell extracts were analyzed by western blotting to detect STAU2 migration and mitotic markers (MPM2 and p-S28-H3). All western blots are representatives of three independently performed experiments that showed similar profiles

extracts were analyzed by western blotting (Fig 7a) and FACS (Fig 7b). Inhibition of CDK1 by any of the four kinase inhibitors prevented the phosphorylation-mediated shift migration of STAU2 in mitosis (Fig 7a). As controls, we showed that CDK1 inhibition prevented the phosphorylation of histone H3, lamina A/C and aurora kinases, known targets of CDK1 (Fig 7a). As further controls, CDK1 inhibition did not allow accumulation of MPM2 or phospho-CDK/MAPK substrates. Importantly, upon these conditions, cells were still in the mitotic

phase as monitored by FACS (Fig 7b). In contrast, none of the other kinase inhibitors changed the phosphorylation status of STAU2. As evidence of their inhibitory effect and specificity, aurora-A and PLK-1 inhibitors abolished aurora-A phosphorylation on threonine288, and aurora-B inhibitors seriously diminished aurora-B threonine232 phosphorylation (Fig 7a).

Similar experiments were reproduced in HeLa cells (Additional file 5). As shown above, CDK1 inhibitors

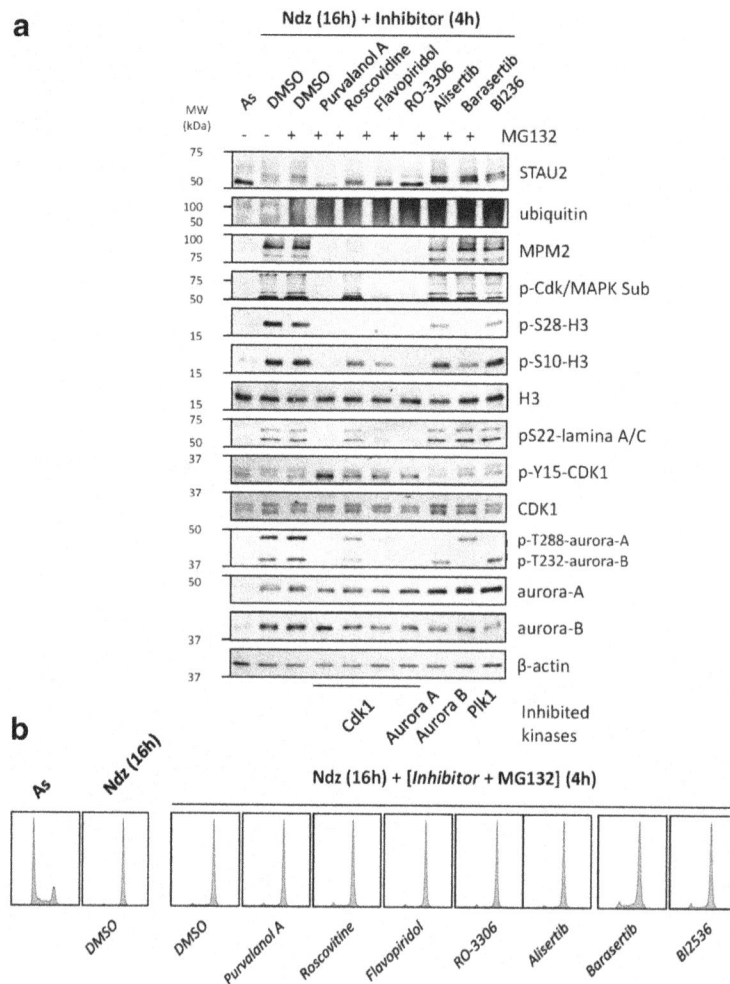

Fig. 7 STAU2 is phosphorylated by Cdk1 during mitosis in hTert-RPE1 cells. **a** hTert-RPE1 cells were incubated with nocodazole (Ndz) for 16 h in the absence (DMSO) or presence of specific kinase inhibitors for 4 h as indicated. Asynchronous (As) cells were used as controls. Cells were treated with the proteasome inhibitor MG132 (+) to keep cells in mitosis. STAU2 migration on SDS-gels was analyzed by western blotting. Cell markers were used to confirm specific inhibition of kinases. **b** Cell synchronization was confirmed by FACS. Western blots and FACS are representatives of three independently performed experiments that showed similar profiles

prevented STAU2 phosphorylation whereas inhibition of aurora A, aurora B or PLK1 had no effect. Additional inhibitors that targeted MAP2K-MEK1/2 (U0126, AZD6244), RSK-1 (BI-D1870), p38MPK (SB203580) and SAPK/JNK (SP600135) were tested in both hTert-RPE1 (Additional file 6A) and HeLa (Additional file 6B) cells. None of these inhibitors affected STAU2 phosphorylation pattern. Inhibition of the MAPKs was monitored with anti-phospho-ERK1/2 (threonine202 and tyrosine204) and through RSK-1 migration pattern.

Discussion

Cell division is a complex process that required a plethora of overlapping mechanisms to precisely control the timing of each step [1]. Post-translational modification of regulatory proteins is a well-documented process that insures spatial and temporal activation and/or inhibition

of protein activity [3, 11]. Among them, protein kinases and phosphatases are at the heart of mechanisms that are essential for progression through mitosis. In this manuscript, we add an additional player to the repertoire of proteins specifically phosphorylated during mitosis. We showed that STAU2 is phosphorylated by a mitotic kinase at the beginning of mitosis and is dephosphorylated as cell exit mitosis. This phosphorylation is ubiquitous, occurs on serine and threonine residues within the tubulin-binding and C-terminal domains of STAU2 and is CDK1-dependant. Unfortunately, the consequence of STAU2 phosphorylation in mitosis is still unclear. However, we excluded the possibility that STAU2 phosphorylation controls cell proliferation, STAU2 protein half-life, interaction with known RNA and protein partners and subcellular localization (see below). One possibility is that, in unstressed cells, STAU2 phosphorylation plays a subtle role that escapes detection but becomes crucial in special conditions (e.g. asymmetric cell division, cell stresses). Alternatively, the putative mechanism that relies on STAU2 phosphorylation might be a backup strategy for the fine tuning of a specific function during cell division. For example, in drosophila, neuroblast asymmetric cell division depends on the differential localization of prospero (pros) mRNA and protein in one of the two progenies. While pros mRNA localization is mediated by Staufen, targeting of Pros protein is not. In Staufen mutant, pros mRNA is no longer localized but still cell fate defects have not been detected in the nervous system, suggesting that Staufen-mediated pros mRNA localization is redundant to Pros protein segregation [57].

Post-translational modifications of STAU2 during mitosis
Our results indicate that STAU2 is specifically phosphorylated during mitosis in human cells. This is reminiscent of the temporal phosphorylation of Xstau proteins during meiotic maturation in Xenopus [33]. It was suggested that phosphorylation of Xstau may alter its RNA-binding activity to allow the release and/or expression of Xstau-bound RNAs. In oocytes, Xstau proteins are transiently phosphorylated by the MAPK pathway. Our pharmacological studies in human cells do not support the possibility that the MAPK pathway phosphorylates STAU2 during mitosis. They rather indicate that cyclin-dependent kinase 1 is involved in this process. We do not exclude the possibility that members of the MAPK pathway may also phosphorylate STAU2 during mitosis but that the process is likely masked by the CDK1-mediated hyper-phosphorylation. Alternatively, phosphorylation of STAU2 by MAPK may occur in other phases of the cell cycle. Indeed, large-scale phosphoproteome analyses and phospho-columns separation (Fig 4d) indicate that STAU2 is also phosphorylated during interphase. Interestingly, STAU2 was shown to harbor a docking site for ERK1/2 [36]. Its presence in the RNA-binding domain inter-region of the protein would be consistent with a role in modulating its RNA binding activity.

CDK1 is the major kinase regulating mitosis. In human, CDK1-cyclin B phosphorylates more than 70 substrates required for mitosis progression [4]. Our results add a member of the RNA-binding proteins family to the list of CDK1 substrates. Previous phosphoproteome analyses identified 10 amino acid residues that are more likely to be phosphorylated in mitoses than during interphase [56]. Our data indicated that phosphorylation of these residues is responsible for the migration shift of STAU2 isoforms during mitosis. Interestingly, our data are consistent with a differential regulation of the P(3) and P(7) clusters and different phosphorylation mechanisms for the two STAU2 isoforms. It is possible that the two isoforms are differently

localized in the cells during mitosis or that they are part of alternative complexes that changes their accessibility for CDK1. However, the exact pattern in the phosphorylated populations and the dynamic and timing of phosphorylation in each complex are still unknown. Many CDK1-mediated phosphorylated residues show the S/T-P-x-K/R consensus sequence [11]. Although none of the phosphorylated residues in the P3 and P7 clusters of STAU2 have a perfect match with the consensus sequence, 4 residues in the P7 clusters and all three residues in the P3 cluster have the S/T-P consensus sequence. It is possible that the 3 other sites contain atypical non-S/T-P consensus motifs as previously described in proteins such as vimentin, desmin and myosin II [58]. We do not exclude the alternative possibilities that other kinases also phosphorylated STAU2 in mitosis but do not significantly contribute to the migration shifts observed in mitosis.

Functions of phosphorylated STAU2 in mitosis
The consequence of STAU2 phosphorylation in mitosis is unclear. Using the phospho-mimetic and phospho-null mutants, we first excluded a role in the RNA-binding capacity of STAU2 (data not shown). This was expected since the phosphorylated residues were mapped within the C-terminal region of the protein, a region that lacks RNA-binding activity. In addition, we showed that STAU2 interaction with ribosomes or the RNA-decay factor UPF1 was not altered by phosphorylation (data not shown), suggesting that its functions in the post-transcriptional regulation of gene expression are not affected. It was previously shown that the human paralog STAU1 interacts with these factors via its TBD [59, 60]. Obviously, STAU2 phosphorylation in TBD does not regulate these processes.

We then excluded a role for phosphorylation in STAU2 stability (Additional file 7). We had noticed a decrease in the amounts of STAU2 in the G_1 phase of the cell cycle as compared to those in G_2, suggesting that STAU2 is partly degraded in mitosis (Fig. 1; Additional file 2A). Phosphorylation could have been used as a signal for protein stability, but this is not the case. STAU2 partial degradation in mitosis likely involved the proteasome since its amount was increased in the presence of the proteasome inhibitor MG132 (Additional file 7). This is reminiscent of STAU1 which is partly degraded during its transit through mitosis as a consequence of its binding to the E3-ubiquitin ligase APC/C and its degradation by the proteasome [55].

We further excluded the possibility that phosphorylation in mitosis changes STAU2 sub-cellular localization (data not shown). We previously showed that STAU2 migrates in dendrites on the microtubule network [29]. Similar results were obtained with STAU1 [61] and the molecular determinant for tubulin-binding was mapped within TBD [60]. Therefore, STAU2 phosphorylation in

TBD might prevent STAU2-microtubule association. Indeed, STAU2 was not found on microtubules during mitosis in hamster BHK and mouse MEF cells [62], although it partly colocalizes with spindle at MI and MII during mouse oocyte meiosis. However, the STAU2 phospho-null mutant was not better in associating with the mitotic spindle than the WT STAU2 or its phospho-mimetic mutant (data not shown).

Finally, it was previously shown that STAU2 may be involved in the control of cell division. Indeed, STAU2 depletion leads to decreased cell proliferation and small eye phenotypes in chicken embryos [50], and increases the number of post-mitotic neurons in rats [47, 48]. Similarly, overexpression of the paralog STAU1 impairs cell proliferation by affecting mitosis entry [55]. Thus, mitosis-specific phosphorylation of STAU2 might be advantageous to regulate some aspects of cell division. However, overexpression of STAU2 phospho-mimetic and phospho-null mutants in the untransformed hTert-RPE1 cells led to the same proliferation pattern as that of the cells that overexpressed the WT protein (Additional file 8).

Conclusions

Altogether, our results indicate that STAU2 is hyperphosphorylated in mitosis suggesting that it may be a novel actor in mitosis regulation and that posttranscriptional regulation of gene expression may be linked to cell cycle pathways in proliferative cells.

Additional files

Additional file 1: Synchronization of hTert-RPE1 cells – Controls for experiments shown in Fig. 1. (A) hTert-RPE1 cells were synchronized by a double thymidine block (DTB) or a nocodazole arrest followed by shake off (Ndz + S.off). Cells were then released in fresh medium for different time periods as indicated. Asynchronous (As) cells were collected as controls. **(B)** Cells were blocked in prometaphase with nocodazole and shake off (Ndz + S.off), released in fresh medium and harvested every 30 min. Cell synchronization was analyzed by FACS. These results are representatives of three independently performed experiments that showed similar profiles. (PDF 132 kb)

Additional file 2: STAU2 is differentially regulated through the cell cycle in HeLa cells. HeLa cells were synchronized by a double thymidine block (DTB) or a nocodazole arrest (Ndz) followed by shake off (S.off). Cells were then released in a fresh medium for different time periods as indicated (Rel (h)). Asynchronous (As) cells were collected as controls. **(A)** Protein extracts from synchronized cells were analyzed by SDS-PAGE and western blotting to investigate STAU2 phosphorylation pattern migration and expression of mitotic markers (MPM2 and cyclins). β-actin was used as loading control. **(B)** As control of synchronization, the percentage of cell population in the G_1, S or G_2/M phases was determined by FACS analysis. Error bars represent the standard deviation. $n = 3$. (PDF 237 kb)

Additional file 3: STAU2 dephosphorylation dynamics was examined in HeLa cells. HeLa cells were blocked in prometaphase with nocodazole (Ndz), released in fresh medium and harvested every 15 min (Rel (min)). **(A)** Extracts from untreated asynchronous (As) and nocodazole-treated cells were analyzed by western blotting. **(B)** The percentage of cell population within the G_1, S or G_2/M phases was determined by FACS. Error bars represent the standard deviation. $n = 3$. (PDF 248 kb)

Additional file 4: STAU2 is phosphorylated in mitosis in all tested cell lines – controls for experiments shown in Fig. 3. Eight cell lines derived from different organs were synchronized (+) by either nocodazole (Ndz) or paclitaxel (Taxol). Cells were collected after a gentle shake off and analyzed by FACS. $n = 3$. (PDF 140 kb)

Additional file 5: STAU2 is phosphorylated by CDK1 during mitosis in HeLa cells. HeLa cells were incubated with nocodazole (Ndz) for 16 h in the absence (DMSO) or presence of specific kinase inhibitors for 4 h as indicated. Asynchronous (As) cells were used as controls. Cells were treated with the proteasome inhibitor MG132 (+) to keep cells in mitosis. STAU2 migration on SDS-gels was analyzed by western blotting. Cell markers were used to confirm specific inhibition of kinases. Western blots are representatives of three independently performed experiments that showed similar profiles. (PDF 422 kb)

Additional file 6: MAPK does not phosphorylated STAU2 in mitosis. hTert-RPE1 **(A)** and HeLa **(B)** cells were incubated with nocodazole (Ndz) for 16 h in the absence (DMSO) or presence of specific kinase inhibitors for 4 h as indicated. Asynchronous (As) cells were used as controls. Cells were treated with the proteasome inhibitor MG132 (+) to keep cells in mitosis. STAU2 migration on SDS-gels was analyzed by western blotting. Cell markers were used to confirm specific inhibition of kinases. Western blots are representatives of three independently performed experiments that showed similar profiles. (PDF 416 kb)

Additional file 7: STAU2 phosphorylation does not modulate protein degradation. hTert-RPE1 cells were infected with retroviruses expressing either empty pMSCV, STAU2^{52}-FLAG$_3$ wild type (WT), phospho-mimetic STAU2^{52}-FLAG$_3$ (P(10)+) or phospho-null STAU2^{52}-FLAG$_3$ (P(10)-). Cells were synchronized in prometaphase with nocodazole and released from the block for 6 h (Ndz + Rel). During release, prometaphase cells were also treated (+) or not (−) with cyclohexamide (CHX) to prevent protein synthesis and with (+) or without (−) the proteasome inhibitor MG132 to prevent protein degradation. Cell extracts were collected and amounts of STAU2 analyzed by western blotting. β-actin was used as loading control. Western blots are representatives of three independently performed experiments that showed similar profiles. (PDF 165 kb)

Additional file 8: STAU2 phosphorylation does not regulate cell proliferation. hTert-RPE1 cells were infected with retroviruses expressing either empty pMSCV (pMSCV), STAU2^{52}-FLAG$_3$ (WT), phospho-mimetic (+) or phospho-null (−) mutants. **(A)** Same amounts of cells were plated and allowed to growth for 5 days in a colony assay. **(B)** Cell proliferation was quantified by crystal violet staining. $n = 3$. **(C)** Western blot analysis indicated that the amounts of each STAU2 overexpressed-protein are similar and slightly above that of the endogenous protein. (PDF 335 kb)

Abbreviations

APC/c: Anaphase-promoting complex/cyclosome; CDK: Cyclin-dependent kinase; CIP: Calf-intestinal phosphatase; DMSO: Dimethyl sulfoxide; dsRBD: Double-stranded RNA-binding domain; DTB: Double-thymidine block; ERK: Extracellular signal-regulated kinase; FACS: Fluorescence-activated cell sorting; GFP: Green-fluorescent protein; HRP: Horseradish peroxidase; MAPK: Mitogen-activated protein kinase; MEK: Mitogen-activated kinase kinase; PBS: Phosphate buffered saline; PLK1: Polo-like kinase 1; Pros: Prospero; RFP: Red-fluorescent protein; RSK: Ribosomal S6 kinase; SAPK/JNK: Stress-activated protein kinase/c-jun kinase; SDS-PAGE: Sodium dodecyl sulfate-polyacrylamide gel electrophoresis; STAU2: Staufen2; TBD: Tubulin-binding domain; YFP: Yellow fluorescent protein; λPP: Lambda protein phosphatase

Acknowledgements

We thank Louise Cournoyer for help in the cell culture and Isabel Gamache for technical assistance.

Funding

This work was supported by a Natural Science and Engineering Research Council of Canada (NSERC) grant (41596–2009) to LDG. EO was supported by a summer studentships from NSERC. The funding body played no role in the design of the study and collection, analysis, and interpretation of data and in writing the manuscript.

Authors' contributions
WV and LDG conceptualized and initiated this study. RB, EO, XZ, CG, SH, SM and WV designed experiments and acquired and analyzed data. RB and LDG wrote the manuscript while others provided editorial comments. All authors have read and approved the final version of the manuscript.

Competing interests
The authors declare that they have no competing interests.

Author details
[1]Present address: UMR PIMIT, Processus Infectieux en Milieu Insulaire Tropical, Université de la Réunion, 97490 Sainte Clotilde, La Réunion, France. [2]Département de biochimie et médecine moléculaire, Faculté de médecine, Université de Montréal, 2900 Edouard Montpetit, Montréal, QC H3T 1J4, Canada.

References

1. Tyson JJ, Novak B. Temporal organization of the cell cycle. Curr Biol. 2008; 18(17):R759–68.
2. Kops GJ, Weaver BA, Cleveland DW. On the road to cancer: aneuploidy and the mitotic checkpoint. Nat Rev Cancer. 2005;5(10):773–85.
3. Lim S, Kaldis P. Cdks, cyclins and CKIs: roles beyond cell cycle regulation. Dev (Cambridge, England). 2013;140(15):3079–93.
4. Malumbres M, Harlow E, Hunt T, Hunter T, Lahti JM, Manning G, et al. Cyclin-dependent kinases: a family portrait. Nat Cell Biol. 2009;11(11):1275–6.
5. Arellano M, Moreno S. Regulation of CDK/cyclin complexes during the cell cycle. Int J Biochem Cell Biol. 1997;29(4):559–73.
6. King RW, Jackson PK, Kirschner MW. Mitosis in transition. Cell. 1994;79(4): 563–71.
7. Morgan DO. Principles of CDK regulation. Nature. 1995;374(6518):131–4.
8. Morgan DO. Cyclin-dependent kinases: engines, clocks, and microprocessors. Annu Rev Cell Dev Biol. 1997;13:261–91.
9. Johnson LN, Lewis RJ. Structural basis for control by phosphorylation. Chem Rev. 2001;101(8):2209–42.
10. Blume-Jensen P, Hunter T. Oncogenic kinase signalling. Nature. 2001; 411(6835):355–65.
11. Ubersax JA, Ferrell JE Jr. Mechanisms of specificity in protein phosphorylation. Nat Rev Mol Cell Biol. 2007;8(7):530–41.
12. Olsen JV, Blagoev B, Gnad F, Macek B, Kumar C, Mortensen P, et al. Global, in vivo, and site-specific phosphorylation dynamics in signaling networks. Cell. 2006;127(3):635–48.
13. Manning G, Whyte DB, Martinez R, Hunter T, Sudarsanam S. The protein kinase complement of the human genome. Science. 2002;298(5600):1912–34.
14. Hanahan D, Weinberg RA. The hallmarks of cancer. Cell. 2000;100(1):57–70.
15. Johnson LN. Protein kinase inhibitors: contributions from structure to clinical compounds. Q Rev Biophys. 2009;42(1):1–40.
16. Noble ME, Endicott JA, Johnson LN. Protein kinase inhibitors: insights into drug design from structure. Science. 2004;303(5665):1800–5.
17. Krause DS, Van Etten RA. Tyrosine kinases as targets for cancer therapy. N Engl J Med. 2005;353(2):172–87.
18. Daub H, Olsen JV, Bairlein M, Gnad F, Oppermann FS, Korner R, et al. Kinase-selective enrichment enables quantitative phosphoproteomics of the kinome across the cell cycle. Mol Cell. 2008;31(3):438–48.
19. Malik R, Lenobel R, Santamaria A, Ries A, Nigg EA, Korner R. Quantitative analysis of the human spindle phosphoproteome at distinct mitotic stages. J Proteome Res. 2009;8(10):4553–63.
20. Dephoure N, Zhou C, Villen J, Beausoleil SA, Bakalarski CE, Elledge SJ, et al. A quantitative atlas of mitotic phosphorylation. Proc Natl Acad Sci U S A. 2008;105(31):10762–7.
21. Glotzer M, Murray AW, Kirschner MW. Cyclin is degraded by the ubiquitin pathway. Nature. 1991;349(6305):132–8.
22. Dynlacht BD. Regulation of transcription by proteins that control the cell cycle. Nature. 1997;389(6647):149–52.
23. Chen HH, Wang YC, Fann MJ. Identification and characterization of the CDK12/cyclin L1 complex involved in alternative splicing regulation. Mol Cell Biol. 2006;26(7):2736–45.
24. Trembley JH, Loyer P, Hu D, Li T, Grenet J, Lahti JM, et al. Cyclin dependent kinase 11 in RNA transcription and splicing. Prog Nucleic Acid Res Mol Biol. 2004;77:263–88.
25. Pyronnet S, Sonenberg N. Cell-cycle-dependent translational control. Curr Opin Genet Dev. 2001;11(1):13–8.
26. Bu X, Haas DW, Hagedorn CH. Novel phosphorylation sites of eukaryotic initiation factor-4F and evidence that phosphorylation stabilizes interactions of the p25 and p220 subunits. J Biol Chem. 1993;268(7):4975–8.
27. Pyronnet S, Imataka H, Gingras AC, Fukunaga R, Hunter T, Sonenberg N. Human eukaryotic translation initiation factor 4G (eIF4G) recruits mnk1 to phosphorylate eIF4E. EMBO J. 1999;18(1):270–9.
28. Raught B, Peiretti F, Gingras AC, Livingstone M, Shahbazian D, Mayeur GL, et al. Phosphorylation of eucaryotic translation initiation factor 4B Ser422 is modulated by S6 kinases. EMBO J. 2004;23(8):1761–9.
29. Duchaine TF, Hemraj I, Furic L, Deitinghoff A, Kiebler MA, DesGroseillers L. Staufen2 isoforms localize to the somatodendritic domain of neurons and interact with different organelles. J Cell Sci. 2002;115(Pt 16):3285–95.
30. Tang SJ, Meulemans D, Vazquez L, Colaco N, Schuman E. A role for a rat homolog of staufen in the transport of RNA to neuronal dendrites. Neuron. 2001;32(3):463–75.
31. Maher-Laporte M, Berthiaume F, Moreau M, Julien LA, Lapointe G, Mourez M, et al. Molecular composition of staufen2-containing ribonucleoproteins in embryonic rat brain. PLoS One. 2010;5(6):e11350.
32. Mallardo M, Deitinghoff A, Muller J, Goetze B, Macchi P, Peters C, et al. Isolation and characterization of Staufen-containing ribonucleoprotein particles from rat brain. Proc Natl Acad Sci U S A. 2003;100(4):2100–5.
33. Allison R, Czaplinski K, Git A, Adegbenro E, Stennard F, Houliston E, et al. Two distinct Staufen isoforms in Xenopus are vegetally localized during oogenesis. RNA. 2004;10(11):1751–63.
34. Goetze B, Tuebing F, Xie Y, Dorostkar MM, Thomas S, Pehl U, et al. The brain-specific double-stranded RNA-binding protein Staufen2 is required for dendritic spine morphogenesis. J Cell Biol. 2006;172(2):221–31.
35. Jeong JH, Nam YJ, Kim SY, Kim EG, Jeong J, Kim HK. The transport of Staufen2-containing ribonucleoprotein complexes involves kinesin motor protein and is modulated by mitogen-activated protein kinase pathway. J Neurochem. 2007;102(6):2073–84.
36. Nam YJ, Cheon HS, Choi YK, Kim SY, Shin EY, Kim EG, et al. Role of mitogen-activated protein kinase (MAPK) docking sites on Staufen2 protein in dendritic mRNA transport. Biochem Biophys Res Commun. 2008;372(4):525–9.
37. Ramasamy S, Wang H, Quach HN, Sampath K. Zebrafish Staufen1 and Staufen2 are required for the survival and migration of primordial germ cells. Dev Biol. 2006;292(2):393–406.
38. Thomas MG, Martinez Tosar LJ, Loschi M, Pasquini JM, Correale J, Kindler S, et al. Staufen recruitment into stress granules does not affect early mRNA transport in oligodendrocytes. Mol Biol Cell. 2005;16(1):405–20.
39. Sanchez-Carbente M, DesGroseillers L. Understanding the importance of mRNA transport in memory. Prog Brain Res. 2008;169:41–58.
40. Lebeau G, Miller LC, Tartas M, McAdam R, Laplante I, Badeaux F, et al. Staufen 2 regulates mGluR long-term depression and Map1b mRNA distribution in hippocampal neuron. Learn Mem. 2011;18(5):314–26.
41. Mikl M, Vendra G, Kiebler MA. Independent localization of MAP2, CaMKIIalpha and beta-actin RNAs in low copy numbers. EMBO Rep. 2011;12(10):1077–84.
42. Miki T, Kamikawa Y, Kurono S, Kaneko Y, Katahira J, Yoneda Y. Cell type-dependent gene regulation by Staufen2 in conjunction with Upf1. BMC Mol Biol. 2011;12:48.
43. Park E, Gleghorn ML, Maquat LE. Staufen2 functions in Staufen1-mediated mRNA decay by binding to itself and its paralog and promoting UPF1 helicase but not ATPase activity. Proc Natl Acad Sci U S A. 2013;110(2):405–12.
44. O'Leary DA, Sharif O, Anderson P, Tu B, Welch G, Zhou Y, et al. Identification of small molecule and genetic modulators of AON-induced dystrophin exon skipping by high-throughput screening. PLoS One. 2009;4(12):e8348.
45. Maher-Laporte M, DesGroseillers L. Genome wide identification of Staufen2-bound mRNAs in embryonic rat brains. BMB Rep. 2010;43(5):344–8.
46. Furic L, Maher-Laporte M, DesGroseillers L. A genome-wide approach identifies distinct but overlapping subsets of cellular mRNAs associated with Staufen1- and Staufen2-containing ribonucleoprotein complexes. RNA. 2008; 14(2):324–35.
47. Vessey JP, Amadei G, Burns SE, Kiebler MA, Kaplan DR, Miller FD. An asymmetrically localized Staufen2-dependent RNA complex regulates maintenance of mammalian neural stem cells. Cell Stem Cell. 2012;11(4):517–28.
48. Kusek G, Campbell M, Doyle F, Tenenbaum SA, Kiebler M, Temple S. Asymmetric segregation of the double-stranded RNA binding protein Staufen2 during mammalian neural stem cell divisions promotes lineage progression. Cell Stem Cell. 2012;11(4):505 16.

49. Bilogan CK, Horb ME. Xenopus staufen2 is required for anterior endodermal organ formation. Genesis. 2012;50(3):251–9.

50. Cockburn DM, Charish J, Tassew NG, Eubanks J, Bremner R, Macchi P, et al. The double-stranded RNA-binding protein Staufen 2 regulates eye size. Mol Cell Neurosci. 2012;51(3–4):101–11.

51. Zhang X, Trepanier V, Beaujois R, Viranaicken W, Drobetsky E, DesGroseillers L. The downregulation of the RNA-binding protein Staufen2 in response to DNA damage promotes apoptosis. Nucleic Acids Res. 2016;44(8):3695–712.

52. Harper JV. Synchronization of cell populations in G1/S and G2/M phases of the cell cycle. Methods Mol Biol. 2005;296:157–66.

53. Martel C, Dugre-Brisson S, Boulay K, Breton B, Lapointe G, Armando S, et al. Multimerization of Staufen1 in live cells. RNA. 2010;16(3):585–97.

54. Vassilev LT. Cell cycle synchronization at the G2/M phase border by reversible inhibition of CDK1. Cell Cycle. 2006;5(22):2555–6.

55. Boulay K, Ghram M, Viranaicken W, Trepanier V, Mollet S, Frechina C, et al. Cell cycle-dependent regulation of the RNA-binding protein Staufen1. Nucleic Acids Res. 2014;42(12):7867–83.

56. Hornbeck PV, Kornhauser JM, Tkachev S, Zhang B, Skrzypek E, Murray B, et al. PhosphoSitePlus: a comprehensive resource for investigating the structure and function of experimentally determined post-translational modifications in man and mouse. Nucleic Acids Res. 2012;40(Database issue):D261–70.

57. Li P, Yang X, Wasser M, Cai Y, Chia W. Inscuteable and Staufen mediate asymmetric localization and segregation of prospero RNA during drosophila neuroblast cell divisions. Cell. 1997;90(3):437–47.

58. Suzuki K, Sako K, Akiyama K, Isoda M, Senoo C, Nakajo N, et al. Identification of non-ser/Thr-pro consensus motifs for Cdk1 and their roles in mitotic regulation of C2H2 zinc finger proteins and Ect2. Sci Rep. 2015;5:7929.

59. Luo M, Duchaine TF, DesGroseillers L. Molecular mapping of the determinants involved in human Staufen-ribosome association. Biochem J. 2002;365(Pt 3):817–24.

60. Wickham L, Duchaine T, Luo M, Nabi IR, DesGroseillers L. Mammalian staufen is a double-stranded-RNA- and tubulin-binding protein which localizes to the rough endoplasmic reticulum. Mol Cell Biol. 1999;19(3):2220–30.

61. Kiebler MA, Hemraj I, Verkade P, Kohrmann M, Fortes P, Marion RM, et al. The mammalian staufen protein localizes to the somatodendritic domain of cultured hippocampal neurons: implications for its involvement in mRNA transport. J Neurosci. 1999;19(1):288–97.

62. Cao Y, Du J, Chen D, Wang Q, Zhang N, Liu X, et al. RNA- binding protein Stau2 is important for spindle integrity and meiosis progression in mouse oocytes. Cell Cycle. 2016;15(19):2608–18.

8

EphA receptors and ephrin-A ligands are upregulated by monocytic differentiation/maturation and promote cell adhesion and protrusion formation in HL60 monocytes

Midori Mukai, Norihiko Suruga, Noritaka Saeki and Kazushige Ogawa* ⓘ

Abstract

Background: Eph signaling is known to induce contrasting cell behaviors such as promoting and inhibiting cell adhesion/spreading by altering F-actin organization and influencing integrin activities. We have previously demonstrated that EphA2 stimulation by ephrin-A1 promotes cell adhesion through interaction with integrins and integrin ligands in two monocyte/macrophage cell lines. Although mature mononuclear leukocytes express several members of the EphA/ephrin-A subclass, their expression has not been examined in monocytes undergoing during differentiation and maturation.

Results: Using RT-PCR, we have shown that EphA2, ephrin-A1, and ephrin-A2 expression was upregulated in murine bone marrow mononuclear cells during monocyte maturation. Moreover, EphA2 and EphA4 expression was induced, and ephrin-A4 expression was upregulated, in a human promyelocytic leukemia cell line, HL60, along with monocyte differentiation toward the classical CD14^{++}CD16$^-$ monocyte subset. Using RT-PCR and flow cytometry, we have also shown that expression levels of αL, αM, αX, and β2 integrin subunits were upregulated in HL60 cells along with monocyte differentiation while those of α4, α5, α6, and β1 subunits were unchanged. Using a cell attachment stripe assay, we have shown that stimulation by EphA as well as ephrin-A, likely promoted adhesion to an integrin ligand-coated surface in HL60 monocytes. Moreover, EphA and ephrin-A stimulation likely promoted the formation of protrusions in HL60 monocytes.

Conclusions: Notably, this study is the first analysis of EphA/ephrin-A expression during monocytic differentiation/maturation and of ephrin-A stimulation affecting monocyte adhesion to an integrin ligand-coated surface. Thus, we propose that monocyte adhesion via integrin activation and the formation of protrusions is likely promoted by stimulation of EphA as well as of ephrin-A.

Keywords: Monocytes, HL60, EphA, Ephrin-A, Cell adhesion, Differentiation

Background

Eph receptors and ephrin ligands are membrane proteins that primarily regulate cell-cell repulsion and adhesion as well as cell adhesion and movement by modulating the organization of the actin cytoskeleton mainly via Rho family GTPases [1]. In mammals, the Eph receptor tyrosine kinase family has 14 members that are divided into the EphA (A1–A8 and A10) and EphB (B1–B4 and B6) subclasses based on the sequence similarity of their extracellular domains. The members of these two receptor subclasses promiscuously bind the ligands of the ephrin-A (A1–A5) and -B (B1–B3) classes, respectively. Ephrin-A ligands are anchored to the plasma membrane through a glycosyl phosphatidylinositol linkage, whereas ephrin-B ligands are a class of transmembrane proteins. Interaction of Eph receptors with ephrin results in bidirectional signaling in both the receptor- and ligand-expressing cells. Forward signaling by Eph mainly depends on autophosphorylation and phosphorylation by other tyrosine kinases as well as by the

* Correspondence: kogawa@vet.osakafu-u.ac.jp
Laboratory of Veterinary Anatomy, Graduate School of Life and Environmental Sciences, Osaka Prefecture University, 1-58 Rinku-Ourai-Kita, Izumisano, Osaka 598-8531, Japan

association of the receptor with various signaling proteins, whereas reverse signaling by ephrin largely depends on Src family kinases [2, 3].

Integrins, a large family of cell adhesion molecules, bind to proteins as ligands in the extracellular matrix and on the cell surface. Integrins are heterodimeric transmembrane proteins composed of α and β integrin subunits, and 18 α subunits and 8 β subunits have been identified in humans to date, thus generating 24 heterodimers [4]. Notably, integrins can transform their conformation from a bent inactive form to an extended closed form (intermediate ligand affinity) and further to an extended open conformation (high ligand affinity) in response to stimulation from other receptors. Integrins play important roles during leukocyte chemotaxis, infiltration, and migration, and conformational changes have been studied intensively for the leukocyte integrin, LFA-1 (αLβ2). In this context, chemokine receptors, upon binding chemokines, rapidly induce integrin activation by activating Rap1 small GTPase, a key regulator of integrins and many other molecules involved in chemokine-driven signaling cascades in leukocytes [5–7]. These signals resulting in integrin activation are termed "inside-out signaling." Once activated and bound to their ligands, integrins generate intracellular signals referred to as "outside-in signaling," which in turn alter various cellular functions such as cell motility and proliferation involving the activation and/or recruitment of signaling molecules such as Src, phosphatidylinositol 3-kinase (PI3K), and Rho family GTPases. Moreover, in their activated state, integrins can regulate their own molecular configuration, such as their clustering and stabilization in focal adhesions, allowing them to continue the regulation of downstream signaling [5–7].

Evidence indicating crosstalk between Eph/ephrin signaling and integrin signaling has accumulated recently [3, 8]. Integrin inside-out and outside-in signaling involves the activation and/or recruitment of upstream/downstream signaling molecules such as Rap1, Rho family GTPases, Src, PI3K, and focal adhesion kinase (FAK), all of which also serve as key players in the downstream cascade mediated by Eph/ephrin signaling [3]. While these receptor-regulated pathways appear to overlap in some ways, to our knowledge, studies investigating the interaction between Eph/ephrin and integrin/integrin ligand signaling on cell adhesion are insufficient, particularly in terms of leukocytes and related cells, and many of the reported investigations are conflicting. For example, in human T cells, β1- and β2 integrin-mediated adhesion to integrin ligands has been shown to be stimulated by ephrin-A activation and inhibited by EphA [9]. However, ephrin-A signaling significantly reduced adhesion to integrin ligand-coated surfaces in addition to impairing

chemokine-mediated trans-endothelial migration in chronic lymphocytic leukemia cells [10], whereas EphA signaling increased β1-integrin-mediated adhesion to an integrin ligand-coated surface in dendritic cells [11]. Thus, additional research is warranted to elucidate whether EphA and ephrin-A promote or inhibit integrin mediated cell adhesion specifically in mononuclear leukocytes (T cells, B cells, and monocytes) since these are known to express several members of the EphA/ephrin-A subclass [12–16]. The goal of this study is to clear up these uncertainties using a monocyte differentiation model cell line.

To our knowledge, EphA and ephrin-A subclass expression has not been examined in monocytes during differentiation even though mature monocytes express several members in the EphA/ephrin-A subclass [12]. The human promyelocytic leukemia cell line, HL60, has been used as a monocyte differentiation model for mechanistic studies, and monocytes differentiated from HL60 (HL60 monocytes) are frequently used as monocyte substitutes [17–19]. Thus, we examined the expression of EphA/ephrin-A in HL60 cells during monocytic differentiation as well as in bone marrow mononuclear cells during monocyte maturation. Recently, using a cell attachment stripe assay, we demonstrated that EphA2, upon stimulation with ephrin-A1, promotes cell adhesion through interactions between integrins and integrin ligands in monocyte/macrophage cell lines and, further, that the ectodomain of truncated EphA2 itself induces cell adhesion [20, 21]. Since monocytes express certain members of EphA as well as ephrin-A, we examined whether ephrin-A stimulation affects cell adhesion to integrin ligand-coated surface in HL60 monocytes. This is the first study demonstrating that certain members of EphA and ephrinA are upregulated during monocyte differentiation/maturation and that ephrin-A, upon stimulation with EphA2, likely promotes adhesion through integrin-integrin ligand interaction in monocytes.

Methods
Animals
Male ICR mice at 7 weeks of age, maintained under standard housing and feeding conditions, were used for isolation of bone marrow mononuclear cells (MNCs). The animal experimentation protocol was approved by the Animal Research Committee of the Osaka Prefecture University.

Isolation of bone marrow mononuclear cells and their differentiation/maturation into adherent monocytes
The femurs, tibias, and humeri were aseptically removed from male ICR mice, and the epiphyses of the bones were dissected. Bone marrow cavities were flushed with ice-cold Hank's balanced salt solution (HBSS; Sigma-

Aldrich, St Louis, MO, USA) using a syringe and a 23-gauge needle to collect bone marrow cells (BMCs). MNCs were isolated by density-gradient centrifugation according to the method described by Graziani-Bowering et al. [22] with some modifications. In brief, BMCs were fractionated by equilibrium density centrifugation on a discontinuous gradient of an iodixanol solution (Opti-Prep; Axis-Shield, Oslo, Norway). A 4 mL mixture containing BMCs and an iodixanol solution at a density of 1.090 g/mL was prepared by dilution with RPMI-1640 medium (Sigma-Aldrich). An iodixanol solution of 4 mL diluted with the medium at a density of 1.080 g/mL was overlaid on the mixture in a 15 mL plastic centrifuge tube followed by addition of 0.5 mL HBSS. The tube was allowed to stand upright at 4 °C for several min and was then centrifuged at $100 \times g$ for 20 min at 4 °C. MNCs fractionated between the iodixanol solution and HBSS were then collected. To remove the adherent cells including mature monocytes and macrophages in this fraction, MNCs at a density of 1×10^6 cells/mL were incubated overnight in a tissue culture dish with RPMI-1640 medium containing 10% heat-inactivated fetal bovine serum (FBS; Nichirei Biosciences, Tokyo, Japan), 100 U/mL penicillin, 100 μg/mL streptomycin (pen/strep; Sigma-Aldrich), and 5 ng/mL murine macrophage colony-stimulating factor (M-CSF; PeproTech, Rocky Hill, NJ, USA). Non-adherent MNCs were then seeded at a density of 3.2×10^5 cells/mL, cultured in medium containing 20 ng/mL M-CSF, and allowed to propagate and differentiate into monocytes. At day 1 after seeding, adherent cells were collected as samples (MC-1d), and at day 2, non-adherent cells were discarded and adherent MNCs were cultured with fresh medium for 3 more days (MC-5d). Adherent MNCs detached from the dish surface by pipetting were collected by centrifugation and used for nonspecific esterase (NSE) staining to identify monocytes and for RT-PCR analyses for the expression of the monocyte differentiation marker CD115 [23, 24] and the undifferentiated myeloid cell marker CD34 [25] to estimate the differentiation states between groups, and among members of the EphA/ ephrin-A subclass.

Differentiation of HL60 into monocytes

The human promyelocytic leukemia cell line, HL60, was obtained from the RIKEN BioResource Center (Ibaraki, Japan), cultured in suspension in RPMI-1640 supplemented with 10% FBS and pen/strep, and maintained in a 5% CO_2 atmosphere at 37 °C.

HL60 cells have been widely used as terminal differentiation models of monocytes, with 1α, 25-dihydroxyvitamin D_3 (VD) and TNFα as inducers of monocytic differentiation. Therefore, HL60 cells were differentiated to monocytes by stimulation with VD and/or TNFα, in accordance with previous studies [17–19]. Cells were

seeded at a concentration of 5×10^4 cells/mL in a tissue culture dish, treated with 50 nM VD (Sigma-Aldrich) dissolved in ethanol, and cultured for 3 days to allow differentiation (VD group). In some dishes, TNFα at 5 ng/mL (Roche Diagnostics, Mannheim, Germany) was added 2 days after VD addition and culture continued for 1 day (VD-TNF group). Control cultures were treated with the same volume of ethanol, reaching less than 0.1% (v/v) of the final volume (control group). At 3 days after seeding, the non-adherent cells and the adherent cells detached from the dish by pipetting were collected by centrifugation. These cells were used for NSE staining to determine the frequency of monocyte differentiation as well as for RT-PCR analyses to compare the expression of monocyte markers among the three groups; CD14, CD16, and CD115 were used as monocyte markers [23, 24] and CD15 was used as a marker of undifferentiated hematopoietic cells [26]. Expression levels of EphA receptors and ephrin-A ligands as well as the various integrin α/ß chains were also determined by RT-PCR in HL60 cells from the control and VD-TNF groups. Moreover, cell surface expression of the integrin α/ß subunit proteins whose expressions were identified by RT-PCR was examined by flow cytometry.

Nonspecific esterase staining analysis

Enzyme cytochemical staining for the fluoride-sensitive NSE enzyme, which has been widely used for identifying monocytes because of its simplicity, was performed to identify the adherent MNCs treated with M-CSF and the HL60 cells treated with VD and/or TNFα as monocytes according to the method described by Li et al. [27] with some modifications. Cell smears on glass slides were dried and fixed with 10% formalin and 60% acetone in PBS for 1 min at 4 °C. After washing with distilled water, the smears were incubated with a mixture of 10 mg/mL 1-naphthyl butyrate (Sigma-Aldrich) in ethylene glycol monomethyl ether (EGME; Sigma-Aldrich) and 0.5 mg/mL Fast Garnet GBC salt (Sigma-Aldrich) in 1/15 M phosphate buffer, pH 6.4 at a ratio of 1:20 for 40 min at 32 °C. Fluoride-sensitivity for NSE activity was verified by incubation in the reaction mixture with 40 mM sodium fluoride (Sigma-Aldrich) [28]. After washing, the cells were mounted in glycerol and photographed under a microscope (IX71; Olympus, Tokyo, Japan). To determine the differentiation efficiencies of HL60 cells treated with VD and/or TNFα, three random field images per smear sample were photographed for each group (control, VD, VD-TNF) using a 10 × objective lens. Cells stained brown to dark brown or almost black were defined as NSE-positive cells, i.e. monocytes, and cells stained yellow or pale brown were defined as NSE-negative cells. We counted 500 cells per image and, in total, 1500 cells per sample in each group. The frequencies of NSE-positive cells in each group were determined from three independent experiments.

RT-PCR analysis

Total RNA was isolated from adherent MNCs from two groups (MC-1d, MC-5d) and the HL60 cells from three groups (control, VD, VD-TNF) using TRIZOL reagent (Invitrogen, Carlsbad, CA, USA). RT-PCR analysis was performed as previously described [29]. Briefly, 1 μg of total RNA was transcribed into first-strand cDNA using M-MLV reverse transcriptase, RNase H$^-$ (Promega, Madison, WI, USA) and oligo (dT)$_{18}$ primers, according to the manufacturer's instructions. For detecting monocyte differentiation/undifferentiation markers (CD14, CD16, CD115/CD15), various α/β integrin chains (α1, α2, α4–6, αD, αL, αM, αX, β1, β2), Ras-related protein 1A (Rap1A), EphA (A1-A8, A10), ephrin-A (A1-A5), and GAPDH, 1 μL of the reaction mix (final 25 μL) was amplified using the reverse-transcribed cDNA as template. The mouse and human primer pairs and the cycle numbers for PCR amplification used in this study are shown in Tables 1 and 2, respectively. The RT reaction was omitted for the negative controls. The amplified mRNA expression levels were determined from three or four independent experiments and were normalized to those of GAPDH in the adherent MNCs between the two groups, to HL60 cells among the three groups, or between the control and VD-TNF groups.

Flow cytometry

We examined cell surface expression of α4, α5, α6, αL, αM, αX, β1, and β2 integrin subunit proteins by flow cytometry because they were clearly expressed, as detected by the RT-PCR analysis in HL60 cells of the control and VD-TNF groups. Cells were prepared at a concentration of 1×10^6 /50 μL in PBS containing 1% bovine serum albumin (BSA; Fraction V, Sigma-Aldrich) and 2 mM EDTA. To avoid non-specific Fc-gamma receptor-mediated binding of fluorochrome-conjugated antibodies, cell suspensions were pretreated with 20 μL of the human Fc receptor-binding inhibitor (20 μL/10^6 cells; Affymetrix, San Diego, CA, USA) for 20 min at 4 °C according to the manufacturer's instructions. To the cell suspensions, we added FITC-conjugated anti-α4 antibody (5 μL/10^6 cells; Miltenyi Biotec, Bergisch Gladbach, Germany), PE-conjugated anti-α5 antibody (5 μL/10^6 cells; Miltenyi Biotec), PE-conjugated anti-α6 antibody (5 μL/10^6 cells; Miltenyi Biotec), FITC-conjugated anti-αL antibody (5 μL/10^6 cells; Miltenyi Biotec), FITC-conjugated anti-αM antibody (0.25 μg /10^6 cells; Tonbo Biosciences, San Diego, CA, USA), APC-conjugated anti-αX antibody (20 μL/10^6 cells; BD Biosciences, San Jose, CA, USA), FITC-conjugated anti-ß1 antibody (5 μL/10^6 cells;

Miltenyi Biotec), or APC-conjugated anti-ß2 antibody (5 μL/10^6 cells; Miltenyi Biotec), and then incubated the cells for 20 min at room temperature. After washing, 50,000 cells were analyzed for their expression characteristics by using a flow cytometer (S3 Cell Sorter; Bio-Rad Laboratories, Hercules, CA, USA). We used cell suspensions pretreated with the human Fc receptor-binding inhibitor as controls. Some of the cell suspensions pretreated with the human Fc receptor-binding inhibitor were treated with an isotype control antibody (APC-conjugated mouse IgG1, 0.5 μg/10^6 cells; Tonbo).

Cell adhesion stripe assay and time-lapse microscopy

Because treatment with VD and TNF was most effective for monocytic differentiation of HL60 cells, we examined the adhesion of HL60 cells in the VD-TNF group to a Matrigel-coated coverslip surface on which EphA2-Fc or ephrin-A1-Fc was adsorbed in stripes according to the method described by Ogawa et al. [30] with some modifications. Briefly, coverslips (15 mm in diameter) were incubated for 3 h in 100 μg/mL poly-L-ornithine (Sigma-Aldrich) in PBS, washed with sterile water, and dried. Comb-shaped silicon masks with parallel teeth and gaps of about 0.48 mm width were then applied to the glass surface. EphA2-Fc (100 μL of 8 μg/mL in HBSS; R&D Systems, Minneapolis, MN, USA), ephrin-A1-Fc (R&D Systems), or human IgG Fc as a control (Fc; OEM Concepts, Inc., Toms River, NJ, USA) was then adsorbed onto the surface for 60 min. The coverslips were washed with HBSS and the masks were removed. After washing, Fc (100 μL of 8 μg/mL in HBSS) followed by Matrigel (100 μL of 40 μg/mL in HBSS; Corning Inc., Tewksbury, MA, USA) was adsorbed on the surface for 60 min. After washing, the coverslips were placed in 6 cm culture dishes with 5 mL RPMI 1640 containing 2% FBS. Cells were plated at a density of 3×10^5 cells/mL and were allowed to adhere for 16 h at 37 °C. Cells were then fixed in 4% paraformaldehyde in PBS for 15 min at room temperature. Phase-contrast images of the fields including both the Fc-chimera protein-adsorbed (test) and the adjacent Fc and Matrigel-adsorbed control regions were acquired using a 4 × and 10 × objective lens (IX71; Olympus). For quantitative analysis of cell density in the test versus the control regions, we counted cell numbers in an area of 0.48 mm × 0.96 mm in each of the test and control regions. This was repeated for each Fc-chimera protein tested. Cell densities were determined from three independent experiments and were normalized to those from the Fc plus Matrigel-only adsorbed control region.

Cell adhesion behaviors on coverslips coated with the EphA2-Fc or with ephrin-A1-Fc in stripes were also analyzed by time-lapse microscopy as previously described [29] with minor modifications. HL60 cells in the VD-

Table 1 Primers and cycle numbers for RT-PCR amplification of mouse mRNAs

Mouse	Primer		Product size (bp)	Annealing temp. (°C)	Cycle number
CD115	Forward	5-GGGACAGCACGAGAATATAG-3′	590	53.0	26
	Reverse	5′-CACTCTGAACTGTGTAGACG-3′			
CD34	Forward	5′-ACAACCACAGACTTCCCCAA-3′	424	59.5	33
	Reverse	5′-ATTGGCCAAGACCATCAGCA-3′			
EphA1	Forward	5′-TCGAGCCTTACGCCAACTAC-3′	466	69	38
	Reverse	5′-CCGATCAGCAGAGCTATTCC-3′			
EphA2	Forward	5′-GAGTGTCCAGAGCATACCCT-3′	549	62.5	37
	Reverse	5′-GCGGTAGGTGACTTCGTACT-3′			
EpA3	Forward	5′-TGTTGGTGCTTGTGTTGCC-3′	440	57	38
	Reverse	5′-CATTTCTTGGTGCGGATGG-3′			
EphA4	Forward	5′-GGGCAGTGAATGGAGTGTCT-3′	448	61.4	38
	Reverse	5′-AGAATGACCACGAGGACCAC-3′			
EphA5	Forward	5′-CTCTGGACGTGCCTTCTC-3′	319	57	38
	Reverse	5′-TCCCCAGTCCTCCAGGAA-3′			
EphA6	Forward	5′-GTGCGGAGCTTGGCTATGTT-3′	373	61.4	38
	Reverse	5′-GAGCTGAAGGTGGTCTAGTG-3′			
EphA7	Forward	5′-ACGGGGGAAGAAACGATGTC-3′	544	61.4	38
	Reverse	5′-GCTTCCTCAAGTGTGGCAAC-3′			
EphA8	Forward	5′-TTGTGGGAGTGGAACTCGCT-3′	632	61.4	38
	Reverse	5′-CCTGTCCGTTCTGGTAATGC-3′			
EphA10	Forward	5′-ACTACCTAGAAACCGAGACC-3′	493	53.0	38
	Reverse	5′-GACACCTTGTAAAACCCTGG-3′			
ephrin-A1	Forward	5′-CATCATCTGCCCACATTACG-3′	463	60.0	37
	Reverse	5′-AGCAGTGGTAGGAGCAATAC-3′			
ephrin-A2	Forward	5′-GGTGAGCATCAACGACTACC-3′	464	61.4	38
	Reverse	5′-CTGACACTAGGAGCCCAGAA-3′			
ephrin-A3	Forward	5′-TCCGCACTACAACAGCTCAG-3′	504	62.5	38
	Reverse	5′-TAGGAGGCCAAGAGCGTCAT-3′			
ephrin-A4	Forward	5′-AAGATTCAGCGCTACACACC-3′	420	61.4	38
	Reverse	5′-GATCCTCCGACTTTGCACAT-3′			
ephrin-A5	Forward	5′-CGTGGAGATGTTGACGCTG-3′	585	57.0	38
	Reverse	5′-GGCTCGGCTGACTCATGTA-3′			
GAPDH	Forward	5′-GACTCCACTCACGGCAAAT-3′	689	57.0	22
	Reverse	5′-TCCTCAGTGTAGCCCAAGAT-3′			

TNF group at a density of 3×10^5 cells/mL (1.5 mL/3.5-cm culture dish) were plated on coverslips coated with EphA2-Fc or ephrin-A1-Fc in stripes and placed in a 3.5-cm culture dish in an incubator (maintained at 37 °C in humidified 5% CO_2/95% air; ONI-INU-F1, Tokai Hit Co., Ltd., Fujinomiya, Japan) installed on the stage of an inverted microscope (IX71, Olympus). Phase contrast images were obtained at 2 min intervals (for 16 h) using a digital camera (DP72, Olympus) controlled by the manufacturer's software (DP2-BSW, Olympus).

Visualization of focal adhesions and F-actin

We examined the formation of focal adhesions and F-actin in the cells after incubation for 16 h on coverslips coated with EphA2-Fc or ephrin-A1-Fc in stripes, as described in the cell adhesion stripe assay, by fluorescence microscopy according to the method of Ogawa et al. [29] with some modifications. Briefly, HL60 cells from the VD-TNF group on the coverslips were fixed with 4% paraformaldehyde in PBS for 15 min at room temperature. The cells were then incubated with 0.02% Triton X-100 in PBS for 15 min at

Table 2 Primers and cycle numbers for RT-PCR amplification of human mRNAs

Human	Primer		Product size (bp)	Annealing temp. (°C)	Cycle number
CD14	Forward	5′-AGAACCTTGTGAGCTGGACGAT-3′	369	61.0	24
	Reverse	5′-GAAAGTGCAAGTCCTGTGGCTT-3′			
CD15	Forward	5′-TTTGGATGAACTTCGAGTCGCC-3′	360	60.0	30
	Reverse	5′-AAGCCAGGTAGAACTTGTAGCG-3′			
CD16	Forward	5′-GACTGAAGATCTCCCAAAGG-3′	668	52.3	24
	Reverse	5′-CCTTCCAGTCTCTTGTTGAG-3′			
CD115	Forward	5′-GTGGTAGAGAGTGCCTACTT-3′	342	53.4	30
	Reverse	5′-CCATATGACGCTTACCTCTG-3′			
integrin α1	Forward	5′-GTCTATCCACGGAGAAATGG-3′	417	50.0	32
	Reverse	5′-CTCACAGAGTCCTGAAAGTC-3′			
integrin α2	Forward	5′-CTACAATGTTGGTCTCCCAG-3′	440	50.0	32
	Reverse	5′-CAACATCTATGAGGGAAGGG-3′			
integrin α4	Forward	5′-GATCATCTTACTGGACTGGC-3′	320	54.8	32
	Reverse	5′-CAGATCTGAGAAGCCATCTG-3′			
integrin α5	Forward	5′-CCAGCCCTACATTATCAGAG-3′	390	50.0	32
	Reverse	5′-GAGATGAGGGACTGTAAACC-3′			
integrin α6	Forward	5′-GGAGATAAACTCCCTGAACC-3′	416	50.0	32
	Reverse	5′-CGAGAATAGCCACTAGGATG-3′			
integrin αD	Forward	5′-ATCAGCAGGCAGGAAGAATC-3′	518	50.1	32
	Reverse	5′-ACCTCGTCTTCTTCTAGCAC-3′			
integrin αL	Forward	5′-GCCCTGGTTTTCAGGAATG-3′	430	53.0	32
	Reverse	5′-CCAATCCCGATGATGTAGC-3′			
integrin αM	Forward	5′-GTGTGATGCTGTTCTCTACG-3′	364	50	32
	Reverse	5′-CTCCATGATTGCCTTGACTC-3′			
integrin αX	Forward	5′-CCAGATCACCTTCTTGGCTAC-3′	523	61.3	32
	Reverse	5′-CTTCAGGGTGAAATCCAGCTC-3′			
integrin β1	Forward	5′-TTCAAGGGCAAACGTGTGAG-3′	459	54.4	28
	Reverse	5′-CCGTGTCCCATTTGGCATTC-3′			
integrin β2	Forward	5′-CAATAAACTCTCCTCCAGGG-3′	522	52.5	28
	Reverse	5′-CAGTACTGCCCGTATATCAG-3′			
Rap1A	Forward	5′-GTACAAGCTAGTGGTCCTTG-3′	371	50.0	29
	Reverse	5′-CAACTACTCGCTCATCTTCC-3′			
EphA1	Forward	5′-TGGTGGAACTTCCTTCGAGA-3′	446	61.2	35
	Reverse	5′-ACAATCCCAAAGCTCCACAC-3′			
EphA2	Forward	5′-CGCCGGCTCTGATGCACCTT-3′	312	62.0	38
	Reverse	5′-TCCTGAGGTGCCCGGAAGAA-3′			
EphA3	Forward	5′-TGCGGTCAGCATCACAACTA-3′	380	59.3	38
	Reverse	5′-TTGCTACTGCCGCTGAAATG-3′			
EphA4	Forward	5′-TGCAGCTTTTGTCATCAGCC-3′	381	60.3	38
	Reverse	5′-TTAGTGACCACGCCTTCCAA-3′			
EphA5	Forward	5′-CTCTGGACGTGCCTTCTC-3′	319	56.6	38
	Reverse	5′-TCCCCAGTCCTCCAGGAA-3′			
EphA6	Forward	5′-TGCGAAGTCCGGGAATTTCT-3′	627	61.6	38
	Reverse	5′-ACGAACAGTGAAGGGGCATT-3′			

Table 2 Primers and cycle numbers for RT-PCR amplification of human mRNAs *(Continued)*

EphA7	Forward	5'-AAAGCTGACCAAGAAGGCGA-3'	311	62.0	38
	Reverse	5'-TCAAACTGCCCCATGATGCT-3'			
EphA8	Forward	5'-ATCACCTACAATGCCGTGTG-3'	341	58.4	38
	Reverse	5'-TACTCCAGGATGATGCCGTT-3'			
EphA10	Forward	5'-CACGTACCAAGTGTGCAATG-3'	646	56.8	38
	Reverse	5'-CAACTGGATACCTTCGCAGA-3'			
ephrin-A1	Forward	5'-GGAACCCAGACCCATAGGAG-3'	825	59.7	38
	Reverse	5'-CCCGTTTTGAGGCTGCTAGG-3'			
ephrin-A2	Forward	5'-TACGTGCTGTACATGGTCAACG-3'	316	58.1	38
	Reverse	5'-GGCTGCTACACGAGTTATTGCT-3'			
ephrin-A3	Forward	5'-TTTACTGCCCGCACTACAACAG-3'	523	59.7	38
	Reverse	5'-AACGTCATGAGGAAGAAGGCGA-3'			
ephrin-A4	Forward	5'-GCCCCGAGACGTTTGCTTTGTA-3'	364	62.5	34
	Reverse	5'-TTGTCGGTCTGAATTGGCACCC-3'			
ephrin-A5	Forward	5'-ACTCTCCAAATGGACCGCTGAA-3'	363	66.0	38
	Reverse	5'-TCAAAAGCATCGCCAGGAGGAA-3'			
GAPDH	Forward	5'-GTCGGAGTCAACGGATTTGG-3'	607	57.2	21
	Reverse	5'-GGATGATGTTCTGGAGAGCC-3'			

room temperature. To visualize focal adhesions, cells were pre-incubated with 1% BSA in PBS for 30 min in a humidified chamber, followed by incubation with an anti-human vinculin monoclonal antibody (hVIN-1, Sigma-Aldrich) at a dilution of 1:200 in 1% BSA-PBS for 60 min at 32 °C. After washing with PBS, the cells were incubated with a mixture of Alexa 488-conjugated goat anti-mouse IgG (5 µg/mL; Molecular Probes, Inc., Eugene, OR, USA) and 165 nM Alexa 568-labeled phalloidin (to visualize F-actin; Invitrogen) in 1% BSA-PBS for 30 min at 32 °C, followed by washing with PBS and mounting with Perma-Fluor (Thermo Fisher Scientific, Fremont, CA, USA). The cells were photographed under a fluorescence microscope (IX71; Olympus).

Statistical analysis
Statistical analyses were performed with the statistical software package Statcel 2 (OMS Publishing Inc., Tokorozawa, Japan). The bar graphs represent means ± SD. Unpaired *t*-test was used to determine the statistical significance of the results. P values less than 0.05 were considered significant.

Results
EphA and ephrin-A are upregulated in bone marrow mononuclear cells during monocytic maturation
M-CSF induces proliferation and differentiation of bone marrow MNCs into the mononuclear phagocytic lineage, wherein the M-CSF receptor signaling is involved in cell adhesion to extracellular matrices [31]. Bone marrow

MNCs, immediately after fractionation by equilibrium density centrifugation, consisted of many non-adherent/NSE-negative cells and some adherent/NSE-positive cells (Fig. 1a). Non-adherent/NSE-negative MNCs were selected and differentiated to monocytes by treatment with M-CSF to examine the expression levels of EphA and ephrin-A during the monocyte maturation process. First, we examined NSE-reactivity and the mRNA expression levels of marker molecules (CD115, a monocyte marker [23, 24], CD34, a myeloid cell marker [25]) in adherent MNCs by semi-quantitative RT-PCR in both groups (MC-1d group: non-adherent MNCs cultured in M-CSF-containing medium which became adherent in one day; MC-5d group: non-adherent MNCs cultured in M-CSF-containing medium which became adherent in two days and were cultured for three more days). All cells from both groups became NSE-positive (Fig. 1a), and the expression level of CD115 in the MC-5d group was elevated 2.80-fold and significantly higher than that in the MC-1d group (P = 0.002), whereas the expression levels of CD34 were similar between both groups (Fig. 1b). These findings indicate that adherent MNCs in these two groups likely belong to the monocyte lineage undergoing maturation and that the cells from the MC-5d group are closer to mature monocytes.

We screened EphA and ephrin-A mRNA expression in the adherent MNCs of two groups by RT-PCR to examine whether these molecules were up- or downregulated during monocytic maturation. EphA2, EphA4, ephrin-A1, ephrin-A2, and ephrin-A4 were detected in the adherent

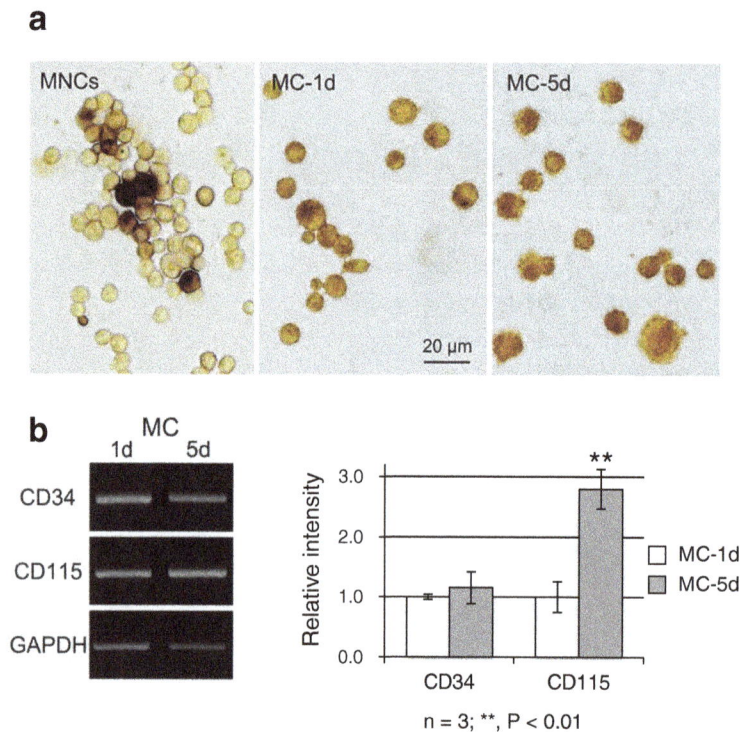

Fig. 1 M-CSF induces differentiation of non-adherent bone marrow MNCs to adherent cells of the monocytic lineage. **a** NSE staining of mouse bone marrow MNCs immediately after fractionation and of adherent MNCs from the MC-1d group (non-adherent MNCs cultured in medium containing 20 ng/mL M-CSF that adhered within one day) and MC-5d group (non-adherent MNCs cultured in medium containing 20 ng/mL M-CSF that adhered within two days and were cultured for three more days). **b** RT-PCR amplification of the myeloid cell marker CD34, and the monocyte marker CD115 in bone marrow MNCs from the MC-1d and MC-5d group. Densitometric quantification of mRNA expression levels from three independent experiments, normalized to GAPDH, is shown as the means ± SD. The expression of CD115 in the MC-5d group is upregulated 2.80-fold ($P = 0.002$) compared to that in the MC-1d group, whereas CD34 expression levels are almost the same between both groups

MNCs of both groups. EphA2 and ephrin-A1 expression was distinctly higher among the EphA and ephrin-A members compared to that of other receptors and ligands, respectively, as observed by stronger band intensities despite one less cycle of PCR amplification (Fig. 2a). EphA7 was detected in the MC-1d group but was decreased to a level less than the detection threshold of PCR amplification in the MC-5d group. This indicates that EphA7 expression was likely downregulated during monocyte maturation. We compared mRNA expression in adherent MNCs between the two groups. The expression levels of EphA2, ephrin-A1, and ephrin-A2 in the MC-5d group were 2.10, 3.66, and 2.56-fold higher, respectively, and were significantly increased compared to those in the MC-1d group ($P = 0.014$, $P = 0.011$, $P = 0.001$; Fig. 2b). There were no significant differences in the expression levels of EphA4 and ephrin-A4 between the two groups, although they tended to increase in the MC-5d group (Fig. 2b).

Treatment with VD plus TNFα efficiently differentiates HL60 cells into monocytes

HL60 cells have been widely used as a terminal differentiation model of monocytes, and VD and TNFα are used to induce monocytic differentiation [17–19]. HL60 cells adhered to tissue culture dishes and frequently ceased cell propagation in the VD group (HL60 cells cultured in VD-containing medium for 3 days), and this was more prominent in the VD-TNF group (HL60 cells cultured in VD-containing medium for 2 days and thereafter in VD plus TNFα for 1 day). Thus, we determined the differentiation efficiencies of HL60 to monocytes in both groups cytochemically, by examining the fluoride-sensitive NSE activity that is specific for monocytes/macrophages. NSE-reactivity showed various color ranges (brown to dark brown or almost black) in the NSE-positive cells of the VD and VD-TNF groups (Fig. 3a). NSE activity in these cells was almost completely inhibited by the addition of 40 mM fluoride to the reaction medium, and the colors turned to yellow or pale brown in almost all cells. In contrast, almost all cells in the control group stained pale brown or yellow and were thus considered NSE-negative. Next, we compared the frequency of NSE-positive cells among the three groups. In the control group, the frequency of the NSE-positive cells that stained light brown in most cases was 1.5 ± 0.4% of the total (Fig. 3b). In the VD and VD-TNF group, the frequencies of NSE-positive

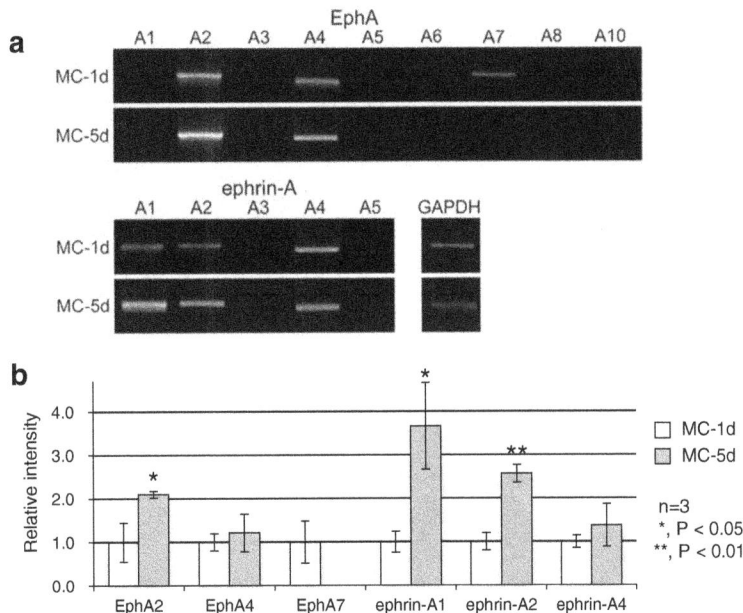

Fig. 2 RT-PCR amplification of EphA and ephrin-A mRNAs from mouse bone marrow MNCs treated with M-CSF. **a** MC-1d represents non-adherent MNCs cultured in medium containing 20 ng/mL M-CSF that adhered within one day, and MC-5d represents non-adherent MNCs cultured in medium containing 20 ng/mL M-CSF that adhered in two days and were cultured for three more days. Compared to other EphAs and ephrin-As, EphA2 and ephrin-A1 required one less cycle of amplification. **b** Densitometric quantification of mRNA expression levels from three independent experiments, normalized to GAPDH, is shown as the mean ± SD. The expression levels of EphA2, ephrin-A1, and ephrin-A2 in the MC-5d group were upregulated by 2.10, 3.66, and 2.56-fold, respectively ($P = 0.014$, $P = 0.011$, $P = 0.001$) compared to that in the MC-1d group, whereas EphA7 expression was downregulated in the MC-5d group and was less than the detection limit after 38 cycles of PCR amplification

cells that were dark brown or almost black in many cases were 46.4 ± 1.5% and 69.5 ± 2.7%, respectively, and were significantly increased compared to those in the control. The frequency of the NSE-positive cells in the VD-TNF group was increased by 1.50-fold and was significantly higher than that in the VD group ($P < 0.001$).

We then examined the mRNA expression levels of marker molecules (CD14, CD16, and CD115, monocyte markers [5, 23, 24]; CD15, a myeloid cell marker [26]) in HL60 cells from the three groups by semi-quantitative RT-PCR. CD14, CD16, and CD115 expression was not detected in the control group. CD16 was not clearly expressed in our RT-PCR analysis even with 30 amplification cycles, whereas CD14 and CD115 were readily detected in the VD and VD-TNF group (Fig. 4a). The expression level of CD14 in the VD-TNF group was increased 1.71-fold, significantly higher than that in the VD group ($P = 0.024$), whereas the expression level of CD115 was similar in the two groups (Fig. 4b). CD15 mRNA was detected in all groups, and its expression level was similar among all three groups (Fig. 4b). NSE staining and RT-PCR analyses showed that treatment with the combination of VD and TNFα was more effective for differentiation of HL60 into monocytes, which could be considered as classical monocytes (CD14^{++}CD16^{-}) [24].

Therefore, we used HL60 cells from the VD-TNF group for further experiments.

EphA2, EphA4, and ephrin-A4 are induced and/or upregulated during differentiation of HL60 cells into monocytes

We screened the mRNA expression of all members of the EphA and ephrin-A groups in HL60 cells from the three groups by RT-PCR to determine whether these receptors and ligands were expressed, and if they were upregulated or downregulated during monocytic differentiation. Among the EphA receptors and ephrin-A ligands, EphA1 and ephrin-A4 were detected in HL60 cells from the control group, whereas EphA1, EphA2, EphA4, and ephrin-A4 were detected in HL60 cells from the VD and VD-TNF groups (Fig. 5a). This indicates that EphA2 and EphA4 expression is likely induced during monocytic differentiation.

We compared the expression of these mRNAs in HL60 cells among the three groups. The expression of EphA1 was similar among the three groups (Fig. 5b). The expression levels of EphA2 and EphA4 induced in the VD and VD-TNF group were not significantly different between the two groups ($P = 0.491$, $P = 0.160$) although expression levels of both receptors in the VD-TNF group tended to increase slightly compared to

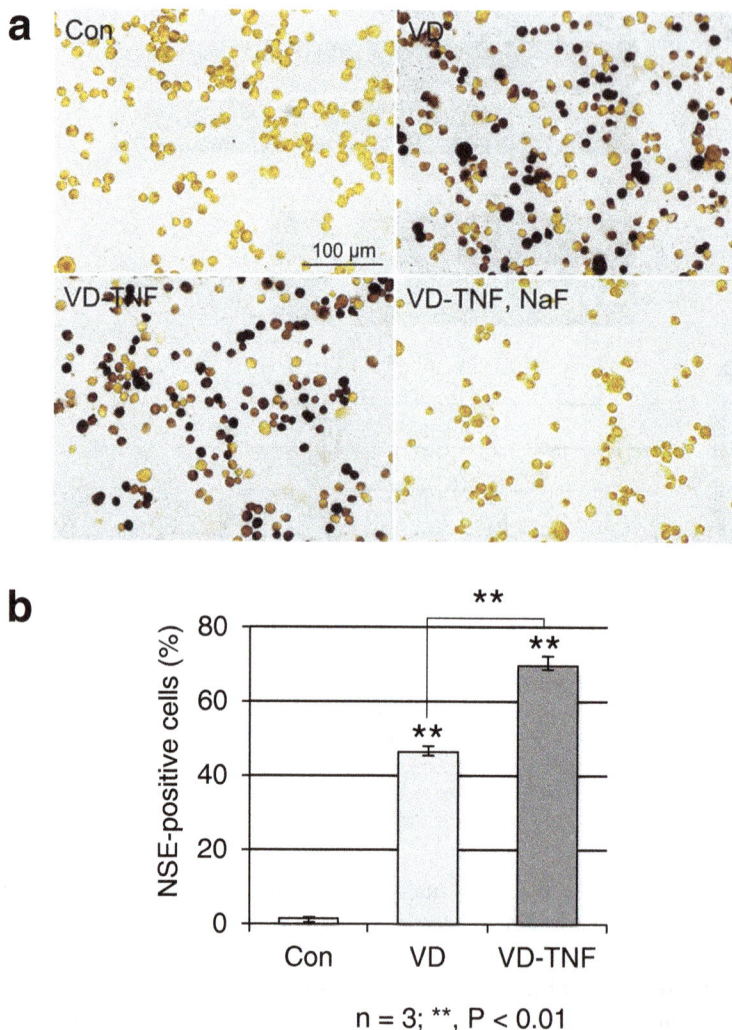

Fig. 3 Treatment with a combination of VD and TNFα effectively induces monocyte differentiation in HL60 cells. **a** NSE staining of HL60 cells treated with the vehicle (control group; Con), 50 nM VD (VD group; VD), and 50 nM VD with 5 ng/mL TNFα (VD-TNF group; VD-TNF). Note that NSE-reactivities in the VD-TNF group become negative by adding 40 mM fluoride (NaF) to the reaction medium, indicating that NSE-positive cells are possibly monocytes. **b** Frequencies of NSE-positive cells in the control, VD, and VD-TNF groups as determined from three independent experiments. Cells stained brown to dark brown or almost black were defined as NSE-positive cells. At random, three field images per smear sample, 500 cells per image, and thus 1500 cells per sample were counted in each group. NSE-positive cells were 46.4 ± 1.5% and 69.5 ± 2.7% in the VD and VD-TNF group, respectively, and are significantly increased compared to those in the control group (1.5 ± 0.4%). The frequency of NSE-positive cells in the VD-TNF group is 1.50-fold greater and statistically significant compared to that in the VD group. *$P < 0.01$

those in the VD group. The expression of ephrin-A4 in the VD-TNF group was elevated 2.08-fold and was significantly higher than that in the control ($P = 0.039$) whereas that in the VD group was not statistically different ($P = 0.060$) compared to that in the control (Fig. 5b). The overall expression patterns of EphA/ephrin-A were similar between the VD and VD-TNF groups. These results show that EphA2, EphA4, and ephrin-A4 expression are induced and/or upregulated during monocytic differentiation in HL60 cells whereas naïve cells express substantial amounts of EphA1 and ephrin-A4.

Integrin expression during monocytic differentiation of HL60 cells

Before examining the adhesive behavior of HL60 cells, we screened for the expression of various α and β integrin subunits (α1, α2, α4–6, αD, αL, αM, αX, β1, β2), and Rap1A in HL60 cells from the control and VD-TNF groups by RT-PCR. These integrins are generally expressed in leukocytes [4, 32], and Rap1A, a small GTPase, is an important regulator of conformational changes in integrins to the high affinity form for generating integrin ligands via external stimuli such as chemokines [5–7].

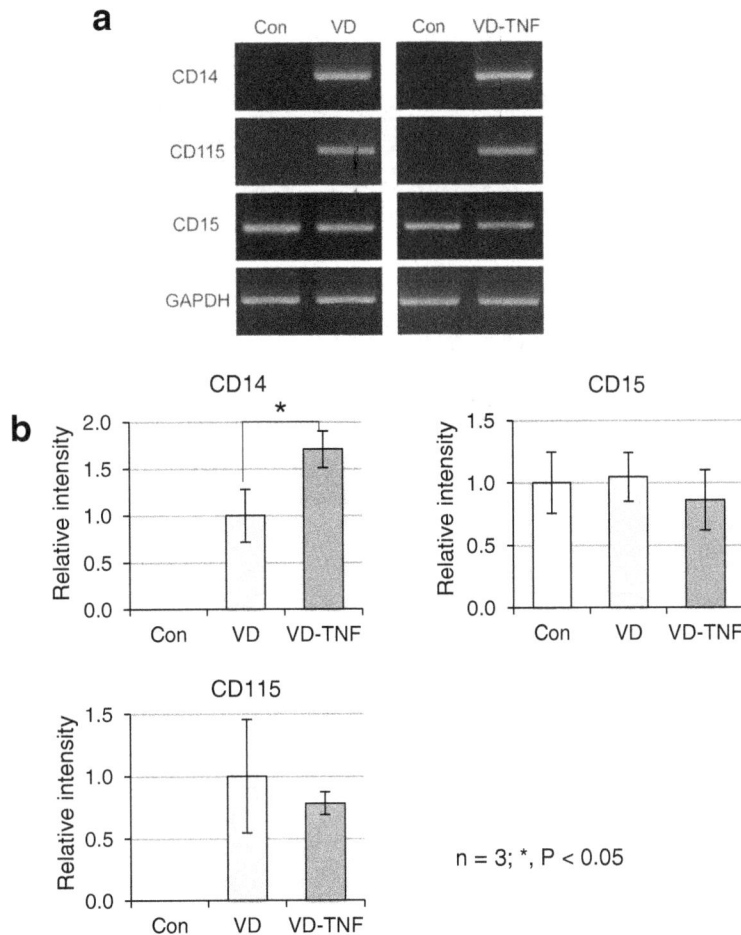

Fig. 4 RT-PCR amplification of monocyte markers and a myeloid cell marker in HL60 cells. **a** Treatment with VD and/or TNFα induces monocyte markers CD14 and CD115 expression in HL60 cells. **b** Densitometric quantification of mRNA expression levels from three independent experiments, normalized to GAPDH, is shown as means ± SD. The expression level of CD14 in the VD-TNF group is 1.71-fold and significantly higher than that in the VD group (P = 0.024) whereas the expression levels of CD115 are similar between both groups

The α1, α2, and αD integrin subunits were not clearly expressed in HL60 cells in our RT-PCR analysis up to 32 amplification cycles although the other examined integrin subunits were detected in 28 or 32 amplification cycles in HL60 cells from both groups. Relatively high expression of α4, α5, α6, αL, β1, and β2 integrin subunits was detected in HL60 cells of the control group. In contrast, high expression of α4, α5, α6, αL, αM, αX, β1, and β2 integrin subunits was detected in HL60 cells from the VD-TNF group (Fig. 6a). The expression levels for α4, α5, α6, and β1 subunits were almost the same between the control and VD-TNF group. In contrast, the expression levels of αL, αM, αX, and β2 subunits in the VD-TNF group were 1.38 ± 0.17, 4.25 ± 0.50, 4.93 ± 0.97, and 2.23 ± 0.27 fold higher than those in the control group, respectively (P = 0.008, P < 0.001, P = 0.004, P = 0.002; Fig. 6b). Moreover, high expression of Rap1A was detected in HL60 cells of both the

control and VD-TNF groups. We also examined cell surface expression of α4, α5, α6, αL, αM, αX, β1, and β2 integrin subunit proteins in HL60 cells of the control and VD-TNF group by flow cytometry. Expression levels of α4, α5, α6, and β1 subunits in HL60 cells were almost the same between the control and VD-TNF group (Fig. 7). In contrast, expression levels of αL, αM, αX, and β2 subunits in the VD-TNF group were obviously higher compared to those in the control, while the αX subunit was not clearly detected in HL60 cells of the control (Fig. 7). Thus relative expression levels of the cell surface proteins tested by flow cytometry were nearly identical to those of the mRNA by RT-PCR in each integrin subunit in HL60 cells between the control and VD-TNF group.

These data indicated that naïve HL60 cells likely express relatively large amounts of α4β1 (CD49d/CD29,

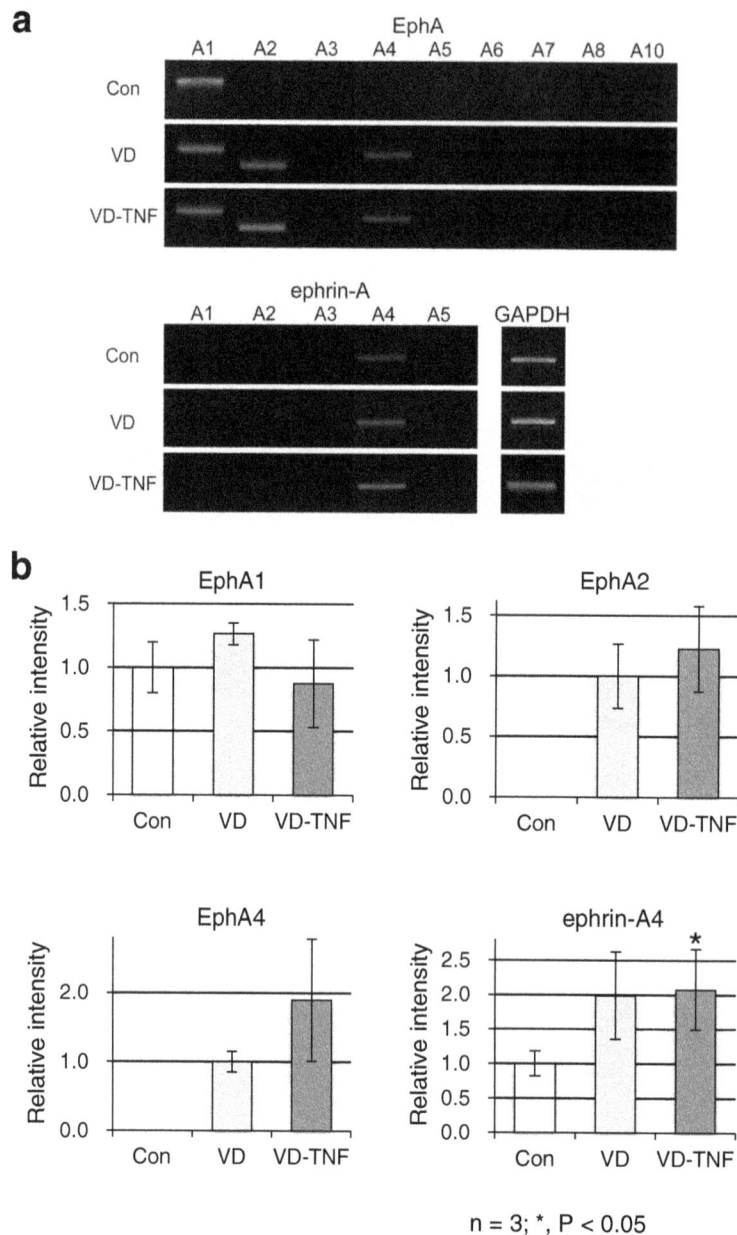

Fig. 5 RT-PCR amplification of EphA and ephrin-A mRNAs in HL60 cells. HL60 cells were treated with the vehicle (control group; Con), 50 nM VD (VD group; VD), and 50 nM VD with 5 ng/mL TNFα (VD-TNF group; VD-TNF). **a** Treatments with VD and/or TNFα induce EphA2 and EphA4 expression in HL60 cells. **b** Densitometric quantification of mRNA expression levels from three independent experiments, normalized to GAPDH, is shown as the mean ± SD. The expression level of EphA1 was similar among the three groups whereas that of ephrin-A4 in the VD-TNF group was elevated 2.08-fold and was significantly higher than that in the control ($P = 0.039$)

VLA-4), $\alpha5\beta1$, $\alpha6\beta1$ and $\alpha L\beta2$ (CD11a/CD18, LFA-1) integrins with substantial levels of $\alpha M\beta2$ (CD11b/CD18, Mac-1) integrin. In contrast, HL60 monocytes likely express relatively large amounts of $\alpha4\beta1$, $\alpha5\beta1$, $\alpha6\beta1$, $\alpha L\beta2$, $\alpha M\beta2$, and $\alpha X\beta2$ (CD11c/CD18) integrins. Moreover, the integrin regulator Rap1A is expressed in HL60 cells from both groups. These results indicated that HL60 monocytes as well as naïve HL60 cells likely possess sufficient adhesive ability towards integrin ligands, including membrane proteins such as ICAM-1 and VCAM-1, and extracellular matrix proteins, including laminin and collagen, based on the ligand specificities of the expressed integrins ($\alpha4\beta1$: VCAM-1, MAdCAM-1, etc.; $\alpha5\beta1$: fibronectin, osteopontin, etc.; $\alpha6\beta1$: laminin, thrombospondin, etc.; $\alpha L\beta2$, $\alpha M\beta2$ and $\alpha X\beta2$: ICAM-1, collagen, etc.) [4, 33].

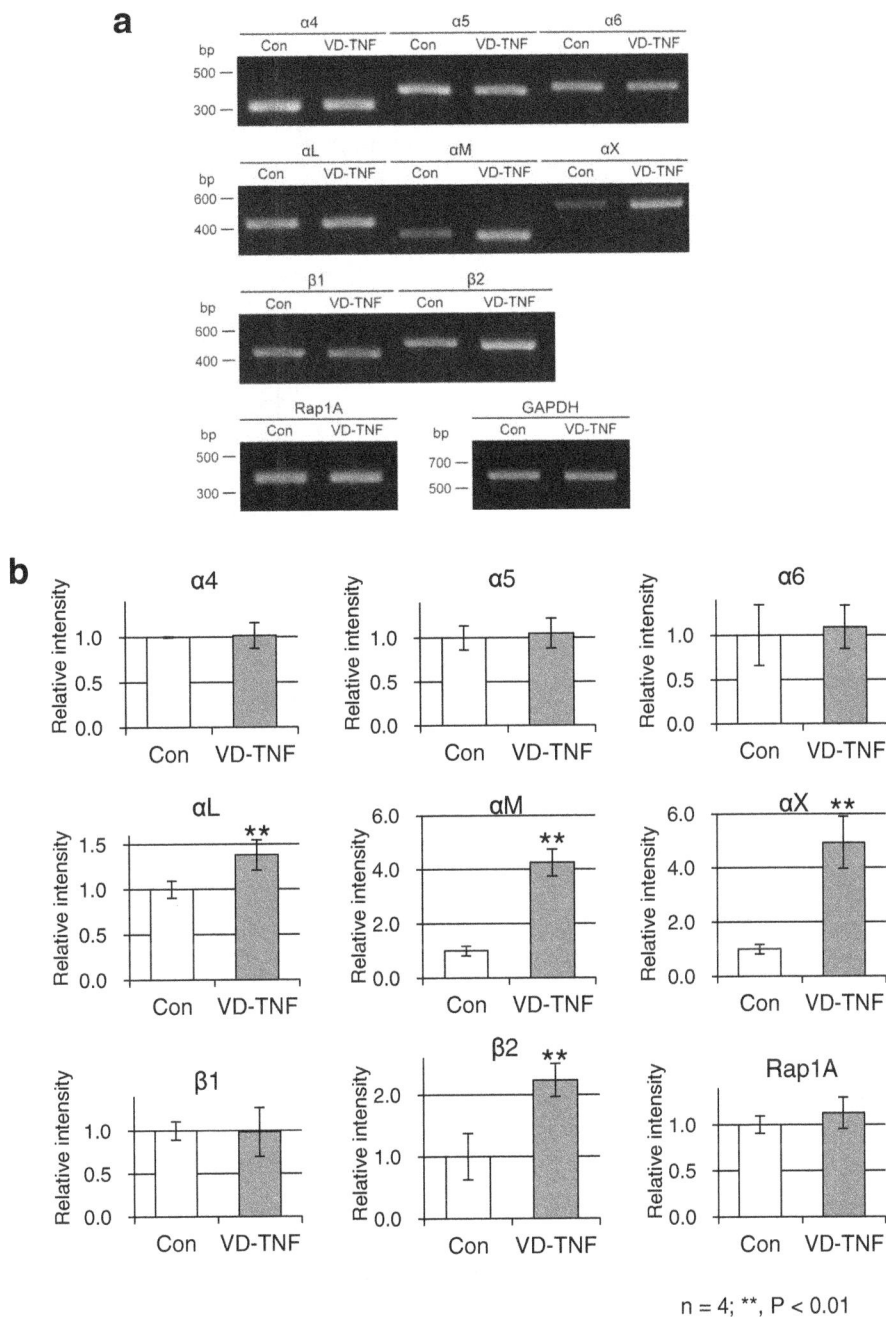

Fig. 6 RT-PCR amplification of integrin subunits and Rap1A mRNAs in HL60 cells. RT-PCR amplification (**a**) and densitometric quantification (**b**) of α4, α5, α6, αL, αM, αX, β1, and β2 integrin subunits as well as Rap1A expression in HL60 cells treated with the vehicle (Con) and 50 nM VD plus 5 ng/mL TNFα (VD-TNF). Data from four independent experiments, normalized to GAPDH, are shown as means ± SD. HL60 cells in the control group likely express relatively large amounts of α4β1, α5β1, and α6β1, αLβ2 integrins and substantial amounts of αMβ2 and αXβ2 integrins, whereas HL60 cells in the VD-TNF group likely express relatively large amounts α4β1, α5β1, α6β1, αLβ2, αMβ2, and αXβ2 integrins. *p < 0.05, **p < 0.01

EphA and ephrin-A activation likely promotes cell adhesion and formation of cytoplasmic protrusions in HL60 monocytes

To examine whether EphA/ephrin-A activation affects integrin-mediated cell adhesion to the extracellular matrix surface in HL60 cells during monocytic differentiation, we compared adhesive properties between HL60 cells of the control and the VD-TNF groups on coverslip surfaces to which EphA2-Fc or ephrin-A1-Fc was adsorbed in stripes and Fc plus Matrigel was overlaid. A test region of the Fc-chimera protein adsorbed surface presented in stripes at certain intervals on the Fc plus Matrigel-adsorbed

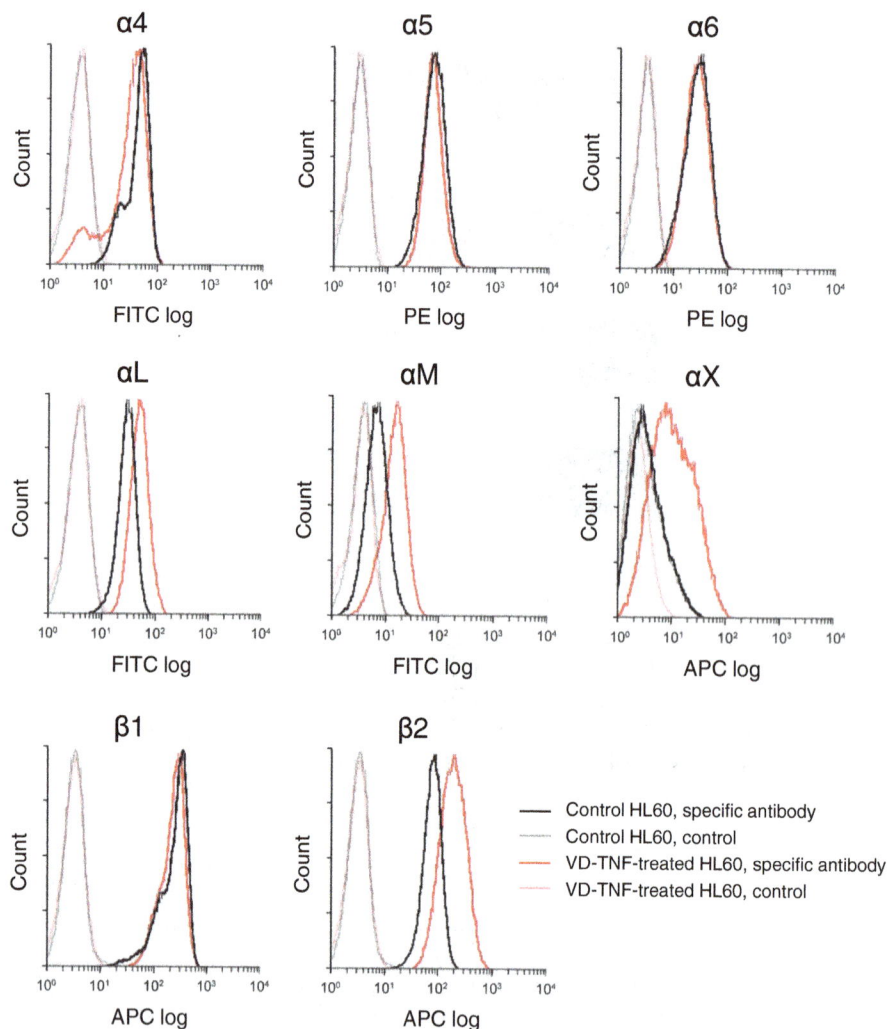

Fig. 7 Cell surface expression of integrin subunit proteins in HL60 cells. Representative histograms from flow cytometric analyses, showing the cell surface expression of α4, α5, α6, αL, αM, αX, β1, and β2 integrin subunits in HL60 cells of the control group (black line, specific antibody; gray line, control) and in the VD-TNF group (red line, specific antibody; pink line, control). For the control in each group, cell suspensions were pretreated with the human Fc receptor-binding inhibitor and with an APC-conjugated isotype control antibody in case of the APC-conjugated anti-αX, anti-β1, and anti-β2 antibody. All integrin subunits are clearly expressed except for the αX integrin subunit in HL60 cells of the control group

coverslip (Fig. 8a). Using phase-contrast microscopy, we found that HL60 cells from the VD-TNF group formed stripes with different cell densities corresponding to the stripes of the Fc-chimera protein-absorbed surface. They preferentially occupied ephrin-A1-Fc-adsorbed as well as EphA2-Fc-adsorbed test surfaces rather than only the Fc plus Matrigel-adsorbed control surface (Fig. 8b). In contrast, HL60 cells of the VD-TNF group were adhered to the surface of control coverslips on which Fc was adsorbed instead of the Fc-chimera protein, but did not form stripes of different cell densities (Fig. 8b). Moreover, HL60 cells of the control group scarcely adhered to the Fc-chimera protein-adsorbed test coverslips or the control coverslips. We calculated the cell densities of HL60 cells

from the VD-TNF group on the Fc-chimera protein-adsorbed region relative to those on the adjacent control region. The relative cell densities on EphA2-Fc and ephrin-A1-Fc-adsorbed regions compared to the control regions were increased 6.4 ± 1.1 and 4.5 ± 0.9 fold (means $\pm$ SD), respectively, and showed significant differences from the controls ($P < 0.001$, $P = 0.002$; Fig. 8c). These results, along with the RT-PCR findings, indicate that EphA2 and/or EphA4 activation induced by ephrin-A1-Fc as well as ephrin-A4 activation induced by EphA2-Fc in HL60 monocytes, likely promote cell adhesion to Matrigel, which contains large amounts of integrin ligands, mainly laminin and type IV collagen. Moreover, adhesiveness to Matrigel and its potentiation

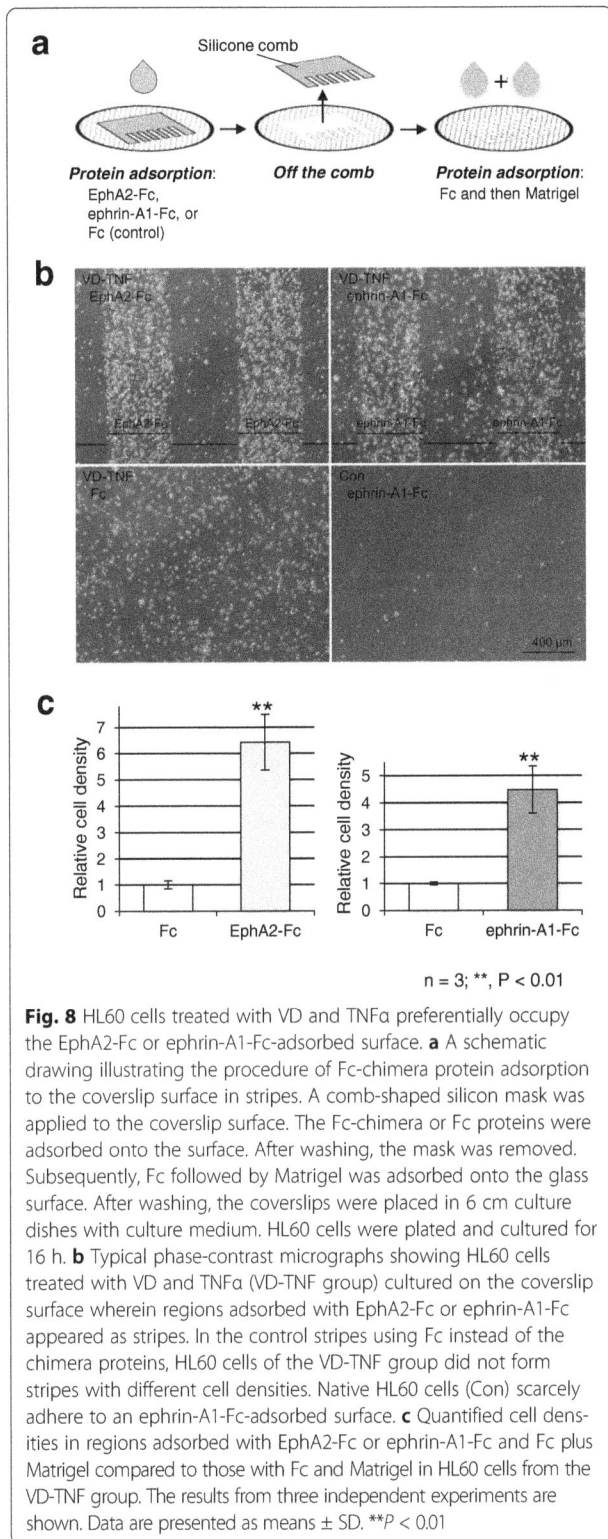

Fig. 8 HL60 cells treated with VD and TNFα preferentially occupy the EphA2-Fc or ephrin-A1-Fc-adsorbed surface. **a** A schematic drawing illustrating the procedure of Fc-chimera protein adsorption to the coverslip surface in stripes. A comb-shaped silicon mask was applied to the coverslip surface. The Fc-chimera or Fc proteins were adsorbed onto the surface. After washing, the mask was removed. Subsequently, Fc followed by Matrigel was adsorbed onto the glass surface. After washing, the coverslips were placed in 6 cm culture dishes with culture medium. HL60 cells were plated and cultured for 16 h. **b** Typical phase-contrast micrographs showing HL60 cells treated with VD and TNFα (VD-TNF group) cultured on the coverslip surface wherein regions adsorbed with EphA2-Fc or ephrin-A1-Fc appeared as stripes. In the control stripes using Fc instead of the chimera proteins, HL60 cells of the VD-TNF group did not form stripes with different cell densities. Native HL60 cells (Con) scarcely adhere to an ephrin-A1-Fc-adsorbed surface. **c** Quantified cell densities in regions adsorbed with EphA2-Fc or ephrin-A1-Fc and Fc plus Matrigel compared to those with Fc and Matrigel in HL60 cells from the VD-TNF group. The results from three independent experiments are shown. Data are presented as means ± SD. **P < 0.01

by ephrin-A and EphA activation likely appears during monocytic differentiation in HL60 cells, although naïve HL60 cells also express substantial amounts of EphA1, ephrin-A4, the integrins targeting the extracellular matrices, as well as Rap1A.

Under phase-contrast microscopy, two types of cells were observed on the EphA2-Fc or ephrin-A1-Fc and Fc plus Matrigel-adsorbed test surfaces, and on the Fc plus Matrigel-adsorbed control surface. The majority of cells were bright and round and a minor proportion of cells was observed to be dark and polymorphic, i.e., spread on the surfaces to some extent (Fig. 9a). Therefore, we then examined the adhesive behavior of HL60 cells from the VD-TNF group on EphA2-Fc or ephrin-A1-Fc and the integrin ligand-adsorbed surfaces by time-lapse microscopy, wherein phase-contrast images were obtained at 2 min intervals for 16 h. Cells were found to behave similarly on both the EphA2-Fc and ephrin-A1-Fc-adsorbed test surfaces. Cell density differences between the two regions, corresponding to the stripes of the Fc-chimera proteins, formed quickly within a few minutes after seeding and remained largely unchanged afterwards as observed by time-lapse microscopy. At a few hours after seeding and thereafter, many cells on the Fc plus Matrigel-adsorbed control regions moved relatively rapidly whereas the cells on the adjacent EphA2-Fc or ephrin-A1-Fc and Fc plus Matrigel-adsorbed test regions were mainly confined to a narrow area. In other words, when the cells moved from the control region to the adjacent test region, most cells abruptly stopped, temporarily spread, and remained in the restricted narrow area as if they were attached or tethered to these surfaces (see Additional file 1: Video S1 and Additional file 2: Video S2). Bright round cells that remained in a narrow area on the test surfaces frequently repeated the extension and retraction of cytoplasmic protrusions towards every direction in the X-Y plane (Fig. 9b). These results indicated that ephrin-A4 activation by EphA2-Fc and EphA2, and/or EphA4 activation by ephrin-A1-Fc likely promotes cell adhesion and induces extension and retraction of cytoplasmic protrusions in HL60 monocytes.

EphA and ephrin-A activation likely promote focal adhesion formation in HL60 monocytes

Because HL60 cells from the VD-TNF group were significantly more adherent to the EphA2-Fc or ephrin-A1-Fc and Fc plus Matrigel-adsorbed surfaces than on the Fc plus Matrigel-adsorbed control surfaces, we examined whether ephrin-A and EphA activation induce morphological changes in focal adhesions and in actin filament organization in HL60 monocytes. Focal adhesions and F-actin were visualized by fluorescence labeling using an anti-vinculin antibody and fluorescence-tagged phalloidin, respectively. These labeling studies revealed dramatic cellular changes on both test surfaces compared to those on the control surface. Cells that extended more widely and frequently, possessed few large

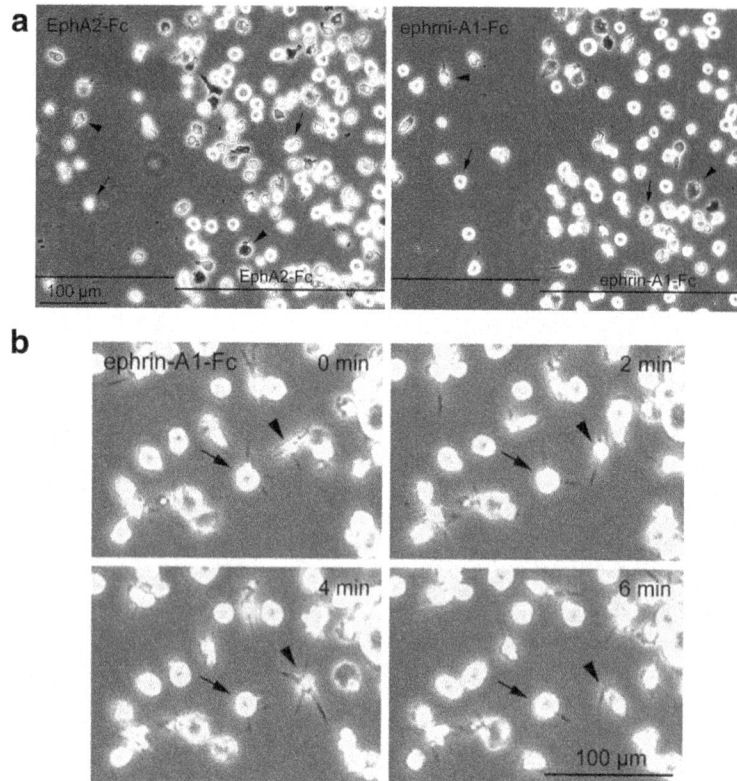

Fig. 9 Typical micrographs of HL60 cells from the VD-TNF group on EphA2-Fc or ephrin-A1-Fc-adsorbed surfaces. **a** HL60 cells from the VD-TNF group were cultured on a coverslip surface on which the regions adsorbed with Fc, EphA2-Fc or ephrin-A1-Fc and Fc plus Matrigel and those with Fc plus Matrigel appeared alternatively as stripes, and phase-contrast images were photographed. Bright/round (arrows) and dark/polymorphic cells (arrowheads) are present as the major and minor populations of cells, respectively, observed not only on the EphA2-Fc or ephrin-A1-Fc-adsorbed test surfaces but also on the control surface. **b** A series of time-lapse micrographs obtained at 2-min intervals in the ephrin-A1-Fc-adsorbed region showing the extension and retraction of a number of cytoplasmic protrusions from the cells (arrows and arrowheads)

protrusions accompanied by prominent F-actin on the Fc-chimera protein-adsorbed test surfaces (Fig. 10a). Focal adhesions labeled by vinculin appeared indistinct in cells on the control surface. In contrast, a large number of clear and prominent focal adhesions, visualized as dot-like structures accompanied by clear spots of F-actin, appeared in the cells on the test surfaces (Fig. 10b). Moreover, doughnut-shaped focal adhesions with F-actin cores were frequently observed in cells on the Fc-chimera protein-adsorbed test surfaces (see also Additional file 3: Fig. S1): they appeared prominently in cells on the ephrin-A1-Fc-adsorbed surface compared to those on EphA2-Fc-adsorbed surface. These results indicated that EphA activation by ephrin-A1-Fc, and ephrin-A activation by EphA2-Fc, likely promote the formation of focal adhesions and F-actin reorganization in HL60 monocytes.

Discussion

Eph receptors and their ligands, ephrins, primarily regulate cell adhesion and movement by modulating actin cytoskeleton organization mainly via Rho family GTPases [1]. Here,

we demonstrated upregulation of EphA2, ephrin-A1, and ephrin-A2 during monocyte maturation in mouse bone marrow MNCs, and the induction of EphA2 and EphA4 as well as upregulation of ephrin-A4 during monocyte differentiation from HL60 cells. Thus, during monocyte differentiation/maturation, functional regulation by EphAs and ephrin-As is possibly acquired as a new feature, likely required for cell adhesion and movement. Several members of the EphA/ephrin-A subclass are known to be expressed in monocytes, macrophages, and monocyte/macrophage cell lines from humans and mice. Sakamoto et al. showed that human peripheral blood monocytes clearly express EphA1, EphA2, EphA4, ephrinA3, ephrinA4, and ephrin-A5 [34]. We have also shown that a human monocytic cell line, U937, expresses EphA1, EphA2, EphA4, ephrin-A1, ephrin-A3, and ephrin-A4; a mouse monocyte/macrophage cell line, J774.1, expresses EphA2, EphA4, ephrin-A1, and ephrin-A4 [20, 21]. These and our present findings indicate that the expression patterns of EphA/ephrin-A subclass members are certainly different between human and murine monocytes/monocyte-related cells, although the expression of EphA2, EphA4, and ephrin-A4 are common to

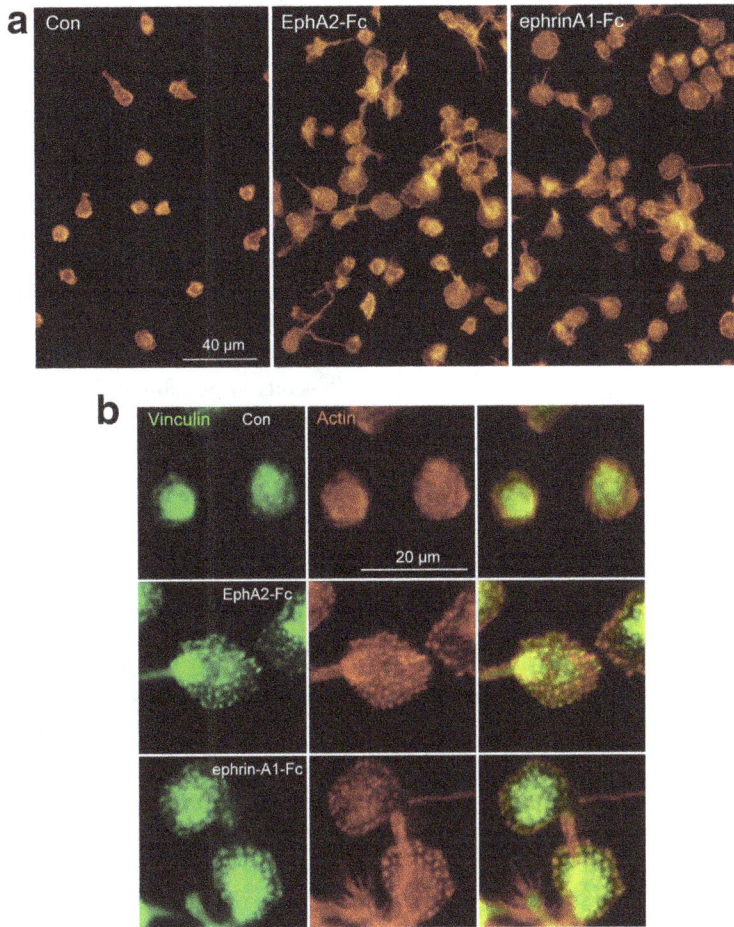

Fig. 10 Representative images of F-actin and vinculin highlighting HL60 cell morphology in the VD-TNF group. HL60 cells from the VD-TNF group were cultured on a coverslip surface on which the regions adsorbed with EphA2-Fc or ephrin-A1-Fc and Fc plus Matrigel and those with Fc plus Matrigel appeared alternatively as stripes. Fluorescence images of F-actin staining (red) and vinculin immunostaining (green) highlighting morphology of cells on the Fc and Matrigel-adsorbed control surface (Con), the EphA2-Fc and Fc plus Matrigel-adsorbed, and the ephrin-A1-Fc and Fc plus Matrigel-adsorbed test surfaces were photographed. **a** Cells extend wider and frequently show a number of large protrusions accompanied with prominent F-actin on the Fc-chimera-protein-adsorbed test surfaces. **b** Prominent dot-like focal adhesions labeled with anti-vinculin antibody were accompanied by spots of F-actin in the cells on the EphA2-Fc-adsorbed surface. Doughnut-shaped focal adhesions accompanied with F-actin in their cores frequently appeared in cells on the ephrin-A1-Fc-adsorbed surface

both. It may not be important to note the difference in EphA/ephrinA expression patterns in monocytes and monocyte-related cells between mice and humans, as the cell surface markers that identify the monocyte subsets are largely different between these species [24]. Because EphA receptors promiscuously bind ephrin-A ligands within the same subclass [2, 8], the induction and upregulation of EphA/ephrin-A during monocyte differentiation and maturation should be highlighted.

We have previously demonstrated that EphA stimulation with ephrin-A1-Fc promotes cell adhesion through interaction with integrins and integrin ligands in the J774.1 monocyte/macrophage cell line and in the U937 monocytic cell line [20, 21]. In the present study, we have demonstrated that HL60 monocytes preferentially migrate and adhere to surfaces that contain ligands for EphA, ephrin-A, and integrins using a cell adhesion stripe assay with alternating stripes of ephrin-A1-Fc (to trigger EphA signaling) or EphA2-Fc (to trigger ephrin-A signaling) and integrin ligands with regions containing integrin ligands alone. HL60 cells formed stripes of different cell densities on ephrin-A1-Fc stripes as well as on EphA2-Fc stripes on the integrin ligand-coated surfaces. Naïve HL60 cells did not adhere to the Fc-chimera protein plus integrin ligand-coated surface, indicating that physical tethering by binding of EphA with ephrin-A1-Fc or by the binding of ephrin-A with EphA2-Fc, is not a likely driving force behind stripe formation because naïve HL60 cells expressed a certain member of the EphA/ephrin-A subclass at substantial levels. Thus, the findings of the cell adhesion stripe assay indicate that the arrest of cellular movement may be driven by EphA as well as ephrin-A signaling in interactions with integrins. To our knowledge, this is the first evidence of ephrin-A promoting cell adhesion to integrin ligands in monocytes. The crosstalk between Eph/ephrin and integrin/integrin ligands has been investigated to a limited extent in blood cells and their related cells. For example, EphA4 activation in human T-cells by ephrin-A1-Fc is shown to inhibit cell adhesion to fibronectin-, VCAM-1-, and ICAM-1-coated surfaces whereas ephrin-A activation by EphA2-Fc promotes cell adhesion to integrin ligand-coated surfaces [9]. Cell adhesion to fibronectin appears to be increased in dendritic cells derived from CD34+ positive progenitors following EphA2 activation with ephrin-A3-Fc, probably through a β1 integrin activation pathway [11]. In chronic lymphocytic leukemia cells, ephrin-A4 activation by EphA2-Fc significantly reduced cell adhesion to fibronectin-, collagen-, laminin-, ICAM-1-, and VCAM-1-coated surfaces [10]. EphA4 was previously shown to be physically associated with the αIIbβ3 integrin in resting platelets, and this association appears to support the stable accumulation of platelets on collagen surfaces even under flow [35]. Thus, we propose that EphA as well as ephrin-A signaling drives cellular movement arrest and promotes integrin-mediated cell adhesion, at least in monocytes and related cells. The present findings of the cell adhesion stripe assay also suggest that ephrin-A and EphA stimulation likely induces inside-out signaling to regulate integrin activation. This is because (1) additives such as chemokines that activate the signaling molecules mediating inside-out signaling by integrins were not used in this assay, thus, most integrins likely remain in their inactive form on the cell surface under this in vitro experimental condition, and (2) Rap1 GTPase, an essential molecule functioning as a molecular switch operating the inside-out signaling of integrins mediated by chemokine/chemokine receptor signaling [5–7], also operates in the Eph and ephrin signaling cascade [2, 3]. Further investigations will be required to determine Rap1 activation in EphA and ephrin-A signaling during monocyte adhesion to integrin ligands.

Furthermore, we have shown that EphA stimulation by ephrin-A1-Fc and ephrin-A4 stimulation by EphA2-Fc likely induce the formation and retraction of cell protrusions composed of F-actin in HL60 monocytes on integrin ligand coated surfaces. We have previously shown that by binding with ephrin-A1-Fc, the truncated EphA2 lacking almost the entire cytoplasmic region, promotes cell spreading and/or elongation along with promoting F-actin formation in U937 and J774.1 cells [20, 21]. It is well accepted that Eph and ephrin primarily regulate cell adhesion and movement by modulating the organization of the actin cytoskeleton mainly via the Rho family GTPases [1]. Thus, EphA and ephrin-A signaling likely regulate the

formation of cell protrusions by modulating the activities of the Rho family GTPases in monocytes. We have also demonstrated that stimulation of both EphA and ephrin-A4 promotes the formation of prominent focal adhesions in HL60 monocytes on an integrin ligand coated surface. Moreover, doughnut-shaped focal adhesions that exhibit structural features similar to podosomes composed of F-actin dots surrounded by vinculin rings as visualized by phalloidin and vinculin staining [36], were frequently induced by EphA stimulation and occasionally by ephrin-A4 stimulation in HL60 monocytes on the integrin ligand-coated surface. Thus, molecular construction of focal adhesions is possibly different between those induced by EphA and ephrin-A signaling.

We have previously demonstrated that endogenous EphA stimulation with ephrin-A1-Fc promotes cell adhesion through interaction with integrins and integrin ligands and that, upon stimulation with ephrin-A1-Fc, the truncated EphA2 lacking almost the entire cytoplasmic region including the kinase domain, potentiates this adhesion and becomes associated with the integrin/integrin-ligand complex in J774.1 and U937 cells that express the truncated EphA2 construct [20, 21]. Here, we have demonstrated that ephrin-A1 stimulation likely promotes cell adhesion to the integrin ligand-coated surface in HL60 monocytes. Bone marrow stromal cells and vascular endothelial cells are known to express several members of EphA and ephrin-A subclasses [37–40]. Therefore, when monocytes encounter stromal and vascular endothelial cells, EphA as well as ephrin-A signaling is activated in monocytes, which possibly regulates integrin activities through several signaling pathways. Further studies are required to fully elucidate the EphA/integrin and the ephrin-A/integrin interactions underlying cell adhesion, migration, and infiltration of monocytes and their related cells.

Conclusions

We have shown that EphA2, ephrin-A1, and ephrin-A2 expression was upregulated in murine bone marrow MNCs during monocyte maturation. In addition, EphA2 and EphA4 expression was induced, and ephrin-A4 expression was upregulated in HL60 cells along with monocyte differentiation. Using a cell attachment stripe assay, we have shown that stimulation of EphA as well as ephrin-A likely promoted adhesion to an integrin ligand-coated surface in HL60 monocytes. Moreover, EphA and ephrin-A stimulation likely promoted the formation of protrusions in HL60 monocytes. Notably, this study is the first analysis of EphA/ephrin-A expression during monocytic differentiation/maturation and of ephrin-A stimulation affecting monocyte adhesion to an integrin ligand-coated surface. Thus, we propose that monocyte adhesion via integrin activation and the formation of protrusions is likely promoted by stimulation of EphA as well as ephrin-A.

Additional files

Additional file 1: Video S1. Cell adhesion behaviors on coverslips coated with EphA2-Fc in stripes. (MOV 7290 kb)

Additional file 2: Video S2. Cell adhesion behaviors on coverslips coated with ephrin-A1-Fc in stripes. HL60 cells in the VD-TNF group at a density of 3×10^5 cells/mL (1.5 mL/3.5-cm culture dish) were plated on coverslips coated with EphA2-Fc or ephrin-A1-Fc in stripes and Fc plus Matrigel overall, and placed in a 3.5 cm culture dish in an incubator (maintained at 37 °C in humidified 5% CO_2/95% air; ONI-INU-F1, Tokai Hit Co., Ltd., Fujinomiya, Japan) installed on the stage of an inverted microscope (IX71, Olympus). Phase contrast images were obtained at 2 min intervals (for 16 h) using a digital camera (DP72, Olympus) controlled by software (DP2-BSW, Olympus). Videos starting from a few min after cell seeding consist of a series of time-lapse micrographs in a field wherein a region adsorbed with EphA2-Fc (Additional file 1: Video S1) or ephrin-A1-Fc (Additional file 2: Video S2) and Fc plus Matrigel (left side) and that with Fc plus Matrigel (right side) appears adjacently. (MOV 7230 kb)

Additional file 3: Figure S1. Immunofluorescence micrographs showing vinculin and paxillin localization in HL60 cell from the VD-TNF group. HL60 cells from the VD-TNF group were cultured on a coverslip surface on which EphA2-Fc or ephrin-A1-Fc and Fc plus Matrigel were adsorbed. To visualize focal adhesions, cells fixed with 4% paraformaldehyde were incubated with 0.02% Triton X-100 in PBS and then with a mixture of an anti-human vinculin mouse monoclonal antibody (hVIN-1, Sigma-Aldrich) at a dilution of 1:200 and an anti-human paxillin rabbit monoclonal antibody (Y113, Abcam, Cambridge, UK) at a dilution of 1:250 in 1% BSA-PBS for 60 min at 32 °C. After washing with PBS, the cells were incubated with a mixture of Alexa 488-conjugated goat anti-mouse IgG (5 μg/mL; Molecular Probes) and Alexa 568-conjugated donkey anti-rabbit IgG (5 μg/mL; Molecular Probes) in 1% BSA-PBS for 30 min at 32°C. After mounting with PermaFluor (Thermo Fisher Scientific), fluorescence images of vinculin (green) and paxillin immunostaining (red) were photographed. (PPTX 976 kb)

Abbreviations
BMCs: Bone marrow cells; BSA: Bovine serum albumin; F-actin: Filamentous actin; FBS: Fetal bovine serum; GAPDH: Glyceraldehyde-3-phosphate dehydrogenase; HBSS: Hank's balanced salt solution; ICAM-1: Intercellular adhesion molecule 1; LFA-1: Lymphocyte function-associated antigen 1; Mac-1: Macrophage-1 antigen; MAdCAM-1: Mucosal vascular addressin cell adhesion molecule 1; M-CSF: Macrophage colony-stimulating factor; MNCs: Mononuclear cells; NSE: Nonspecific esterase; p: Probability; pen/strep: Penicillin and streptomycin; Rap1: Ras-related protein 1; RT-PCR: Reverse transcription polymerase chain reaction; SD: Standard deviation; TNF: Tumor necrosis factor; VCAM-1: Vascular cell adhesion molecule 1; VD: 1α, 25-dihydroxy-vitamin D_3; VLA-4: Very late antigen-4

Acknowledgements
The authors thank Dr. Kikuya Sugiura for technical assistance in the isolation of MNCs from bone marrow. The authors also thank Editage (https://www.editage.jp) for English language editing.

Funding
This work was supported by a Grant-in-Aid for Scientific Research from the Japan Society for the Promotion of Science (to K.O.; Nos. 24,580,429, 15 K07769).

Authors' contributions
KO designed the study; MM, NS, NS, and KO performed the experiments; MM and KO analyzed the data; KO and MM wrote the paper. All authors read and approved the final manuscript.

Competing interests
The authors declare that they have no competing interests.

References
1. Noren NK, Pasquale EB. Eph receptor-ephrin bidirectional signals that target Ras and rho proteins. Cell Signal. 2004;16(6):655–66.
2. Pasquale EB. Eph-ephrin bidirectional signaling in physiology and disease. Cell. 2008;133(1):38–52.
3. Pasquale EB. Eph receptors and ephrins in cancer: bidirectional signalling and beyond. Nat Rev Cancer. 2010;10(3):165–80.
4. Takada Y, Ye X, Simon S. The integrins. Genome Biol. 2007;8(5):215.
5. Hogg N, Patzak I, Willenbrock F. The insider's guide to leukocyte integrin signalling and function. Nat Rev Immunol. 2011;11(6):416–26.
6. Abram CL, Lowell CA. The ins and outs of leukocyte integrin signaling. Annu Rev Immunol. 2009;27:339–62. doi:10.1146/annurev.immunol.021908.132554.
7. Ley K, Laudanna C, Cybulsky MI, Nourshargh S. Getting to the site of inflammation: the leukocyte adhesion cascade updated. Nat Rev Immunol. 2007;7(9):678–89.
8. Pasquale EB. Eph receptor signalling casts a wide net on cell behaviour. Nat Rev Mol Cell Biol. 2005;6(6):462–75.
9. Sharfe N, Nikolic M, Cimpeon L, Van De Kratts A, Freywald A, Roifman CM. EphA and ephrin-A proteins regulate integrin-mediated T lymphocyte interactions. Mol Immunol. 2008; 45(5):1208–1220. Epub 2007 Nov 1205.
10. Trinidad EM, Ballesteros M, Zuloaga J, Zapata A, Alonso-Colmenar LM. An impaired transendothelial migration potential of chronic lymphocytic leukemia (CLL) cells can be linked to ephrin-A4 expression. Blood. 2009; 114(24):5081–90.
11. de Saint-Vis B, Bouchet C, Gautier G, Valladeau J, Caux C, Garrone P. Human dendritic cells express neuronal Eph receptor tyrosine kinases: role of EphA2 in regulating adhesion to fibronectin. Blood. 2003;102(13):4431–40.
12. Sakamoto A, Kawashiri M, Ishibashi-Ueda H, Sugamoto Y, Yoshimuta T, Higashikata T, et al. Expression and function of ephrin-B1 and its cognate receptor EphB2 in human abdominal aortic aneurysm. Int J Vasc Med. 2012; 2012:127149.
13. Sharfe N, Freywald A, Toro A, Roifman CM. Ephrin-A1 induces c-Cbl phosphorylation and EphA receptor down-regulation in T cells. J Immunol. 2003;170(12):6024–32.
14. Holen HL, Shadidi M, Narvhus K, Kjosnes O, Tierens A, Aasheim HC. Signaling through ephrin-A ligand leads to activation of Src-family kinases, Akt phosphorylation, and inhibition of antigen receptor-induced apoptosis. J Leukoc Biol. 2008;84(4):1183–1191. doi:10.1189/jlb.1207829. Epub 1202008 Jul 1207821.
15. Aasheim HC, Munthe E, Funderud S, Smeland EB, Beiske K, Logtenberg T. A splice variant of human ephrin-A4 encodes a soluble molecule that is secreted by activated human B lymphocytes. Blood. 2000;95(1):221–30.
16. Aasheim HC, Terstappen LW, Logtenberg T. Regulated expression of the Eph-related receptor tyrosine kinase Hek11 in early human B lymphopoiesis. Blood. 1997;90(9):3613–22.
17. Koffel R, Meshcheryakova A, Warszawska J, Hennig A, Wagner K, Jorgl A, et al. Monocytic cell differentiation from band-stage neutrophils under inflammatory conditions via MKK6 activation. Blood. 2014;124(17):2713–2724. doi: 10.1182/blood-2014-2707-588178. Epub 582014 Sep 588111.
18. Collins SJ. The HL-60 promyelocytic leukemia cell line: proliferation, differentiation, and cellular oncogene expression. Blood. 1987;70(5):1233–44.
19. Trinchieri G, Kobayashi M, Rosen M, Loudon R, Murphy M, Perussia B. Tumor necrosis factor and lymphotoxin induce differentiation of human myeloid cell lines in synergy with immune interferon. J Exp Med. 1986;164(4):1206–25.
20. Saeki N, Nishino S, Shimizu T, Ogawa K. EphA2 promotes cell adhesion and spreading of monocyte and monocyte/macrophage cell lines on integrin ligand-coated surfaces. Cell Adhes Migr. 2015;9(6):469–82.
21. Konda N, Saeki N, Nishino S, Ogawa K. Truncated EphA2 likely potentiates cell adhesion via integrins as well as infiltration and/or lodgment of a monocyte/macrophage cell line in the red pulp and marginal zone of the mouse spleen, where ephrin-A1 is prominently expressed in the vasculature. Histochem Cell Biol. 2017;147(3):317–39. doi:10.1007/s00418-016-1494-8.
22. Graziani-Bowering GM, Graham JM, Filion LG. A quick, easy and inexpensive method for the isolation of human peripheral blood monocytes. J Immunol Methods. 1997;207(2):157–68.
23. Yu W, Chen J, Xiong Y, Pixley FJ, Dai XM, Yeung YG, et al. CSF-1 receptor structure/function in MacCsf1r−/− macrophages: regulation of proliferation, differentiation, and morphology. J Leukoc Biol. 2008;84(3):852–63.
24. Shi C, Pamer EG. Monocyte recruitment during infection and inflammation. Nat Rev Immunol. 2011;11(11):762–74. doi:10.1038/nri3070.

25. Okuno Y, Iwasaki H, Huettner CS, Radomska HS, Gonzalez DA, Tenen DG, et al. Differential regulation of the human and murine CD34 genes in hematopoietic stem cells. Proc Natl Acad Sci U S A. 2002;99(9):6246–51.

26. Gadhoum SZ, Sackstein R. CD15 expression in human myeloid cell differentiation is regulated by sialidase activity. Nat Chem Biol. 2008;4(12):751–7.

27. Li CY, Lam KW, Yam LT. Esterases in human leukocytes. J Histochem Cytochem. 1973;21(1):1–12.

28. Stadnyk AW, Befus AD, Gauldie J. Characterization of nonspecific esterase activity in macrophages and intestinal epithelium of the rat. J Histochem Cytochem. 1990;38(1):1–6.

29. Ogawa K, Takemoto N, Ishii M, Pasquale EB, Nakajima T. Complementary expression and repulsive signaling suggest that EphB receptors and ephrin-B ligands control cell positioning in the gastric epithelium. Histochem Cell Biol. 2011;136(6):617–36.

30. Ogawa K, Saeki N, Igura Y, Hayashi Y. Complementary expression and repulsive signaling suggest that EphB2 and ephrin-B1 are possibly involved in epithelial boundary formation at the squamocolumnar junction in the rodent stomach. Histochem Cell Biol. 2013;140(6):659–75.

31. Pixley FJ, Stanley ER. CSF-1 regulation of the wandering macrophage: complexity in action. Trends Cell Biol. 2004;14(11):628–38.

32. Luo BH, Carman CV, Springer TA. Structural basis of integrin regulation and signaling. Annu Rev Immunol. 2007;25:619–47.

33. Lahti M, Heino J, Kapyla J. Leukocyte integrins alphaLbeta2, alphaMbeta2 and alphaXbeta2 as collagen receptors–receptor activation and recognition of GFOGER motif. Int J Biochem Cell Biol. 2013;45(7):1204–11.

34. Sakamoto A, Sugamoto Y, Tokunaga Y, Yoshimuta T, Hayashi K, Konno T, et al. Expression profiling of the ephrin (EFN) and EPH receptor (EPH) family of genes in atherosclerosis-related human cells. J Int Med Res. 2011;39(2):522–7.

35. Prevost N, Woulfe DS, Jiang H, Stalker TJ, Marchese P, Ruggeri ZM, et al. Eph kinases and ephrins support thrombus growth and stability by regulating integrin outside-in signaling in platelets. Proc Natl Acad Sci U S A. 2005; 102(28):9820–5.

36. Linder S, Wiesner C. Tools of the trade: podosomes as multipurpose organelles of monocytic cells. Cell Mol Life Sci. 2015;72(1):121–35.

37. Ting MJ, Day BW, Spanevello MD, Boyd AW. Activation of ephrin a proteins influences hematopoietic stem cell adhesion and trafficking patterns. Exp Hematol. 2010;38(11):1087–98.

38. Ogawa K, Pasqualini R, Lindberg RA, Kain R, Freeman AL, Pasquale EB. The ephrin-A1 ligand and its receptor, EphA2, are expressed during tumor neovascularization. Oncogene. 2000;19(52):6043–52.

39. Pandey A, Shao H, Marks RM, Polverini PJ, Dixit VM. Role of B61, the ligand for the Eck receptor tyrosine kinase, in TNF-alpha-induced angiogenesis. Science (New York, NY) 1995;268(5210):567–569.

40. Funk SD, Yurdagul A Jr, Albert P, Traylor JG Jr, Jin L, Chen J, et al. EphA2 activation promotes the endothelial cell inflammatory response: a potential role in atherosclerosis. Arterioscler Thromb Vasc Biol. 2012;32(3):686–95.

9

Honokiol improved chondrogenesis and suppressed inflammation in human umbilical cord derived mesenchymal stem cells via blocking nuclear factor-κB pathway

Hao Wu[1], Zhanhai Yin[1], Ling Wang[2], Feng Li[1] and Yusheng Qiu[1*]

Abstract

Background: Cartilage degradation is the significant pathological process in osteoarthritis (OA). Inflammatory cytokines, such as interleukin-1β (IL-1β), activate various downstream mediators contributing to OA pathology. Recently, stem cell-based cartilage repair emerges as a potential therapeutic strategy that being widely studied, whereas, the outcome is still far from clinical application. In this study, we focused on an anti-inflammatory agent, honokiol, which is isolated from an herb, investigated the potential effects on human umbilical cord derived mesenchymal stem cells (hUC-MSCs) in IL-1β stimulation.

Methods: Second passage hUC-MSCs were cultured for multi-differentiation. Flow cytometry, qRT-PCR, von Kossa stain, alcian blue stain and oil red O stain were used for characterization and multi-differentiation determination. Honokiol (5, 10, 25, 50 μM) and IL-1β (10 ng/ml) were applied in hUC-MSCs during chondrogenesis. Analysis was performed by MTT, cell apoptosis evaluation, ELISA assay, qRT-PCR and western blot.

Results: hUC-MSC was positive for CD73, CD90 and CD105, but lack of CD34 and CD45. Remarkable osteogenesis, chondrogenesis and adipogenesis were detected in hUC-MSCs. IL-1β enhanced cell apoptosis and necrosis and activated the expression of caspase-3, cyclooxygenase-2 (COX-2), interleukin-6 (IL-6) and matrix metalloproteinase (MMP)-1, −9, 13 in hUC-MSCs. Moreover, the expression of SRY-related high-mobility group box 9 (SOX-9), aggrecan and col2α1 was suppressed. Honokiol relieved these negative impacts induced by IL-1β and suppressed Nuclear factor-κB (NF-κB) pathway by downregulating expression of p-IKKα/β, p-IκBα and p-p65 in dose-dependent and time-dependent manner.

Conclusions: Honokiol improved cell survival and chondrogenesis of hUC-MSCs and inhibited IL-1β-induced inflammatory response, which suggested that combination of anti-inflammation and stem cell can be a novel strategy for better cartilage repair.

Keywords: Osteoarthritis, Honokiol, Interleukin-1β, Mesenchymal stem cell, Cartilage repair

* Correspondence: yusheng.qiu@mail.xjtu.edu.cn
[1]Department of Orthopaedics, The First Affiliated Hospital, College of Medicine, Xi'an Jiaotong University, Xi'an 710061, People's Republic of China
Full list of author information is available at the end of the article

Background

Articular cartilage has limited and insufficient ability to self-regeneration once being damaged. Great efforts have been made consistently so far, whereas, no effective therapeutic approach was claimed to fully repair damaged articular cartilage. Potential therapies based on multi-differentiation characteristics of mesenchymal stem cells (MSCs) and for cartilage regeneration were widely studied recently. MSCs can also secret various growth factors, including transforming growth factor (TGF), insulin-like growth factor (IGF), hepatocyte growth factor (HGF), fibroblast growth factor (FGF) and vascular endothelial growth factor (VEGF), which promoting cell proliferation and angiogenesis in different tissues [1, 2] and preventing cell apoptosis from trauma, oxidation stress, radiation and chemicals [3]. MSCs can be obtained from various mature tissues, such as bone marrow [4], adipose tissue [5], and synovium [6]. Fetus tissues contains abundant MSCs [7], such as umbilical cord blood, placenta [8, 9] and umbilical cord matrix [10]. Human umbilical cord derived mesenchymal stem cell (hUC-MSC) was isolated from human umbilical cord and possess several advantages such as vast source, easy isolation, stable multi-differentiation capacity. However, many studies indicated that the application of MSCs in in vitro osteoarthritis (OA) models led to fibrotic cartilage instead of hyaline cartilage [11], the reasons remained unclear.

Interleukin-1β (IL-1β) is generally regarded as one of the main initiators in OA, which being produced by different types of cell including chondrocytes, macrophages and synovial fibroblasts [12, 13]. Importantly, IL-1β activate inflammatory pathways resulting in a vicious circle of articular cartilage damage rather than cartilage regeneration. Cell survival and chondrogenic potential of MSCs should be negatively affected once being transplanted in such inflammatory environment, but whether anti-inflammation treatment would be an improvement has not been examined. Although anti-inflammation has been studied for decades, significant pathological effects of cytokines on OA still deserve adequate attention.

Anti-inflammation therapy has a long history in OA treatment, but traditional agents, such as dexamethasone and Non-Steroidal Anti-Inflammatory Drugs (NSAIDs), have different adverse effects that limit their long-term application for chronic inflammatory diseases [14, 15]. Honokiol is a natural biphenolic compound purified from a traditional Chinese medicine called *Magnolia officinalis*. Its anti-inflammation function and less adverse effects has been reported [16, 17]. One of the therapeutic targets of honokiol is Nuclear Factor-κB (NF-κB) pathway [18], which regulates important downstream signals in inflammatory process [19]. Thus, honokiol has the potential to be a promising anti-inflammatory drug for different inflammatory diseases including OA.

In this study, we investigated the cell survival and chondrogenesis of hUC-MSCs and the effects of honokiol on hUC-MSCs during chondrogenic process in IL-1β stimulation. Our data may provide useful information and a novel approach for articular cartilage regeneration.

Methods

Cell isolation and culture of hUC-MSCs

All human umbilical cords (n = 6; gestational ages, 39-40 weeks) were obtained from the First Affiliated Hospital of Xi'an Jiaotong University and the study was approved by the institutional review board of the First Affiliated Hospital of Xi'an Jiaotong University. Human umbilical cords were kept in phosphate buffered saline (PBS) with penicillin in ice-box after collection and delivered to our lab as soon as possible. Then umbilical cords were cut into 5 cm sections and all veins and arteries were removed. After being washed by PBS, sections were minced into $1mm^3$ cubes and placed on a petri dish with the same intervals. These cubes were incubated with Liberase Enzyme Blends™ (Roche, Switzerland) at 37 °C for 30 min to dissolve remaining tissue, then transferred into tubes, cell suspensions were centrifuged at 4 °C and 1500 rpm for 5 min, cells were washed by PBS and re-suspended with α-MEM (Gibco, USA). We counted cell number with a hemocytometer, then cells were cultured on petri dishes with a density of 6×10^3 cells/cm^2 in incubator at 37 °C and under 5% CO_2. Culture medium was composed of high glucose α-MEM (Gibco, USA), 10% fetal bovine serum (FBS; Gibco, USA), 100 U/ml penicillin and 0.1 mg/ml streptomycin and replaced twice a week. When cell confluence reached 80%, we used 0.25% trypsin to detach cells from petri dishes and passaged cells at 1: 4 dilutions. Cells were passaged for 3 times, each passage was collected and re-suspended with culture medium. After cell number count, cell density was adjusted to 1×10^6 cells/ml for reservation.

Multi-differentiation of hUC-MSCs

Cells from 2nd passage were cultured in differentiation culture medium for multi-differentiation (Table. 1) [20]. For chondrogenesis, cells were cultured as pellet. Cells were detached by 0.25% trypsin, then transferred into a 15 ml conical tube for centrifugation at room temperature, 1000 rpm for 5 min. The supernatant was removed and cells were re-suspended with 5 ml chondrogenic differentiation medium and then counted. We adjusted the cell density to 1×10^6 cells/tube and centrifuged cells at room temperature, 1000 rpm for 5 min. The supernatant was discarded and 1 ml chondrogenic differentiation medium was added to each tube, then all

Table 1 Differentiation culture medium

Osteogenesis	Low glucose a-MEM (Gibco, USA)
	10% (*v/v*) fetal bovine serum (Gibco, USA)
	0.1uM dexamethasone (Sigma-Aldrich, USA)
	10 mM β-glycerolphosphate (Sigma-Aldrich, USA)
	0.2 mM ascorbic acid (Sigma-Aldrich, USA)
Chondrogenesis	Low glucose a-MEM (Gibco, USA)
	0.1 uM dexamethasone (Sigma-Aldrich, USA)
	50 µg/mL AsA, 100 µg/mL sodium pyruvate (Sigma-Aldrich, USA)
	40 µg/mL proline (Sigma-Aldrich, USA)
	10 ng/mL TGF-1 (Invitrogen, USA)
	50 mg/mL ITS$^+$ premix (Becton Dickinson, USA)
Adipogenesis	Low glucose a-MEM (Gibco, USA)
	0.5 mM 3-isobutyl-1-methylxanthine (IBMX, Sigma-Aldrich, USA)
	1 µM hydrocortisone (Sigma-Aldrich, USA)
	0.1 mM indomethacin (Sigma-Aldrich, USA)
	10% rabbit serum (Sigma-Aldrich, USA)

tubes were cultured in cell incubator at 37 °C and under 5% CO_2. For osteogenesis or adipogenesis, each well in 6-well plates was added with 2 ml osteogenic differentiation medium or adipogenic differentiation medium and 200 µl high density cells suspension (a density of 1×10^6 cells/ml), the final cell density in each well was 2×10^4 cells/cm^2. All groups were cultured for 2 weeks and culture medium was replaced twice a week. At the end of 2nd week, all cells were collected for subsequent analysis.

Flow cytometry

The cells were cultured in high glucose α-MEM (Gibco, USA), 10% FBS (Gibco, USA), 100 U/ml penicillin and 0.1 mg/ml streptomycin for 1 week, culture medium was replaced twice a week. Cells were collected and counted, then re-suspended with α-MEM, we adjusted the cell density at 1×10^6 cells/ml. Each 0.1 ml sample was incubated with 20 µl of CD73-PE, CD90-FITC, CD34-PE, CD45-PE, CD105-PE (Santa Cruz, USA) at 4 °C for 1 h. Antibody binding was analyzed by flow cytometry (Becton Dickson, USA).

Multi-differentiation staining

After 2 weeks' culture, cells in osteogenesis and adipogenesis group were fixed with 2.5% glutaraldehyde for 15 min. Cells pellets in chondrogenesis group were fixed with 4% paraformaldehyde for 1 h, then embedded in paraffin and cross sectioned. Slices were stained with hematoxylin and eosin (Beyotime, China). Von Kossa staining (GENMED, China), oil red O staining (Beyotime, China) and alcian blue staining (Beyotime, China) were conducted by using commercial stain kits. All staining results were photographed and analyzed by using microscope (Olympus IX35, Japan).

qRT-PCR

qRT-PCR were applied to determine gene markers expression in cell differentiation. The total RNAs were extracted from cells by Trizol (Invitrogen, USA) according to the manufacturer's instructions. 1 µg RNA was used to synthesize cDNA using a reverse transcription reagents kit (Roche, Switzerland). The qRT-PCR was carried out with the following protocol and conducted with Applied Biosystems 7500 Fast (Applied Biosystems, USA). The qRT-PCR system performed with the following temperature profile: 50 °C for 2 min, 95 °C for 2 min, then 40 cycles of 95 °C for 3 s and 40 cycles of 60 °C for 30 s. All data was analyzed by using $2^{-\Delta\Delta CT}$ method [21]. GADPH was used as control. All primer sequences were showed in Table 2.

Cell viability

Cell viability was evaluated by 3-(4,5- dimethylthiazol-2-yl)-2,5-diphenyltetrazolium bromide (MTT; Sigma-Aldrich, USA). 2nd passage cells were cultured in 96-wells culture plate with a density of 2×10^4 cells/cm^2. Honokiol (Sigma-Aldrich, USA) was dissolved in dimethyl sulfoxide (DMSO; Sigma-Aldrich, USA) for 24 h in advance and added in cell culture plate with gradient concentration (0, 5, 10, 25, 50 µM). After being washed by PBS, cells were incubated with MTT (0.2 mg/ml) at 37 °C for 4 h. Then culture medium was replaced by DMSO and culture plate was shake for 10 min at room temperature. Each well was determined by microreader (SpectraMax i3, Molecular Devices, USA) at 550 nm.

ELISA assay for IL-6 and COX-2 production determination

2nd passage cells were cultured with a density of 2×10^4 cells/cm^2 in 6-wells plate with chondrogenic medium, control group was cultured in common culture medium. IL-1β group was treated with IL-1β (10 ng/ml; Sigma-Aldrich, USA), honokiol group were treated with both honokiol (5, 10, 25 µM) and IL-1β (10 ng/ml) for 24 h. IL-6 and COX-2 production was determined by using commercial ELISA kit (R&D System, USA). All plates were read at 460 nm by a microreader (SpectraMax i3, Molecular Devices, USA).

Inflammatory stimulation and honokiol treatment

2nd passage hUC-MSCs were randomly divided into 4 groups, control group was cultured in common medium. Remaining groups were cultured as pellets for chondrogenesis, the method was previously described. IL-1β group was treated with IL-1β (10 ng/ml), honokiol group was treated with both IL-1β (10 ng/ml) and honokiol (25 µM). Chondrogenesis group was cultured in chondrogenic medium without any additions. All groups were cultured for 2 weeks and culture mediums were replaced

Table 2 Primer sequences for qRT-PCR

Gene	Forward sequence 5'-3'	Reverse sequence 5'-3'
GADPH	TGTTGCCATCAATGACCCCTT	CTCCACGACGTACTCAGCG
ALP	CCACGTCTTCACATTTGGTG	AGACTGCGCCTGGTAGTTGT
RUNX-2	AGTGGACGAGGCAAGAGTTTC	CCTTCTGGGTTCCCGAGGT
SOX-9	CTTCCGCGACGTGGACAT	GTTGGGCGGCAGGTACTG
Aggrecan	ACAGCTGGGGACATTAGTGG	GTGGAATGCAGAGGT GGTTT
Col2α1	GCCTGGTGTCATGGGTTT	GTCCCTTCTCACCAGCTTTG
CEBP	AGGAACACGAAGCACGATCAG	CGCACATTCACATTGCACAA
FABP4/aP2	TACTGGGCCAGGAATTTGAC	GGACACCCCCATCTAAGGTT
Caspase-3	AGAACTGGACTGTGGCATTGAG	GCTTGTCGGCATACTGTTTCAG
MMP-1	AGTGACTGGGAAACCAGATGCTGA	GCTCTTGGCAAATCTGGCCTGTAA
MMP-9	GCGGAGATTGGGAACCAGCTGTA	GACGCGCCTGTGTACACCCACA
MMP-13	TGCTGCATTCTCCTTCAGGA	ATGCATCCAGGGGTCCTGGC

twice a week. At the end of 2nd week, cells were collected for subsequently analysis.

Apoptosis analysis

Cell apoptosis and necrosis of hUC-MSCs were assessed by Hoechst 33,342 and propidium iodide (PI) staining using a commercial kit (Beyotime, China) and evaluated by fluorescence microscopy (Olympus IX35, Japan). Results were analyzed by ImageJ (NIH, USA).

Immunofluorescent staining

All slides were rinsed with PBS, and fixed with 2.5% glutaraldehyde for 15 min, followed by three times PBS washing, then treated with 0.3% Triton X-100 for 0.5 h and 1% BSA in PBS for 1 h. Rabbit anti-collagen II monoclonal antibody (1:100, Abcam, UK) was incubated at 4 °C overnight, followed by Alexa flour 488 conjugated secondary antibody (Molecular Probes, USA) incubation at 37 °C for 1 h. Cells were counterstained with DAPI for 10 min and analyzed by fluorescence microscopy (IX53, Olympus, Japan).

Western blot analysis

To investigate the pattern of suppressive effects of honokiol on p-IKKα/β, p-IκBα, p-p65, 2nd passage hUC-MSCs were grouped as previously described in inflammatory stimulation part and treated in two different ways. One is that honokiol group was cultured with gradient dose of honokiol (5, 10, 25 μM) and IL- β (10 ng/ml) for 2 weeks. The other one is that honokiol group was cultured with honokiol (25 μM) and IL- β (10 ng/ml) for 1, 3, 7 and 14 days respectively. We determined the expression of p-IKKα/β, p-IκBα, p-p65 at each preset time point using western blot, the protocol was briefly described as followed: cells were detached by 10% trypsin and washed by PBS, then lysed by lysis buffer containing protease inhibitors (TianGen, China). The total protein concentration was determined by the Bicinchoninic acid assay (BCA assay; Bio-Rad, USA).

Protein extracts were heated for denaturation at 100 °C for 5 min and a 12% sodium dodecyl sulfate polyacrylamide gel electrophoresis (SDS-PAGE; Bio-Rad, USA) was used for electrophoretic separation of proteins. Proteins were transferred to a PVDF membrane (Millipore, USA). The membrane was blocked with 5% non-fat dried milk in TBST buffer (0.1 M Tris-HCl and 0.1% Tween-20, pH = 7.5) for 1 h and probed with β-actin (1: 1000, Santa Cruz Biotechnology, USA), p-IKKα/β, p-IκBα, p-p65 (1: 500, Santa Cruz Biotechnology, USA), Horseradish peroxidase-conjugated anti-rabbit was used as the secondary antibody (1: 1000, Jackson Immunoresearch, USA). The detection was performed by the Thermo-Scientific Pierce ECL Western blotting substrate (Thermo-Fisher Scientific, USA). Images were scanned by Tanon-410 automatically gel imaging system (Shanghai Tianneng Corporation, China), all samples were normalized to the internal control β-actin and the optical density were determined by ImageJ (NIH, USA).

Statistical analysis

Cells obtained from 6 donors were made into a mixture for this study. All results were presented as mean ± SD and analyzed with two-tailed Student's t –test and one-way analysis of variance (ANOVA). Statistical analysis was conducted by SPSS 23.0 for Mac (IBM Inc., USA); $p < 0.01$ was considered as significant.

Results

Flow cytometry

The cell surface markers for MSCs were analyzed by flow cytometry, the results showed that cells expressed high levels of CD73, CD90 and CD105, but were lack of CD34 and CD45 expression (Fig. 1).

Histochemical staining

After 2 weeks' culture, each differentiation group was evaluated by alcian blue staining, oil red O staining and von

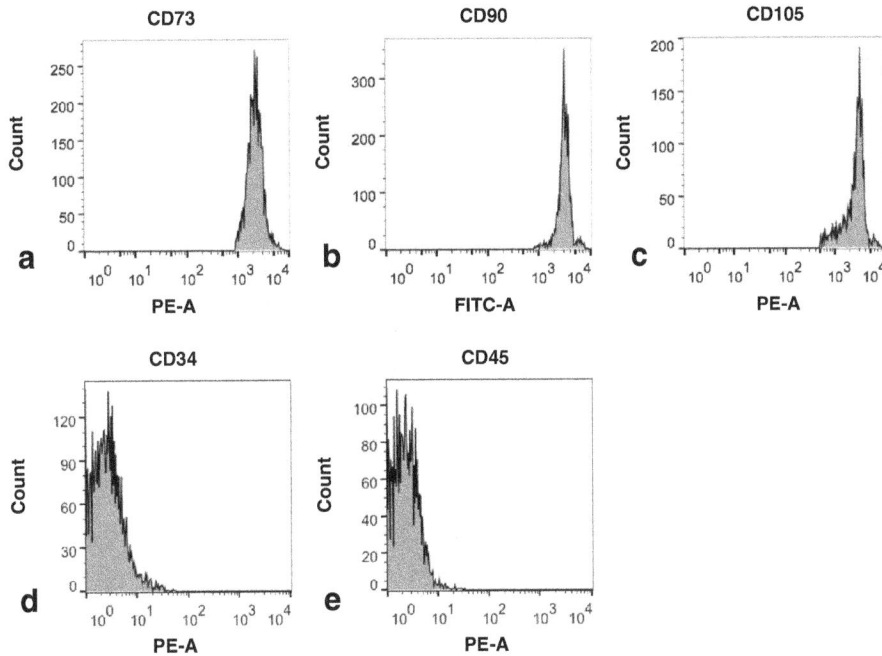

Fig. 1 Flow cytometry analysis of cell surface marker expression on hUC-MSCs. hUC-MSCs were cultured in high glucose α-MEM containing 10% FBS, 100 U/ml penicillin and 0.1 mg/ml streptomycin and were passaged for two times. Then cells were analyzed by flow cytometry (n = 3). The data presented here were representative of all sample obtained in 6 donors. **a**: CD73, **b**: CD90, **c**: CD105, **d**: CD34, **e**: CD45

Kossa staining respectively. Our results showed vary degrees of positive staining results in differentiation groups. Control group cells remained the shape of MSCs with negative staining marks (Fig. 2a), In chondrogenesis group, cells were cultured in pellets, positive alcian blue staining was detected in slices (Fig. 2b). In osteogenesis group, von Kossa stain-positive nodules were formed (Fig. 2c). In adipogenic group, positive oil red O stain-cells were widely detected (Fig. 2d).

qRT-PCR
We also evaluated key genes of differentiation in hUC-MSCs. Our results revealed that the expression of SOX-9, col2α1, aggrecan in chondrogenesis group was 8.6 ($p = 9.65 \times 10^{-5}$), 3 ($p = 2 \times 10^{-3}$) and 10 ($p = 6.4 \times 10^{-4}$) folds of control respectively (Fig. 2e), The expression of osteogenic markers, Alkaline phosphatase (ALP) and Runt-related transcription factor 2 (RUNX-2) in osteogenesis group had approximate 7.7 ($p = 6.79 \times 10^{-4}$) and 2 ($p = 2.26 \times 10^{-4}$) folds of control (Fig. 2f). Additionally, the expression of adipogenic markers, CCAAT-enhancer-binding proteins (CEBP) and fatty acid-binding protein-4 (FABP4/aP2) in adipogenesis group, were highly expressed and had almost 18.3 ($p = 5.83 \times 10^{-6}$) and 165 ($p = 2.02 \times 10^{-5}$) folds of control separately (Fig. 2g).

The effects of honokiol on cell viability in hUC-MSCs
MTT assay was introduced here to investigate the effect of different concentration of honokiol on cell viability.

The results showed that the gradient dose (5, 10, 25, 50 μM) of honokiol didn't have remarkable cytotoxic effect on hUC-MSCs until the concentration increased to 50 μM (Fig. 3a). During the culture of hUC-MSCs, SOX-9, Aggrecan and Col2α1 expression were evaluated in different passages of cells to investigate the prime cell passage for chondrogenesis, the results indicated that three markers were highly expressed in the 2nd passage cells (Additional file 1: Figure S1).

The effects of honokiol on IL-6 and COX-2 production in hUC-MSCs
According to previous results, the relatively safe dose of honokiol in vitro ranged from 5 to 25 μM. To discover the prime dose of honokiol for anti-inflammation in vitro, hUC-MSCs were treated with honokiol (5, 10, 25 μM) under IL-1β (10 ng/ml) stimulation for 24 h. The ELISA results showed that honokiol suppressed IL-6 and COX-2 production in hUC-MSCs in a dose-dependent way (Fig. 3b/c). Especially, IL-6 and COX-2 production in honokiol group was still higher than control group, but production in honokiol group (25 μM) were only 34.9% ($p = 2.94 \times 10^{-11}$) and 11.6% ($p = 2.42 \times 10^{-13}$) of IL-1β group respectively.

Cell survival of hUC-MSCs in IL-1β stimulation
25 μM was selected for applied concentration of honokiol in subsequent experiments according to our study. A commercial fluorescence staining kit was used to evaluate cell

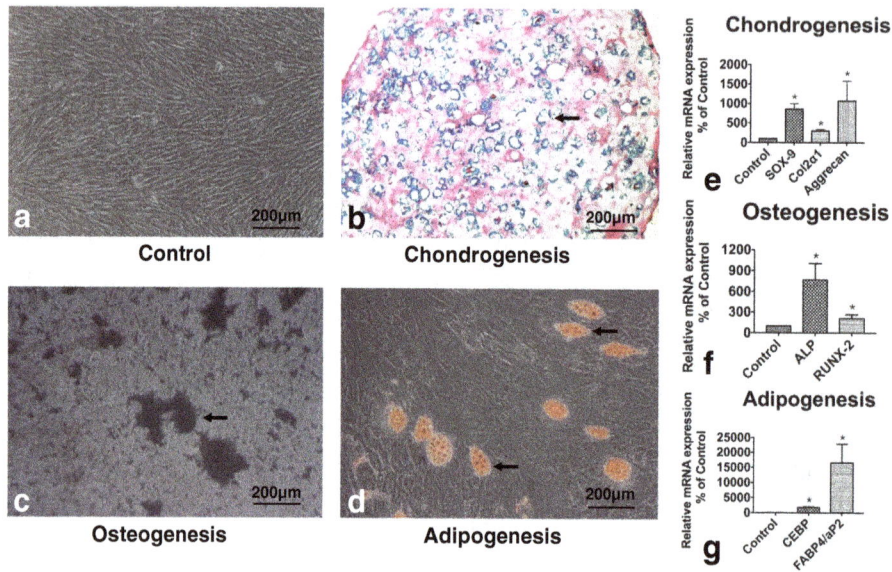

Fig. 2 Determination of multi-differentiation of hUC-MSCs. 2nd passage hUC-MSCs were used for multi-differentiation determination. For osteogenesis or adipogenesis, cells were cultured in 6-wells plate with osteogenic or adipogenic medium for 2 weeks. Chondrogenesis was conducted in pellet culture for 2 weeks ($n = 6$). **a**: control group. **b**: Chondrogenesis group cells. Hematoxylin-eosin and alcian blue dual-staining (arrow indicated the positive alcian blue staining). c: Osteogenesis group cells (arrow indicated the positive von Kossa staining). **d**: Adipogenesis group cells (arrow indicated the positive oil red O staining). The results showed here were representative of all samples from 6 donors. The expression of differentiation related genes was evaluated by qRT-PCR. **e**: Chondrogenesis related genes. f: Osteogenesis related genes. g: Adipogenesis related genes. Data was analyzed by using the $2^{-\Delta\Delta CT}$ method. All qRT-PCR results were presented as mean ± SD ($n = 9$); p* < 0.01 versus control

apoptosis and necrosis. Control and chondrogenesis group cells showed low degree of apoptosis and necrosis (Fig. 4a/b), which had 8.99% apoptotic cells and 0.91% necrotic cells in control group and 7.47% and 0.82% in chondrogenesis group (Fig. 4e/f). Cell apoptosis and necrosis were enhanced in IL-1β group (Fig. 4c), the apoptotic and necrotic cells were 19.71% ($p = 3.85 \times 10^{-9}$) and 15.63% ($p = 3.72 \times 10^{-9}$) in IL-1β group compared with the low percentage in control and chondrogenesis group (Fig. 4e/f). Apoptotic and necrotic cell percentages were improved in honokiol group, which were 12.75% ($p = 9.6 \times 10^{-6}$) and 2.5% ($p = 1.96 \times 10^{-8}$) respectively (Fig. 4d) and much lower than IL-1β group (Fig. 4e/f). qRT-PCR was used for

evaluating the expression of caspase-3 (Fig. 4g), which is one of the most important regulators and markers for apoptosis. Results showed that the expression of caspase-3 in IL-1β group was 7.5 folds of control ($p = 5.15 \times 10^{-9}$), chondrogenesis group had an equal level with control ($p = 0.72361$). In honokiol group, the expression were only 1.45 folds of control ($p = 1.28 \times 10^{-4}$) and 19% of IL-1β group ($p = 3.06 \times 10^{-9}$).

Maintenance of Chondrogenic potential of hUC-MSCs

Chondrogenic potential of hUC-MSCs is the key for cartilage regeneration, especially in IL-1β stimulation. As described previously, SOX-9, aggrecan and col2α1 were

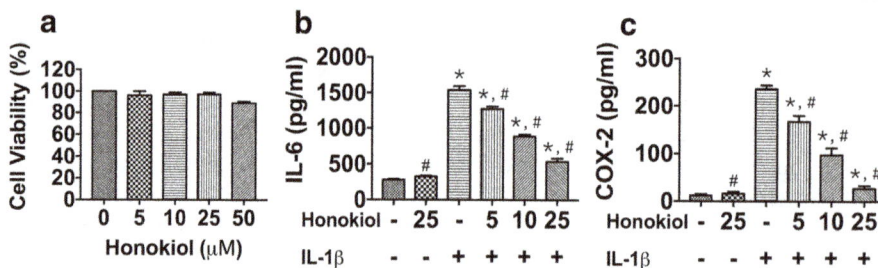

Fig. 3 Cell viability and the production of IL-6 and COX-2 in a gradient dose of honokiol. 2nd passage hUC-MSCs were cultured in high glucose α-MEM containing 10% FBS, 100 U/ml penicillin and 0.1 mg/ml streptomycin. A gradient dose of honokiol (0, 5, 10, 25, 50 μM) with or without IL-1β (10 ng/ml) were added for incubation for 24 h. **a**: Cell viability determination. **b**, **c**: IL-6 and COX-2 production. All results were presented as mean ± SD ($n = 6$); p* < 0.01 versus control; p# < 0.01 versus IL-1β alone (10 ng/ml IL-1β without honokiol)

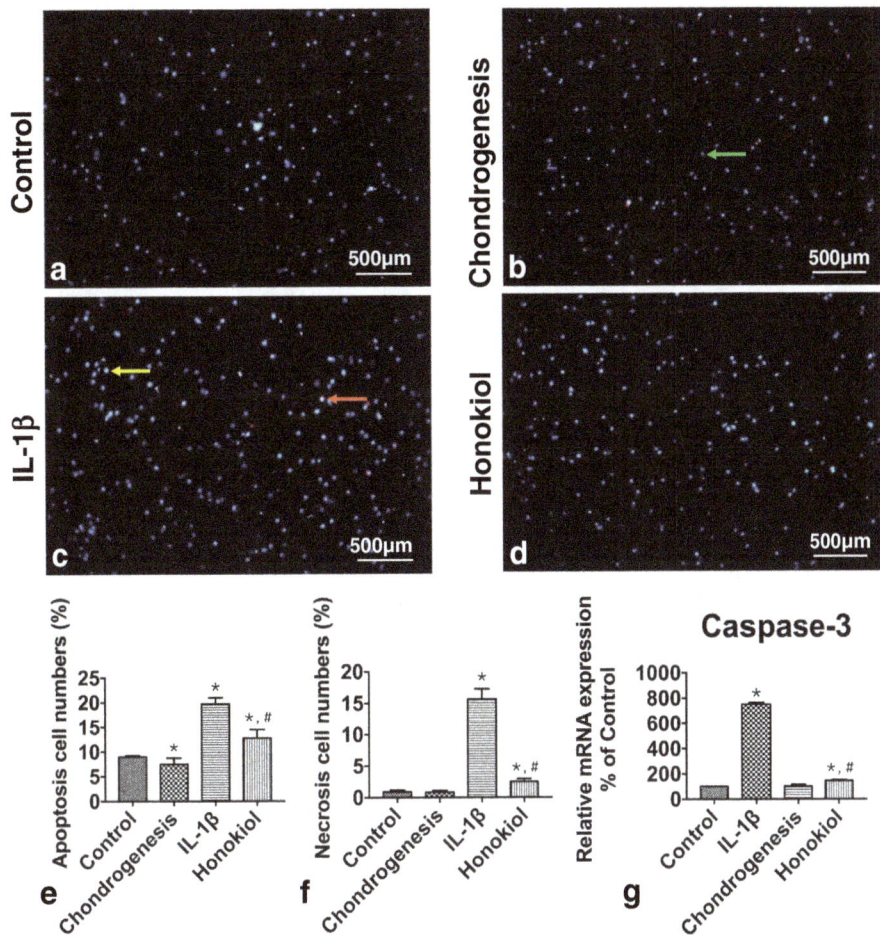

Fig. 4 The effects of honokiol on apoptosis and necrosis of hUC-MSCs. 2nd passage hUC-MSCs were cultured in 4 groups, **a**: Control, **b**: Chondrogenesis, **c**: IL-1β, **d**: Honokiol for 2 weeks. At the end of 2nd week, cells were stained with Hoechst 33,342 and PI for apoptosis and necrosis analysis. Green arrow indicated the normal cell, yellow arrow indicated the apoptotic cell and red arrow indicated the necrotic cell. The apoptotic cell number and necrotic cell number were analyzed by ImageJ, all data were presented as mean ± SD ($n = 3$), $p^* < 0.01$ versus control; $p\# < 0.01$ versus IL-1β. **e**: apoptotic cells percentage, **f**: necrotic cells percentage. Caspase-3 expression was determined by qRT-PCR. **g**: caspase-3 expression. Gene expression data were analyzed by using the $2^{-\Delta\Delta CT}$ method. All qRT-PCR results were presented as mean ± SD ($n = 9$); $p^* < 0.01$ versus control; $p\# < 0.01$ versus IL-1β

selected as markers for chondrogenesis and cartilage ECM formation (Fig. 5a/b/c). In chondrogenesis group, elevations of expression in SOX-9, aggrecan and col2α1 were noticed, which were 1.8-. 1.7-, 3.7-fold of control respectively ($p = 1.6 \times 10^{-5}$, 6.1×10^{-4}, 1.3×10^{-5}). However, the expression level was 1.1-, 1.2- and 2-fold of control ($p = 0.37067$, 0.27118, 0.04465) in IL-1β group. The expressions were upregulated to varied extent in honokiol group, which were 1.3-, 1.1-, 1.3- fold of IL-1β group respectively ($p = 9.604 \times 10^{-3}$, 8.0956×10^{-3}, 3.363×10^{-3}). Immunofluorescent staining was also used for the evaluation of col2α1 expression in different groups. The results indicated that the expression of col2α1in control and IL-1β group remained a low level compared with chondrogenesis group and honokiol group, moreover, the expression in honokiol was lower than in chondrogenesis group (Fig. 6).

Cartilage ECM degradation and inflammation activation of hUC-MSCs

ECM is the main content of cartilage and its degradation is enhanced by cell apoptosis and inflammation. In cartilage repair, ECM provides a suitable environment for cells to proliferate and differentiate. MMPs play a very important role in ECM degradation in many cell types. MMP-1, −9, −13 were assessed in our study (Fig. 5d/e/f). Results showed that the expression of MMP-1, −9, −13 in IL-1β group was highly elevated, which were 77-, 1.4- and 13-fold of control respectively ($p = 2.8 \times 10^{-6}$, 2.15×10^{-3}, 1.1×10^{-4}). The expression was almost at the same level in control and chondrogenesis group ($p = 0.12229$, 0.97183, 0.33852). Expressions were inhibited in honokiol group, which were only 24%, 87% and 39% of IL-1β group ($p = 2.8 \times 10^{-5}$, 0.07168, 6.21×10^{-3}).

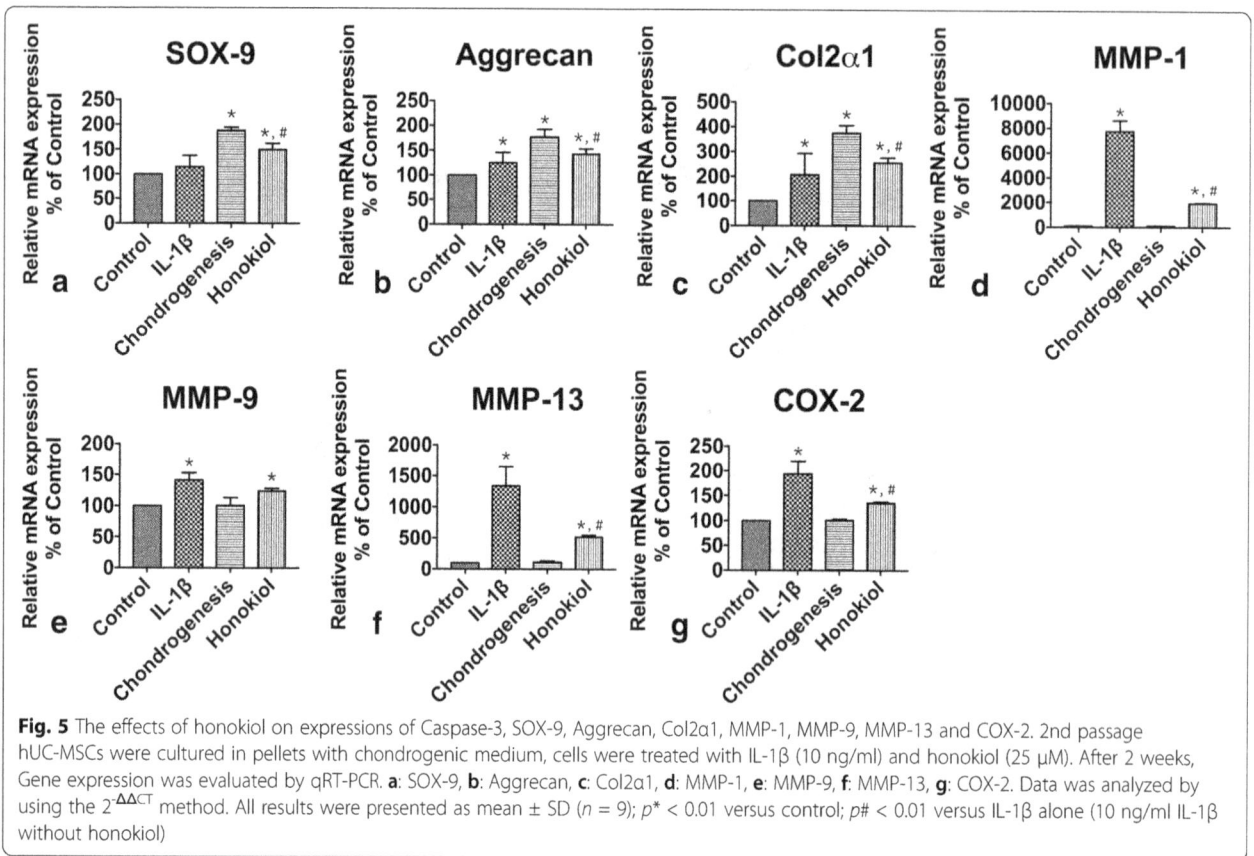

Fig. 5 The effects of honokiol on expressions of Caspase-3, SOX-9, Aggrecan, Col2α1, MMP-1, MMP-9, MMP-13 and COX-2. 2nd passage hUC-MSCs were cultured in pellets with chondrogenic medium, cells were treated with IL-1β (10 ng/ml) and honokiol (25 μM). After 2 weeks, Gene expression was evaluated by qRT-PCR. **a**: SOX-9, **b**: Aggrecan, **c**: Col2α1, **d**: MMP-1, **e**: MMP-9, **f**: MMP-13, **g**: COX-2. Data was analyzed by using the $2^{-\Delta\Delta CT}$ method. All results were presented as mean ± SD (n = 9); p^* < 0.01 versus control; $p\#$ < 0.01 versus IL-1β alone (10 ng/ml IL-1β without honokiol)

COX-2 is the central regulators and effectors in inflammation relating to many clinical symptoms in OA (Fig. 5g), result showed no statistical difference in expression of COX-2 between control and chondrogenesis group (p = 0.30602), but expression was 1.9-fold of control (p = 1.22 × 10^{-3}) in IL-1β group, but had a 30% decline in honokiol group compared with IL-1β group (p = 1.6 × 10^{-6}).

Honokiol inhibited NF-κB pathway activation in hUC-MSCs

NF-κB pathway is a significant cellular and molecular regulative network in inflammatory diseases, whose effects on OA initiation and progression can't be underestimated. High expression of p-IKKα/β, p-IκBα and p-p65 is applicative marker for NF-κB pathway activation. The western blot results demonstrated that honokiol suppressed the expression of p-IKKα/β, p-IκBα and p-p65 in a dose-dependent way, the expression of p-IKKα/β, p-IκBα and p-p65 in hUC-MSCs dropped along with the increase of honokiol concentration (Fig. 7a/b/c). Moreover, during 2 weeks chondrogenesis, the expression of p-IKKα/β, p-IκBα and p-p65 in IL-1β group remained a high level in contrast to control and chondrogenesis group. With a slight upregulation in first 3 days, the expression of p-IKKα/β, p-IκBα and p-p65 in hUC-MSCs

declined from the 3rd day, which indicating that honokiol suppressed the expression of p-IKKα/β, p-IκBα and p-p65 in a time-dependent manner in long-term application (Fig. 8a/b/c).

Discussion

OA is an inflammatory disease and characterized with pain and cartilage degradation. Pro-inflammation cytokines, such as IL-1β, play important roles in different stages of OA. Anti-inflammation is considered as one effective therapy for OA symptoms, but cartilage repair or cartilage regeneration is the fundamental solution for OA. Stem cell-based therapeutic strategy for cartilage repair is widely studied in recent years. Various types of stem cell were considered as the promising candidates for cartilage regeneration. hUC-MSCs have its advantages such as vast source and easy isolation among them. However, the stem cell-based cartilage regeneration didn't achieve a satisfied therapeutic effect in different studies [11]. Anti-inflammation is the main therapy for OA in clinical practice, however, the combination of stem cell-based therapy and anti-inflammation remains a new concept for cartilage regeneration. Despite of various traditional anti-inflammation agents, honokiol, which is an extract from traditional Chinese medicine, was introduced in this work. Anti-inflammation, anti-

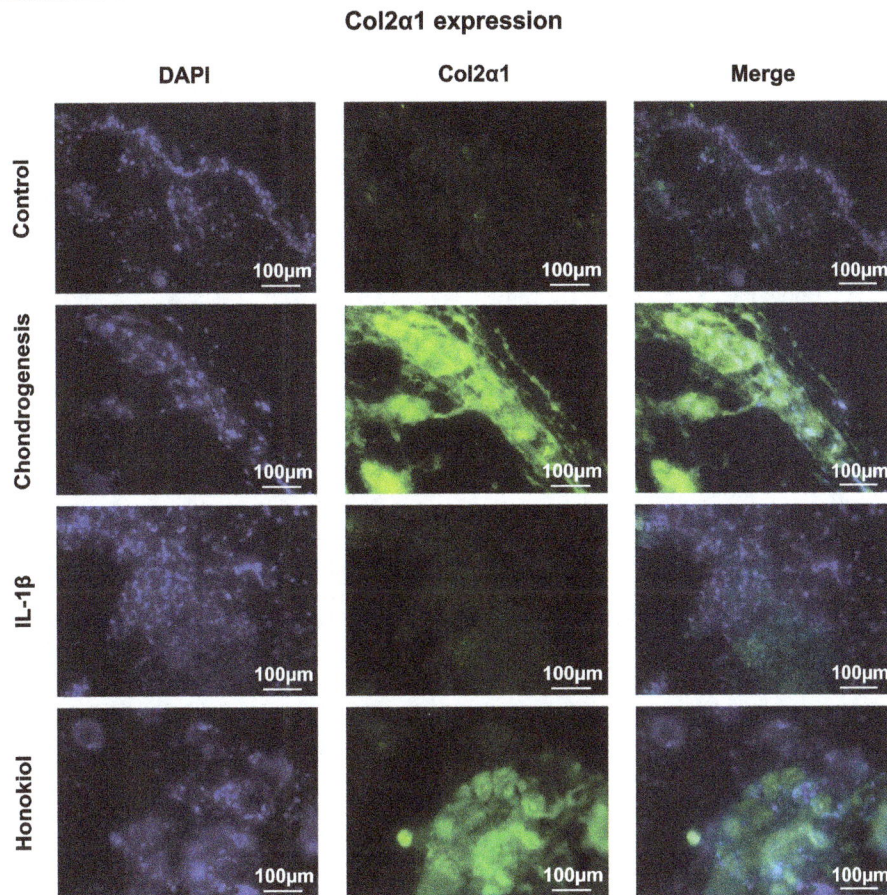

Fig. 6 Expression of Col2α1. Immunofluorescent staining results of Col2α1 were representative of the cells obtained in 6 donors. DAPI stain (blue) indicated the cells, col2α1 stain (green) indicated the positive expression of col2α1 ($n = 3$)

oxidation and other pharmacological features were reported recently [18, 22], thus, honokiol may prove to be a potential candidate for OA therapy. Studies indicated that MSCs being applied in OA models usually formed fibrotic cartilage instead of hyaline cartilage [11]. Therefore, it is easy to hypothesize that inflammation has the negative effect on MSCs once being transplanted in an inflammatory environment, inflammation may be one of major reasons for the unsuccessful MSCs-based cartilage repair. In our study, we demonstrated that IL-1β suppressed the cell survival, chondrogenesis and ECM degradation in hUC-MSCs, but honokiol relieved the negative effects by partly blocking the activation of NF-κB pathway. These findings reported here are the first of this kind of study according to our knowledge. It has been well-documented that OA joints express high level of pro-inflammatory cytokines, such as IL-1β and TNF-α [13], our data indicated anti-inflammation and protective effect of honokiol on hUC-MSCs may provide a novel thought for stem cell-based cartilage repair. In addition, hUC-MSCs were positive for certain cell surface markers of MSCs and performed successful

chondrogenesis, osteogenesis and adipogenesis. The fulfillment of the criteria for stem cell therapy [23] ensured the isolation of MSCs from human umbilical cords.

IL-1β is one of the pro-inflammatory cytokines getting involved with various inflammatory diseases. IL-1β is highly expressed in damaged articular cartilage during the initiation and progression of OA, the detection of IL-1β in synovium and articular fluid had been noticed as well [24]. IL-1β can stimulate COX-2 to upregulate other inflammatory mediators. Other pro-inflammatory cytokines, such as IL-6, also can be upregulated by IL-1β [13]. IL-1β activates the expression of MMPs contributing to ECM degradation of cartilage [25]. The relevance of IL-1β and elevated level of human chondrocyte apoptosis, enhances the synthesis of aggrecanase in human chondrocytes and synovial fibroblasts during OA was also reported [26]. The inhibition of synthesis of col2α1 and proteoglycan by IL-1β results in catabolism of cartilage tissue [27]. Our study indicated that honokiol inhibited the production of IL-6 and COX-2 in a dose-dependent manner in hUC-MSCs. Moreover, IL-1β can bind to IL-1 receptor and other receptors, then activate

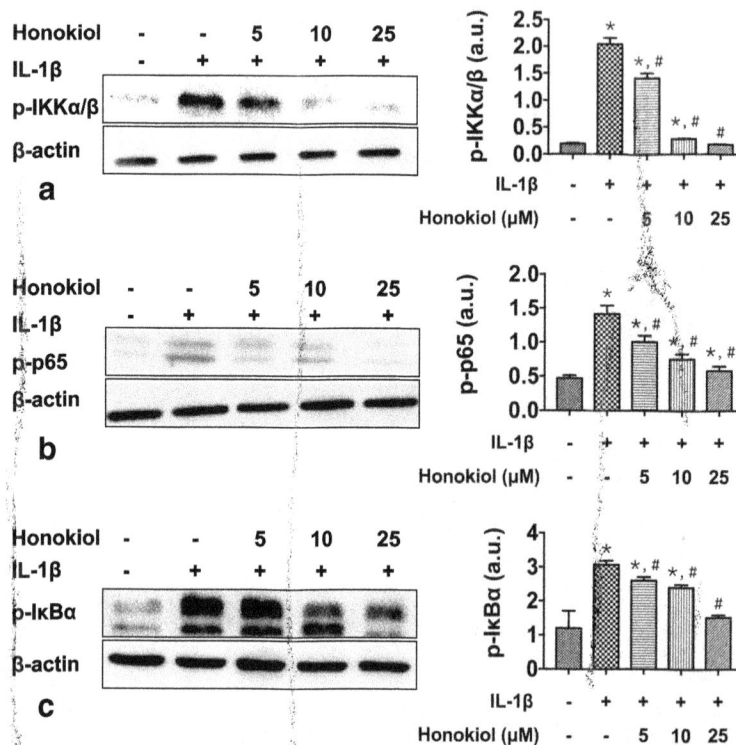

Fig. 7 Honokiol inhibited NF-κB pathway in hUC-MSCs in a dose-dependent manner. 2nd passage hUC-MSCs were cultured in pellets with chondrogenic medium, cells were treated with IL-1β (0 or 10 ng/ml) and honokiol (0, 5, 10, 25 μM). After 2 weeks, the expression of (**a**): p-IKKα/β, (**b**): p-p65, (**c**): p-IκBα were evaluated by western blot. Integrated density values were analyzed by ImageJ and normalized to β-actin, all results were presented as mean ± SD ($n = 9$); $p*< 0.01$ versus control; $p\# < 0.01$ versus IL-1β alone (10 ng/ml IL-1β without honokiol)

downstream cellular and molecular signals such as NF-κB pathway and Mitogen-activated protein kinases (MAPK) pathway to induce the cascades activation of caspases, which is the key process in apoptosis. Caspase-3 acts as the effector in apoptosis. We reported that IL-1β induced a significant apoptosis and necrosis in hUC-MSCs, the expression of caspase-3 was highly up-regulated in this process. Apoptosis and necrosis were remarkably inhibited by honokiol, which was clearly presented in Fig. 4.

Chondrogenesis is a sophisticated process being associated with many genes and cellular signals. SOX-9, an important regulator in the early stage of chondrogenesis, can regulate downstream genes related to chondrogenesis and ECM synthesis, such as col2α1 and aggrecan [28–31]. In normal chondrogenic process, MSCs differentiate into progenitor cells, then into chondrocytes, synthesizing col2α1 and aggrecan, but in some cases, especially inflammation, normal chondrocytes undergo hypertrophy leading to the production of collagen type I and XI, which are main ECM contents of fibrotic cartilage [32, 33]. IL-1β suppressed the expression of SOX-9, aggrecan and col2α1 according to our results, which indicated an unsatisfied chondrogenesis of hUC-MSCs. Instead, the expression of the three proteins were up-

regulated by honokiol and recovered to some extent, honokiol showed the capacity of maintaining the chondrogenic potential of hUC-MSCs in some ways.

Cartilage ECM provide the fundamental support for cell survival, proliferation and chondrogenesis of hUC-MSCs. Aggrecan and col2α1 are two main components of ECM in hyaline cartilage, both can be inhibited by IL-1β as we described previously. IL-1β not only suppresses ECM synthesis, but also enhances ECM degradation through activation of certain proteins. The main enzymes responsible for ECM degradation are MMPs, a large protein family containing various proteases. MMP-1, 9, 13 were assessed and showed an excessive synthesis in IL-1β group, but the upregulation was partly blocked by honokiol. The inhibition of MMP-9 wasn't as effective as MMP-1, 13 by honokiol. Interestingly, the substrates of MMP-1 include collagen type I, II, III, VII, VIII, X and gelatin, as to MMP-13, collagen type IV, IX, XIV are also included. However, the substrates of MMP-9 only contained gelatin, collagen type IV and V [34]. Thus, MMP-1, 13 may serve a more crucial role than MMP-9 in cartilage degradation as the main collagen type in articular cartilage ECM is collagen type II. It was indicated that honokiol was an appropriate anti-

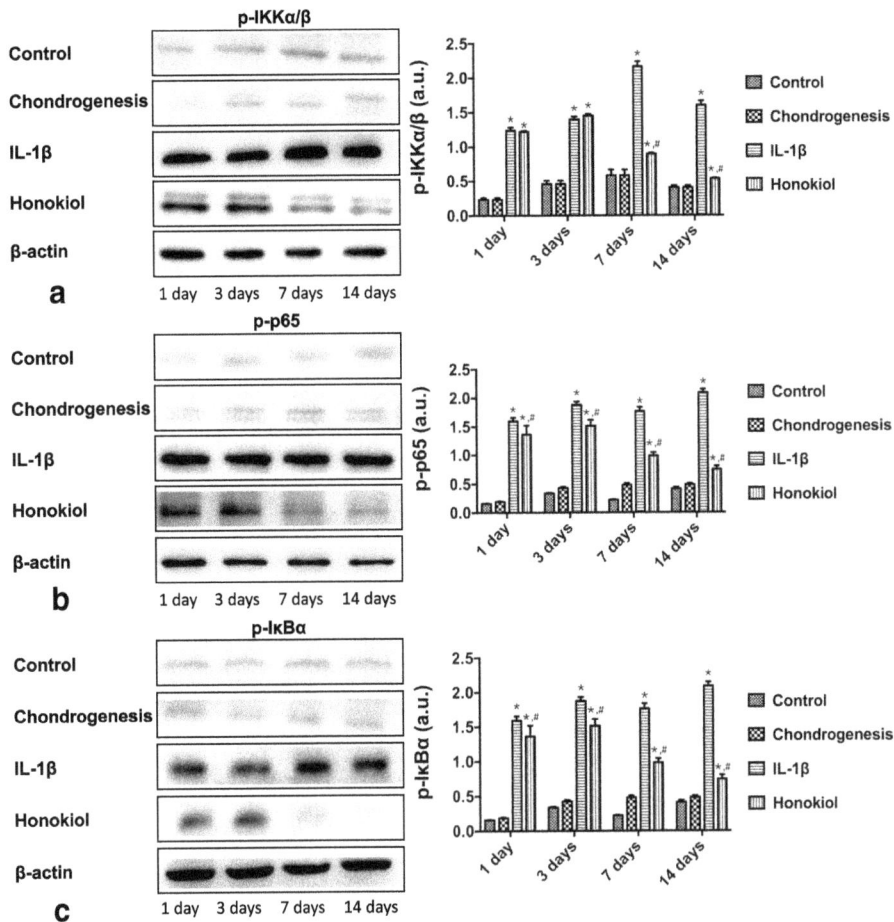

Fig. 8 Honokiol inhibited NF-κB pathway in hUC-MSCs in a time-dependent manner. 2nd passage hUC-MSCs were cultured in pellets with chondrogenic medium, cells were treated with IL-1β (10 ng/ml) and honokiol (25 μM). At 1st day, 3rd day, 7th day and 14th day, the expression of (a): p-IKKα/β, (b): p-p65, (c): p-IκBα were evaluated by western blot. Integrated density values were analyzed by ImageJ and normalized to β-actin, all results were presented as mean ± SD ($n = 9$); $p*$ < 0.01 versus control; $p\#$ < 0.01 versus IL-1β alone (10 ng/ml IL-1β without honokiol)

inflammation agent for hUC-MSCs to survive and differentiate in IL-1β stimulation.

The inhibition of NF-κB pathway is one vital process in anti-inflammation of honokiol, which has been proved in several types of mature cells. NF-κB pathway regulates various cellular signals in inflammation and has great therapeutic significance in OA [19], p-IKKα/β, p-IκBα and p-p65 are the phosphorylated forms of three important proteins in NF-κB pathway. Once NF-κB is activated by the binding of IL-1β and its receptors, these proteins are phosphorylated and translocated from cytoplasm to nucleus. The high level of the protein phosphorylation is a symbol for NF-κB activation. Our study reported that p-IKKα/β, p-IκBα and p-p65 were elevated by IL-1β stimulation as expected and the expression of p-IKKα/β, p-IκBα and p-p65 in hUC-MSCs were suppressed by honokiol in both dose-dependent and time-dependent manner. As the concentration of IL-1β applied in our study was much higher than in OA joint, the anti-

inflammation and protective effects of honokiol on hUC-MSCs may be more efficient in long-term in vivo application with an accurate and sustained release-control of honokiol. OA is characterized with chronic pathological process and low concentration of IL-1β, our colleagues are focusing on an animal model to test the hypothesis and preliminary data will be published in near future.

Conclusions

The expectation of regenerating or reconstructing cartilage defect by simple application of MSCs is proved less effective. One possible reason is the intensive focus on stem cells (the seeds) but ignoring cell living environment (the soil). In brief, IL-1β induced apoptosis in hUC-MSCs followed by ECM degradation, synthesis down-regulation, inflammation activation and cytokines secretion. We demonstrated that honokiol can significantly improve cell survival, maintain chondrogenic potential and ECM synthesis in hUC-MSCs by inhibiting NF-κB activation in dose-dependent and time-dependent way. Given the complex inflammation regulatory networks, cell survival, chondrogenesis and ECM production in hUC-MSCs didn't recover or maintain at the normal level, however, the combination of anti-inflammation and hUC-MSCs may be a novel strategy for cartilage regeneration. In vitro study was only a primary verification of our hypothesis, more attention has been focused on sustained-release of anti-inflammatory agent, MSCs and tissue engineering scaffolds to construct a cartilage regenerative complex. These findings will provide a new thought for cartilage repair.

Additional file

Additional file 1: Figure S1. Prime passage number of hUC-MSCs. hUC-MSCs from 1st, 2nd, 3rd passage were cultured in chondrogenic medium as pellets for 2 weeks. The expression of SOX-9, col2α1 and aggrecan was evaluated by qRT-PCR. Data was analyzed by using the $2^{-\Delta\Delta CT}$ method. All results were presented as mean ± SD ($n = 9$); $p^* < 0.01$ versus P1. (PPTX 206 kb)

Abbreviations

ALP: Alkaline phosphatase; CEBP: CCAAT-enhancer-binding proteins; COX-2: Cyclooxygenase-2; ELISA: Enzyme linked immunosorbent assay; FABP4/aP2: Fatty acid-binding protein-4; FGF: Fibroblast growth factor; GAPDH: Glutaraldehyde phosphate dehydrogenase; HGF: Hepatocyte growth factor; hUC-MSCs: Human umbilical cord derived messechymal stem cells; IGF: Insulin-like growth factor; IL-1β: Interleukin-1β; IL-6: Interleukin-6; IκBα: Inhibitor of nuclear factor κB α; MMP: Matrix metalloproteinase; MSCs: Mesenchymal stem cells; NF-κB: Nuclear factor-κB; OA: Osteoarthritis; PBS: Phosphate buffered saline; p-IKKα/β: Inhibitor of nuclear factor κB kinase α/β; qRT-PCR: Quantitative real-time polymerase chain reaction; RUNX-2: Runt-related transcription factor 2; SOX-9: SRY-related high-mobility group box 9; TGF: Transforming growth factor; VEGF: Vascular endothelial growth factor

Acknowledgements
We acknowledged the kind support from Department of obstetrics, the first affiliated hospital, college of medicine, Xi'an Jiaotong University.

Funding
This work is supported by National Natural Science Foundation of China (No. 81371943) and Fundamental Research Funds for the Central Universities of China (No. 30801173).

Authors' contributions
The author met all the following conditions: (1) substantial contribution to design (HW, ZY, YQ, FL), (2) preparation of cell materials (HW, LW, FL), acquisition and analysis of data (HW, LW), and drafting the article or revising it critically for important intellectual content (HW, LW, ZY, FL, YQ), (3) final approval of the version to be published (HW, ZY, LW, FL YQ). All authors read and approved the manuscript.

Competing interests
The authors declared that they have no competing interests.

Author details
[1]Department of Orthopaedics, The First Affiliated Hospital, College of Medicine, Xi'an Jiaotong University, Xi'an 710061, People's Republic of China. [2]Center for Biomedical Engineering and Regenerative Medicine, Frontier Institute of Science and Technology, Xi'an Jiaotong University, Xi'an 710049, People's Republic of China.

References

1. Pers Y-M, Ruiz M, Noël D, Jorgensen C. Mesenchymal stem cells for the management of inflammation in osteoarthritis: state of the art and perspectives. Osteoarthr Cartil. 2015;23:2027–35.
2. Rodrigues M, Griffith LG, Wells A. Growth factor regulation of proliferation and survival of multipotential stromal cells. Stem Cell Res Ther. 2010;1:32.
3. Rehman J, Traktuev D, Li J, Merfeld-Clauss S, Temm-Grove CJ, Bovenkerk JE, et al. Secretion of angiogenic and antiapoptotic factors by human adipose stromal cells. Circulation. 2004;109:1292–8.
4. Castro-Malaspina H, Gay RE, Resnick G, Kapoor N, Meyers P, Chiarieri D, et al. Characterization of human bone marrow fibroblast colony-forming cells (CFU-F) and their progeny. Blood. 1980;56:289–301.
5. Zuk PA, Zhu M, Mizuno H, Huang J, Futrell JW, Katz AJ, et al. Multilineage cells from human adipose tissue: implications for cell-based therapies. Tissue Eng. 2001;7:211–28.
6. de Sousa EB, Casado PL, Moura Neto V, Duarte MEL, Aguiar DP. Synovial fluid and synovial membrane mesenchymal stem cells: latest discoveries and therapeutic perspectives. Stem Cell Res Ther. 2014;5:112.
7. Campagnoli C, Roberts IA, Kumar S, Bennett PR, Bellantuono I, Fisk NM. Identification of mesenchymal stem/progenitor cells in human first-trimester fetal blood, liver, and bone marrow. Blood. 2001;98:2396–402.
8. Lee OK. Isolation of multipotent mesenchymal stem cells from umbilical cord blood. Blood. 2004;103:1669–75.
9. Park Y-B, Seo S, Kim J-A, Heo J-C, Lim Y-C, Ha C-W. Effect of chondrocyte-derived early extracellular matrix on chondrogenesis of placenta-derived mesenchymal stem cells. Biomed Mater Bristol Engl. 2015;10:35014.
10. Wang H-S, Hung S-C, Peng S-T, Huang C-C, Wei H-M, Guo Y-J, et al. Mesenchymal stem cells in the Wharton's jelly of the human umbilical cord. Stem Cells. 2004;22:1330–7.
11. Caldwell KL, Wang J. Cell-based articular cartilage repair: the link between development and regeneration. Osteoarthr Cartil. 2015;23:351–62.
12. Sandell LJ, Aigner T. Articular cartilage and changes in arthritis. An introduction: cell biology of osteoarthritis. Arthritis Res. 2001;3:107–13.
13. Haseeb A, Haqqi TM. Immunopathogenesis of osteoarthritis. Clin Immunol Orlando Fla. 2013;146:185–96.
14. Curtis JR, Westfall AO, Allison J, Bijlsma JW, Freeman A, George V, et al. Population-based assessment of adverse events associated with long-term glucocorticoid use. Arthritis Rheum. 2006;55:420–6.
15. Harirforoosh S, Asghar W, Jamali F. Adverse effects of nonsteroidal Antiinflammatory drugs: an update of gastrointestinal, cardiovascular and renal complications. J Pharm Pharm Sci. 2014;16:821–47.
16. Chen YJ, Tsai KS, Chan DC, Lan KC, Chen CF, Yang RS, et al. Honokiol, a low molecular weight natural product, prevents inflammatory response and cartilage matrix degradation in human osteoarthritis chondrocytes. J Orthop Res Off Publ Orthop Res Soc. 2014;32:573–80.
17. Vaid M, Sharma SD, Katiyar SK. Honokiol, a phytochemical from the Magnolia plant, inhibits photocarcinogenesis by targeting UVB-induced inflammatory mediators and cell cycle regulators: development of topical formulation. Carcinogenesis. 2010;31:2004–11.
18. Weng TI, Wu HY, Kuo CW, Liu SH. Honokiol rescues sepsis-associated acute lung injury and lethality via the inhibition of oxidative stress and inflammation. Intensive Care Med. 2011;37:533–41.
19. Roman-Blas JA, Jimenez SA. NF-κB as a potential therapeutic target in osteoarthritis and rheumatoid arthritis. Osteoarthr Cartil. 2006;14:839–48.
20. Vater C, Kasten P, Stiehler M. Culture media for the differentiation of mesenchymal stromal cells. Acta Biomater. 2011;7:463–77.
21. Livak KJ, Schmittgen TD. Analysis of relative gene expression data using real-time quantitative PCR and the 2−ΔΔCT method. Methods. 2001;25:402–8.
22. Zhang P, Liu X, Zhu Y, Chen S, Zhou D, Wang Y. Honokiol inhibits the inflammatory reaction during cerebral ischemia reperfusion by suppressing NF-κB activation and cytokine production of glial cells. Neurosci Lett. 2013; 534:123–7.
23. Dominici M, Blanc KL, Mueller I, Slaper-Cortenbach I, Marini F, Krause D, et al. Minimal criteria for defining multipotent mesenchymal stromal cells. The International Society for Cellular Therapy position statement. Cytotherapy. 2006;8:315–7.
24. Moos V, Fickert S, Müller B, Weber U, Sieper J. Immunohistological analysis of cytokine expression in human osteoarthritic and healthy cartilage. J Rheumatol. 1999;26:870–9.

25. Lefebvre V, Peeters-Joris C, Vaes G. Modulation by interleukin 1 and tumor necrosis factor alpha of production of collagenase, tissue inhibitor of metalloproteinases and collagen types in differentiated and dedifferentiated articular chondrocytes. Biochim Biophys Acta. 1990;1052:366–78.

26. Fan Z, Bau B, Yang H, Soeder S, Aigner T. Freshly isolated osteoarthritic chondrocytes are catabolically more active than normal chondrocytes, but less responsive to catabolic stimulation with interleukin-1beta. Arthritis Rheum. 2005;52:136–43.

27. Goldring MB, Birkhead J, Sandell LJ, Kimura T, Krane SM. Interleukin 1 suppresses expression of cartilage-specific types II and IX collagens and increases types I and III collagens in human chondrocytes. J Clin Invest. 1988;82:2026–37.

28. Lefebvre V, Huang W, Harley VR, Goodfellow PN, de Crombrugghe B. SOX9 is a potent activator of the chondrocyte-specific enhancer of the pro alpha1(II) collagen gene. Mol Cell Biol. 1997;17:2336–46.

29. Sekiya I, Tsuji K, Koopman P, Watanabe H, Yamada Y, Shinomiya K, et al. SOX9 enhances aggrecan gene promoter/enhancer activity and is up-regulated by retinoic acid in a cartilage-derived cell line, TC6. J Biol Chem. 2000;275:10738–44.

30. Zhang P, Jimenez SA, Stokes DG. Regulation of human COL9A1 gene expression. Activation of the proximal promoter region by SOX9. J Biol Chem. 2003;278:117–23.

31. Kou I, Ikegawa S. SOX9-dependent and -independent transcriptional regulation of human cartilage link protein. J Biol Chem. 2004;279:50942–8.

32. Steck E, Bertram H, Abel R, Chen B, Winter A, Richter W. Induction of intervertebral disc-like cells from adult mesenchymal stem cells. Stem Cells Dayt Ohio. 2005;23:403–11.

33. Winter A, Breit S, Parsch D, Benz K, Steck E, Hauner H, et al. Cartilage-like gene expression in differentiated human stem cell spheroids: a comparison of bone marrow-derived and adipose tissue-derived stromal cells. Arthritis Rheum. 2003;48:418–29.

34. Verma RP, Hansch C. Matrix metalloproteinases (MMPs): chemical–biological functions and (Q)SARs. Bioorg Med Chem. 2007;15:2223–68.

10

Small molecule modulation of splicing factor expression is associated with rescue from cellular senescence

Eva Latorre[1], Vishal C. Birar[2], Angela N. Sheerin[2], J. Charles C. Jeynes[3], Amy Hooper[1], Helen R. Dawe[4], David Melzer[1], Lynne S. Cox[5], Richard G. A. Faragher[2], Elizabeth L. Ostler[2*] and Lorna W. Harries[1*]

Abstract

Background: Altered expression of mRNA splicing factors occurs with ageing in vivo and is thought to be an ageing mechanism. The accumulation of senescent cells also occurs in vivo with advancing age and causes much degenerative age-related pathology. However, the relationship between these two processes is opaque. Accordingly we developed a novel panel of small molecules based on resveratrol, previously suggested to alter mRNA splicing, to determine whether altered splicing factor expression had potential to influence features of replicative senescence.

Results: Treatment with resveralogues was associated with altered splicing factor expression and rescue of multiple features of senescence. This rescue was independent of cell cycle traverse and also independent of SIRT1, SASP modulation or senolysis. Under growth permissive conditions, cells demonstrating restored splicing factor expression also demonstrated increased telomere length, re-entered cell cycle and resumed proliferation. These phenomena were also influenced by ERK antagonists and agonists.

Conclusions: This is the first demonstration that moderation of splicing factor levels is associated with reversal of cellular senescence in human primary fibroblasts. Small molecule modulators of such targets may therefore represent promising novel anti-degenerative therapies.

Keywords: Alternative splicing, Ageing, Resveratrol, Senescence, Fibroblasts

Background

Messenger RNA (mRNA) processing has been implicated as a key determinant of lifespan. Splicing factor expression is dysregulated in the peripheral blood of aging humans, where they are the major functional gene ontology class whose transcript patterns alter with advancing age [1] and in senescent primary human cells of multiple lineages [2]. Splicing factor expression is also an early determinant of longevity in mouse and man [3], and in both species these changes are likely to be functional, since they are associated

with alterations in splice site usage for many genes [1–3]. Recent data suggests that modification of the levels of SFA-1, a core component of the spliceosome, influences lifespan in *C. elegans* through interaction with TORC1 machinery [4]. Diseases for which age is a significant risk factor including Alzheimer's disease [5], Parkinson's disease [6] and cancer [7] are also marked by major changes in the isoform repertoires, highlighting the importance of correct splicing for health throughout the life course. Thus, the loss of fine-tuning of gene expression in ageing tissues and the resulting failure to respond appropriately to intrinsic and extrinsic cellular stressors has the potential to be a major contributor to the increased physiological frailty seen in aging organisms [8].

The splicing process is regulated on two levels. Firstly, constitutive splicing is carried out by the core spliceosome, which recognises splice donor and acceptor sites that define

---

* Correspondence: E.Ostler@brighton.ac.uk; L.W.Harries@exeter.ac.uk
Eva Latorre and Vishal C. Birar are co-first authors.
Richard G. A. Faragher, Elizabeth L. Ostler and Lorna W. Harries are co-senior authors.
[2]School of Pharmacy and Biomolecular Sciences, University of Brighton, Cockcroft Building, Moulsecoomb, Brighton BN2 4GJ, UK
[1]Institute of Biomedical and Clinical Sciences, University of Exeter Medical School, University of Exeter, Barrack Road, Exeter, Devon EX2 5DW, UK
Full list of author information is available at the end of the article

introns and exons. However, fine control of splice site usage is orchestrated by a complex interplay between splicing regulator proteins such as the Serine Arginine (SR) class of splicing activators and the heterogeneous ribonucleoprotein (hnRNP) class of splicing repressors. Splicing activators bind to exon and intron splicing enhancers (ESE, ISE), and splicing inhibitors to intron and exon splicing silencers (ESS, ISS). Splice site usage relies on the balance between these factors and occurs in a concentration-dependent manner [9–11]. Other aspects of information transfer from DNA to protein, such as RNA export and mRNA stability are also influenced by splicing factors [12]. Intriguingly, in addition to their splicing roles, many splicing factors have non-canonical additional functions regulating processes relevant to ageing. For example, hnRNPK, hnRNPD and hnRNPA1 have been shown to have roles in telomere maintenance [13–15], hnRNPA1 regulates the stability of SIRT1 mRNA transcripts [16] and hnRNPA2/B1 is involved in maintenance of stem cell populations [17]. Splicing factor expression is known to be dysregulated in senescent cells of multiple lineages [2] and it is now well established that the accumulation of senescent cells is a direct cause of multiple aspects of both ageing and age-related disease in mammals [18].

Senescent cells accumulate progressively through life in a variety of mammalian species [15], and premature senescence is a hallmark of many human progeroid syndromes. Conversely, dietary restriction, which increases longevity, retards the accumulation of senescent cells. Most compellingly, deletion of senescent cells in transgenic mice improves multiple aspects of later life health and extends lifespan [19]. The mechanisms by which senescent cells mediate these deleterious effects are complex but include factors such as ectopic calcification in the case of vascular smooth muscle cells [20] and secretion of pro-inflammatory cytokines, the well-known Senescence Associated Secretory Phenotype (SASP) [21]. These observations suggest that an interrelationship may exist between well characterised mechanisms of ageing, such as cellular senescence, and the RNA splicing machinery where the mechanistic relationship to ageing remains largely correlational.

In contrast to the situation with core spliceosomal proteins such as SFA-1, perturbation of a single splicing regulator by standard molecular techniques such as knockdown or overexpression is unlikely to be informative for assessment of effects on ageing and cell senescence, since ageing is characterised by co-ordinate dysregulation of large modules of splicing factors [1, 2]. Splice site choice is also dependent on the balance between more than a hundred splicing activator and splicing inhibitor regulatory proteins, which differ from splice site to splice site and from tissue to tissue [9, 10]. Thus experimental tools capable of co-ordinately modulating the expression of multiple components simultaneously are required to address the potential effects of the dysregulation of large numbers of splicing factors that we note during the ageing process. Small molecules such as resveratrol have been reported to influence splicing regulatory factor expression in transformed cell lines such as HEK293 and HeLa [22], although it is not yet known whether this is a direct or indirect effect. Unfortunately, resveratrol has multiple biological effects, including a reduction of pro-inflammatory cytokine expression [23] as well as its canonical activity against SIRT1 [24] thus a 'clean' assessment of the effects of moderation of splicing factor levels on cell physiology cannot be achieved using this compound alone.

We have overcome this limitation through development of a novel library of resveratrol-related compounds (resveralogues) which are all capable of either directly or indirectly influencing the expression of multiple splicing factors of both *SRSF* and *HNRNP* subtypes, whilst exhibiting differential activity against SIRT1 and SASP. Treatment of senescent human fibroblasts from different developmental lineages with any of these novel molecules shifts expression patterns of multiple splicing factors to those characteristic of much earlier passage cells. This change occurs regardless of cell cycle traverse and is associated with a marked decrease in key biochemical and molecular biomarkers of senescence without any significant alteration in levels of apoptosis. Elevated splicing factor expression is also associated with elongation of telomeres, and in growth permissive conditions, these previously senescent populations show significant increases in growth fraction (as measured by Ki67 staining) and in absolute cell number, indicating cell cycle reentry. The mechanisms by which 'rejuvenation' occurs are independent both of SIRT1 activation, or effects on the SASP. Thus, molecules that modulate RNA splicing patterns, either directly or indirectly, may have the potential to delay or reverse cellular senescence with consequent positive impact on human health span.

Results

Synthesis of novel resveralogues

Resveratrol (RSV) has been reported to extend lifespan in various model organisms through activation of the NAD-dependent protein deacetylase, SIRT1 [24], while replenishment of NAD^+ improves lifespan and health span in ATM^- worms and mice [25]. We therefore set out to rationally design a panel of novel resveratrol-like compounds (Fig. 1a) with the goal of identifying compounds that could restore splicing factor expression to levels comparable with those seen in young cells, but with differing effects on SIRT1 activation and the senescence-associated secretory phenotype (SASP) to allow assessment of molecular mechanism. Synthesis of the backbone was achieved as previously reported [26], with additional functionality and diversity achieved via

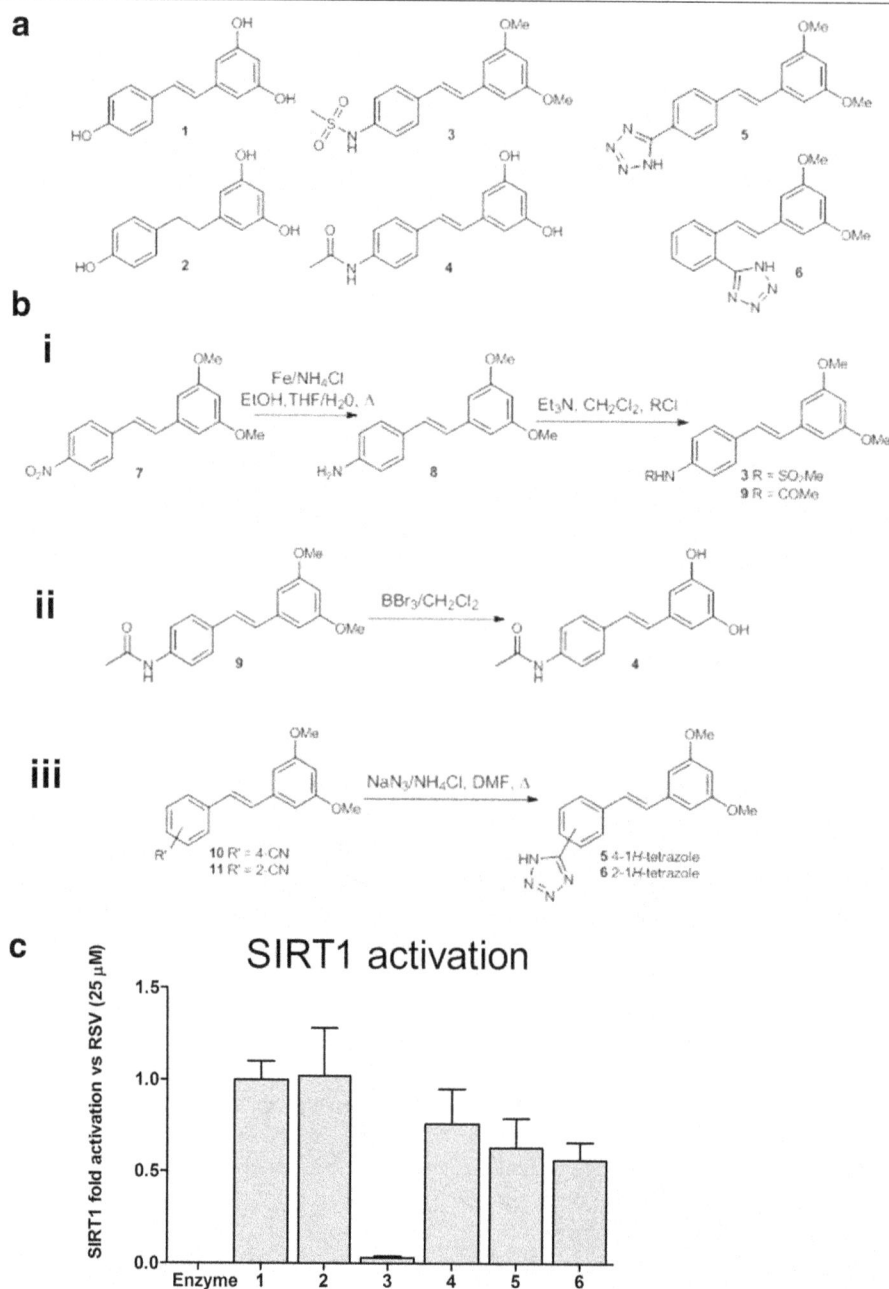

Fig. 1 Synthesis and characterisation of novel resveralogues. **a** Structures of resveralogues 1–6. Compounds are: **1** resveratrol, **2** resveratrol's primary metabolite, dihydroresveratrol, **3** (*E*)-N-(4-(3,5-dimethoxystyryl) phenyl)methanesulfonamide, **4** (*E*)-N-(4-(3,5-dihydroxystyryl)phenyl)acetamide, **5** (*E*)-5-(4-(3,5-dimethoxystyryl)phenyl)-1*H*–tetrazole and **6** (*E*)-5-(2-(3,5-dimethoxystyryl)phenyl)-1*H*–tetrazole. **b** Scheme of synthesis of compounds **3–6** (see Methods for details). **c** Fluorescence determination of SIRT1 activity in vitro in the presence of 25 µM each compound, normalised against resveratrol (1) and vehicle only control (0). Data are presented as fold change (mean ± SD) in activity normalised to enzyme-only (0) and resveratrol (1), such that 0 represents no activation, and 1.0 indicates activation equivalent to that observed with resveratrol 1. The experiment was carried out in 3 replicates. The numbers on the X axis (**1–6**) refer to the identity of each resveralogue as indicated above. Uncertainty was calculated by subjecting the standard deviation of the control, Resveratrol and compound data to combination using standard methods for propagation of uncertainty [49]

functional group interconversion (Fig. 1b). Compounds were chosen for further analysis based on (i) structural novelty and low cytotoxicity (ii) differential SIRT1 activation activity (iii) differential effects on the suppression of SASP components and (iv) previously observed increases in the Ki67 positive fraction of MRC5 cultures at 5 µM. We also included the parent compound (resveratrol) and a major metabolite (dihydroresveratrol).

SIRT1 activation is significantly altered following side chain modification of resveratrol

Since RSV has been suggested to exert its pro-longevity effects predominantly through activation of SIRT1, we first tested the ability of our novel compounds to activate SIRT1 in an ex vivo enzyme assay (Fig. 1c), with data normalised against activity detected on treatment with resveratrol (RSV, 1). While dihydroresveratrol (Fig. 1a, 2) displayed SIRT1-activation activity equivalent to that of resveratrol, the four novel analogues (3–6) displayed a range of activities from zero (compound 3) to around 75% of control levels (compound 4) (Fig. 1c). These marked differences in SIRT1 activation by the novel resveralogues (compared with RSV and DHRSV) therefore allow us to probe SIRT1-dependence of any biological effects.

Impact of resveralogues on the senescence-associated secretory phenotype

We then set out to determine if treatment with resveratrol or the novel resveralogues had an impact on the senescence-associated secretory phenotype (SASP) in senescent cultures of human fibroblasts (NHDF). The levels of multiple cytokines including key SASP components (IL6, IL8, TNFα, IL2, IL1β, IL-12p70, IL10, INFγ and GMCSF) were determined in senescent NHDF by ELISA (Fig. 2). Although each of the compounds altered cytokine profiles to some extent (Fig. 2, see also Additional file 1: Table S1), there was no consistent pattern with which this occurred. Resveratrol 1 was the only compound to reduce

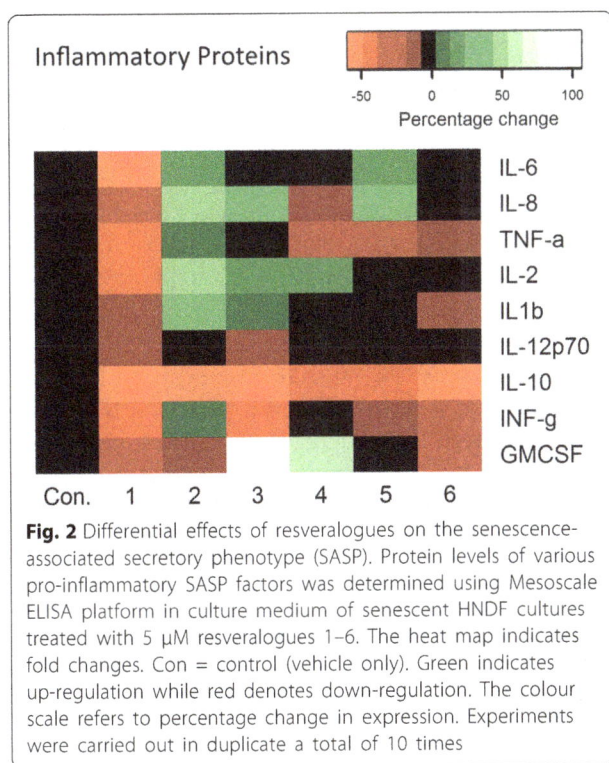

Fig. 2 Differential effects of resveralogues on the senescence-associated secretory phenotype (SASP). Protein levels of various pro-inflammatory SASP factors was determined using Mesoscale ELISA platform in culture medium of senescent HNDF cultures treated with 5 μM resveralogues 1–6. The heat map indicates fold changes. Con = control (vehicle only). Green indicates up-regulation while red denotes down-regulation. The colour scale refers to percentage change in expression. Experiments were carried out in duplicate a total of 10 times

the levels of multiple cytokines including the key SASP mediators IL-6 and IL-8 as well as IL2, TNFα and IFNγ, consistent with previous reports [27]. By contrast, dihydroresveratrol (2) treatment significantly elevated levels of IL-8 and several other inflammatory mediators, whilst 3–6 had variable impact on the expression of the SASP proteins assayed. The only cytokine showing a consistent reduction in level in response to all 6 compounds was IL-10 (Fig. 2, Additional file 1: Table S1).

Splicing factor expression and splicing patterns of senescence-associated genes are restored in senescent cultures of fibroblasts following treatment with resveralogues

To establish whether RSV and the novel resveralogues could influence splicing regulators, we first measured splicing factor expression by qRT-PCR in senescent cultures of human fibroblasts (NHDF) following 24 h treatment with 5 μM of compounds 1–6. Consistent with previous studies in HEK293 cells [22], we find that resveratrol (1) treatment increased levels of both splicing activators (SRSF transcripts) and inhibitors (HNRNP transcripts) (Fig. 3a). Importantly, novel resveratrol analogues also partially restored levels of both splicing activator and inhibitor transcripts (Fig. 3a, Additional file 2: Table S2). The level of restoration of splicing regulator expression in treated cells was similar to levels previously reported in early passage fibroblasts [2]. This reversal of the age-related decline in splicing factor expression was present for compounds with no discernible SIRT activity (compound 3) as well as those that elevated IL6 and IL8 levels (compounds 2 and 5), indicating that the action of splicing factors is independent of SIRT1 and the SASP.

We then asked whether this restoration of a 'youthful' complement of splicing factors is biologically relevant. To do this, we examined the alternative splicing profiles of key genes involved in cellular senescence in senescent NHDF cultures treated with each of the compounds (Fig. 3b). In some cases, it was not possible to distinguish an effect on splicing from effects on transcription, since multiple isoforms were affected with the same directionality. For example, both p14ARF and p16INK4A isoforms of the CDKN2A gene, which increases with cellular senescence [28], were down-regulated in response to treatment with most resveralogues. However, in other cases, only one isoform was affected; the expression of the pro-apoptotic p21b isoform, but not the consensus isoform p21a of the CDKN1A gene was altered, demonstrating an effect on splicing. Similarly, increased expression of the CHK1S isoform of the CHK1 gene, which induces mitosis [29], but not the consensus CHK1 isoform which does not, was seen (Additional file 2: Table S2). SIRT1 mRNA expression was upregulated by treatment with the novel

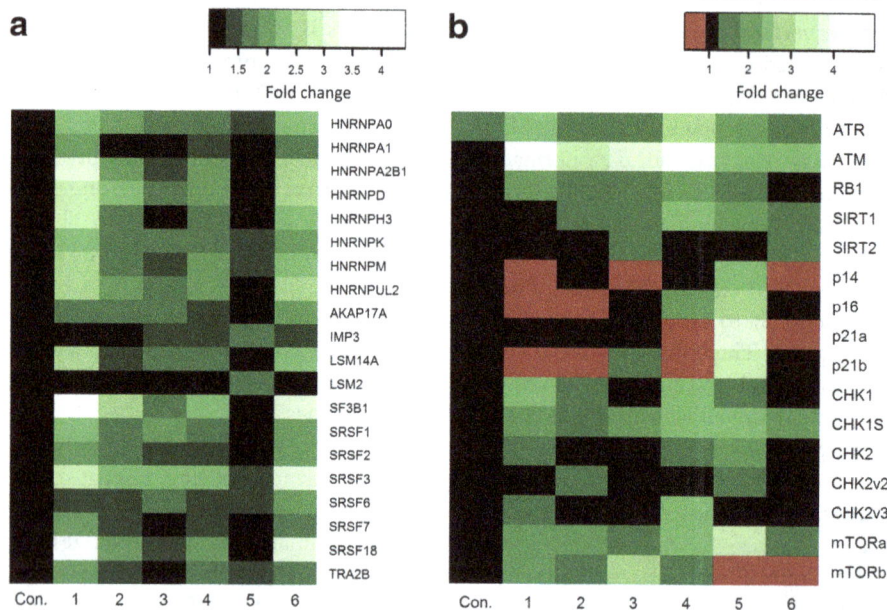

Fig. 3 Splicing factor regulators are elevated following treatment with resveratrol analogues. Changes to mRNA levels in HNDF cells in response to treatment with 5 μM resveratrol (**1**) or resveralogues **2–6** determined by quantitative reverse transcription PCR. **a** Expression of splicing factor regulatory genes (**b**) Isoform-specific transcripts of genes associated with senescence and/or DNA damage responses. Con = control (vehicle only). Green indicates up-regulated genes, red denotes down-regulated genes. The colour scale refers to fold-change in expression. Data are derived from duplicate testing of 3 biological replicates

resveralogues but not RSV itself. A major regulator of cell proliferation and potential driver of senescence is mTOR: inhibition of mTORC1 by rapamycin increases longevity in animal models [30], while mTORC inhibition can reverse multiple phenotypes of cell senescence [31]. We found elevated expression of both the *mTORα* and β isoforms, which regulate cell metabolism and cell proliferation respectively [32], on treatment with resveralogues **1–4** (Fig. 3b, Additional file 2: Table S2), though mTORβ was suppressed on exposure of cells to resveralogues **5** and **6**. Overall, the changes in alternatively-expressed isoforms following resveralogue treatment are consistent with a shift towards a more proliferation-competent repertoire.

Treatment of senescent cells with resveralogues is associated with reduction in biomarkers of senescence

To assess whether restored splicing factor expression was associated with rescue from cellular senescence, we treated senescent cultures of normal human diploid fibroblasts from three genetically distinct cell strains (NHDF and HF043 dermal fibroblasts and MRC5 lung fibroblasts) for 24 h with compounds **3–6**, compared with RSV (**1**) and DHRSV (**2**) and measured transcript levels of senescence biomarkers *CD248* and *CDKN2A* (encoding p16$^{INK4/6}$) by quantitative reverse transcription PCR, normalised against the *IDH3B*, *GUSB* and *PPIA* endogenous control genes, which we have found

to be stable in response to senescence and ageing in our previous work [1, 2]. Stability of control genes to resveralogue treatment was verified empirically. While we observed differences between the cell lineages, there was an overall significant decrease in *CDKN2A* and *CD248* molecular markers of senescence compared with vehicle-only control cell populations (Fig. 4a), which was most marked for the foreskin fibroblast line HF043. To further assess senescence, we examined levels of senescence-associated β galactosidase (SA β-Gal). The percentage of NHDF cells staining positive for SA β-Gal decreased from ~75 to ~25%, compared with much lower levels (~7%) in younger cells at PD25 (Fig. 4b), and similar highly significant reductions in SA β-Gal reactivity were seen in senescent cultures of MRC5 and HF043 fibroblasts (Fig. 4b). These reductions in senescence markers were still evident in NHDF cells 4 weeks after initial treatment and larger reductions occurred following repeated treatments at 48 h intervals (Additional file 3: Figure S1). We conclude therefore that senescence markers are markedly diminished upon resveralogue treatment. Given that compound **3** (which does not activate SIRT1) has very similar effects on these senescence biomarkers compared with resveratrol and other resveralogues with variable SIRT-activation activity (**4, 5, 6**), we can conclude that the decrease in senescence biomarker expression on resveralogue treatment can occur independently of SIRT1 activation.

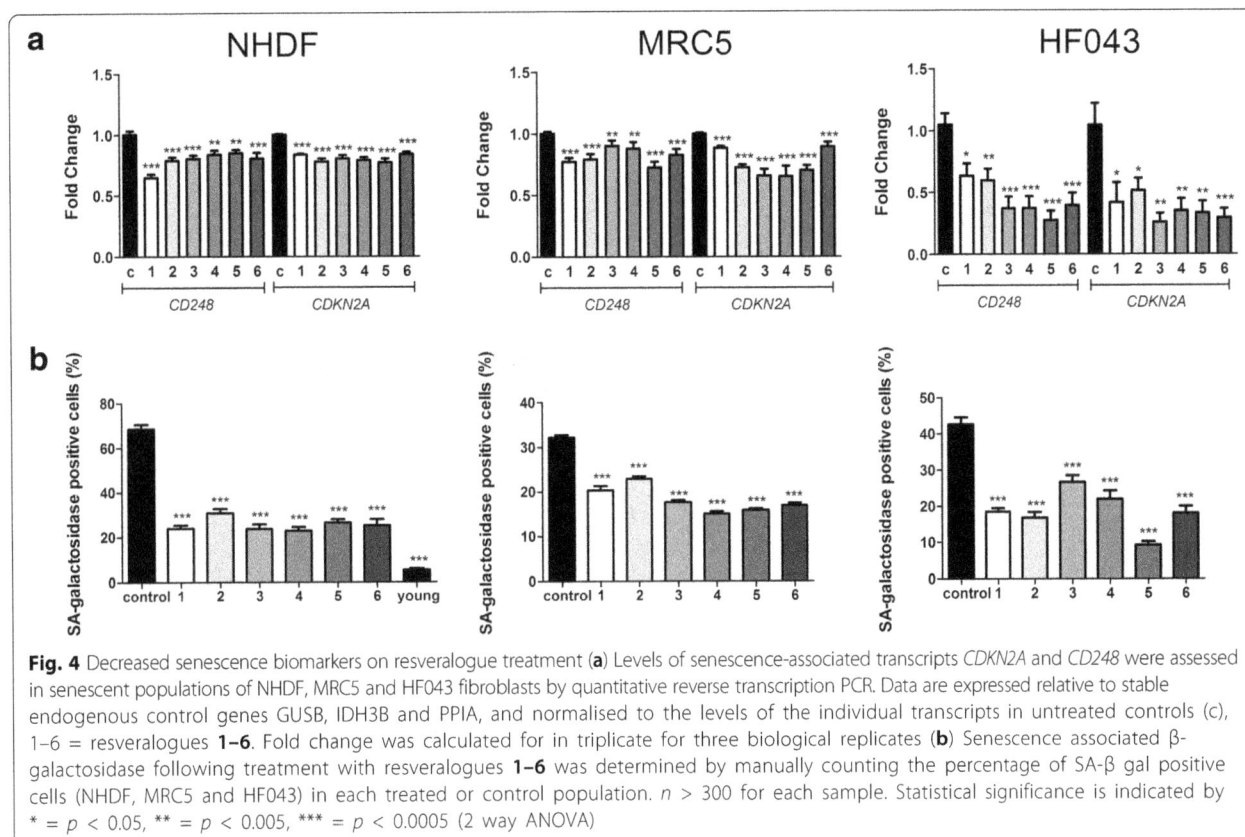

Fig. 4 Decreased senescence biomarkers on resveralogue treatment (**a**) Levels of senescence-associated transcripts *CDKN2A* and *CD248* were assessed in senescent populations of NHDF, MRC5 and HF043 fibroblasts by quantitative reverse transcription PCR. Data are expressed relative to stable endogenous control genes GUSB, IDH3B and PPIA, and normalised to the levels of the individual transcripts in untreated controls (c), 1–6 = resveralogues **1–6**. Fold change was calculated for in triplicate for three biological replicates (**b**) Senescence associated β-galactosidase following treatment with resveralogues **1–6** was determined by manually counting the percentage of SA-β gal positive cells (NHDF, MRC5 and HF043) in each treated or control population. $n > 300$ for each sample. Statistical significance is indicated by * = $p < 0.05$, ** = $p < 0.005$, *** = $p < 0.0005$ (2 way ANOVA)

Treatment of senescent cells with resveralogues is associated with re-entry of cell cycle

While decreases in senescence biomarkers may be beneficial in alleviating some of the detrimental effects of senescent cells, it is the loss of proliferative capacity of senescent cell populations that is likely to lead to stem cell exhaustion and loss of tissue function/frailty with increasing age [33]. We therefore also assessed cell proliferation and re-entry into the proliferative cell cycle. Initially, using live cell imaging of senescent NHDF cells treated with resveratrol for up to 92 h, we found that some cells within this population showed clear evidence of mitosis within a little as 17.5 h after treatment (Additional file 4: Figure S2). We therefore assessed whether senescent populations of three different fibroblasts lines (NHDF, MRC5 and HF043) could undergo mitosis following treatment with the novel compounds. Remarkably, treatment with even very low doses (5 μM) of the resveralogues led to significant increases (up to 0.6 population doublings) in total cell numbers over only 24 h of drug exposure, while vehicle-only controls remained proliferation-arrested (Fig. 5a). Increases in cell number strongly suggest that a significant proportion of cells in the non-cycling senescent population have been induced to re-enter the mitotic cell cycle.

Cell proliferation kinetics are altered in treated cells

To further probe this potential induction of proliferation, the proliferation kinetics of these cultures were determined by immunocytochemical and catalytic histochemical measurement of the levels of the proliferation marker, Ki67, and the senescence marker, SA β-Gal, respectively. Compounds **1–6** induced a consistent increase in the Ki67 positive fraction of cells in senescent NHDF cultures from ~20% of nuclei to ~40%, whereas levels in younger cells at PD25 were > 90% (Fig. 5b), consistent with the findings of increased cell numbers and mitotic figures following drug administration (Fig. 5a, Additional file 4: Figure S2 and data not shown). Since the increased number of cells staining for the proliferation marker Ki67 correlates inversely with the decreased numbers staining for SA-β gal (see Fig. 4b), we suggest that cells have exited senescence to enter the cell cycle.

Treatment of senescent cells with resveralogues is associated with telomere elongation

Telomere shortening is perhaps the best known trigger of cellular senescence. Several splicing factors have been previously demonstrated to unwind telomeres and activate telomerase and could thus potentially lengthen telomeres [13, 14, 34]. We therefore measured telomere length by qPCR in NHDF cells treated with 5 μM

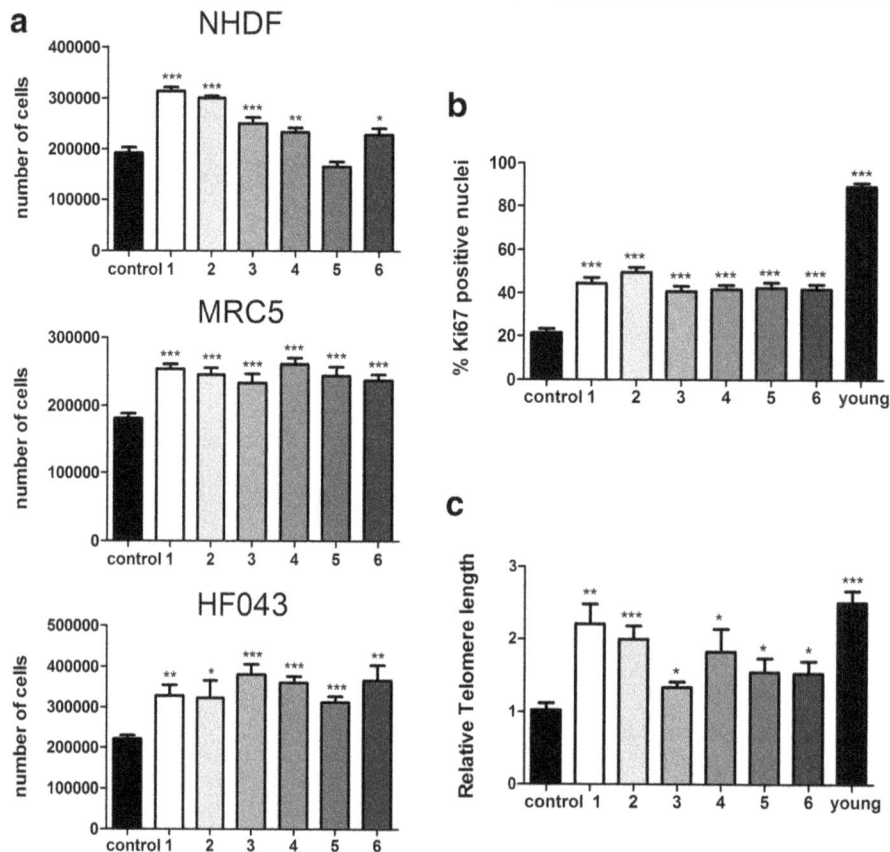

Fig. 5 Increased proliferation of senescent cell populations following resveralogue treatment. **a** Cell numbers of NHDF, MRC5 and HF043 fibroblast populations following treatment with resveralogues **1–6**. Experiments were carried out in triplicate for three biological replicates and *** represents $p < 0.001$ (2 way ANOVA). **b** Proliferation index was assessed for control and treated NHDFs, as well as younger (PD25) cells as assessed by Ki67 immunofluorescence (> 400 nuclei counted per sample, *** $p < 0.001$ by 2 way ANOVA). **c** Telomere length was quantified by qPCR relative to the *36B4* endogenous control and normalised to telomere length in vehicle-only controls, younger passage cells (PD25) and in cells treated with compounds **1–6**. Experiments were carried out in triplicate for three biological replicates. Statistical significance is indicated by * = $p < 0.05$, ** = $p < 0.005$, *** = $p < 0.0005$ (2 way ANOVA)

resveratrol or resveralogues for 24 h, relative to telomere length in untreated cells. We found that cells treated with resveratrol or any of the novel resveralogues had telomeres that were 1.3–2.4 times longer than vehicle-only controls, compared with younger cells at PD25, which showed telomeres 2.6 times longer than untreated senescent cells (Fig. 5c).

Changes in splicing factor expression and senescence markers are not effects of cell proliferation

To determine whether the changes in splicing factor expression were a cause or consequence of renewed cell proliferation, we measured splicing factor expression and selected senescence markers under low serum conditions, which would induce proliferating cells to enter quiescence. Unsurprisingly, serum-starved cultures demonstrated no increase in cellular proliferation in response to resveralogue treatment, as determined by lack of an observable increase in cell numbers (Fig. 6a) or

Ki67 index (Fig. 6b) in treated cells. However effects on both senescence markers (Fig. 6c) and splicing factor expression (Fig. 6d) were still observed, indicating that the effects on senescence and splicing factor expression were independent of proliferation. Uncoupling rescue from proliferation also allows us to quantify more precisely the percentage of cells in which senescence has been reversed from the dilution effect of increased cell number. The number of 'reverted' cells is ~15%, which is similar to the levels we had predicted based on the cell proliferation kinetics.

Decrease in senescent cell fraction is not due to selective death of senescent cells

To exclude the possibility that the decrease in the percentage of senescent cells following treatment resulted from selective cell death of non-proliferating cells, cytotoxicity was assessed using an assay for extracellular lactate dehydrogenase (LDH); this intracellular enzyme is

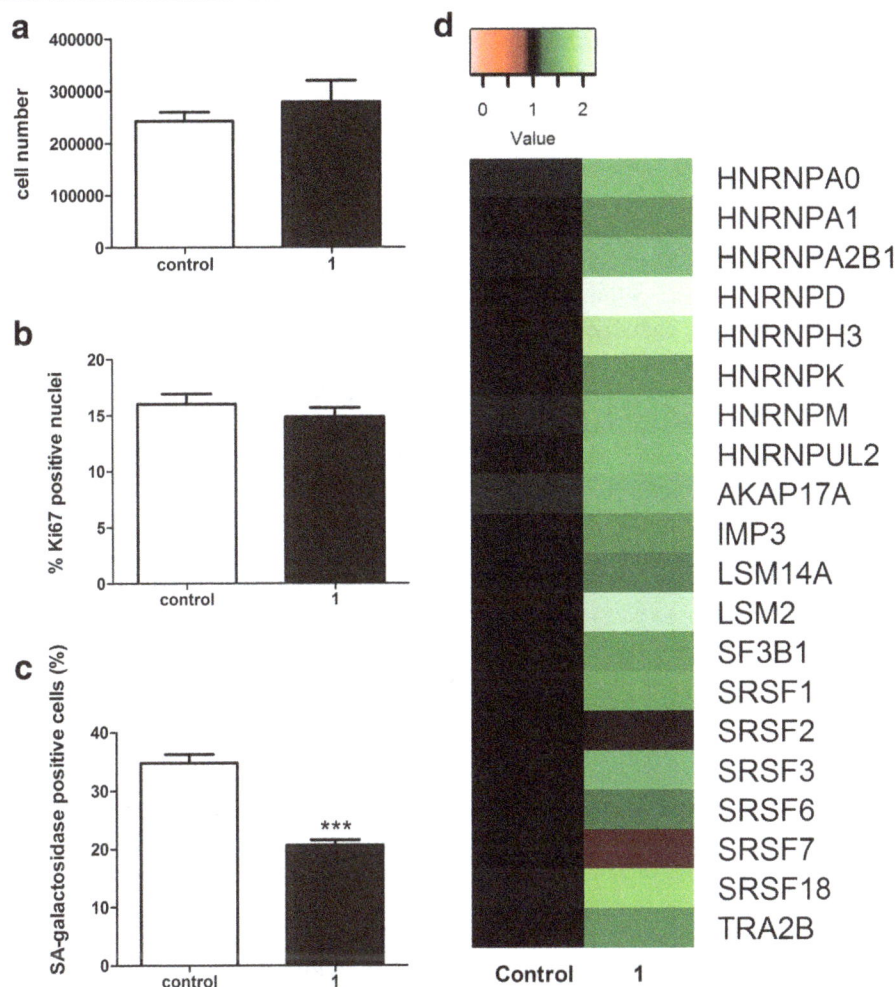

Fig. 6 Effects of resveratrol treatment in cells grown under serum starvation conditions. **a** Cell numbers of NHDF fibroblasts following treatment with 5 μM resveratrol for 24 h under conditions of serum starvation. Experiments were carried out in triplicate for three biological replicates. (2 way ANOVA). **b** Proliferation index was assessed for NHDF fibroblasts following treatment with 5 μM resveratrol for 24 h under conditions of serum starvation as assessed by Ki67 immunofluorescence (> 400 nuclei counted per sample). **c** Senescence associated β-galactosidase following NHDF fibroblasts following treatment with 5 mM resveratrol for 24 h under conditions of serum starvation was determined by manually counting the percentage of SA-β gal positive cells in each treated or control population. *n* > 300 for each sample. Statistical significance is indicated by *** = *p* < 0.0005 (2 way ANOVA). **d** Changes to splicing factor mRNA levels in NHDF fibroblasts following treatment with 5 μM resveratrol for 24 h under conditions of serum starvation determined by qRTPCR. Control = vehicle only. Green indicates up-regulated genes, red denotes down-regulated genes. The colour scale refers to fold-change in expression. Data are derived from duplicate testing of 3 biological replicates

only released into the culture medium upon cell death. In all cases, cells treated with the novel compounds released lower levels of LDH than those treated with RSV (at doses up to 100 μM) in comparison with vehicle only controls (Additional file 5: Figure S3); compound **6** in particular showed very low levels of LDH release. These results demonstrate low cytotoxicity of dihydroresveratrol and all four novel resveralogues.

While necrotic cell death was not detected, it was important to rule out selective loss of senescent cells by apoptosis. Levels of apoptosis in senescent NHDF cultures treated with resveralogues **1–6** were determined by both TUNEL and by Caspase 3 and 7 assays (Additional file 5:

Figure S3B and C). No increases in levels of apoptosis were observed in the resveralogue-treated cultures compared with vehicle-only control treatments, suggesting that the increased proliferation on resveralogue treatment was not a consequence of selective death of non-proliferating cells within the population.

ERK agonists and antagonists influence cellular senescence and splicing factor expression

ERK signalling has previously been reported to be influenced by resveratrol [35, 36]. ETS-1, a transcription factor downstream of ERK activation has also been reported to regulate the expression of *TRA2B*, an important splicing regulator [37]. To investigate the potential interplay between resveratrol and ERK signalling on splicing factor expression and cellular senescence phenotypes, we treated senescent NHDF cells with low dose (1 μM or 10 μM) of trametinib, a well-characterised signalling inhibitor that inhibits the ERK signalling pathway. Treatment of senescent cells with trametinib resulted in a robust decrease in the proportion of senescent cells in the culture, which was apparent at 1 μM and 10 μM, but not at 20 μM. Such dose effects are not uncommon in signalling pathways due to interconnectivity with other signalling pathways and autoregulation (Additional file 6: Figure S4A). Conversely, treatment with the ERK agonist ceramide resulted in a comparable increase in the senescent cell fraction after 24 h. Notably, the effect of ceramide was negated by the addition of 5 μM of any of the novel resveralogues (Additional file 6: Figure S4B). Trametinib also restored splicing factor expression to profiles consistent with earlier passage in a manner similar to that observed with the resveralogues (Additional file 7: Figure S5).

Discussion

We have generated a panel of novel molecules based on the small molecule resveratrol, to determine whether alteration to regulators of mRNA processing could influence cellular senescence phenotypes in human fibroblasts of different lineages. Treatment of senescent cultures of cells from different genetic backgrounds with these novel molecules was associated with an increase in the expression of multiple splicing factors, to levels consistent with those seen in early passage cells [2], although at present it is not clear whether these are direct or indirect effects. Treatment with all 6 resveralogues also resulted in a decline in the senescent cell fraction, along with changes to the splicing patterns of genes involved in cell senescence to a profile indicative of much 'younger' cells. Our evidence suggests cells have also re-entered the cell cycle, as determined by an increase in markers of cell division with concurrent increases in cell number. Finally, in accordance with the reported role of some splicing factors on telomere accessibility and telomerase activity [13–15], telomere length was lengthened in treated cells, consistent with a 'resetting' of the telomere clock. The absence of any elevation of either necrotic (LDH release) or apoptotic cell death (TUNEL and caspase), also excludes the possibility that the dramatic decline in the senescent fraction results from selective killing of senescent cells by resveralogues.

Disruption to splicing factor transcript expression levels is known to be a major feature of ageing in humans [1] and also in senescent human primary cell lines of multiple lineages which have undergone in 'ageing' by repeated culture in vitro [2]. Splicing factor expression is also associated with lifespan in humans and also in mice, where their expression appears to be early-life determinants of longevity [3].

Recent data adds weight to this hypothesis, since abolition of the core splicing factor 1 (SFA-1) alone was to reduce lifespan in *C.elegans* by interaction with the TORC1 pathway [4]. Splicing factors are also known to be drivers of cell proliferation [38–40], through effects both on splicing patterns, and through their non-canonical roles in telomere maintenance [13–15, 17, 34]. Telomere maintenance is critical in permitting cell proliferation; restoration of hTERT allows prematurely senescing human Werner syndrome fibroblasts to proliferate with kinetics of wild type cells [41]. Splicing factors hnRNPK and hnRNPD interact with the hTERT promoter while knockdown of hnRNPD notably reduces transcription of the telomerase gene [13, 14]. Additionally, hnRNPA1 is required for telomere maintenance in multiple species and has been proposed to facilitate the access of telomerase to the telomere [15].

The question of whether senescence drives splicing changes, or whether splicing alterations are causative of cell senescence in different species, tissues and points in the life course is a challenging and multifaceted one. The conventional approaches to answer this question are intractable in this system, since there are over 100 splicing factors involved in regulation of splicing, with exon usage determined by the balance of activators and inhibitors at each individual splice site [10]. The pattern and dosage of splicing factors involved will also differ from splice site to splice site and from tissue to tissue. There is also redundancy between splicing factors, both in terms of regulation of splicing, and also in their non-canonical roles - at least 6 hnRNPs and some SRSF proteins are known to have effects on telomere structure or telomerase activity. However, in this initial study the issue of causality can be distilled down to whether the changes in splicing factor expression we observe on resveralogue treatment drive rescue from senescence or are a consequence of re-entry into cell cycle. The observation that the alteration in splicing factor expression and the decrease in numbers of senescent cells occurs when proliferation is blocked provides evidence to suggest effects we note may lie upstream in the causal pathway (Fig. 5).

Presently, it is not possible to attribute specific splicing changes to alterations in the levels of specific splicing factors. Splice site choice is governed by the balance of activators and inhibitors at individual splice sites, and the binding sites are short and degenerate [10]. Similarly, in some cases, it is not possible to determine whether the effects on expression we note are transcriptional or due to splicing on the basis that expression changes of alternative isoforms share directionality. However, for other genes, the effect is confined to specific isoforms, clearly indicating an effect on splice site choice. Another caveat to our work is that we have assessed expression changes at the level of mRNA only. This is due to the inherent difficulty in culturing sufficient quantities of senescent cells to allow large scale protein analysis.

At present, the specific mechanism(s) by which resveralogues may influence splicing factor expression and senescence phenotypes in our work are not clear. Resveratrol has previously been demonstrated to have beneficial effects on senescence phenotypes through other pathways such as SIRT1 activity [24] and also through effects on the senescence-associated secretory phenotype (SASP) [23]. Our data suggest that resveralogues can influence splicing factor expression and cell division in senescent cultures independently of SIRT1 activity, since one compound, molecule 3 has no discernible SIRT1 activity (Fig. 1), despite an induction of *SIRT1* at the mRNA level. The action of resveratrol on SIRT1 is at the level of enzyme activation. In the case of compound 3, although there appears to be an effect on transcription, this compound is not able to activate the translated protein. Our data are also consistent with earlier studies in siRNA SIRT1 knockout cells which demonstrated that the effect of resveratrol on splicing factor expression occurs irrespective of SIRT1 activity [22]. Similarly, although resveralogues 1–6 display very similar effects on splicing factor expression, ability to supress the SASP varies widely (as shown in Fig. 2). Indeed, treatment with compound 2 significantly elevates levels of IL-8, one of the canonical cytokines that causes paracrine senescence (alongside IL-6). The only consistent change to cytokine levels that we detect is a reduction in IL-10, which is not growth suppressive.

Resveratrol has been reported to modulate the ERK pathway [35, 36]. ERK signalling has previously been suggested as a potential regulator of splicing factor expression [37]. Indeed, ETS1, a downstream target of ERK signalling has previously been reported to regulate the expression of TRA2B [37]. ERK inhibition has also been demonstrated to suppress cellular senescence [42] and to influence lifespan in animal models [43]. Our data are consistent with these observations, since alterations to ERK signalling with ERK antagonists was also associated with altered splicing factor expression and senescence phenotypes. ERK agonists were also able to ameliorate the effects of resveratrol on both phenotypes. At present, however, we cannot state definitively that this is the primary mode of induction of these effects, given the context and cell type dependence of ERK signalling, and the existence of crosstalk with other pathways. Interpretation of data are also made more complicated by the observation that even a population of senescent cells derived from a single 'young' culture is actually fairly heterogeneous, consisting of deeply senescent, newly senescent and pre-senescent cells. Within a senescent cell culture, there are also several routes by which those cells may have become senescent. These include replicative senescence, mitochondrial senescence, oncogene-induced senescence, paracrine senescence and autocrine

senescence. At the present time, it is unclear whether all subpopulations respond to resveralogue treatment equivalently, or whether cells that have become senescent via different routes respond equivalently to resveralogues.

There is already considerable interest in the development of drugs that can attenuate senescence for eventual human use. Notable successes have come from overcoming apoptosis in senescence using Bcl-2 inhibitors [44], and by modifying mTORC signalling using rapamycin and other rapalogues or ATP mimetics specific for the mTOR kinase active site [31]. SIRT1 is also a current target for drug design and for nutraceutical interventions and is known also to be activated by resveratrol. We suggest that focusing on SIRT1 activity alone may be misleading and that other pathways activated by resveralogues may be more important in alleviating senescence and improving health outcomes in later life. The renewal of proliferation we observe upon resveralogue treatment obviously raises questions about the potential cancer risk attached to such treatment, should it eventually be employed in a clinical setting. We propose that the renewed proliferation arises from a transient increase in telomerase activity brought about by the induction of specific splicing factor proteins, and that the growth is still regulated. This is in accordance with observations that treatment with resveratrol has been suggested to have a protective effect against cancer in both humans and rodent models [45, 46].

Conclusions

During the ageing process, both senescent and non-senescent cells lose a degree of response to cellular stressors. The upstream causes of this are as yet unclear, but may include changes in genes controlling alternative splicing; a major regulator of gene expression which ensures genomic plasticity. Here, we provide evidence that treatment with novel analogues of the stilbene compound resveratrol is associated not only with restoration of splicing factor expression but also with amelioration of multiple cellular senescence phenotypes in senescent human primary fibroblasts. At present, the precise mechanisms behind these observations are unclear, but may involve both the restoration of a more 'youthful' pattern of alternative splicing, and also effects of specific splicing factors on telomere maintenance. We propose therefore that splicing factors, and the upstream drivers of splicing factor expression may prove promising as druggable targets to ameliorate ageing phenotypes and hold promise as anti-degenerative compounds effective in human cells in the future.

Methods
Synthesis of novel resveralogues
Resveratrol (Sigma Aldrich, UK; 1) was used to synthesise dihydroresveratrol 2 as reported previously [47]. (*E*)-N-(4-(3,5-Dimethoxystyryl)phenyl) methanesulfonamide 3 was

synthesised from the previously reported nitro-substituted analogue **7** via an Fe/NH_4Cl reduction to give amine **8** [26], followed by sulfonylation with methanesulfonyl chloride (Fig. 1b). The corresponding amide **9** was also prepared from **8**, by acylation with acetylchloride. The product **9** was subjected to demethylation (BBr_3, CH_2Cl_2) to give the target compound (E)-N-(4-(3,5-dihydroxystyryl)phenyl)acetamide **4**. (E)-5-(4-(3,5-dimethoxystyryl)phenyl)-1H–tetrazole **5** and the isomeric 2-1H-tetrazole analogue **6** were prepared directly via acid-catalysed cycloaddition with azide ion from the 4- and 2-cyanostilbenes [26] (**10** and **11** respectively). (Fig. 1b) Details of the synthesis, purification and characterisation of the resveralogues are given in Additional file 8.

Determination of SIRT1 enzyme activation

SIRT1 enzyme activity was measured by using the SIRT1 Fluorometric drug discovery kit (Cayman Chemicals, Michigan, USA) according to the manufacturer's instructions. This assay is a standard direct fluorescent screening assay for SIRT1 ex-vivo and is essentially a variant of the well-known "fluor de lys" system. For determination of the relative capacity of each resveralogue to activate the enzyme, 25 μM solution of each compound ($n = 3$) was preincubated with the enzyme and co-factors before measurement of activity. Quantification was achieved by measuring output at $\lambda_{ex} = 360$ nm and $\lambda_{em} = 460$ nm. Each plate included background measurements and enzyme-only controls. Data are presented as fold change (mean ± sd) in activity normalised to enzyme-only and resveratrol **1**, such that 0 represents no activation, and 1.0 indicates activation equivalent to that observed with resveratrol **1**.

Determination of cytotoxicity of resveralogue library

A commercial LDH release assay (Pierce LDH Cytotoxicity Assay Kit) was used to determine cell death. Briefly, MRC5 cells (at population doubling (PD) = 45) were seeded in 24 well plates at 1.3×10^5 cells/cm^2 and allowed to recover from trypsinisation for 24 h then exposed to each of the resveralogues (3 biological replicates × 3 concentrations; 10, 50 and 100 μM) for a further 24 h. 50 μl of media from each well was then mixed with an equal volume of LDH assay reaction mixture and incubated at room temperature in the dark for 30 min. 50 μl of LDH assay stop solution was added to each well and the absorbance of the solution was measured by spectrophotometry at 490 nm. Complete lysis and vehicle only positive and negative controls were included. Data are presented as mean (+/–standard deviation) % of the total lysis control.

Culture of human primary fibroblasts (NHDF, MRC-5 and HF043)

Fibroblast cell strains of three genetic backgrounds and two lineages were used in this study: normal human dermal fibroblasts (NHDF; Heidelburg, Germany), human

diploid foetal lung fibroblasts (MRC-5; Coriell Institute for Medical Research) and neonatal foreskin fibroblasts (HF043; Dundee Cell Products, UK). Standard culture conditions were a seeding density of 6×10^4 cells/cm^2 in media (C-23020, Promocell, Heidelburg, Germany) containing 1% penicillin and streptomycin, and a fibroblast-specific supplement mix consisting of foetal calf serum (3% v/v), recombinant fibroblast growth factor (1 ng/ml) and recombinant human insulin (5 μg/ml) (Promocell, Heidelburg, Germany). For the assays requiring senescent cultures, cells were counted and equal numbers of cells seeded at each passage until the growth of the culture slowed to less than 0.5 PD/week as previously described [2] (this occurred at PD = 64 (NHDF), 65 (MRC-5) and 64 (HF043). Viable cell numbers were determined at each passage by trypan blue staining. For cultures grown under serum starvation conditions, cells were maintained in DMEM (Sigma Aldrich, Dorset, UK) supplemented with 0.1% of serum and 1% penicillin and streptomycin in the absence of fibroblast-specific supplement, for 24 h prior to treatment.

Quantification of secretion of key cytokines

NHDF cells from a senescent culture were seeded at 6×10^4 cells/cm^2 in a 6 well plate in serum-free media, and after 10 days were treated with 5 μM of each of **1–6** for 24 h. Cell supernatants were then harvested and stored at –80 °C. Levels of 9 cytokines (GMCSF, IFNγ, IL1β, IL2, IL6, IL8, IL10, IL-12p70, and TNFα) in cell supernatants from treated and vehicle-only control cells were determined using the K15007B MesoScale Discovery multiplex ELISA immunoassay (MSD, Rockville, USA) in 11 replicates. Proteins were quantified relative to a standard curve using a Sector Imager SI-6000 according to the manufacturer's instructions. Data are presented as mean (+/–SEM).

Expression profiling of splicing factor expression in cultures of senescent cells

NHDF cells were seeded at 6×10^4 cells/cm^2 in 6 well plates, allowed to grow for 10 days then treated with 5 μM of each compound for 24 h in 3 biological replicates, with vehicle only controls (DMSO). Resveratrol **1** acute treatment was at an initial dose of 5 μM, followed by culture without further treatment for 4 weeks. For chronic treatment regimes, resveratrol (or DMSO vehicle) was added once every 48 h during 4 weeks. 20 splicing factor transcripts that associated with age and replicative senescence in our previous work [1, 2] were selected a priori for assessment here. (Assay identifiers are available on request). RNA was extracted by using 1 ml of TRI reagent ® (Life Technologies, Foster City USA) according to the manufacturer's instructions. Total RNA (100 ng) was reverse transcribed in 20 μl reactions using the Superscript III VILO kit (Life

Technologies, Foster City, USA). Transcript expression was then quantified in triplicate for each biological replicate using TaqMan Low Density Array (TLDA) on the ABI-Prism 7900HT platform. Cycling conditions were 1 cycle each of 50 °C for 2 min, 94.5 °C for 10 min and then 40 cycles of 97 °C for 30 s and 57.9 °C for 1 min. The reaction mixes included 50 μl TaqMan Fast Universal PCR Mastermix (Life Technologies, Foster City, USA), 30 μl dH2O and 20 μl cDNA template. 100 μl reaction mixture was dispensed into the TLDA card chamber and centrifuged twice for 1 min at 1000 rpm to ensure correct distribution of solution to each well. Transcript expression was assessed by the Comparative Ct approach, relative to the *IDH3B*, *GUSB* and *PPIA* endogenous control genes, selected on the basis of empirical evidence for stability with age in our earlier microarray data [1] and with cellular senescence in our earlier work [2]. Transcript expression was expressed relative to the level of splicing factor expression in vehicle treated control cells.

Assessment of total gene expression and alternative splicing for senescence-related genes

To assess gene expression and splicing, NHDF cells were seeded at 6×10^4 cells/cm^2 in 6 well plates and after 10 days were treated with 5 μM of each compound for 24 h in 3 biological replicates. Target transcripts included the known age-related genes *CDKN2A*, *CDKN1A*, *TP53*, *MTOR*, *CHK1* and *CHK2*. Probes specific to particular isoforms or groups of isoforms were designed to unique regions of the transcripts in question. Assays were validated by standard curve analysis of 7 serial 1:2 dilutions of pooled cDNA and proved robust and sensitive with an average efficiency of −3.4 and an average r^2 for reproducibility between replicates of 0.87. PCR reactions contained 2.5 μl TaqMan Universal Mastermix (no AMPerase) (Applied Biosystems, Foster City, USA), 0.9 μM each primer, 0.25 μM probe and 0.5 μl cDNA reverse transcribed as above in a total volume of 5 μl in a 384 well plate. PCR conditions were a single cycle of 95 °C for 10 min followed by 40 cycles of 95 °C for 15 s and 60 °C for 1 min. We also measured the total expression of the *ATR*, *ATM*. *RB1*, *SIRT1* and *SIRT2* genes. Probe and primer details are available on request. Each biological replicate was tested in triplicate. Isoform-specific and total expression changes were examined for statistical significance by two way ANOVA analysis using SPSS v.22 (IBM, USA).

Catalytic histochemical determination of SA β-gal positive fraction

Senescence marker SA β-Gal was assayed in triplicate using a commercial kit (Sigma Aldrich, UK); according to manufacturer's instructions, with a minimum of 400 cells assessed per replicate.

qRTPCR measurement of transcripts of senescence associated genes

Molecular markers of senescence (*CDKN2A* and *CD248* transcript levels) were measured by qRTPCR relative to the *GUSB* and *PPIA* endogenous control genes, on the ABI Prism 7900HT platform. PCR conditions and analysis were as previously described [2].

Live cell capture microscopy

For live capture microscopy, cells were seeded at a density of 5×10^4 cells per 35 mm glass bottomed dish (World Precision Instruments, USA) in 2 ml of media. They were then imaged on a Leica Axiovert inverted environmental microscope with heated chamber (37 °C) and CO$_2$ capabilities. An image was taken every 10 min over the course of 92 h. A 20× objective was used with a 30 mS shutter speed and 10% light intensity from a widefield white light source for each image giving optimal contrast and minimal light exposure. Images were analysed using Leica LAS X software.

Determination of cell proliferation

Senescent cultures of each strain were seeded at 6×10^4 cells/cm^2 into 6-well plates and cultured for 10 days then treated with 5 μM of each compound for 24 h. Cell counts in three replicates of treated and vehicle-only cultures were carried out manually following trypsinisation and suspension of cells and are presented as mean (+/−SEM).

Immunocytochemical determination of Ki67 positive fraction

Proliferation index was assessed by using Ki67 staining on NHDF cells. Cells were seeded at 1×10^4 cells/coverslip and after 10 days were treated with 5 μM of each compound for 24 h in 3 biological replicates. Cells were fixed for 10 min with 4% PFA and permeabilized with 0.025% Triton and 10% serum in PBS for 1 h. Cells were then incubated with a rabbit monoclonal anti-Ki67 antibody (ab16667, Abcam, UK) at 1:200 overnight at 4 °C followed by FITC-conjugated secondary goat anti-rabbit (1:400) for 1 h, and nuclei were counterstained with DAPI. Coverslips were mounted on slides in DAKO fluorescence mounting medium (S3023; Dako). The proliferation index was determined by counting the percentage of Ki67 positive cells from at least 400 nuclei from each biological replicate at 400× magnification under a Leica D4000 fluorescence microscope.

Assessment of apoptosis using TUNEL assay

Terminal DNA breakpoints in situ 3 - hydroxy end labeling (TUNEL) was to quantify levels of apoptosis in NHDF cells. Cells were seeded at 1×10^4 cells/cm^2 in 6 well plates and after 10 days were treated with 5 μM of each compound for 24 h in 3 biological replicates. The TUNEL assay was performed with Click-iT® TUNEL Alexa Fluor® 488 Imaging Assay kit (Thermofisher, UK) following the manufacturer's instructions. Negative and positive (DNase1) controls were also performed. The apoptotic index was determined by counting the percentage of positive cells from at least 400 nuclei from each biological replicate at 400× magnification.

Assessment of apoptosis by assessment of Caspase 3 and 7 activity

Caspase-3 and-7 activities were assessed as secondary measures of apoptosis. Cells were seeded (1000 cells per well) in a white-walled 96-well plate and then treated with 5 μM of each compound for 24 h in 11 biological replicates alongside vehicle-only controls. Caspase-3 and -7 activities in the supernatants were then measured by Caspase-Glo 3/7 assay (Promega, Madison, WI, USA) following the manufacturer's instructions. Luminescence was measured by using a BMG Pherastar FSX.

Moderation of ERK signalling pathway with inhibitors and agonists

The role of ERK signalling in reversal of senescence was investigated using agonists (ceramide) and inhibitors (trametinib) of the ERK pathway. Cells from a senescent culture were seeded at 6×10^4 cells/cm^2 in a 6 well plate in serum free media, and after 10 days were treated with 1-20 μM of the ERK inhibitor trametinib (LC laboratories, Woburn, USA for 24 h hours, or with the ERK agonist N-Acetyl-D-sphingosine (C2-ceramide; Sigma Aldrich, UK) at 20 μM for 24 or 120 h. To examine the role of ERK signalling in resveralogue-induced rescue of senescence, HNDF cells were treated with 20 μM of the ERK agonist C2-ceramide as above, but with the addition of 5 μM resveralogue for 24 h.

Assessment of telomere length in resveratrol treated cells

DNA was extracted from 2×10^5 NHDF cells treated with 5 μM resveralogue for 24 h, using the PureLink® Genomic DNA Mini Kit (Invitrogen™/Thermo Fisher, MA, USA) according to the manufacturer's instructions. DNA quality and concentration was checked by Nanodrop spectrophotometry (NanoDrop/Thermo Fisher, MA, USA). Relative telomere length was assessed by a modified qPCR protocol [48]. PCR reactions contained 1 μl EvaGreen (Solis Biodyne, Tartu, Estonia), 2 μM each primer and 25 ng DNA in a total volume of 5 μl in a 384 well plate. PCR conditions were a single cycle of 95 °C for 15 min followed by 45 cycles of 95 °C for 10 s, 60 °C for 30 s and 72 °C for 1 min. Telomere length was calculated using the comparative Ct approach relative to the *36B4* housekeeping gene and normalised to the quantification from untreated cells. Three biological replicates were tested and each was assessed in triplicate.

Statistical analysis

Unless otherwise indicated, differences between treated and vehicle-only control cultures were assessed for statistical significance by two way ANOVA analysis using SPSS v.22 (IBM, USA).

Additional files

Additional file 1: Table S1. Changes in inflammatory proteins following treatment with resveratrol analogues. (DOCX 13 kb)

Additional file 2: Table S2. Splicing factor expression and changes in alternative splicing following treatment with resveratrol analogues. (DOCX 19 kb)

Additional file 3: Figure S1. Changes in biochemical and molecular markers of cellular senescence following chronic or repeated treatment with resveratrol. (TIFF 154 kb)

Additional file 4: Figure S2. Live cell capture image following resveratrol treatment. (TIFF 243 kb)

Additional file 5: Figure S3. Level of necrosis and apoptosis following treatment with resveratrol analogues. (TIFF 230 kb)

Additional file 6: Figure S4. The effect of manipulation of the ERK pathway with chemical inhibitors and agonists on cellular senescence. (TIFF 193 kb)

Additional file 7: Figure S5. The effect of ERK inhibition on splicing factor expression. (TIFF 143 kb)

Additional file 8: Synthesis and characterisation of resveralogues. (PDF 3019 kb)

Abbreviations

DHRSV: Dihydroresveratrol; ESE: Exon splicing enhancer; ESS: Exon splicing silencer; HF043: Neonatal foreskin fibroblast; hnRNP: heterogeneous ribonucleoprotein; ISE: Intron splicing enhancer; ISS: Intron splicing silencer; LDH: Lactate dehydrogenase; MRC5: Human diploid foetal lung; NHDF: Normal human dermal fibroblast; RSV: Resveratrol; SA β-Gal: Senescence-associated β galactosidase; SASP: Senescence associated secretory phenotype; SR: Serine arginine class; TLDA: TaqMan low density array

Acknowledgements

The authors would like to thank Nicola Jeffery and Ben Lee for technical assistance, and Luke Pilling for help in preparing the heat maps.

Funding

This work was supported by The Dunhill Medical Trust [grant number: R386/1114] to Lorna Harries, a studentship to Vishal Birar from the University of Brighton and Glenn Foundation for Medical Research personal Awards to Richard Faragher and Lynne Cox. Work in Lynne Cox's lab is also supported by BBSRC grant [BB/M006727/1].

Authors' contributions

EL carried out the majority of the experiments and reviewed the manuscript. VB carried out the synthesis of the novel compounds. AS contributed to the chemical synthesis of the compounds. JCCJ carried out the live cell microscopy experiments. AH provided technical help for the ERK inhibitor and agonist experiments. HRD oversaw and interpreted the apoptosis assays. DM reviewed the manuscript. LSC supplied senescent fibroblasts and interpreted the data. RGAF interpreted the data, contributed to project direction and reviewed the manuscript, EO managed the chemical aspects of the study interpreted the data and reviewed the manuscript. LWH managed the molecular aspects of the study, interpreted the data and wrote the manuscript. All authors read and approved the final manuscript.

Competing interests

The authors declare that they have no competing interests.

Author details

[1]Institute of Biomedical and Clinical Sciences, University of Exeter Medical School, University of Exeter, Barrack Road, Exeter, Devon EX2 5DW, UK. [2]School of Pharmacy and Biomolecular Sciences, University of Brighton, Cockcroft Building, Moulsecoomb, Brighton BN2 4GJ, UK. [3]Centre for Biomedical Modelling and Analysis, University of Exeter, Exeter, Devon EX2 5DW, UK. [4]College of Life and Environmental Sciences, University of Exeter, Exeter, Devon EX4 4QD, UK. [5]Department of Biochemistry, University of Oxford, Oxford OX1 3QU, UK.

References

1. Harries LW, Hernandez D, Henley W, Wood AR, Holly AC, Bradley-Smith RM, Yaghootkar H, Dutta A, Murray A, Frayling TM, et al. Human aging is characterized by focused changes in gene expression and deregulation of alternative splicing. Aging Cell. 2011;10:868–78.
2. Holly AC, Melzer D, Pilling LC, Fellows AC, Tanaka T, Ferrucci L, Harries LW. Changes in splicing factor expression are associated with advancing age in man. Mech Ageing Dev. 2013;134:356–66.
3. Lee BP, Pilling LC, Emond F, Flurkey K, Harrison DE, Yuan R, Peters LL, Kuschel G, Ferrucci L, Melzer D, Harries LW. Changes in the expression of splicing factor transcripts and variations in alternative splicing are associated with lifespan in mice and humans. Aging Cell. 2016;15:903–13.
4. Heintz C, Doktor TK, Lanjuin A, Escoubas CC, Zhang Y, Weir HJ, Dutta S, Silva-Garcia CG, Bruun GH, Morantte I, et al. Splicing factor 1 modulates dietary restriction and TORC1 pathway longevity in C. Elegans. Nature. 2016;541:102–6.
5. Lee VM, Goedert M, Trojanowski JQ. Neurodegenerative tauopathies. Annu Rev Neurosci. 2001;24:1121–59.
6. Beyer K, Ariza A. Alpha-Synuclein posttranslational modification and alternative splicing as a trigger for neurodegeneration. Mol Neurobiol. 2013;47:509–24.
7. Wojtuszkiewicz A, Assaraf YG, Maas MJ, Kaspers GJ, Jansen G, Cloos J. Pre-mRNA splicing in cancer: the relevance in oncogenesis, treatment and drug resistance. Expert Opin Drug Metab Toxicol. 2015;11:673–89.
8. Stegeman R, Weake VM. Transcriptional signatures of aging. J Mol Biol. 2017;429:2427–37.
9. Smith CW, Valcarcel J. Alternative pre-mRNA splicing: the logic of combinatorial control. Trends Biochem Sci. 2000;25:381–8.
10. Cartegni L, Chew SL, Krainer AR. Listening to silence and understanding nonsense: exonic mutations that affect splicing. Nat Rev Genet. 2002;3:285–98.
11. Ramanouskaya TV, Grinev VV. The determinants of alternative RNA splicing in human cells. Mol Gen Genomics. 2017;
12. Muller-McNicoll M, Botti V, de Jesus Domingues AM, Brandl H, Schwich OD, Steiner MC, Curk T, Poser I, Zarnack K, Neugebauer KM. SR proteins are NXF1 adaptors that link alternative RNA processing to mRNA export. Genes Dev. 2016;30:553–66.
13. Kang X, Chen W, Kim RH, Kang MK, Park NH. Regulation of the hTERT promoter activity by MSH2, the hnRNPs K and D, and GRHL2 in human oral squamous cell carcinoma cells. Oncogene. 2009;28:565–74.
14. Pont AR, Sadri N, Hsiao SJ, Smith S, Schneider RJ. mRNA decay factor AUF1 maintains normal aging, telomere maintenance, and suppression of senescence by activation of telomerase transcription. Mol Cell. 2012;47:5–15.
15. Sikora E, Bielak-Zmijewska A, Mosieniak G. Cellular senescence in ageing, age-related disease and longevity. Curr Vasc Pharmacol. 2014;12:698–706.
16. Wang H, Han L, Zhao G, Shen H, Wang P, Sun Z, Xu C, Su Y, Li G, Tong T, Chen J. hnRNP A1 antagonizes cellular senescence and senescence-associated secretory phenotype via regulation of SIRT1 mRNA stability. Aging Cell. 2016;15:1063–73.
17. Choi HS, Lee HM, Jang YJ, Kim CH, Ryu CJ. Heterogeneous nuclear ribonucleoprotein A2/B1 regulates the self-renewal and pluripotency of human embryonic stem cells via the control of the G1/S transition. Stem Cells. 2013;31:2647–58.
18. van Deursen JM. The role of senescent cells in ageing. Nature. 2014;509:439–46.
19. Baker DJ, Childs BG, Durik M, Wijers ME, Sieben CJ, Zhong J, Saltness RA, Jeganathan KB, Verzosa GC, Pezeshki A, et al. Naturally occurring p16(Ink4a)-positive cells shorten healthy lifespan. Nature. 2016;530:184–9.
20. Burton DG, Giles PJ, Sheerin AN, Smith SK, Lawton JJ, Ostler EL, Rhys-Williams W, Kipling D, Faragher RG. Microarray analysis of senescent vascular smooth muscle cells: a link to atherosclerosis and vascular calcification. Exp Gerontol. 2009;44:659 65.
21. Coppe JP, Patil CK, Rodier F, Sun Y, Munoz DP, Goldstein J, Nelson PS, Desprez PY, Campisi J. Senescence-associated secretory phenotypes reveal cell-nonautonomous functions of oncogenic RAS and the p53 tumor suppressor. PLoS Biol. 2008;6:2853–68.
22. Markus MA, Marques FZ, Morris BJ. Resveratrol, by modulating RNA processing factor levels, can influence the alternative splicing of pre-mRNAs. PLoS One. 2011;6:e28926.
23. Fuggetta MP, Bordignon V, Cottarelli A, Macchi B, Frezza C, Cordiali-Fei P, Ensoli F, Ciafre S, Marino-Merlo F, Mastino A, Ravagnan G. Downregulation of proinflammatory cytokines in HTLV-1-infected T cells by Resveratrol. J Exp Clin Cancer Res. 2016;35:118.
24. Sinclair DA, Guarente L. Small-molecule allosteric activators of sirtuins. Annu Rev Pharmacol Toxicol. 2014;54:363–80.
25. Fang EF, Kassahun H, Croteau DL, Scheibye-Knudsen M, Marosi K, Lu H, Shamanna RA, Kalyanasundaram S, Bollineni RC, Wilson MA, et al. NAD+ replenishment improves lifespan and Healthspan in ataxia Telangiectasia models via Mitophagy and DNA repair. Cell Metab. 2016;24:566–81.
26. Birar VC, Sheerin AN, Milkovicova J, Faragher RG, Ostler EL. A facile, stereoselective, one-pot synthesis of resveratrol derivatives. Chem Cent J. 2015;9:26.
27. Chung EY, Kim BH, Hong JT, Lee CK, Ahn B, Nam SY, Han SB, Kim Y. Resveratrol down-regulates interferon-gamma-inducible inflammatory genes in macrophages: molecular mechanism via decreased STAT-1 activation. J Nutr Biochem. 2011;22:902–9.
28. Aguilo F, Zhou MM, Walsh MJ. Long noncoding RNA, polycomb, and the ghosts haunting INK4b-ARF-INK4a expression. Cancer Res. 2011;71:5365–9.
29. Pabla N, Bhatt K, Dong Z. Checkpoint kinase 1 (Chk1)-short is a splice variant and endogenous inhibitor of Chk1 that regulates cell cycle and DNA damage checkpoints. Proc Natl Acad Sci U S A. 2012;109:197–202.
30. Harrison DE, Strong R, Sharp ZD, Nelson JF, Astle CM, Flurkey K, Nadon NL, Wilkinson JE, Frenkel K, Carter CS, et al. Rapamycin fed late in life extends lifespan in genetically heterogeneous mice. Nature. 2009;460:392–5.
31. Walters HE, Deneka-Hannemann S, Cox LS. Reversal of phenotypes of cellular senescence by pan-mTOR inhibition. Aging (Albany NY). 2016;8:231–44.
32. Panasyuk G, Nemazanyy I, Zhyvoloup A, Filonenko V, Davies D, Robson M, Pedley RB, Waterfield M, Gout I. mTORbeta splicing isoform promotes cell proliferation and tumorigenesis. J Biol Chem. 2009;284:30807–14.
33. Zhou S, Greenberger JS, Epperly MW, Goff JP, Adler C, Leboff MS, Glowacki J. Age-related intrinsic changes in human bone-marrow-derived mesenchymal stem cells and their differentiation to osteoblasts. Aging Cell. 2008;7:335–43.
34. Ford LP, Wright WE, Shay JW. A model for heterogeneous nuclear ribonucleoproteins in telomere and telomerase regulation. Oncogene. 2002;21:580–3.
35. Chen H, Jin ZL, Xu H. MEK/ERK signaling pathway in apoptosis of SW620 cell line and inhibition effect of resveratrol. Asian Pac J Trop Med. 2016;9:49–53.
36. Tillu DV, Melemedjian OK, Asiedu MN, Qu N, De Felice M, Dussor G, Price TJ. Resveratrol engages AMPK to attenuate ERK and mTOR signaling in sensory neurons and inhibits incision-induced acute and chronic pain. Mol Pain. 2012;8:5.
37. Kajita K, Kuwano Y, Kitamura N, Satake Y, Nishida K, Kurokawa K, Akaike Y, Honda M, Masuda K, Rokutan K. Ets1 and heat shock factor 1 regulate transcription of the transformer 2beta gene in human colon cancer cells. J Gastroenterol. 2013;48:1222–33.
38. Kerins JA, Hanazawa M, Dorsett M, Schedl T. PRP-17 and the pre-mRNA splicing pathway are preferentially required for the proliferation versus meiotic development decision and germline sex determination in Caenorhabditis Elegans. Dev Dyn. 2010;239:1555–72.
39. Anczukow O, Rosenberg AZ, Akerman M, Das S, Zhan L, Karni R, Muthuswamy SK, Krainer AR. The splicing factor SRSF1 regulates apoptosis and proliferation to promote mammary epithelial cell transformation. Nat Struct Mol Biol. 2012;19:220–8.
40. He Y, Brown MA, Rothnagel JA, Saunders NA, Smith R. Roles of heterogeneous nuclear ribonucleoproteins a and B in cell proliferation. J Cell Sci. 2005;118:3173–83.
41. Davis T, Singhrao SK, Wyllie FS, Haughton MF, Smith PJ, Wiltshire M, Wynford-Thomas D, Jones CJ, Faragher RG, Kipling D. Telomere-based proliferative lifespan barriers in Werner-syndrome fibroblasts involve both p53-dependent and p53-independent mechanisms. J Cell Sci. 2003;116:1349–57.
42. Blagosklonny MV. Aging-suppressants: cellular senescence (hyperactivation) and its pharmacologic deceleration. Cell Cycle. 2009;8:1883–7.

11

A novel interaction between kinase activities in regulation of cilia formation

Nicole DeVaul[2], Katerina Koloustroubis[1], Rong Wang[1] and Ann O. Sperry[1*]

Abstract

Background: The primary cilium is an extension of the cell membrane that encloses a microtubule-based axoneme. Primary cilia are essential for transmission of environmental cues that determine cell fate. Disruption of primary cilia function is the molecular basis of numerous developmental disorders. Despite their biological importance, the mechanisms governing their assembly and disassembly are just beginning to be understood. Cilia growth and disassembly are essential events when cells exit and reenter into the cell cycle. The kinases never in mitosis-kinase 2 (Nek2) and Aurora A (AurA) act to depolymerize cilia when cells reenter the cell cycle from G_0.

Results: Coexpression of either kinase with its kinase dead companion [AurA with kinase dead Nek2 (Nek2 KD) or Nek2 with kinase dead AurA (AurA KD)] had different effects on cilia depending on whether cilia are growing or shortening. AurA and Nek2 are individually able to shorten cilia when cilia are growing but both are required when cilia are being absorbed. The depolymerizing activity of each kinase is increased when coexpressed with the kinase dead version of the other kinase but only when cilia are assembling. Additionally, the two kinases act additively when cilia are assembling but not disassembling. Inhibition of AurA increases cilia number while inhibition of Nek2 significantly stimulates cilia length. The complex functional relationship between the two kinases reflects their physical interaction. Further, we identify a role for a PP1 binding protein, PPP1R42, in inhibiting Nek2 and increasing ciliation of ARPE-19 cells.

Conclusion: We have uncovered a novel functional interaction between Nek2 and AurA that is dependent on the growth state of cilia. This differential interdependence reflects opposing regulation when cilia are growing or shortening. In addition to interaction between the kinases to regulate ciliation, the PP1 binding protein PPP1R42 directly inhibits Nek2 independent of PP1 indicating another level of regulation of this kinase. In summary, we demonstrate a complex interplay between Nek2 and AurA kinases in regulation of ciliation in ARPE-19 cells.

Keywords: Cilia, AurA, Nek2, PP1, PPP1R42

Background

Primary cilia are microtubule-based organelles that protrude from the cell membrane to receive and transduce environmental signals. Interference with the formation and/or stability of cilia disrupts signaling pathways essential for normal development and maintenance of the differentiated state ([1–3]; as reviewed in [4, 5]). A diverse collection of developmental disorders, ciliopathies, stem directly from disruption of ciliary function and display a wide range of abnormalities from cystic kidney to obesity (as reviewed in [6, 7]). Virtually all cells form primary cilia, which are structurally analogous to flagella. Cilia assemble when cells exit the cell cycle. As cells reenter the cell cycle and begin to proliferate, cilia disassemble, the basal body detaches from the plasma membrane, and centrosomes duplicate to form the mitotic spindle. Although the inventory of proteins that constitute cilia is increasing, the mechanisms regulating their formation and disassembly are just beginning to be defined.

Cilia display regulated growth and retraction when entering and exiting the cell cycle. Cilia grow as cells enter G_0 and are absorbed prior to reentry into the cell cycle. Two important regulators of cilia absorption are the kinases Nek2 (NIMA related kinase 2) and AurA (Aurora A). Nek2 is a serine/threonine kinase that is localized to the distal portion of the mother centriole and functions in both cilia

* Correspondence: sperrya@ecu.edu
[1]Anatomy and Cell Biology, East Carolina University, Brody School of Medicine, Greenville, NC, USA
Full list of author information is available at the end of the article

shortening and centrosome duplication [8, 9]. Nek2 controls cilia disassembly; depletion of Nek2 causes an increase in the number of ciliated cells [8]. Nek2 has been linked directly to disruption of left-right asymmetry, a biological consequence of cilia dysfunction [10]. Nek2 also regulates intraflagellar transport (IFT) through phosphorylation of the kinesin KIF24 to stimulate cilia depolymerization [11]. Additionally, Nek2 induces centrosome separation prior to cell division. Overexpression of active Nek2 causes premature splitting of centrosomes [12] due to phosphorylation and destabilization of centrosomal linker proteins [9].

AurA, another serine/threonine kinase, is essential for maintenance of cilia length and cilia retraction prior to cell cycle reentry in diverse organisms [13, 14] as well as centrosome maturation, duplication, and spindle assembly (as reviewed by [15]). Pioneering work in *Chlamydomonas reinhardtii* provided the first indication that AurA regulates the length of the flagellum of this biflagellate alga [16, 17]. AurA is localized to and activated at the basal body of cilia when cilia disassemble. Overexpression of AurA in ciliated mammalian cells induces cilia disassembly through activation of a tubulin deacetylase [13]. Like Nek2, AurA participates in preparation of centrosomes for cell division (reviewed in [18–20]).

PP1, a serine/threonine phosphatase, is a common regulator of both kinases in control of centrosome separation prior to spindle formation at mitosis; however, its role in cilia biogenesis has not been investigated [19–22]. PP1 activity is itself regulated by both positive and negative regulatory subunits. The negative regulator PPP1R2 (I2) inhibits PP1 activity in both centrosome separation and cilia acetylation and stabilization [19, 23]. We have previously identified a PP1 binding protein, PPP1R42 that is involved in centrosome separation [24]; however, its role in ciliation is not known.

Our study provides evidence that Nek2 and AurA interact differentially depending on cilia growth status. We demonstrate that Nek2 and AurA interact on several levels. They appear to share positive and negative factors to enhance or inhibit depolymerization activity when cilia are disassembling or assembling, respectively. Nek2 and AurA act independently when cilia are growing but both are required to depolymerize cilia. Furthermore, we demonstrate that these two kinases act additively to depolymerize cilia when cilia are growing and are independently involved in cilia number and length. These findings represent a novel functional interaction between two kinases involved in cilia disassembly. In addition, we identify inhibition of Nek2 by PPP1R42, a PP1 binding protein, which is independent of PP1.

Results

Requirement for kinase activity is dependent on cilia growth state

We investigated the interaction between AurA and Nek2 by overexpressing the kinases and their kinase dead counterparts either alone or in combination in cells either growing cilia after serum starvation or absorbing cilia after reintroduction of serum (Fig. 1). The kinase dead versions of Nek2 and AurA have been shown to localize to the centrosome and to have a dominant negative effect on endogenous kinase function by sequestering substrates and upstream regulators of the kinases (Dr. Andrew Fry, personal communication and [12, 25, 26]). Expressed protein is maintained throughout the time course of treatment (Additional file 1: Figure S1) with a transfection efficiency of 90% on average (Additional file 2: Figure S2) and cells show little toxicity after transfection. These experiments examine a time window between formation of cilia after cell division and before the approximate onset of the next division. The effect of experimental manipulation on cilia in cell populations has precedent and results of such studies have expanded our knowledge of cilia biology [13, 27–29].

We first compared cilia number and length in cells expressing either kinase active or kinase dead versions of AurA and Nek2 in cells assembling or disassembling cilia to determine if the effect on cilia depolymerization differed depending on cilia growth status (Fig. 1, compare Empty (control) vs AurA, and Empty (control) vs Nek2). Previous studies have shown that AurA and Nek2 induce cilia depolymerization [11, 13]; therefore, we expected that cells transfected with the active kinases would be less ciliated than control. This was true for cells assembling cilia (Fig. 1A, left panel, AurA vs control, $p = 0.008$; Fig. 1B left panel, Nek2 vs control, $p = 0.0009$) but not for cells disassembling cilia. Only Nek2 transfected cells were significantly less ciliated than control (Fig. 1B, right panel, $p = 0.03$). There was no significant difference in cilia length between active kinase and control (Fig. 1C, AurA vs control; Fig. 1D, Nek2 vs control). The stronger effect of the active kinase compared to control in cells assembling cilia may reflect a shift in the balance from assembling to disassembling catalyzed by the active kinases. The active kinases are not as effective at further depolymerizing cilia when cilia are already disassembling (Fig. 1A and B, right panels).

We expected that cells transfected with the kinase dead enzymes would have more and longer cilia compared to control and wild-type (compare Fig. 1 control and AurA vs AurAKD, and control and Nek2 vs Nek2KD). However, the kinase dead enzymes of AurA and Nek2 did not increase cilia number compared to control either when cilia are assembling or disassembling which may suggest that ciliation is at a maximum and the kinase dead version cannot further stimulate cilia formation above control levels

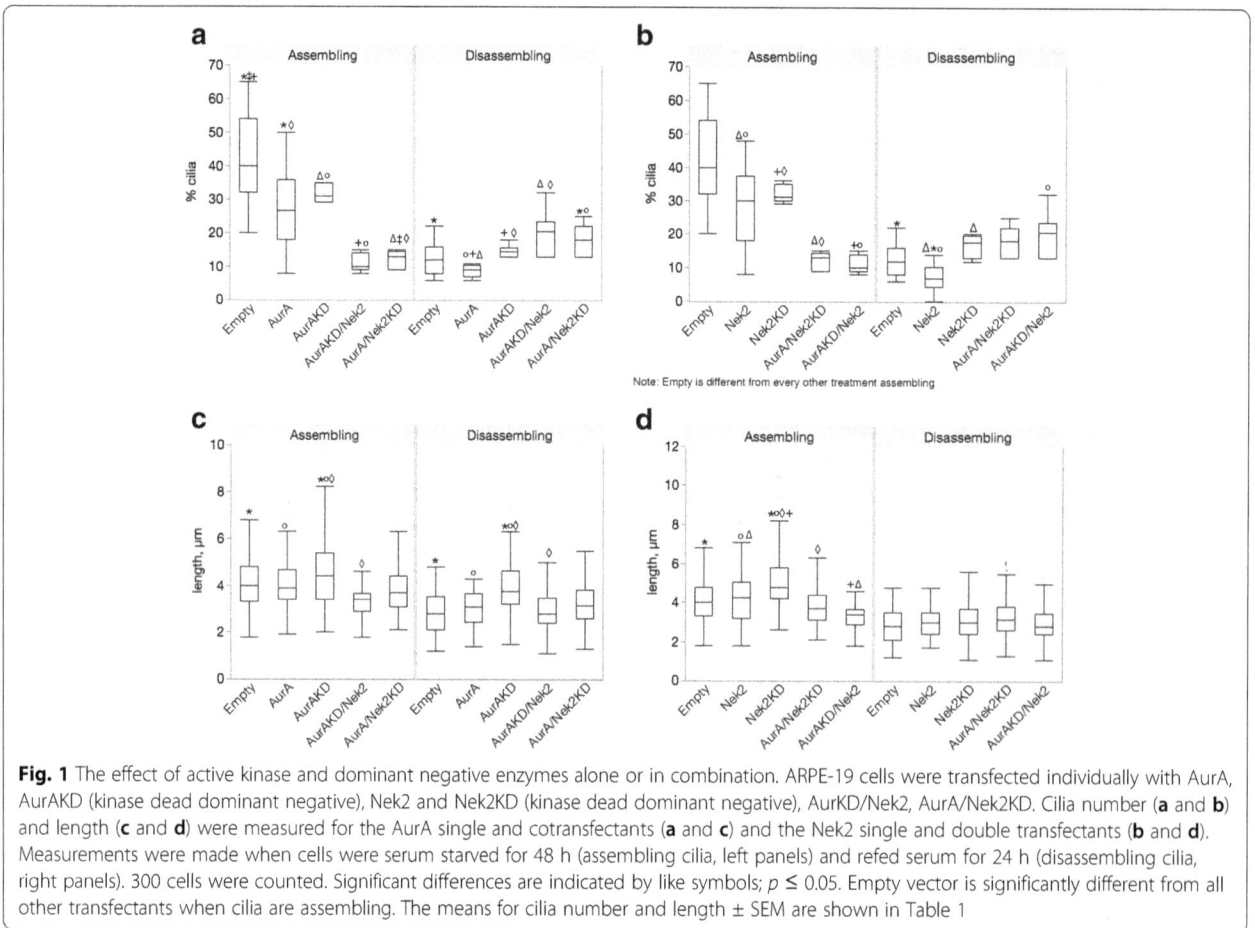

Fig. 1 The effect of active kinase and dominant negative enzymes alone or in combination. ARPE-19 cells were transfected individually with AurA, AurAKD (kinase dead dominant negative), Nek2 and Nek2KD (kinase dead dominant negative), AurKD/Nek2, AurA/Nek2KD. Cilia number (**a** and **b**) and length (**c** and **d**) were measured for the AurA single and cotransfectants (**a** and **c**) and the Nek2 single and double transfectants (**b** and **d**). Measurements were made when cells were serum starved for 48 h (assembling cilia, left panels) and refed serum for 24 h (disassembling cilia, right panels). 300 cells were counted. Significant differences are indicated by like symbols; $p \leq 0.05$. Empty vector is significantly different from all other transfectants when cilia are assembling. The means for cilia number and length ± SEM are shown in Table 1

(Fig. 1A and B). The KD was significantly different from the active kinase only when cilia were disassembling (Fig. 1A, right panel, AurA vs AurAKD, $p = 0.047$; Fig. 1B, right panel Nek2 vs. Nek2KD, $p = 0.004$) suggesting that the observed destabilization of cilia by the active kinase is dependent on its kinase activity when cilia are disassembling but not assembling.

We also measured cilia length in cells transfected with the kinase dead enzyme and compared the values to wild-type and control cells (Fig. 1C and D). Cilia were significantly longer in cells transfected with the kinase dead compared to control when cilia were assembling (Fig. 1C, left panel, AurAKD vs control, $p = 0.01$; Fig. 1D left panel, Nek2KD vs control, $p < 0.0001$) and for AurKD when cilia were disassembling (Fig. 1C, right panel, AurAKD vs control, $p < 0.0001$). Cilia were also significantly longer when the KD was compared to the wild-type enzyme in cells assembling (Fig. 1C, left panel, AurA vs AKD, $p = 0.02$; Fig. 1D, left panel, Nek2 vs Nek2KD, $p < 0.0001$) or disassembling cilia (Fig. 1C, right panel, AurA vs AKD, $p = 0.0002$). There was no significant difference in length between cells expressing Nek2KD and Nek2 during disassembly (right panel, Fig. 1D). The KDs influenced cilia length but not cilia number suggesting that cilia length

may be controlled by a different mechanism than initiation of cilia growth.

One catalytically active kinase is sufficient to destabilize cilia when cilia are assembling but not disassembling

We next compared cilia number and length in cells expressing the kinase dead to those coexpressing the kinase dead with the wild-type companion kinase (compare AurAKD/Nek2 vs AurAKD; AurA/Nek2KD vs Nek2KD in Fig. 1) to determine whether the ability to depolymerize cilia requires both catalytically active kinases. If both active kinases are required for depolymerization, cilia would remain long and numerous in the double transfectants. Alternatively, the kinase dead enzymes could indirectly affect the activity of the wild-type partner enzyme by sequestering necessary factors. If the kinase dead protein appropriates negative factors that restrict kinase activity in cells assembling cilia, the companion kinase would be activated and cilia would disassemble. If, on the other hand, the KD sequesters positive factors that stimulate kinase activity in cells disassembling cilia, the kinase would be less able to depolymerize cilia and they would remain long and numerous. This model depends on AurA and Nek2 sharing a common set of upstream regulators.

Table 1 Summary of percentage cilia and length for transfections

	Percentage cilia			Length, µm		
	Cycling	Assemb	Disassemb	Cycling	Assemb	Disassemb
Empty	1.7 ± 0.8	42.1 ± 3.7	12.6 ± 1.4	1.7 ± 0.4	4.1 ± 0.1	2.9 ± 0.1
AurA	0	26.5 ± 2.8	8.8 ± 0.9	0	4.1 ± 0.1	3.1 ± 0.2
Nek2	0.5 ± 0.5	27.8 ± 2.3	7.2 ± 1.9	0.9 ± 0.5	4.2 ± 0.1	3.0 ± 0.1
AurA/Nek2	ND	13.3 ± 1.0	15.5 ± 1.6	ND	7.2 ± 0.2	7.0 ± 0.3
AurAKD	ND	31.7 ± 1.8	13.6 ± 1.3	ND	4.5 ± 0.2	4.0 ± 0.1
Nek2KD	ND	32.2 ± 1.2	16.8 ± 1.8	ND	5.3 ± 0.2	3.2 ± 0.2
AurAKD/Nek2KD	ND	34.5 ± 2.7	25.6 ± 1.5	ND	14.0 ± 0.8	10.1 ± 0.3
AurA/Nek2KD	ND	12 ± 1.3	18 ± 2.0	ND	3.9 ± 0.1	3.3 ± 0.1
AurAKD/Nek2	ND	11 ± 1.0	19.6 ± 2.0	ND	3.4 ± 0.2	3.0 ± 0.1
R42	9.1 ± 1.3	47.2 ± 2.9	22.3 ± 3.0	4.2 ± 0.3	4.3 ± 0.1	3.4 ± 0.2
R42/AurA	1 ± 0.4	24.8 ± 2.1	7.2 ± 1.0	2.6 ± 0.7	4.3 ± 0.1	2.7 ± 0.1
R42/Nek2	0.8 ± 0.4	18.4 ± 2.6	12.5 ± 2.6	2.2 ± 0.8	4.0 ± 0.1	3.0 ± 0.1
OT siRNA	4.8 ± 1.3	46 ± 4.6	26.3 ± 8.6	2.6 ± 0.2	4.0 ± 0.1	2.9 ± 0.1
R42 siRNA	3.4 ± 0.7	41 ± 2.2	24 ± 4.0	2.7 ± 0.2	3.6 ± 0.1	3.4 ± 0.2
Empty-MLN	ND	26.4 ± 3.1	13.2 ± 1.3	ND	6.4 ± 0.4	5.3 ± 0.4
Empty + MLN	ND	33.4 ± 1.7	18.3 ± 3.3	ND	6.8 ± 0.4	4.8 ± 0.4
Nek2-MLN	ND	25.2 ± 2.3	10.6 ± 1.5	ND	5.9 ± 0.4	4.8 ± 0.4
Nek2 + MLN	ND	36.2 ± 2.0	17.5 ± 1.4	ND	6.3 ± 0.3	4.8 ± 0.4
Empty-rac	ND	46 ± 3.7	17.6 ± 2.0	ND	4.6 ± 4.7	5.3 ± 0.6
Empty + rac	ND	43 ± 5.8	27.7 ± 2.3	ND	10.8 ± 1.2	6.4 ± 0.6
AurA-rac	ND	33 ± 4.4	14.3 ± 2.3	ND	5.3 ± 0.7	3.8 ± 0.3
AurA + rac	ND	48.4 ± 5.6	28 ± 5.0 l	ND	7.4 ± 0.9	11.8 ± 1.3

The mean percentage of cilia number and length ± SEM is shown for each transfection

Cells cotransfected with AurAKD/Nek2 or Nek2KD/AurA had significantly fewer and shorter cilia than the corresponding kinase dead when cilia were assembling (Fig. 1). Cells co-expressing AurAKD/Nek2 were significantly less ciliated than cells transfected with AurAKD alone and the control (AurAKD/Nek2 vs AurAKD, $p = 0.02$, Fig. 1A, left panel; AurAKD/Nek2 vs control, $p < 0.0001$). Cells co-expressing AurA/Nek2KD were significantly less ciliated than cells expressing Nek2KD alone and control (Fig. 1B, left panel, AurA/Nek2KD vs Nek2KD alone, $p = 0.01$; AurA/Nek2KD vs control, $p < 0.0001$) when cells are assembling cilia. Similarly, the cilia of cells co-expressing AurAKD/Nek2 were shorter than cells expressing AurAKD alone (Fig. 1C, left panel, $p < 0.0001$) and those co-expressing AurA/Nek2KD were shorter than cells expressing Nek2KD alone and control (Fig. 1D, left panel AurA/Nek2KD vs Nek2KD, $p < 0.0001$, AurA/Nek2KD vs control, $p = 0.002$).

Ciliation was not reduced in the double transfectants in cells disassembling cilia. Cilia number of the cotransfectants was significantly increased (compare AurAKD/Nek2 vs AurA KD; AurA/Nek2KD vs Nek2KD, Fig. 1A and B, right panels). The active enzymes could not reduce cilia number in the presence of the KD of the partner kinase (Fig. 1A and B, right panels). Cells coexpressing AurAKD/Nek2 were significantly more ciliated than those expressing AurAKD alone but not different from control (Fig. 1A, right panel, $p = 0.02$). AurAKD/Nek2 were significantly shorter than AurAKD alone but not different from control (Fig. 1C, right panel). There was no significant difference in cilia number or length between cells expressing AurA/Nek2KD and Nek2KD during disassembly. Together, these data demonstrate that the wild-type kinases can destabilize cilia in the presence of the catalytically inactive companion kinase when cilia are assembling but not disassembling. These data do not discriminate between a differential requirement for both kinases depending on cilia growth, or activation/inhibition of the companion kinase due to sequestration of common negative factors when cilia are assembling or positive factors when cilia are disassembling.

The KD proteins enhance the activity of the partner wild-type kinase but only when cilia are assembling

To determine whether the KD might directly or indirectly affect the activity of the wild-type partner kinase, we compared cilia number and length between cells transfected with either wild-type kinase alone or the double transfectants (compare AurA/Nek2KD vs AurA; AurAKD/Nek2 vs Nek2, Fig. 1). Cells expressing AurA/Nek2KD were significantly less ciliated than cells expressing AurA alone or control when cilia are growing (Fig. 1A, left panel, AurA/Nek2KD vs AurA, $p = 0.002$; AurA/Nek2KD vs control, $p < 0.0001$). Similarly, AurAKD enhanced the depolymerization activity of Nek2 (Fig. 1B, left panel, AurAKD/Nek2 vs Nek2, $p = 0.002$). Only cells transfected with AurAKD/Nek2 displayed a decrease in cilia length compared to Nek2 during cilia assembly (Fig. 1D, left panel, $p = 0.0001$).

In contrast, coexpression with the kinase dead companion kinase did not result in increased depolymerization when cilia were disassembling compared to wild-type alone (Fig. 1A and B, right panels). Instead of stimulating depolymerization, the catalytically inactive kinase suppressed cilia depolymerization by the companion kinase. Cells dissembling cilia and coexpressing AurA/Nek2KD were significantly more ciliated than cells expressing AurA alone and control (Fig. 1A, right panel AurA/Nek2KD vs AurA, $p = 0.002$; AurA/Nek2KD vs control, $p = .02$). Cells coexpressing AurAKD/Nek2 were significantly more ciliated than Nek2 alone and control (Fig. 1B, right panel, AurAKD/Nek2 vs Nek2, $p < 0.0004$; AurKD/Nek2 vs control, $p = 0.007$). Cilia length was not significantly different in cotransfectants compared to wild-type or control.

Our results support differential effects of the KDs on the wild-type kinases depending on whether cilia are assembling or disassembling. Coexpression of the dominant negative enzyme with the wild-type companion kinase stimulates activity of the other kinase when cilia are growing but inhibits its activity when cilia are disassembling. We conclude from these experiments that the dominant negative enzymes sequester factors that inhibit depolymerization by the kinases when cilia are assembling and factors that stimulate depolymerization when cilia are disassembling. This result supports the existence of regulatory proteins common to the two kinases.

AurA and Nek2 activities are additive when cilia are growing

We next compared cells cotransfected with AurA/Nek2 or with AurAKD/Nek2KD (shown as AKD/NKD in the graph) to their respective single transfectants (Fig. 2). We predict that the double wild type transfectants would increase depolymerization compared to the single transfectants and that the double kinase dead would inhibit depolymerization more than the single KD transfectants. This was true for cells assembling cilia (Fig. 2A, left panel). AurA and Nek2 acted additively in cells assembling cilia to reduce the percentage of ciliated cells in cotransfected cells; cilia were significantly reduced in

AurA/Nek2 double transfectants compared to single transfectants and control when cilia were assembling (Fig. 2A, left panel, AurA/Nek2 vs AurA or Nek2, $p < 0.0001$; AurA/Nek2 vs control, $p < 0.0001$). However, when cilia were disassembling, cilia were more numerous in double transfectants but this difference was not statistically significant nor different from control (Fig. 2A, right panel, $p = 0.077$ Nek2/AurA vs Nek2). These results suggest that their kinase activities are additive when cilia are assembling not disassembling suggesting that they have different downstream effectors that stimulate depolymerization independently of one another. The double KD increased ciliation in cells disassembling cilia compared to control (Fig. 2A right panel; AKD/AKD vs control, $p < 0.0001$) and compared to the KD of each kinase alone (AKD/NKD vs AKD, $p < 0.0001$; AKD/NKD vs NKD, $p < 0.0001$). No significant difference was observed during assembly. The effects of the double KD to stimulate ciliation suggests that AurA and Nek2, in addition to sharing activators as demonstrated by our wt/KD double transfections (Fig. 1), must have distinct coregulators as the KDs have additive effects to supresses depolymerization.

The effect of the double wild-type transfectants on cilia length was unexpected. Instead of shorter cilia as we would predict by the reduced cilia number in these samples (Fig. 2A), the double wild-type transfectants had significantly longer cilia compared to the single transfectants and controls ($p < 0.0001$) when cilia were either assembling ($p < .0001$) or disassembling ($p < 0.0001$) (Fig. 6B). One possibility is that the wild-type enzymes inhibit one another in control of cilia length. Alternatively, they could each regulate downstream negative effectors that control cilia length.

The effect on cilia length of the double KD was particularly striking (Fig. 2B). Cilia of double KD transfectants were extremely long compared to control and the single KD transfectants ($p < 0.0001$ in all cases). Cilia were about three times longer in cells transfected with AKD/NDK compared to controls when cilia were assembling and disassembling. The surprising finding that the double KD causes extremely long cilia (Fig. 2C) suggests to us that repression of multiple pathways is necessary to maintain cilia length. This is consistent with the additive effects of the KD enzymes.

Nek2 and AurA form a complex in ARPE-19 cells

To determine whether Nek2 and AurA can interact with one another in cultured cells we coexpressed AurA and Nek2 in ARPE-19 cells and immunoprecipitated the complex with antibodies to either Nek2 or AurA followed by interrogation of the complex with antibodies specific for the other kinase. Complexes immunoprecipitated with AurA contain Nek2 and reciprocally (Fig. 3A).

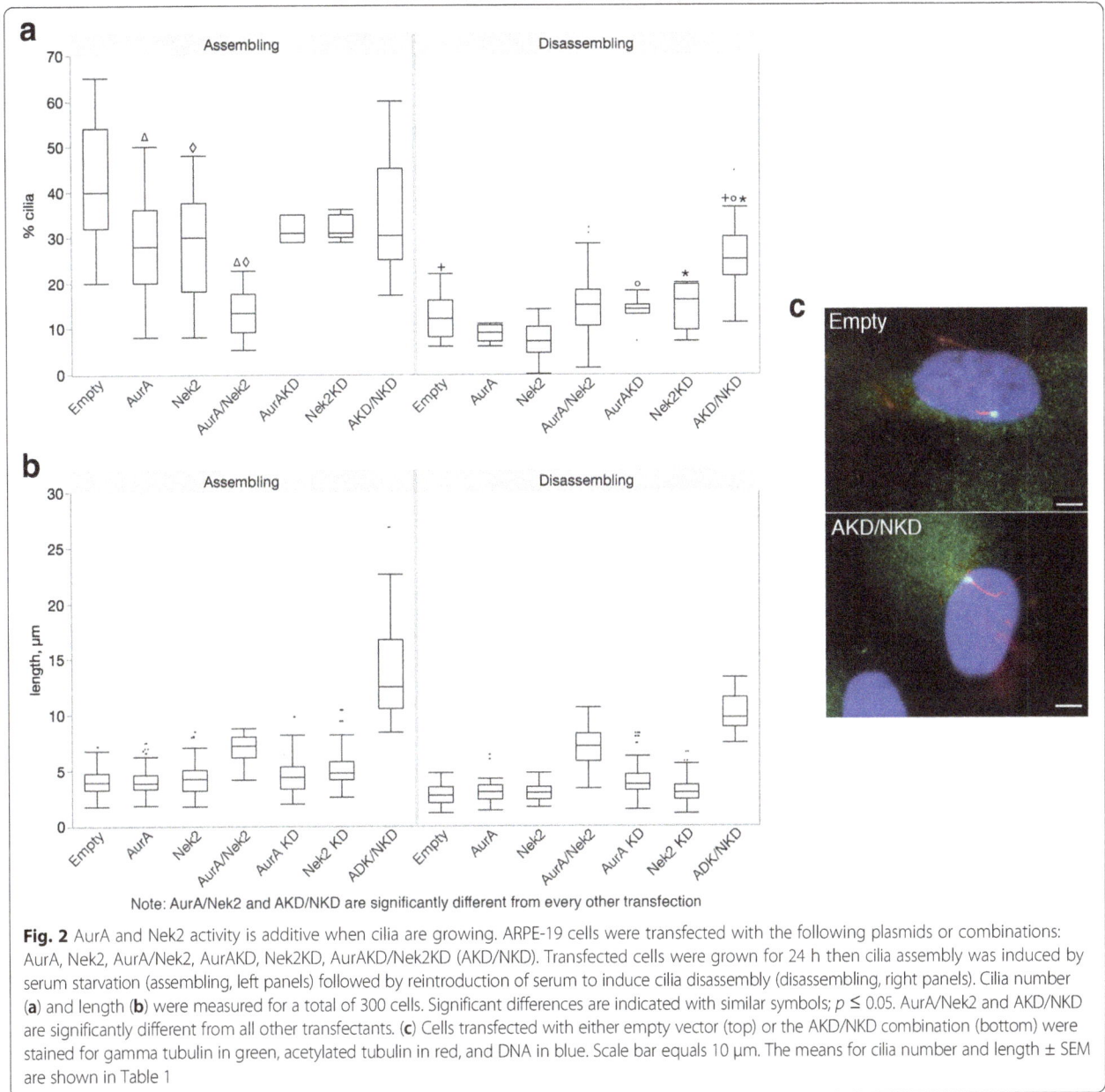

Fig. 2 AurA and Nek2 activity is additive when cilia are growing. ARPE-19 cells were transfected with the following plasmids or combinations: AurA, Nek2, AurA/Nek2, AurAKD, Nek2KD, AurAKD/Nek2KD (AKD/NKD). Transfected cells were grown for 24 h then cilia assembly was induced by serum starvation (assembling, left panels) followed by reintroduction of serum to induce cilia disassembly (disassembling, right panels). Cilia number (**a**) and length (**b**) were measured for a total of 300 cells. Significant differences are indicated with similar symbols; $p \leq 0.05$. AurA/Nek2 and AKD/NKD are significantly different from all other transfectants. (**c**) Cells transfected with either empty vector (top) or the AKD/NKD combination (bottom) were stained for gamma tubulin in green, acetylated tubulin in red, and DNA in blue. Scale bar equals 10 μm. The means for cilia number and length ± SEM are shown in Table 1

The endogenous kinases are also found associated with one another (Fig. 3B). This suggests that the kinases may directly interact to regulate one another during cilia assembly and disassembly. This is consistent with the results described in Fig. 1 where we show that the kinases functionally interact to disassemble cilia. Alternatively, an intermediary protein could be responsible for the effects we observe.

AurA inhibition increases cilia number

To determine whether the wild type kinases effect the activity of the other and whether the kinases have distinct roles in ciliation, we compared cilia number and length in cells transfected with the active kinase while inhibiting the partner kinase with a small molecule inhibitor (Fig. 4). If AurA affects Nek2 when cilia are assembling, release of this action by inhibition of AurA's kinase activity will alter Nek2 catalyzed cilia depolymerization resulting in fewer and shorter cilia. Instead, incubation of cells assembling cilia and expressing Nek2 with the small molecule inhibitor of Aurora A, MLN8237 (MLN) [30], resulted in an increase in the percentage of ciliated cells (Fig. 4A, left panel, Nek2 + MLN vs Nek2 − MLN $p = 0.001$). We also observed this effect in cells disassembling their cilia (Fig. 4A, right panel, Nek2 + MLN vs Nek2 − MLN, $p = 0.04$). However, this appears to be an effect of MLN alone since this difference was also seen when

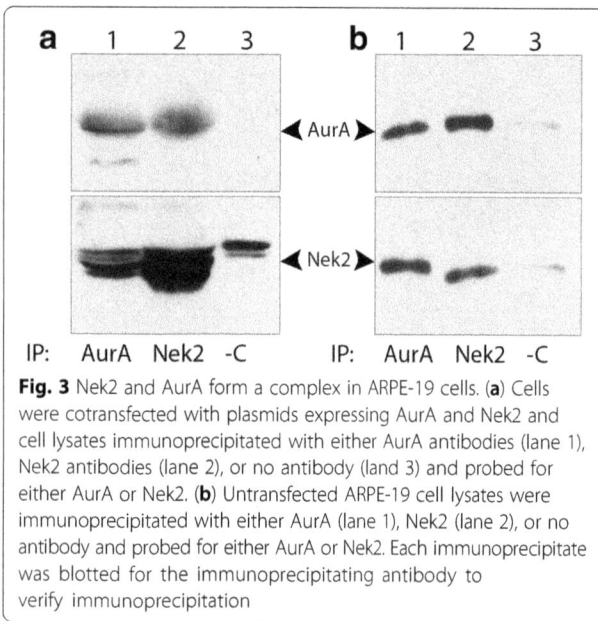

Fig. 3 Nek2 and AurA form a complex in ARPE-19 cells. (**a**) Cells were cotransfected with plasmids expressing AurA and Nek2 and cell lysates immunoprecipitated with either AurA antibodies (lane 1), Nek2 antibodies (lane 2), or no antibody (land 3) and probed for either AurA or Nek2. (**b**) Untransfected ARPE-19 cell lysates were immunoprecipitated with either AurA (lane 1), Nek2 (lane 2), or no antibody and probed for either AurA or Nek2. Each immunoprecipitate was blotted for the immunoprecipitating antibody to verify immunoprecipitation

comparing treated control transfected cells with untreated cells (Fig. 4A, empty – MLN vs empty + MLN, p = 0.01). Treatment of Nek2 transfected cells with MLN did not effect cilia length when cilia were either assembling or disassembling (Fig. 4B). Therefore, any effect of inhibition of AurA on Nek2 activity is masked by the general effect of the AurA inhibitor on cilia number. Our results demonstrate that AurA has a more significant role in control of cilia nucleation than length.

Inhibition of Nek2 significantly lengthens cilia when cilia are assembling and disassembling

We next treated cells transfected with either empty vector or AurA with the Nek2 inhibitor rac-CCT 250863 (termed here as rac, Fig. 5) [31]. Treatment of control transfected cells had no effect on ciliation when cilia were assembling or disassembling. However, cells overexpressing AurA and treated with rac were significantly more ciliated than untreated AurA overexpressing cells when cells were assembling or disassembling (Fig. 5A, p=0.005 in both cases). Therefore, inhibition of Nek2 counteracts the effect of AurA indicating that Nek2 may be upstream of AurA or in a parallel pathway in control of cilia number. The effect of the Nek2 inhibitor on cilia length was striking. Cilia were significantly longer in AurA transfected cells inhibited for Nek2 when cells were both assembling or disassembling (Fig. 5B, AurA – rac vs AurA + rac, assembling, p = 0.001, disassembling, p < 0.0001). Inhibition of Nek2 increased the length of cilia in control transfected cells when cilia were assembling their cilia (p ≤ 0.0001). Examples of long cilia are shown in panels C-E (Fig. 5). In many cells with multiple centrosomes, likely a consequence of Nek2

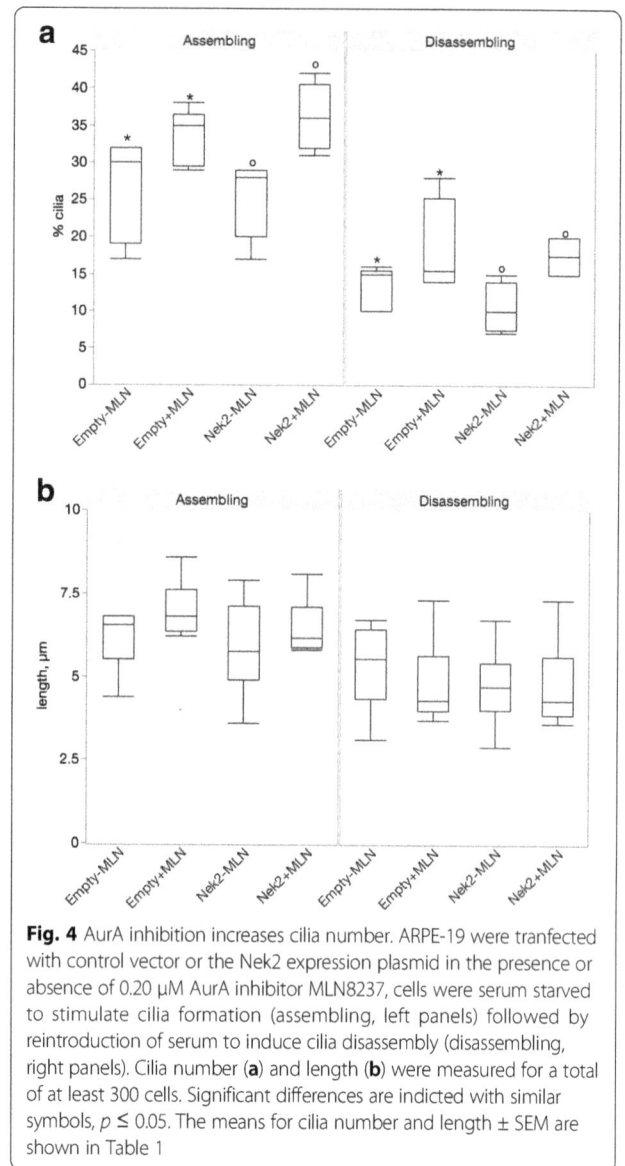

Fig. 4 AurA inhibition increases cilia number. ARPE-19 were tranfected with control vector or the Nek2 expression plasmid in the presence or absence of 0.20 µM AurA inhibitor MLN8237, cells were serum starved to stimulate cilia formation (assembling, left panels) followed by reintroduction of serum to induce cilia disassembly (disassembling, right panels). Cilia number (**a**) and length (**b**) were measured for a total of at least 300 cells. Significant differences are indicted with similar symbols, p ≤ 0.05. The means for cilia number and length ± SEM are shown in Table 1

inhibition, we observed more than one cilia in an individual cell (Fig. 5E).

The effect on cilia length in cells assembling cilia appeared solely due to Nek2 inhibition as there was no difference between control cells treated with inhibitor and AurA transfectants treated with inhibitor; however, in cells disassembling cilia, cilia were significantly longer in cells inhibited for Nek2 and transfected with AurA than treated cells transfected with empty vector or untreated cells expressing AurA (Fig. 5B, right panel, p = 0.0001 for both). As seen for cilia number, Nek2 inhibition counteracts the effect of AurA to destabilize cilia in cells disassembling cilia. We conclude that Nek2 has a profound effect on cilia length and its action is independent of AurA expression in cells assembling cilia but AurA overexpression further inhibits cilia depolymerization

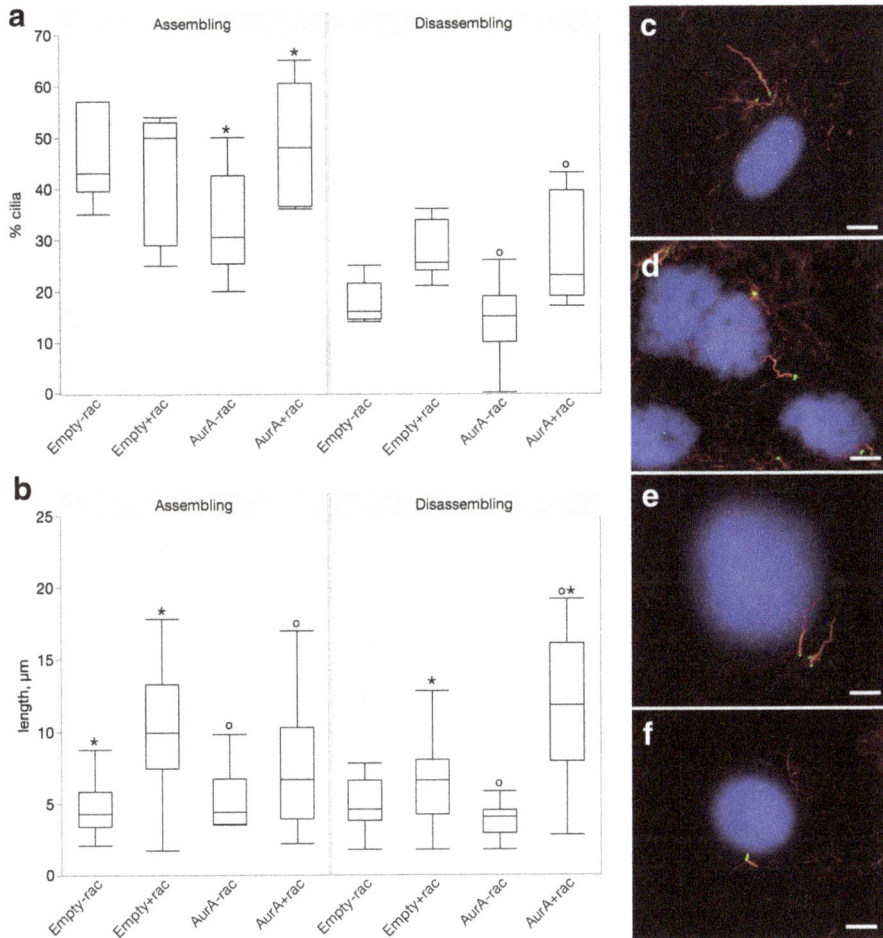

Fig. 5 Inhibition of Nek2 significantly lengthens cilia when cilia are assembling and disassembling. 10 μM of the Nek2 inhibitor (rac-CCT250863) was added to the AurA transfections, cells were serum starved to stimulate cilia formation (assembling, left panels) followed by reintroduction of serum to induce cilia disassembly (disassembling, right panels). Cilia number (**a**) and length (**b**) were measured for a total of at least 300 cells. (**c-e**) show typical examples of long cilia in cells serum starved and exposed to the Nek2 inhibitor and stained for gamma tubulin (green), acetylated tubulin (red) and DNA (blue); (**f**) shows a control cell. Scale bar equals 10 μm. The means for cilia number and length ± SEM are shown in Table 1

when cells are already disassembling their cilia. In this case, Nek2 may activate AurA because in its absence AurA is less able to depolymerize cilia when cilia are disassembling.

R42 overexpression increases the number and length of cilia

We have previously identified PPP1R42 as a binding partner for PP1 in the testes that participates in centrosome dynamics [24]. PPP1R42 (R42) is associated with activated PP1 and localized to the base of flagella in spermatids and cilia in ARPE-19 cells [24, 32]. PP1 is known to regulate both AurA and Nek2 in centrosome separation [19–22], therefore we sought to determine whether R42 could regulate ciliation in ARPE-19 cells and whether any effect was dependent on cilia growth. R42 was overexpressed in actively dividing cells (cycling), and cells where cilia were assembling or disassembling

(Fig. 6). R42 overexpression was confirmed by western blot (Additional file 1: Figure S1D) and indirect immunofluorescence (Additional file 2: Figure S2). Transfection with empty FLAG vector served as negative control. Cells overexpressing R42 were significantly more ciliated compared to the negative control when cells were actively cycling or when cilia were disassembling but not when cilia were assembling (Fig. 6A). There was a 5-fold increase in the percentage of ciliated cells in the cycling cell population overexpressing R42 compared to control (Fig. 6A, left panel, $p = 0.04$); and an 83% increase when cilia were disassembling (Fig. 6A, right panel, $p = 0.03$).

Cilia length was also responsive to R42 overexpression when cells were cycling or disassembling cilia, but not when cilia were assembling. Cilia were 100% longer than control cells (Fig. 6B, left panel, $p < 0.0001$) when R42 was overexpressed in a population of dividing cells and

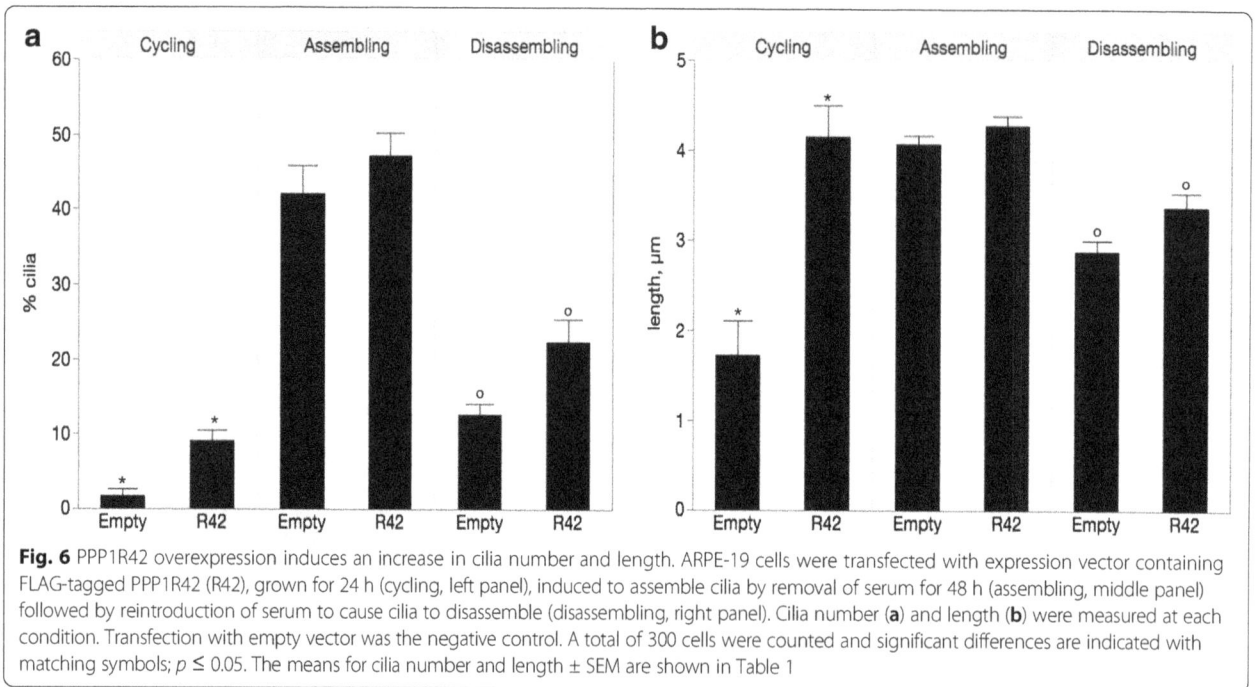

Fig. 6 PPP1R42 overexpression induces an increase in cilia number and length. ARPE-19 cells were transfected with expression vector containing FLAG-tagged PPP1R42 (R42), grown for 24 h (cycling, left panel), induced to assemble cilia by removal of serum for 48 h (assembling, middle panel) followed by reintroduction of serum to cause cilia to disassemble (disassembling, right panel). Cilia number (**a**) and length (**b**) were measured at each condition. Transfection with empty vector was the negative control. A total of 300 cells were counted and significant differences are indicated with matching symbols; $p \leq 0.05$. The means for cilia number and length ± SEM are shown in Table 1

17% longer than control (Fig. 6B, right panel, $p = 0.03$) when cilia were disassembling. Like our results with AurA and Nek2, we see different effects of overexpression on cilia depending on growth status. R42 overexpression does not increase cilia number and length when cilia are assembling, suggesting that cilia number and length are at a maximum in these cells, as we concluded for the KD coexpressions (Fig. 1). However, our results with the double KD (Fig. 2) demonstrated that cilia can become very long when both AurA and Nek2 are affected.

Depletion of PPP1R42 reduces cilia number and length and produces deformed cilia

Protein levels of R42 were reduced with an siRNA pool to examine the effect of this PP1 binding protein on cilia (Additional file 1: Figure S1C and Fig. 7). We have used this approach successfully to show that R42 functions in centrosome dynamics [24]. Cilia length and number were measured in treated cells that were actively dividing, assembling or disassembling cilia. R42 knockdown caused a decrease in ciliated cells but this difference was significant only when cilia were growing (Fig. 7A, p=0.001, middle panel). R42 knockdown significantly decreased cilia length compared to control when cilia were growing; however, cilia were longer in R42 depleted cells that were disassembling cilia ($p = 0.001$ and $p = 0.006$, respectively, Fig. 7B). We note that overexpression of R42 also increased length of cilia instead of decreasing it. This may be because R42 is a multifunctional protein binding to PP1 and Nek2 and these enzymes may have

complex interactions in control of cilia length. Eleven percent of all cells assembling cilia displayed deformed cilia while a smaller fraction of cilia (1.7%) were deformed when cilia were disassembling (Fig. 7A, red bars). Deformed cilia were never seen in control cells (Fig. 7F). The malformed cilia demonstrated a range of shapes from curled into a sphere (Fig. 7B and D) to hook-shaped (Fig. 7C). This difference between cells assembling or disassembling their cilia is similar to our kinase/kinase dead coexpression studies above that revealed different effects on cilia number and length depending on cilia growth status.

R42 interacts with Nek2 and inhibits its activity in vitro

To determine whether R42 interacts with these kinases during cilia assembly and disassembly, we coexpressed R42 with each kinase in cycling cells or in cells with assembling or disassembling cilia, then immunoprecipitated with an R42 antibody and probed for the presence of Nek2 (Fig. 8A, top) or AurA (Fig. 8A, bottom). R42 associates with both Nek2 and AurA during all growth conditions. Notably, R42 interacts with different isoforms of Nek2 depending on the growth status of cilia. Both isoforms interact equally with R42 in dividing cells while the complex with the faster migrating Nek2 species is more abundant when cilia are assembling (Fig. 8A, SS left panel) and the complex with the slower migrating Nek2 species is more abundant when cilia are disassembling (Fig. 8A, +S left panel). Alternative splicing of the *nek2* gene results in isoforms Nek2A, Nek2B, and Nek2C (42 to 48kD) [33]. We used the Nek2A

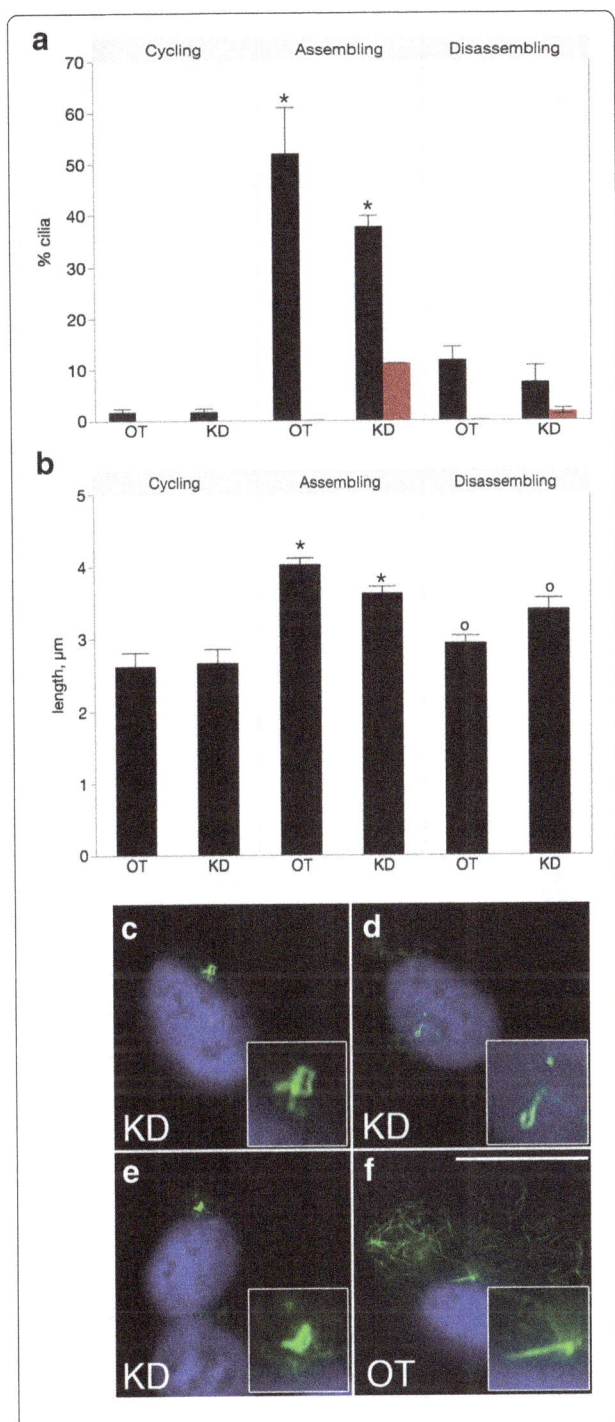

Fig. 7 PPP1R42 depletion reduces cilia number and length and produces abnormal cilia. (**a**) ARPE-19 cells were treated with either PPP1R42 targeting siRNA (knockdown, KD) or with off-target siRNA (off target, OT) and grown for 24 h (cycling, left panel), then cilia assembly was induced by serum starvation (assembling, middle panel) following reintroduction of serum to induce cilia disassembly (disassembling, right panel). Cilia number and length (**a** and **b**) were measured in each condition, 300 cells were counted. Significant differences are indicated with similar symbols; $p \leq 0.05$. Percentage distorted cilia is shown by the red bars in panel A. (**c-e**) ARPE-19 cells were treated with R42 KD siRNA for 24 h and three representative cells are shown with cilia stained green with anti-acetylated tubulin antibody (enriched in cilia) and nuclei blue with DAPI. Insets show deformed cilia at higher magnification. (**f**) A representative control cell treated with off-target siRNA. Scale bar = 10 μm. The means for cilia number and length ± SEM are shown in Table 1

expression plasmid provided by the Fry lab for our experiments; therefore, only one splice variant is overexpressed in cells. Treatment of the R42 IP with λ-phosphatase collapses all bands to the faster migrating species supporting phosphorylation as responsible for the observed isoforms (Fig. 8A, right). Immunoprecipitation of endogenously expressed proteins isolated from untransfected cells showed that R42 interacts with both kinases in vivo (Fig. 8B). To determine whether R42 has any effect on Nek2 activity, we conducted kinase assays after incubating Nek2 with recombinant R42 at different concentrations. We observed a dose dependent inhibition of Nek2 activity by R42 indicating direct interaction and inhibition of Nek2 by R42 (Fig. 8C). R42 also interacts with PP1 [32]; therefore, we wanted to determine whether R42 might also directly affect PP1 activity. We conducted phosphatase assays using recombinant PP1 and R42, along with the PP1 inhibitor PPP1R2 (Inhibitor-2; [34]). Recombinant R42 did not affect PP1 activity whereas PPP1R2 was able to repress phosphatase activity, as has already been shown by others (Additional file 3: Figure S3) [35]. R42 may target PP1 to the centrosome for interaction with kinases.

Nek2 and AurA counteract the effects of R42 overexpression on cilia

We have shown that R42 inhibits Nek2 directly and is a PP1 binding protein [32]. Both Nek2 and AurA kinases induce cilia disassembly and are negatively regulated by PP1 [8, 13, 19, 20, 24, 29, 36, 37]. To determine whether R42 could reverse the depolymerizing activity of Nek2 or AurA, we coexpressed either kinase with R42 (Fig. 9, Additional file 1: Figure S1A). Under every growth condition, R42 was not able to significantly increase the low percentage of ciliated cells induced by either kinase alone (Fig. 9A). A similar effect on cilia length was seen; R42 coexpression with either kinase did not lengthen cilia (Fig. 9B).

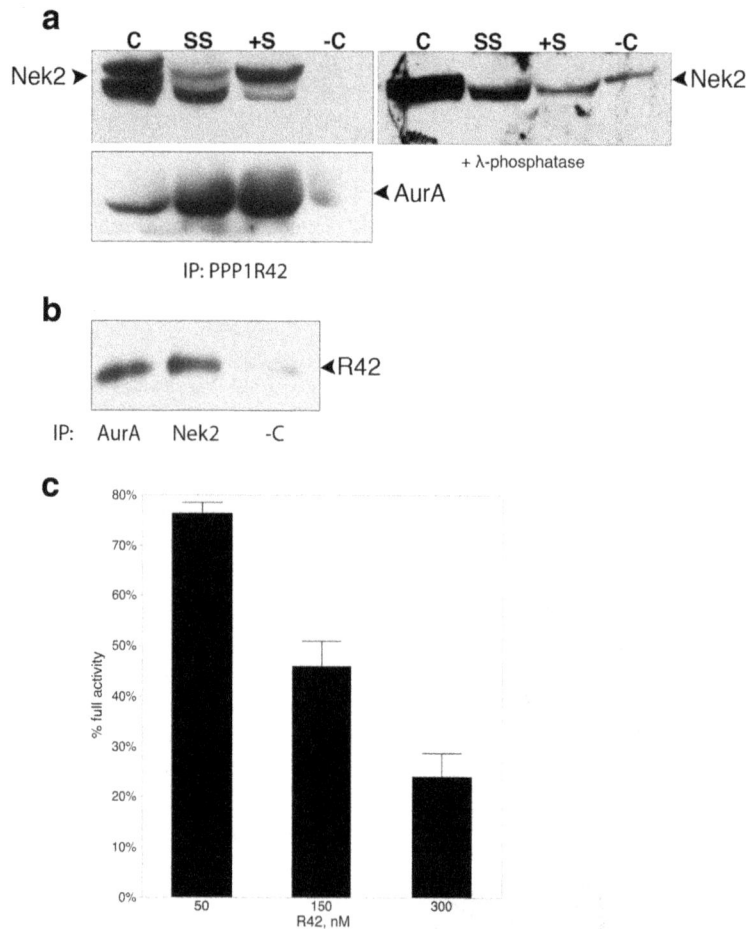

Fig. 8 PPP1R42 interacts with Nek2 and inhibits Nek2 activity in vitro. (**a**) PPP1R42 was overexpressed with either Nek2 or AurA in ARPE-19 cells, grown for 24 h (C, cycling) then the cells were induced to assemble cilia (SS, serum starved) followed by disassembly (+S, serum added back). Protein complexes were immunoprecipitated with anti-PPP1R42 antibody, or no antibody as negative control (−C), and the blots probed for Nek2 or AurA. Immunoprecipitates were treated with λ-phosphatase and probed with Nek2 antibody (**a**, right panel). (**b**) Protein complexes were immunoprecipitated from untransfected ARPE-19 cells with anti-AurA and anti-Nek2 antibodies and probed for PPP1R42. (**c**) Nek2 kinase activity was assayed in the presence of 50, 150 and 300 nM recombinant R42 protein and activity calculated as a percentage of untreated Nek2 activity

Discussion

The mechanisms governing cilia shortening are not well understood. AurA and Nek2 destabilize cilia prior to entry of cells into G_0 [8, 10, 11, 13, 38–41]. The molecules governing this disassembly are depicted in the cartoon in Fig. 10. AurA is activated by multiple intersecting pathways including NEDD9 (Hef1), a scaffolding protein associated with focal adhesions [13] and trichoplein which activates AurA to prevent aberrant cilia formation in proliferating cells [38]. An additional upstream activator, pitchfork, activates AurA at the embryonic node [39]. Nedd9 also regulates Nek2 but inhibits its activity rather than activating this enzyme [42]. AurA and Nek2 also have unique and overlapping downstream effectors. Both kinases can activate HDAC6 to depolymerize microtubules; however, Nek2 can also activate the kinesin KIF24, a member of the family of depolymerizing kinesins, to stimulate cilia depolymerization independent of AurA [11, 13]. Inhibition or overexpression of Nek2 does not affect phosphorylation and activation of AurA at the centrosome, suggesting that Nek2 is downstream of AurA or in a parallel pathway to stimulate cilia absorption [8, 10]. This result examined kinase activation prior to mitosis and not cilia behavior. Our results that Nek2 inhibition can reverse the destabilization effect of AurA to regulate cilia number and length (Fig. 5) indicate that Nek2 may act upstream or in a parallel pathway to control cilia.

In these studies, we provide evidence that AurA and Nek2 functionally and physically interact in cilia reabsorption in ARPE-19 cells and that this interaction varies with the growth state of cilia. The wild-type enzyme can destabilize cilia in the presence of the dominant negative partner enzyme but only when cilia are assembling not disassembling (Fig. 1). One interpretation of this result is that one enzyme is dispensable when cilia are growing but not shortening. This may

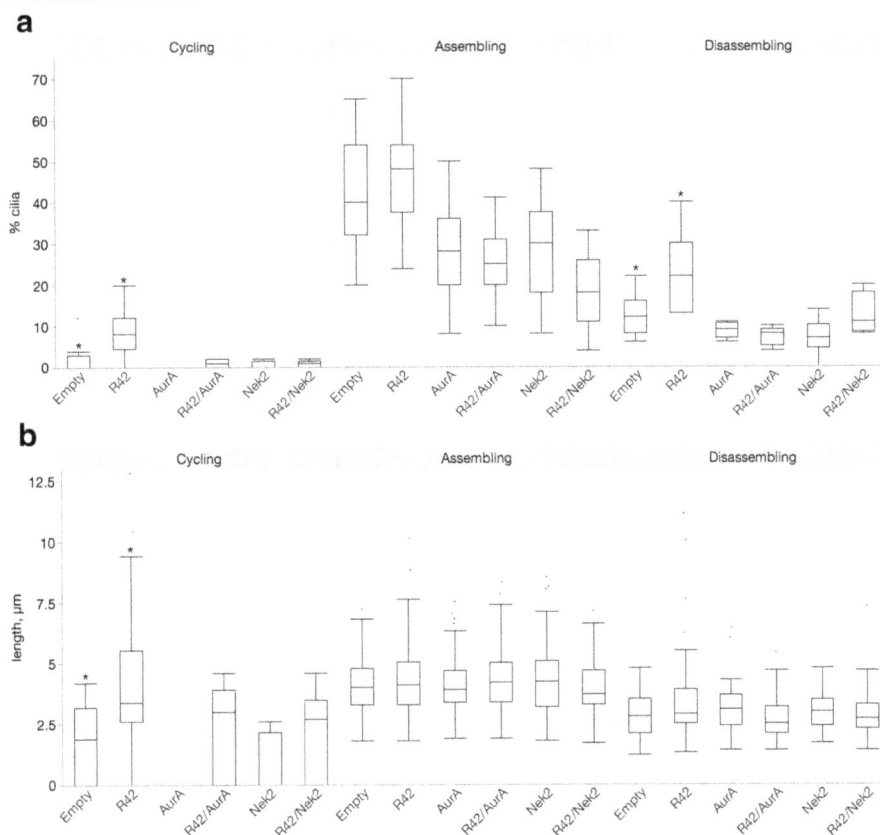

Fig. 9 AurA and Nek2 counteract the stimulatory effect of PPP1R42 on cilia. ARPE-19 cells coexpressing PPP1R42 (R42) with either kinase (R42/Nek2; R42/AurA), or expressing either kinase alone were grown for 24 h (cycling, left panel), then cilia assembly was induced by serum starvation (assembling, middle panel) following by reintroduction of serum to induce cilia to disassemble (disassembling, right panel). Transfection with empty vector served as negative control. Cilia number (**a**) and length (**b**) were measured at each condition; 300 cells were counted. Significant differences are indicated with asterisks; $p \leq 0.05$. The means for cilia number and length ± SEM are shown in Table 1

reflect the fact that the kinases are not as active when cilia are assembling and therefore overexpression of only one kinase is sufficient to shift the balance from increased to reduced ciliation. When cilia are shortening, both enzymes are required to further shorten cilia suggesting that the rapid depolymerization necessary for reentry into the cell cycle may require maximal activation of multiple pathways through activation of several downstream effectors by both Nek2 and AurA.

Alternatively, the dominant negative could have an indirect effect on the activity of the partner kinase. We predict that the kinases are restrained by negative factors when cilia are growing to prevent premature destabilization of cilia in G_0. The dominant negative proteins sequester these negative factors from the partner kinase thereby activating it to prematurely destabilize cilia. Similarly, we predict that the kinases are activated in cells disassembling cilia to efficiently depolymerize cilia prior to reentry into the cell cycle, therefore, competition for positive factors by the dominant negative partner kinase inhibits the other kinase and blocks depolymerization.

This latter explanation is supported by our experiments comparing cells expressing the wild-type kinases to those coexpressing the kinase dead with the wild-type partner kinase (Fig. 1). The dominant negative of either kinase potentiated depolymerization compared to the wild-type kinase alone when cilia were growing and countered depolymerization when cilia were shortening. This is consistent with competition of the dominant negative protein for negative factors when cilia are assembling and positive factors when cilia are disassembling. One conclusion from this finding is that AurA and Nek2 must share common activators and inhibitors of cilia disassembly. However, only reciprocal regulators have been reported. AurA at the basal body is positively regulated by the focal adhesion scaffolding protein Nedd9 to induce cilia disassembly when cells reenter the cell cycle [13]. Conversely, Nedd9 inhibits Nek2 activity to disassemble cilia and increase pericentriolar material at the centrosome [42]. The opposing activity of Nedd9 on Nek2 and AurA is counterintuitive to their demonstrated roles in disassembling cilia and would not likely

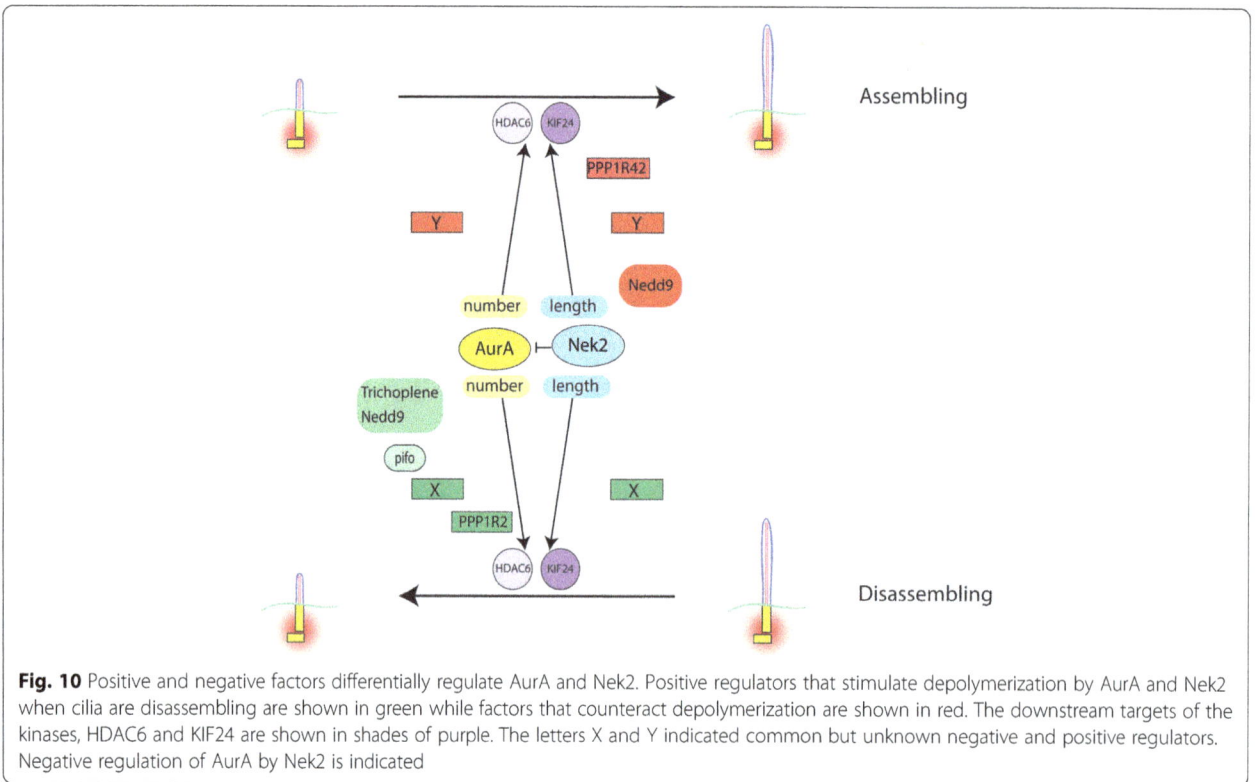

Fig. 10 Positive and negative factors differentially regulate AurA and Nek2. Positive regulators that stimulate depolymerization by AurA and Nek2 when cilia are disassembling are shown in green while factors that counteract depolymerization are shown in red. The downstream targets of the kinases, HDAC6 and KIF24 are shown in shades of purple. The letters X and Y indicated common but unknown negative and positive regulators. Negative regulation of AurA by Nek2 is indicated

counteract one another [8, 10, 11, 13]. It is important to note that the report concerning Nek2 inhibition by Nedd9 was not obtained from cells disassembling cilia but from cycling cells and may reflect differential regulation in cycling vs disassembling cells which is consistent with our work [42]. No common activators or inhibitors of AurA and Nek2 have been identified, however, we would predict that associated positive regulatory factors would be enriched in cells disassembling cilia and negative regulators in cells assembling cilia. This functional interaction is consistent with our finding that the kinases are associated with one another (Fig. 3) and the work of others demonstrating that the two kinases colocalize to the basal body [8, 42].

Although Nek2 and AurA appear to share regulatory proteins in common, their inhibition has different effects on cilia number and length. We show that AurA plays an important role in regulating cilia number (Fig. 4) while Nek2 inhibition significantly impacts cilia length (Fig. 5). This suggests that nucleation of cilia is inhibited by AurA in cells either assembling or disassembling cilia while Nek2 functions primarily to maintain the steady state length of cilia. Our finding contrasts with a previous result that flagella length in *Clamydomonas reinhardtii* is regulated by an AurA-like protein [16]. However, our experiments examine cilia length under dynamic rather than steady state conditions that may involve different mechanisms of regulation. Additionally, we show that Nek2

activates AurA in control of cilia number because Nek2 inhibition inhibits depolymerization by AurA (Fig. 5). However, our results with the double wild-type transfectants indicate that the functions of the kinases are overlapping. Cells cotransfected with active AurA and Nek2 had significantly fewer cilia compared to each kinase transfected separately when cilia were assembling; however, cilia length was increased in the double wild-type transfection. When cilia are disassembling and the kinases are activated, expression of both dominant negative kinases suppresses depolymerization and increases both cilia number and length above the single KD alone but only when cilia are disassembling. This suggests that the mechanism of AurA and Nek2 in regulation of cilia length differs from their control over cilia number.

A particularly intriguing finding was our observation that cilia were extremely long in cells transfected with AKD/NKD (Fig. 2). HDAC6 is a downstream effector of both Nek2 and AurA, therefore we would expect that axonemal microtubules might be stabilized in the double KD and that this effect would be augmented compared to the single KD transfectants [10, 13]. Furthermore, the depolymerizing kinesin KIF24 is activated by Nek2 to ensure cilia disassembly and the Nek2-KIF24 pathway is temporally and functionally distinct from the AurA-HDAC6 pathway [11]. We would expect interruption of both pathways to be additive. Ciliary growth is mediated by intraflagellar transport (IFT) via anterograde motors that move protein and membrane to the ciliary tip and retrograde motors that return cargo to the cilia base [43]. Cilia length is in a state of "dynamic stability" with assembly in balance with disassembly [44–46]. One explanation of our result is that by increasing the length of the microtubule "track" upon which motors transport cargos to the ciliary tip, the cilia is extended beyond its steady state length. This indicates that the length of the microtubule track is one of the limiting factors for IFT. Cilia length in *Chlamydomonas reinhardtii* is regulated by phosphorylation of CALK, an AurA-like protein. Inhibition of CALK activity increases flagella length to twice their normal length [47]. In our system, Nek2 has a more profound effect on cilia length than does AurA.

Nek2 is proposed to act as a switch between cilia growth and resorption in the establishment of left right asymmetry [10]. Increased amounts of Nek2 shift the balance to cilia depolymerization while decreased Nek2 causes centrosome defects; therefore, it is proposed that Nek2 promotes cilia biogenesis and homeostasis. We have identified a new inhibitor of Nek2, the PP1 binding protein PPP1R42. R42 inhibits the activity of Nek2 in vitro and binds to Nek2 in cultured cells. This is consistent with our finding that R42 overexpression increases ciliation presumably by inhibiting cilia depolymerization by Nek2. We conclude that R42 is partially responsible for inhibition of Nek2 in addition to dephosphorylation by PP1 [20, 22]. Our data establishes that different phosphorylated isoforms of Nek2 interact with PPP1R42 depending on cilia growth state and further supports our proposal that Nek2 has different activities when cilia are assembling or disassembling.

We have previously shown that R42 interacts with PP1; however, R42 does not directly affect PP1 activity in vitro (Additional file 3: Figure S3). R42 may target PP1 to Nek2 to facilitate dephosphorylation and inhibition of Nek2; experiments are planned to determine whether a ternary complex exits containing Nek2, PP1, and R42.

Conclusions

Figure 10 presents a model for cilia assembly (top) and disassembly (bottom) based on this work and that of others. In this diagram, positive factors that stimulate depolymerization are shown in green and negative regulators in red and flank either AurA or Nek2. Factors identified by others are Nedd9, trichoplein, pitchfork (pifo) and PP1R2 [10, 13, 38, 48]. HDAC6 and KIF24 are downstream targets of AurA and Nek2 that act to destabilize the microtubules of the axoneme [11, 13]. The work described here makes several unique contributions. First, we propose that AurA and Nek2 are activated and inhibited by positive (green) and negative (red) regulators which control AurA and Nek2 activity differently depending on the growth state of cilia. We

predict that depolymerization is restrained by negative factors that are active when cilia are assembling (top) and that polymerization is enhanced by positive factors when cilia are disassembling (bottom). Second, AurA and Nek2 exist in a complex and share positive and negative regulators, indicated by unknown factors X and Y. This conclusion is supported by our data that the kinase dead of one enzyme affects the activity of the other presumably by sequestering regulatory factors (Fig. 1). Third, AurA and Nek2 have distinct effects on cilia number and length; AurA inhibition increases cilia number with no effect on cilia length while Nek2 inhibition increases cilia length (Figs. 4 and 5).

Fourth, the depolymerizing ability of AurA and Nek2 is additive when cilia are assembling. In addition, the double wild-type transfectants had longer but fewer cilia (Fig. 2). Finally, we show that PPP1R42, a PP1 binding protein, binds to and inhibits Nek2 in vivo and increases ciliation when cilia are shortening but not growing. This dependence on cilia growth status mirrors our results with the kinases. PPP1R42 may actively promote assembly when cilia are disassembling or prevent cilia from shortening. We also show that PPP1R42 does not activate PP1 suggesting that its inhibition of Nek2 is independent of its interaction with PP1.

This work establishes a complex web of regulation of cilia depolymerization that involves both positive and negative effectors that are differentially regulated in cells assembling cilia compared to disassembling cilia. We have also identified a new negative regulator of Nek2 which does not act through PP1 but binds directly to the kinase and represents a candidate molecule for differential modulation in cells assembling or disassembling cilia.

Methods
Cell culture and nucleic acid transfection

Human retinal pigmented epithelial cells (ARPE-19) were obtained from American Type Tissue Collection (Manassas, VA) and grown in DMEM-F12 media supplemented with 10% fetal bovine serum and 1% penicillin-streptomycin as directed by the supplier. To drive cells into quiescence, cells were washed three times in PBS and grown in culture media without fetal bovine serum for 48 h. To release cells from the quiescence, 10% serum was added back to the culture media for 24 h. To study the effect of protein overexpression on cilia growth and retraction, cells were transfected with expression vectors encoding PPP1R42, AurA, AurA kinase dead mutant, Nek2, or Nek2 kinase dead mutant, and 24 h later serum was removed from the media to halt growth and drive cells into G_0 to stimulate cilia growth. Following incubation in serum free media for 48 h, serum was added to the growth media to induce cilia disassembly and cells were observed for cilia number and length 24 h later. Protein expression was confirmed by western analysis (Additional file 1: Figure S1).

Overexpression of PPP1R42 in ARPE-19 cells was performed using a PPP1R42-FLAG expression vector constructed by insertion of the PPP1R42 cDNA downstream and in frame with the FLAG tag of 3XFLAG CMV-14 (Sigma-Aldrich; St. Louis, MO). Overexpression of Nek2 was performed using a Nek2-myc expression vector while inhibition of Nek2 activity was achieved using a Nek2-K37R-myc expression vector, a dominant negative form of Nek2 (kind gifts of Dr. Andrew Fry, [20]). Overexpression of AurA was performed using an AurA-myc expression vector while inhibition of AurA kinase activity was achieved using an AurA-K162R-myc expression vector, a dominant negative form of AurA (kind gifts of Dr. Erich Nigg, [26]). Cells were transfected using Lipofectamine 2000® (Life Technologies; Lincoln, NE) according to the manufacturer's recommendations. Briefly, cells were transfected for 24 h and overexpression was confirmed by western blot (Additional file 1: Figure S1). 90–95% transfection efficiency of the FLAG plasmids was determined using immunofluorescence. Depletion of PPP1R42 in ARPE-19 cells was performed using the ON-TARGETplus SMARTpool siRNA LOC286187 from Dharmacon/Thermo Scientific (Pittsburgh, PA) as previously described [24]. Cells were transfected using the Lipofectamine-RNAiMAX reagent (Life Technologies; Lincoln, NE) according to manufacturer's recommendations. Briefly, cells were transfected for 48 h with a final concentration of 20 μM siRNA. Following incubation, knockdown was confirmed by western blot (Additional file 1: Figure S1).

The AurA inhibitor MLN8237 (0.25 μm; Selleckchem, Houston, TX) was added to cells transfected with the Nek2 plasmid, grown for 24 h, serum removed to induce cilia formation for 48 h and serum reintroduced to induce cilia disassembly. This procedure was repeated with the AurA plasmid and 10 μm Nek2 inhibitor (rac-CCT 250863, Tocris Biosciences, Bristol, UK).

Western blot

ARPE-19 total cell lysates were prepared as previously described [24]. Briefly, cells were resuspended in cell lysis buffer (50 mM Tris, pH 7.5, 1 mM EDTA, 1 mM EGTA, and 1% NP-40) with protease and phosphatase inhibitors. Protein was separated using SDS-PAGE and proteins were transferred from the gel to polyvinylidene difluoride (PVDF) membrane (BIO-RAD Laboratories; Hercules, CA). PPP1R42-FLAG was detected using anti-FLAG antibody (1:1000; F1804; Sigma-Aldrich; St. Louis, MO) and confirmed with anti-human PPP1R42 antibody (1:1000; HPA028628; Sigma-Aldrich; St. Louis, MO). Nek2-myc and AurA-myc were detected using anti-myc antibody (1:1000; TA325701; Origene Technologies;

Rockville, MD). Immune complexes bound to the membrane were detected with horseradish peroxidase-conjugated donkey secondary antibody (711–035-152; Jackson ImmunoResearch Inc.; West Grove, PA) and developed with SuperSignal® West Pico Chemiluminescent Substrate according to directions of the manufacturer (Thermo Fisher Scientific; Asheville, NC).

Indirect immunofluorescence

ARPE-19 cells were grown on coverslips, fixed and permeabilized with methanol at –20 °C for 10 min, and then non-specific sites were blocked by incubation with 3% BSA in Tris-buffered saline and Triton X-100 (TBST) (20 mM Tris, pH 7.5, 150 mM NaCl, 2 mM EGTA, 0.1% Triton X-100) for 30 min. The cells were incubated with anti-acetylated-tubulin antibody to detect primary cilia (1:1000; T6793; Sigma-Aldrich; St. Louis, MO). Acetylated tubulin is a well-accepted marker for primary cilia. Cells were then incubated with FITC-conjugated donkey anti-mouse secondary antibody (1:200; 715–095-150; Jackson ImmunoResearch Inc.; West Grove, PA). DNA was stained with 4',6-diamidino-2-phenylindole (DAPI) incorporated into the mounting media (Vector Labs; Burlingame, CA). The intracellular localization of proteins was observed with a Nikon E600 fluorescence microscope, Pan Fluor 100× objective (N.A. 0.5–1.3) or Pan Fluor 40× objective (N.A. 0.75), fit with appropriate filters and images captured with an Orca II CCD camera, model C4742–95 (Hamamatsu; Middlesex, NJ) and Metamorph image analysis and acquisition software (Molecular Devices; Sunnyvale, CA, USA). Images were exported to Photoshop (Adobe; San Jose, CA) and only linear adjustments to brightness and/or contrast were performed.

Morphometric and statistical analysis

Captured images of cells containing cilia, verified by costaining with anti-acetlyated tubulin and anti-gamma tubulin for the centrosome (Thermo Scientific, PA5–34815; Asheville, NC) were captured by Metamorph and enlarged to visualize cilia clearly. The length of cilia was obtained using the line tool calibrated for the 100X objective. For each treatment, 300 cells were measured. The data for cilia quantification and length are expressed as box and whiskers plots for Figs. 1-2, 4-5 and 10 and as mean ± SEM for Figs. 6 and 7. The differences between groups were analyzed using the unpaired Student's t-test. A p-value of ≤0.05 was considered significant.

Coimmunoprecipitation

Protein complexes were collected by immunoprecipitation. Briefly, affinity purified antibody to PPP1R42 was incubated with precleared cell lysate (1 mg protein) followed by anti-rabbit IgG beads. After transfer to membrane, immunoprecipated proteins were detected with anti-Nek2 (1:500; sc-33,167; Santa Cruz Biotechnology; Dallas, TX) or anti-AurA antibodies (1:500; PC742; EMD Millipore, Billerica, MA), and Veriblot anti-rabbit HRP (Abcam; Cambridge, MA). Use of the Veriblot secondary prevents detection of IgG heavy chain. Negative control for coimmunoprecipitation was precleared lysate incubated with no antibody.

Kinase assay

Kinase assays were conducted in kinase buffer (5 mM MOPS, pH 7.2, 2.5 mM β-glycerophosphate, 1 mM EGTA, 0.4 mM EDTA, 5 mM $MgCl_2$, 0.05 mM DTT) with 1 μg myelin basic protein as substrate and 5 μCi/μl [^{32}P] ATP. 50 nM Nek2 (Thermo Fisher; Waltham, MA) was incubated with varying concentrations of recombinant PPP1R42 (Biomatik; Wilmington, DE). After the designated time, reactions were terminated by spotting onto phosphocellulose P82 paper, washed extensively with 1% phosphoric acid, and the trapped radioactivity measured by scintillation counting.

Phosphatase assay

Inhibition or activation of PP1 by R42 and R2 was accomplished using the fluorescence based RediPlate96© enzcheck serine/threonine phosphatase assay kit from Fisher Scientific (Pittsburgh, PA). Appropriate amounts of recombinant R2 or R42 (Biomatik; Wilmington, DE) were mixed in reaction buffer containing 2 mM DTT and 200 μM $MnCl_2$ and added to wells containing the fluorescent phosphatase substrate. After incubation at 30 °C for 30 min, fluorescence measured at excitation/emission 358/452 nm.

Additional files

Additional file 1: Figure S1. Protein expression in transfected cells. In all cases cell lysates were prepared from transfected cells that were grown for 24 h, serum starved for 48 h followed by reintroduction of serum for 24 h. Proteins were resolved by SDS-PAGE, transferred to membrane and probed with the indicated antibodies. (A) Cells were transfected with the indicated expression plasmids or combinations and proteins probed with anti-FLAG (R42) or anti-myc (AurA and Nek2). The AurA and Nek2 plasmids were verified by sequencing, we were unable to resolve these proteins using this gel system; however, both AurA and Nek2 have been reported as doublets in PAGE [13, 49]. Untransfected cells served as negative control (−C). (B) Cells were transfected with the indicated expression plasmids or combinations and proteins probed with anti-myc (AurA and Nek2). Untransfected cells served as negative control (−C). All panels for each section were exposed to film for the same length of time. (C) Cells were treated with off target (OT) or PPP1R42 (R42) targeting siRNA (KD) and membrane probed with anti-R42 and anti-actin. (D) Cells were transfected with R42-FLAG tagged vector or empty vector (−C) and proteins probed with anti-FLAG. Expressed proteins are maintained in the cell throughout the course of the experiment with reduction when cells are metabolically inactive after starvation. Blots were probed for actin as a loading control. (TIFF 5922 kb)

Additional file 2: Figure S2. Expression plasmids transfect ARPE-19 at high efficiency. ARPE-19 cells were transfected with plasmids expressing either FLAG tagged R42 or myc tagged kinase constructs and grown for 24 h in complete media. Cells were stained with anti-FLAG or anti-myc antibody and detected with Alex Fluor 594 secondary antibody (red). Nuclei were stained with DAPI (blue) (A). 100 cells were counted for each condition and the efficiency of transfection for all constructs was about 90%. Scale bars equal 10 μm. (B) Proteins lysates from cells transfected with either Nek2, AurA, Nek2KD, AurAKD, and R42, were separated by SDS-PAGE transferred to membrane and probed with the appropriate antibodies. Proteins from untransfected cells were loaded to indicate the level of endogenous protein (Ne, Ae, and R42e). (TIFF 16425 kb)

Additional file 3: Figure S3. PPP1R42 does not enhance PP1 activity in vitro. Recombinant PP1 (USBiologicals; Salem, MA) was incubated with varying concentrations of recombinant R2 (A) or R42 (B) (Biomatik; Wilmington, DE) and phosphatase activity measured as described in Materials and Methods. (TIFF 2235 kb)

Abbreviations

AurA KD: Aurora A kinase dead; AurA: Aurora A; KD: kinase dead, knock down; Nek2 KD: Never in mitosis kinase-2 kinase dead; Nek2: Never in mitosis kinase-2; OT: Off target; PPP1R2 or R2: Phosphoprotein phosphatase 1 regulatory subunit 2 or Inhibitor-2; PPP1R42 or R42: Phosphoprotein phosphatase 1 regulatory subunit 42

Acknowledgements
The authors wish to thank Dr. Andrew Fry and Dr. Erich Nigg for gifts of the expression plasmids used in this work.

Funding
This work was supported by grant HD080151 (AOS) from the National Institutes of Health; however, this funding body had no input into the design collection analysis, and interpretation, and in writing the manuscript.

Authors' contributions
The authors contributed to the work in the following ways: ND conducted transfections followed by immunofluorescence and cilia morphometrics and drafted the manuscript. RW conducted transfections followed by immunofluorescence and cilia morphometrics, western blot, immunoprecipitation and phosphatase assays, KK conducted transfections. AOS conceived and oversaw the experiments, wrote the manuscript and conducted kinase assays. All authors read and approved the final manuscript.

Competing interests
The Authors declare they have no competing interests.

Author details
[1]Anatomy and Cell Biology, East Carolina University, Brody School of Medicine, Greenville, NC, USA. [2]Laboratory of Biochemistry and Genetics, National Institute of Diabetics and Digestive and Kidney Diseases, National Institutes of Health, Bethesda, MD, USA.

References

1. Schneider L, Clement CA, Teilmann SC, Pazour GJ, Hoffmann EK, Satir P, Christensen ST. PDGFRalphaalpha signaling is regulated through the primary cilium in fibroblasts. Curr Biol. 2005;15(20):1861–6.
2. Simons M, Gloy J, Ganner A, Bullerkotte A, Bashkurov M, Kronig C, Schermer B, Benzing T, Cabello OA, Jenny A, Mlodzik M, Polok B, Driever W, Obara T, Walz G. Inversin, the gene product mutated in nephronophthisis type II, functions as a molecular switch between Wnt signaling pathways. Nat Genet. 2005;37(5):537–43.
3. Huangfu D, Liu A, Rakeman AS, Murcia NS, Niswander L, Anderson KV. Hedgehog signalling in the mouse requires intraflagellar transport proteins. Nature. 2003;426(6962):83–7.
4. Veland IR, Awan A, Pedersen LB, Yoder BK, Christensen ST. Primary cilia and signaling pathways in mammalian development, health and disease. Nephron Physiol. 2009;111(3):p39–53.

5.   Basten SG, Giles RH. Functional aspects of primary cilia in signaling, cell cycle and tumorigenesis. Cilia. 2013;2(1):6.

6.   Quinlan RJ, Tobin JL, Beales PL. Modeling ciliopathies: primary cilia in development and disease. Curr Top Dev Biol. 2008;84:249–310.

7.   Badano JL, Mitsuma N, Beales PL, Katsanis N. The ciliopathies: an emerging class of human genetic disorders. Annu Rev Genomics Hum Genet. 2006;7:125–48.

8.   Spalluto C, Wilson DI, Hearn T. Nek2 localises to the distal portion of the mother centriole/basal body and is required for timely cilium disassembly at the G2/M transition. Eur J Cell Biol. 2012;91(9):675–86.

9.   Fry AM, Mayor T, Meraldi P, Stierhof YD, Tanaka K, Nigg EA. C-Nap1, a novel centrosomal coiled-coil protein and candidate substrate of the cell cycle-regulated protein kinase Nek2. J Cell Biol. 1998;141(7):1563–74.

10. Endicott SJ, Basu B, Khokha M, Brueckner M. The NIMA-like kinase Nek2 is a key switch balancing cilia biogenesis and resorption in the development of left-right asymmetry. Development. 2015;142(23):4068–79.

11. Kim S, Lee K, Choi JH, Ringstad N, Dynlacht BD. Nek2 activation of Kif24 ensures cilium disassembly during the cell cycle. Nat Commun. 2015;6:8087.

12. Fry AM, Meraldi P, Nigg EA. A centrosomal function for the human Nek2 protein kinase, a member of the NIMA family of cell cycle regulators. EMBO J. 1998;17(2):470–81.

13. Pugacheva EN, Jablonski SA, Hartman TR, Henske EP, Golemis EA. HEF1-dependent aurora a activation induces disassembly of the primary cilium. Cell. 2007;129(7):1351–63.

14. Pan J, Wang Q, Snell WJ. An aurora kinase is essential for flagellar disassembly in Chlamydomonas. Dev Cell. 2004;6(3):445–51.

15. Marumoto T, Zhang D, Saya H. Aurora-a - a guardian of poles. Nat Rev Cancer. 2005;5(1):42–50.

16. Luo M, Cao M, Kan Y, Li G, Snell W, Pan J. The phosphorylation state of an aurora-like kinase marks the length of growing flagella in Chlamydomonas. Curr Biol. 2011;21(7):586–91.

17. Cao M, Li G, Pan J. Regulation of cilia assembly, disassembly, and length by protein phosphorylation. Methods Cell Biol. 2009;94:333–46.

18. Wang G, Jiang Q, Zhang C. The role of mitotic kinases in coupling the centrosome cycle with the assembly of the mitotic spindle. J Cell Sci. 2014;127(Pt 19):4111–22.

19. Eto M, Elliott E, Prickett TD, Brautigan DL. Inhibitor-2 regulates protein phosphatase-1 complexed with NimA-related kinase to induce centrosome separation. J Biol Chem. 2002;277(46):44013–20.

20. Helps NR, Luo X, Barker HM, Cohen PT. NIMA-related kinase 2 (Nek2), a cell-cycle-regulated protein kinase localized to centrosomes, is complexed to protein phosphatase 1. Biochem J. 2000;349(Pt 2):509–18.

21. Katayama H, Zhou H, Li Q, Tatsuka M, Sen S. Interaction and feedback regulation between STK15/BTAK/aurora-a kinase and protein phosphatase 1 through mitotic cell division cycle. J Biol Chem. 2001;276(49):46219–24.

22. Mi J, Guo C, Brautigan DL, Larner JM. Protein phosphatase-1alpha regulates centrosome splitting through Nek2. Cancer Res. 2007;67(3):1082–9.

23. Wang H, Brautigan DL. A novel transmembrane Ser/Thr kinase complexes with protein phosphatase-1 and inhibitor-2. J Biol Chem. 2002;277(51):49605–12.

24. DeVaul N, Wang R, Sperry AO. PPP1R42, a PP1 binding protein, regulates centrosome dynamics in ARPE-19 cells. Biol Cell. 2013;105(8):359–71.

25. Faragher AJ, Fry AM. Nek2A kinase stimulates centrosome disjunction and is required for formation of bipolar mitotic spindles. Mol Biol Cell. 2003;14(7):2876–89.

26. Meraldi P, Nigg EA. Centrosome cohesion is regulated by a balance of kinase and phosphatase activities. J Cell Sci. 2001;114(Pt 20):3749–57.

27. Dere R, Perkins AL, Bawa-Khalfe T, Jonasch D, Walker CL. Beta-catenin links von Hippel-Lindau to aurora kinase a and loss of primary cilia in renal cell carcinoma. J Am Soc Nephrol. 2015;26(3):553–64.

28. Ou Y, Ruan Y, Cheng M, Moser JJ, Rattner JB, van der Hoorn FA. Adenylate cyclase regulates elongation of mammalian primary cilia. Exp Cell Res. 2009;315(16):2802–17.

29. Plotnikova OV, Nikonova AS, Loskutov YV, Kozyulina PY, Pugacheva EN, Golemis EA. Calmodulin activation of aurora-a kinase (AURKA) is required during ciliary disassembly and in mitosis. Mol Biol Cell. 2012;23(14):2658–70.

30. Manfredi MG, Ecsedy JA, Chakravarty A, Silverman L, Zhang M, Hoar KM, Stroud SG, Chen W, Shinde V, Huck JJ, Wysong DR, Janowick DA, Hyer ML, Leroy PJ, Gershman RE, Silva MD, Germanos MS, Bolen JB, Claiborne CF, Sells TB. Characterization of alisertib (MLN8237), an investigational small-molecule inhibitor of aurora a kinase using novel in vivo pharmacodynamic assays. Clin Cancer Res. 2011;17(24):7614–24.

31. Innocenti P, Cheung KM, Solanki S, Mas-Droux C, Rowan F, Yeoh S, Boxall K, Westlake M, Pickard L, Hardy T, Baxter JE, Aherne GW, Bayliss R, Fry AM, Hoelder S. Design of potent and selective hybrid inhibitors of the mitotic kinase Nek2: structure-activity relationship, structural biology, and cellular activity. J Med Chem. 2012;55(7):3228–41.

32. Wang R, Sperry AO. PP1 forms an active complex with TLRR (lrrc67), a putative PP1 regulatory subunit, during the early stages of spermiogenesis in mice. PLoS One. 2011;6(6):e21767.

33. Hames RS, Fry AM. Alternative splice variants of the human centrosome kinase Nek2 exhibit distinct patterns of expression in mitosis. Biochem J. 2002;361(Pt 1):77–85.

34. Huang FL, Glinsmann WH. Separation and characterization of two phosphorylase phosphatase inhibitors from rabbit skeletal muscle. Eur J Biochem. 1976;70(2):419–26.

35. Huang FL, Glinsmann W. A second heat-stable protein inhibitor of phosphorylase phosphatase from rabbit muscle. FEBS Lett. 1976;62(3):326–9.

36. Ohashi S, Sakashita G, Ban R, Nagasawa M, Matsuzaki H, Murata Y, Taniguchi H, Shima H, Furukawa K, Urano T. Phospho-regulation of human protein kinase aurora-a: analysis using anti-phospho-Thr288 monoclonal antibodies. Oncogene. 2006;25(59):7691–702.

37. Hilton LK, Gunawardane K, Kim JW, Schwarz MC, Quarmby LM. The kinases LF4 and CNK2 control ciliary length by feedback regulation of assembly and disassembly rates. Curr Biol. 2013;23(22):2208–14.

38. Inoko A, Matsuyama M, Goto H, Ohmuro-Matsuyama Y, Hayashi Y, Enomoto M, Ibi M, Urano T, Yonemura S, Kiyono T, Izawa I, Inagaki M. Trichoplein and aurora a block aberrant primary cilia assembly in proliferating cells. J Cell Biol. 2012;197(3):391–405.

39. Kinzel D, Boldt K, Davis EE, Burtscher I, Trumbach D, Diplas B, Attie-Bitach T, Wurst W, Katsanis N, Ueffing M, Lickert H. Pitchfork regulates primary cilia disassembly and left-right asymmetry. Dev Cell. 2010;19(1):66–77.

40. Plotnikova OV, Pugacheva EN, Golemis EA. Aurora a kinase activity influences calcium signaling in kidney cells. J Cell Biol. 2011;193(6):1021–32.

41. Inaba H, Goto H, Kasahara K, Kumamoto K, Yonemura S, Inoko A, Yamano S, Wanibuchi H, He D, Goshima N, Kiyono T, Hirotsune S, Inagaki M. Ndel1 suppresses ciliogenesis in proliferating cells by regulating the trichoplein-aurora a pathway. J Cell Biol. 2016;212(4):409–23.

42. Pugacheva EN, Golemis EA. The focal adhesion scaffolding protein HEF1 regulates activation of the aurora-a and Nek2 kinases at the centrosome. Nat Cell Biol. 2005;7(10):937–46.

43. Kozminski KG, Johnson KA, Forscher P, Rosenbaum JL. A motility in the eukaryotic flagellum unrelated to flagellar beating. Proc Natl Acad Sci U S A. 1993;90(12):5519–23.

44. Keeling J, Tsiokas L, Maskey D. Cellular mechanisms of ciliary length control. Cell. 2016;5(1)

45. Stephens RE. Synthesis and turnover of embryonic sea urchin ciliary proteins during selective inhibition of tubulin synthesis and assembly. Mol Biol Cell. 1997;8(11):2187–98.

46. Marshall WF, Rosenbaum JL. Intraflagellar transport balances continuous turnover of outer doublet microtubules: implications for flagellar length control. J Cell Biol. 2001;155(3):405–14.

47. Cao M, Meng D, Wang L, Bei S, Snell WJ, Pan J. Activation loop phosphorylation of a protein kinase is a molecular marker of organelle size that dynamically reports flagellar length. Proc Natl Acad Sci U S A. 2013;110(30):12337–42.

48. Satinover DL, Leach CA, Stukenberg PT, Brautigan DL. Activation of aurora-a kinase by protein phosphatase inhibitor-2, a bifunctional signaling protein. Proc Natl Acad Sci U S A. 2004;101(23):8625–30.

49. Ha Kim Y, Yeol Choi J, Jeong Y, Wolgemuth DJ, Rhee K. Nek2 localizes to multiple sites in mitotic cells, suggesting its involvement in multiple cellular functions during the cell cycle. Biochem Biophys Res Commun. 2002;290(2):730–6.

12

A specific FMNL2 isoform is up-regulated in invasive cells

Christine Péladeau, Allan Heibein, Melissa T. Maltez, Sarah J. Copeland and John W. Copeland[*]

Abstract

Background: Formins are a highly conserved family of cytoskeletal remodeling proteins. A growing body of evidence suggests that formins play key roles in the progression and spread of a variety of cancers. There are 15 human formin proteins and of these the Diaphanous-Related Formins (DRFs) are the best characterized. Included in the DRFs are the Formin-Like proteins, FMNL1, 2 & 3, each of which have been strongly implicated in driving tumorigenesis and metastasis of specific tumors. In particular, increased FMNL2 expression correlates with increased invasiveness of colorectal cancer (CRC) in vivo and for a variety of CRC cell-lines in vitro. FMNL2 expression is also required for invasive cell motility in other cancer cell-lines. There are multiple alternatively spliced isoforms of FMNL2 and it is predicted that the encoded proteins will differ in their regulation, subcellular localization and in their ability to regulate cytoskeletal dynamics.

Results: Using RT-PCR we identified four FMNL2 isoforms expressed in CRC and melanoma cell-lines. We find that a previously uncharacterized FMNL2 isoform is predominantly expressed in a variety of melanoma and CRC cell lines; this isoform is also more effective in driving 3D motility. Building on previous reports, we also show that FMNL2 is required for invasion in A375 and WM266.4 melanoma cells.

Conclusions: Taken together, these results suggest that FMNL2 is likely to be generally required in melanoma cells for invasion, that a specific isoform of FMNL2 is up-regulated in invasive CRC and melanoma cells and this isoform is the most effective at facilitating invasion.

Keywords: Formins, Actin, Invasion, Metastasis, FMNL2, Melanoma, Colorectal cancer

Abbreviations: CRC, Colorectal cancer; DAD, Diaphanous Autoregulatory domain; DID, Diaphanous inhibitory domain; DRF, Diaphanous-Related Formin; FH, Formin homology; FMNL, Formin-like; MT, Microtubules; qPCR, quantitative PCR; RT-PCR, Reverse transcription PCR; SRF, Serum Response Factor

Background

Metastasis initiates with the migration of individual cells, or cellular collectives, from the tumor into the surrounding stroma [1]. Individually invading tumor cells use either a mesenchymal or amoeboid mode of 3-D migration. In this context, mesenchymal invasion requires low actomyosin contractility, greater cell-substrate attachment, and proteolytic digestion of the extracellular matrix. In contrast, amoeboid invasion requires high contractility, low adhesion and is protease-independent [2, 3]. A specific cell-type utilizes a specific mode of invasion, however, cells will switch from mesenchymal to amoeboid migration when mesenchymal motility is inhibited [4]. In vivo, amoeboid invasion is very rapid and is used by breast cancer cells to invade the stroma [5]. In vitro, a variety of colorectal and melanoma cancer cell lines adopt this mode of migration [3, 6]. Both 2D and 3D cell motility is driven by the coordinated regulation of actin dynamics and is dependent upon a variety of actin remodeling proteins.

Formin homology proteins are a highly conserved family of cytoskeletal remodeling proteins distinguished by the presence of two functional domains, formin homology 1 (FH1) and FH2. FH1 is proline-rich and a ligand for the small actin-binding protein profilin [7]. FH2 directly nucleates actin polymerization, inducing the formation of long, unbranched actin filaments (F-actin);

* Correspondence: John.copeland@uottawa.ca
Department of Cellular and Molecular Medicine, Faculty of Medicine, University of Ottawa, 451 Smyth Road, Ottawa, ON K1H 8M5, Canada

some FH2 domains also bind and bundle F-actin [8–11]. FH2 is a dual purpose domain that is also able to regulate microtubule (MT) stabilization [12–14].

Given their dramatic effects on cytoskeletal dynamics, it is not surprising that a number of formins have been shown to play a role in metastasis and invasion [1]. In particular, the FMNL sub-group of Diaphanous-Related Formins have each been shown to be required for migration and invasion by transformed cells. FMNL1 promotes proliferation and motility of leukemia cells [15–17]. FMNL3 is required for invasion in PC3 prostate cancer cells and down-regulation of FMNL3 expression is associated with suppression of metastasis [18–21]. Increased FMNL2 expression correlates with increased invasiveness in CRC cell-lines [22–24] and FMNL2 is consistently highly expressed across the NCI60 panel of melanoma cell-lines [25]. In patient samples increased FMNL2 expression also correlates directly with increased CRC metastasis [23, 24]. FMNL2 is required in both MDA-MB-435 for amoeboid cell invasion in vitro and B16-F1 melanoma cells where it cooperates with the Arp2/3 complex for lamellipodial extension [6, 26]. More recently, FMNL2 has been shown to participate in the regulation of cell-cell and cell-substrate adhesions [27, 28].

The ability of FMNL2 to govern cytoskeletal dynamics is regulated by an autoinhibitory interaction between its C-terminal Diaphanous Autoregulatory Domain (DAD) and N-terminal Diaphanous Inhibitory Domain (DID) [11]. Inhibition is relieved by binding of either active RhoC or cdc42 to the FMNL2 GTPase Binding Domain (GBD) [6, 26]. Deletion of either DID or DAD is sufficient to render FMNL2 constitutively active [11]. As with other formins, constitutively active derivatives of FMNL2 are able to induce F-actin accumulation and MT acetylation in vivo [11, 12] as well as bind and bundle actin filaments and accelerate F-actin polymerization in vitro [11, 26]. F-actin bundling by FMNL2 is dependent on FH2 and an actin-binding WH2 motif immediately C-terminal to the FH2 domain [11]. The FMNL2 protein is targeted to the plasma membrane by N-myristoylation at Gly2 [29], although there are conflicting reports as to whether or not this modification is required for FMNL2-dependent cell migration [22, 26].

Many formin genes undergo alternative splicing of their mRNA and the resulting isoforms have significant impact on the regulation, localization and function of the encoded proteins [30–34]. Similarly, up-regulation of a specific splice form of the cytoskeletal remodeling proteins Mena and palladin potentiates the invasive phenotype of transformed cells [35–38]. There are at least 14 predicted alternative splice forms of FMNL2 affecting regions that are likely to impact on the activity, regulation and subcellular localization of the encoded proteins. We show here that there are at least four of the

predicted FMNL2 isoforms expressed in CRC and melanoma cells and that a novel isoform is preferentially up-regulated in invasive cells. We also show that invasion by A375 and WM266.4 melanoma cells is FMNL2-dependent suggesting that FMNL2 might be generally required for 3D invasion in this cell-type. All four FMNL2 isoforms are able to rescue invasion in FMNL2-depleted A375 cells, but the "invasive" isoform is significantly better than the other three.

Methods

Cell culture

U2OS osteosarcoma human cells were a gift from Dr. Laura Trinkle-Mulcahy and were grown in DMEM 10 % FBS at 37 °C, in 5 % CO_2. All other cell-lines were obtained from the American Type Culture Collection (ATCC) and were maintained as recommended by the supplier.

Plasmids

The FMNL2 isoforms described in this study correspond to the following accession numbers. ITM: NP_443.137.2, YHY: Q96PY5, PMR: XP_005246322, TQS: XP_005246320.1. Full length (FL) FMNL2 isoform cDNA were assembled in a step-wise fashion into the pEF-plink2-mCherry vector. Briefly, the alternative 3′ ends of each FMNL2 isoform was amplified by polymerase chain reaction (PCR) from cDNA generated from SW620 cell mRNA. The conserved 5′ portion of the FMNL2 cDNA was amplified from KIAA1902 (Kazusa Project, Japan). FMNL2 FH1 + FH2 derivatives were subcloned into pEF.NBRSS from the full-length constructs. For rescue, FMNL2 full-length isoforms were cloned into pLVX-IRES-mCherry lentivirus vector. All PCR reactions were performed using standard techniques and the Phusion High Fidelity DNA Polymerase (Thermo scientific).

Serum Response Factor (SRF) assay

SRF reporter gene assays were performed as described previously (Copeland and Treisman [39]). Briefly, NIH 3T3 fibroblasts were seeded on 6-well plates at 125 000 cells/well 1 day prior to transfection. The cells were transfected using polyethylenimine transfection reagent (PEI) with the indicated expression plasmids (0.1ug) and the reporter constructs p3D.A.Luc (50 ng/well) and pMLVLacZ (250 ng/well). After 5 h, the cells were placed in low serum medium (0.5 % FBS in DMEM). Cells were harvested the next day and lysed in 1x Reporter Lysis buffer; luciferase assays were performed according to the supplied protocol (Promega) and read in an LMAXII Luminometer (Molecular Devices). The activation of luciferase was standardized to an SRF-VP16 control. A β-galactosidase (β-gal) assay is performed in parallel as a transfection efficiency control. The cell lysates are also subjected to a sodium dodecyl sulfate polyacrylamide gel (SDS-PAGE) and immunoblotted to detect FMNL2 expression levels.

RNA extraction, RT-PCR, and qPCR

The indicated cell-lines were grown to 70–80 % confluence in 10 cm dishes. Total RNA was harvested using the RNeasy mini kit (Qiagen). Total RNA concentration was determined by reading the OD_{280}. RNA quality was assessed on denaturing formaldehyde/agarose MOPS gel and visualization of distinct 18S and 28S ribosomal subunits. cDNA were synthesized using MuLV reverse transcriptase and the GeneAmp RNA PCR kit (Applied Biosystems). PAW109 RNA serves as a positive RT-PCR control and no RNA (ddH_2O) served as negative control. Characterization of the FMNL2 isoforms in A375, SW620 and SW480 cancer cell lines was accomplished by performing a PCR reaction [2 mM MgCL2 solution, 1X PCR buffer I, ddH_2O, 2.5U/100 µL AmpliTaq DNA Polymerase, 20 µL purified cancer cell cDNA and 0.15 µM of both forward and reverse FMNL2 or control specific primers]. PCR products were separated on 1 % agarose gels. Amplified bands were isolated with Gen Elute Gel extraction kit (Sigma) and directly sequenced. These sequences were used to design primers for qPCR.

Real time PCR analysis was performed to determine absolute expression levels of FMNL2 isoforms using the following primers. ITM forward 5′-GCCATTGAAGA TATTATCACAGATC-3′, rev 5′-AACTTGCGTTCTGT TAATGGTG-3′, amplicon 117 bp; YHY 5′-CTGAAGA CTGTGCCCTTTACTGCT-3′, 5′-CCTGTTCTCACTG AGGAATACCATTAC-3′, amplicon 87 bp; PMR 5′-CA TTGAAGATATTATCACAGCCTTA-3′, 5′-GAGGATC TTAGAAACCAACCATA-3′, amplicon 87 bp; TQS 5′-G ATATTATCACAGCCTTAAAGAAGAAT-3′, 5′-TGGTG GAGGATACACAGAGCT-3′, amplicon 106 bp; pan-FM NL2 5′-GCTCCTCCCTTAGCACCT-3′, 5′-GCCAATCA AGACGAAGTTCAGA-3′, amplicon 127 bp; β-actin 5′-G CACCACACCTTCTACAATGAG-3′, 5′-GACCCAGATC ATGTTTGAGACC-3′, amplicon 122 bp. Standard curves were established using FMNL2 FH1 + FH2 + C templates corresponding to each isoform. A logarithmic series of template dilutions were generated (10^2, 10^3, 10^4, 10^5, 10^6 copies per tube). A standard curve for β-actin template was also generated as a control. The experimental design included standards, no template controls and unknowns. Standards contained: 5 µL FMNL2 isoform template, 5 µL of specific FMNL2 isoform primer (final concentration: 300 ng/well/primer) and 10 µL of the SYBR Green PCR Premix (QuantiTect; QIAGEN) used according to the manufacturer's instructions. Unknown reactions consisted of 5 µL cDNA (harvested from cancer cell lines) (0.05 µg/ tube), 5 µL primers and 10 µL SYBR Green qPCR Premix. No template control reactions were 5 µL of primers, 10 µL

of SYBR Green mix and 5 µL ddH_2O. 5-carboxy-X-Rhodamine (ROX) is also included in the SYBR Green qPCR Premix and serves as a passive loading reference dye. The qPCR reactions were run on an Mx 3005P qPCR instrument (Stratagene). Standard curves for each isoform were used to determine the relationship between the fluorescent signal (the CT value) and absolute copy number of each FMNL2 isoforms in each sample.

Immunofluorescence

NIH-3T3 fibroblast cells were seeded at a density of 125 000 cells/well in 6-well plates containing acid-washed coverslips and 10 % DBS DMEM medium. The cells were transfected using PEI (1 mg/mL) as above. 24 h later, the cells were fixed with 4 % paraformaldehyde (PFA) in PBS for 15 min, washed three times with 1x PBS (5 min) and permeabilized for 30 min with 0.3 % Triton X-100, 10 % horse serum in 1XPBS. Fixed cells were incubated at room temperature with primary antibodies in 0.03 % Triton X-100 plus 5 % horse serum 1X PBS for 1 h. Cells were washed three times in 1X PBS and incubated with the appropriate secondary antibody and Fluorescein-Phalloidin (1:100, Molecular Probes) in 0.03 % Triton X-100 plus 5 % horse serum 1X PBS at room temperature for 1 h. The cells were washed three times in 1X PBS and once in ddH2O and mounted on slides with Vectashield mounting medium with DAPI (Vector Laboratories). Images were captured on a Zeiss Axio Imager Z1 fluorescence microscope, using a 63X Plan Apochromat objective and an AxioCam HRm camera. Optical sections were obtained with the apotome2. Images were processed using Axiovision software and figures assembled using Adobe Photoshop.

Filopodia formation in transfected cells was assessed visually by immunofluorescence. Transfected cells expressing myr-mCherryFP or FMNL2-mCherry were identified by virtue of the mCherryFP tag and F-actin visualized by fluorescein-phalloidin as above. Cells with an obvious overproliferation of filopodia (>25 filopodia/cell) compared to either non-transfected neighbours or myr-mCherryFP expressing control cells were counted and the result expressed as a percent of transfected cells.

Western blotting

Cells were grown to 70–80 % confluence in 10 cm dishes, washed with 1XPBS and harvested in 1xSDS buffer. Lysates were subjected to SDS-PAGE and immunoblotted with the indicated antibodies. Chemiluminescence was used for detection using the western HRP substrate reagent (Millipore) and visualized on a GE Image Quant LAS4010 Imaging System. Blots were stripped in 0.1 M Glycine, 0.5 % SDS (pH2.5) for 1 h. Blots were re-blocked and probed with mouse-anti-α-tubulin (clone DM1A, T9026, Sigma) to assess loading. Chemiluminescence and visualization is performed as described previously. Pan-FMNL2 antiserum was raised in chicken (Cedarlane) using the isolated FMNL2-FH2 protein (codons 599–1045) expressed in *E.coli* and purified as previously described [11]. All FMNL2 antisera were affinity purified using standard protocols [40]. Affinity purified anti-FMNL3 antibody was described previously [41].

FMNL2 siRNA

A375 or WM266.4 melanoma cells were seeded in six well plates or 3.5 cm dishes (Corning) at a density of 125 000 cells/well. The following day, cells were transfected (DharmaFECT #1, Thermo Scientific) with control or FMNL2 siRNA duplexes (TriFECTa Dicer-Substrate RNAi Kit, Integrated DNA Technologies) as directed by the manufacturer. The siRNA duplex targeted the 3′UTR of FMNL2 (5′-CCUGUUCAGAUUAAUCAAAGCAATA-3′). A non-specific universal negative control duplex (Integrated DNA Technologies) was used for all siRNA knockdown experiments. This control duplex does not recognize any sequences in human, mouse or rat transcriptomes (5′-CGUUAAUCGCGUAUAAUAAGAGUAT-3′). Following transfection, cells were incubated at 37 °C (5 % CO_2) for 48 h. A fluorescent TYE 563 DS control was used to verify transfection efficiency. After 48 h, cells are harvested and the lysates subjected to immunoblotting to detect FMNL2 expression levels.

2-D migration assay

A375 melanoma cells were seeded in six well plates or 3.5 cm petri dish (Corning) at a density of 125,000 cells/well. The following day, the cells were transfected with siRNA; after 48 h 100,000 A375 cells were added to each chamber of an ibidi wound insert in a 3.5 cm petri dish (Ibidi). The outside of the insert was filled with 1.5 ml of DMEM 10 % FBS. In parallel, cells were also seeded in duplicate to assess knockdown efficiency by immunobloting. The next day, the insert was removed to generate the wound and the plate was gently washed with 10 % FBS DMEM to remove any floating cells. Wound closure was monitored for 48 h by live imaging on a Zeiss Axiovert 200 microscope (10x objective, phase 1) in a controlled environment (5 % CO2, 37 °C). The percent wound closure was calculated by measuring the distance of the gap at three points using Northern Eclipse Software (NES, Empix Imaging, Mississauga, Ontario, Canada).

Virus production and transduction

FMNL2 cDNA were cloned into the lentiviral vector pLVX-IRES-mCherry for virus production. Briefly, 10 plates (15 cm) of 293 T cells at 70 % confluence were transfected with 96.85 µg of the FMNL2 pLVX-IRES-mCherry construct, 53.95 µg of the envelop plasmid (pMD2G coding for VSV-G envelope), 99.15 µg of the packaging plasmid psPAX2 using PEI. Virus was collected from the medium supernatant every day for the next 48 h. The virus was concentrated and titrated to determine the multiplicity of infection (MOI). For rescue experiments, A375 melanoma cells were seeded at a density of 125 000 cells/well, in a six well plate or in a 3.5 cm petri dish with a coverslip. The next day, the cells were transfected with siRNAs and incubated for 24 h before infection with the FMNL2 expressing lentiviral vectors using a *multiplicity of infection (MOI)* of 10. The cells were left for another 24 h before seeding for an invasion assay. Efficiency of knockdown and re-expression of FMNL2 was assessed by immunoblotting and immunofluorescence on parallel samples.

Transwell invasion assay

6.5 mm/8 µm transwell inserts (Costar #3422) were coated with 100 µL of 1:1 DMEM:growth factor reduced Matrigel (#356230) and allowed to polymerize for 1 h at 37 °C. The polymerized transwells were flipped upside down and the underside coated with 50 µL of a 1:20 matrigel: DMEM solution to facilitate cell adhesion and placed back into the 37 °C incubator for 1 h. 7000 A375 cells in 70 µL of DMEM (0.5 % FBS) were seeded on the underside of the inserts and allowed to adhere for 3 h at 37 °C. The inserts were then flipped into a 24 well plate containing 600 µL of DMEM (0.5 % FBS) on the bottom. 200 µL of 20 % FBS DMEM was carefully added to the top of the insert. Cells were left to invade through the matrigel at 37 °C (5 % CO_2) for 72 h. The cells were then fixed with CSK buffer (containing 8 % paraformaldehyde) for 30 min, washed with PBS and then permeabilized with 0.3 % Triton X-100 for another 30 min. The cells were stained with 50 µg/mL of Propidium Iodide overnight at 4 °C. The transwells were then gently washed in PBS, rinsed in ddH$_2$O, and then placed onto a 35 mm Mat-Tek coverslip culture dish with 15 µL Vectashield (Vector Labs). Cells were imaged on a Zeiss LSM.Pascal using a 40x NA1.30 oil objective. 100 optical sections were taken (50 slices before the transwell membrane and 50 slices into the matrigel) with the Z step set at Nyquist sampling frequency (~0.5 µM/slice) and total number of cells counted in each section to determine invasion frequency.

Results

There are 14 human FMNL2 isoforms listed in the NCBI protein database at present. These isoforms are generated by alternative splicing of their mRNA and the exons affected encode both functional and regulatory domains in the FMNL2 protein. Therefore we wished to determine which FMNL2 isoforms are expressed in cell-types relevant to the study of the role of FMNL2 in invasion and metastasis. A series of overlapping primer pairs was designed to span the entire FMNL2 coding sequence with attention paid to regions bridging the predicted alternative splice sites. These primers were used to amplify individual sections of the entire FMNL2 coding sequence using cDNA prepared from SW480 to SW620 colorectal cancer, and A375 melanoma, cell-lines. The resulting amplified fragments were gel-purified and sequenced directly. From this analysis only four FMNL2 alternatively spliced isoforms were detected and these varied only at the 3′ end of the coding sequence (see Experimental Procedures for accession numbers). Based on sequence analysis, additional primer pairs were designed to confirm these results directly and to amplify these sequences specifically (Fig. 1a). The proteins encoded by the alternatively spliced mRNA differ only in the region C-terminal to the autoregulatory DAD domain and are designated ITM, PMR, YHY, and TQS for the final three C-terminal amino acid residues of each protein (Fig. 1b). It should be noted that additional alternatively spliced versions of FMNL2 are listed in the NCBI database, however, all of these splicing events are predicted to affect other regions of the coding sequence, that is, there are no additional 3′ splice variants predicted that encode additional alternative C-terminal tails. No other splice variants were detected in our assays.

In our initial RT-PCR analysis we noted apparent variations in the relative abundance of the different FMNL2 splice forms (Fig. 1a). Therefore we wished to determine the expression levels of each isoform in a panel of relevant cell-lines chosen either for utilizing an amoeboid mode of migration, or for their tissue of origin (CRC and melanoma). Primary HUVECs, primary melanocytes, A431 carcinoma and U2OS osteosarcoma cells were also included for comparison. Primers unique to each alternatively spliced isoform were designed for qPCR analysis and amplification efficiency was calibrated using the relevant cloned cDNA. Validated primer sets were obtained for the linear amplification of total FMNL2, ITM, YHY and TQS, however, we were unable to generate primers for the specific linear amplification of PMR. The cloned cDNA were also used to generate standard curves that allowed the derivation of absolute copy number for each isoform in the various cell-lines. Consistent with previous reports [23], we confirmed that FMNL2 expression is elevated in the metastatic SW620 cell-line when compared to parental SW480 cells (Fig. 1c). We also detected similar elevated levels of expression in LS174T CRC cells, as well as A375 and WM266.4 melanoma cells (Fig. 1c). We were unable to detect FMNL2 mRNA in U2OS cells and only low levels of FMNL2 mRNA in primary HUVECs. Surprisingly, the total level of FMNL2 expression in primary melanocytes

was similar to that in the melanoma cell-lines. Specific, individual measurement of the separate FMNL2 isoforms indicated that TQS was essentially the sole isoform expressed in A375, WM266.4 and SW620 cells and the major isoform in LS174T and A431 cells. In primary melanocytes, however, TQS accounts for less than 50 % of total FMNL2 mRNA. We were unable to assess levels of PMR expression in this assay. Nevertheless, a comparison of total FMNL2 with the sum of the individual isoforms suggests that, with the notable exception of primary melanocytes, PMR does not contribute significantly to total FMNL2 mRNA levels.

We next wished to confirm the results of our qPCR assays at the protein level. Total FMNL2 levels were assessed using a pan-FMNL2 antibody recognizing an invariant region in the N-terminus. We also generated polyclonal antibodies raised against peptides corresponding to the unique C-terminal tails of each of the four FMNL2 isoforms (Fig. 1d). These antibodies were affinity purified and used for immunoblot analysis of lysates from the panel of cell-lines used in Fig. 1c. The pan-FMNL2 immunoblot (Fig. 1e, top panel) largely confirmed the qPCR results, with highest levels of FMNL2 expression in primary melanocytes and metastatic melanoma (A375 and WM266.4) and CRC (LS174T and SW620) cell-lines. FMNL2 protein was not detected in U2OS cell lysates. The isoform specific immunoblots also showed that only primary melanocytes express a mix of the four FMNL2 isoforms; significant levels of YHY and PMR are not detected in the other cell-lines and only low levels of ITM are detected across this panel. In contrast, the TQS specific immunoblot shows that it is the major isoform expressed in A375, SW620 and WM266.4 cell-lines.

Amino acid residues C-terminal to the DAD motif are predicted to modulate formin autoregulation [9] and modify FH2 activity [42]. To compare activity of the FMNL2 isoforms, we generated full-length cDNAs encoding each of the four splice variants. FMNL proteins are predicted to be N-myristoylated [26, 29, 33] and this modification may play some role in targeting these proteins to the plasma membrane. To assess the effects of N-myristoylation we generated FMNL2 derivatives with either N-terminal myc epitope tags or C-terminal mCherryFP tags. These were expressed in NIH 3T3 cells by transient transfection and their subcellular localization and effects on cell morphology and the actin cytoskeleton were assessed by immunofluorescence (Fig. 2). The C-terminal tagged derivatives were recruited to the plasma membrane where they induced extensive filopodia formation (Fig. 2a, c). This was not the case with the N-terminal tagged proteins, which were distributed diffusely throughout the cytoplasm.

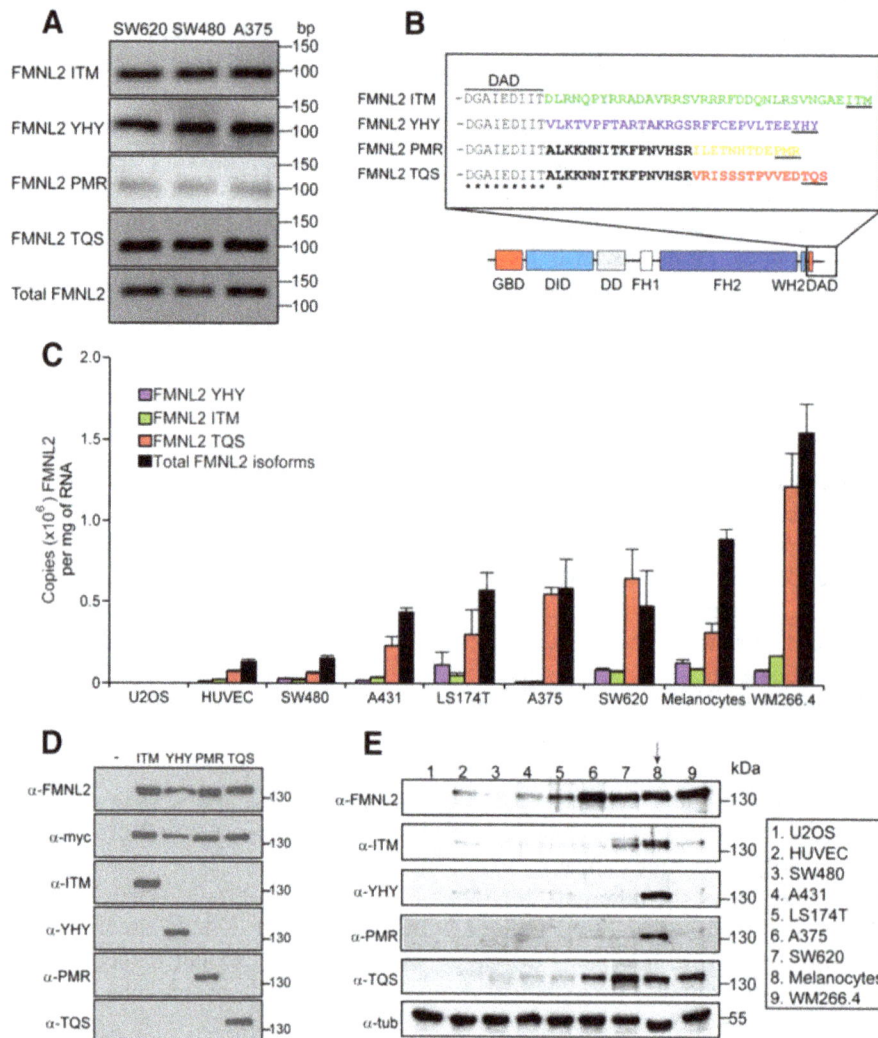

Fig. 1 A specific FMNL2 isoform is over-represented in invasive cells. **a** RT-PCR was performed on total RNA harvested from SW480, SW620 and A375 cells. Specific primers were used to demonstrate the presence of four alternatively spliced FMNL2 isoforms identified in an initial RT-PCR analysis (see Results). **b** The four alternatively spliced mRNA are predicted to encode proteins differing only at amino acid residues C-terminal to the DAD domain. **c** qPCR was used to assess the levels of expression of each isoform in comparison to total FMNL2 in the indicated colorectal and melanoma cell-lines as well as in non-transformed primary endothelial cells and melanocytes. The TQS isoform was predominant in all the invasive cell-lines that were assessed. We were unable to obtain reliable qPCR data for the PMR isoform. **d** Isoform specific antibodies were raised against peptides corresponding to the unique domain of each C-terminal tail. Full-length derivatives of the indicated FMNL2 isoforms bearing N-terminal myc epitope tags were expressed in U2OS cells by transient transfection and the resulting cell lysates were immunoblotted with the indicated antibodies. Total FMNL2 protein was detected using an antibody directed against FMNL2 codons 599–1045. Equivalent samples were loaded on a separate gel and probed with α-myc antibodies. The α-FMNL2 blot was stripped and re-probed with α-TQS antibody; the α-myc blot was stripped and re-probed with α-PMR antibody. Equivalent samples were loaded on separate gels to detect ITM and YHY **e** Immunoblot analysis using antibodies directed against the C-terminal tails of each of the predicted FMNL2 isoforms confirms the qPCR analysis. Total FMNL2 protein was detected using pan FMNL2 antibody shown in 1D. Isoform specific antibodies were raised against peptides corresponding to the unique domain of each C-terminal tail. Equivalent samples were loaded on separate gels to eliminate concerns regarding incomplete stripping. The pan-FMNL2 blot was stripped and re-probed with anti α-tubulin for a loading control

Expression of N-terminal tagged versions of ITM, YHY or TQS did not induce filopodia formation above the background levels seen in control cells while expression of PMR was only sufficient to induce filopodia formation in a minority of cells (Fig. 2b, c). We also compared activity of N- and C-terminal tagged proteins using an SRF reporter gene assay. This reporter gene is activated by the MRTF/SRF transcription pathway in response to depletion of cellular pools of G-actin and is an indirect, sensitive and quantitative measure of changes in actin dynamics [39]. Neither N- nor C-terminal tagged full-length derivatives of the FMNL2 isoforms induced

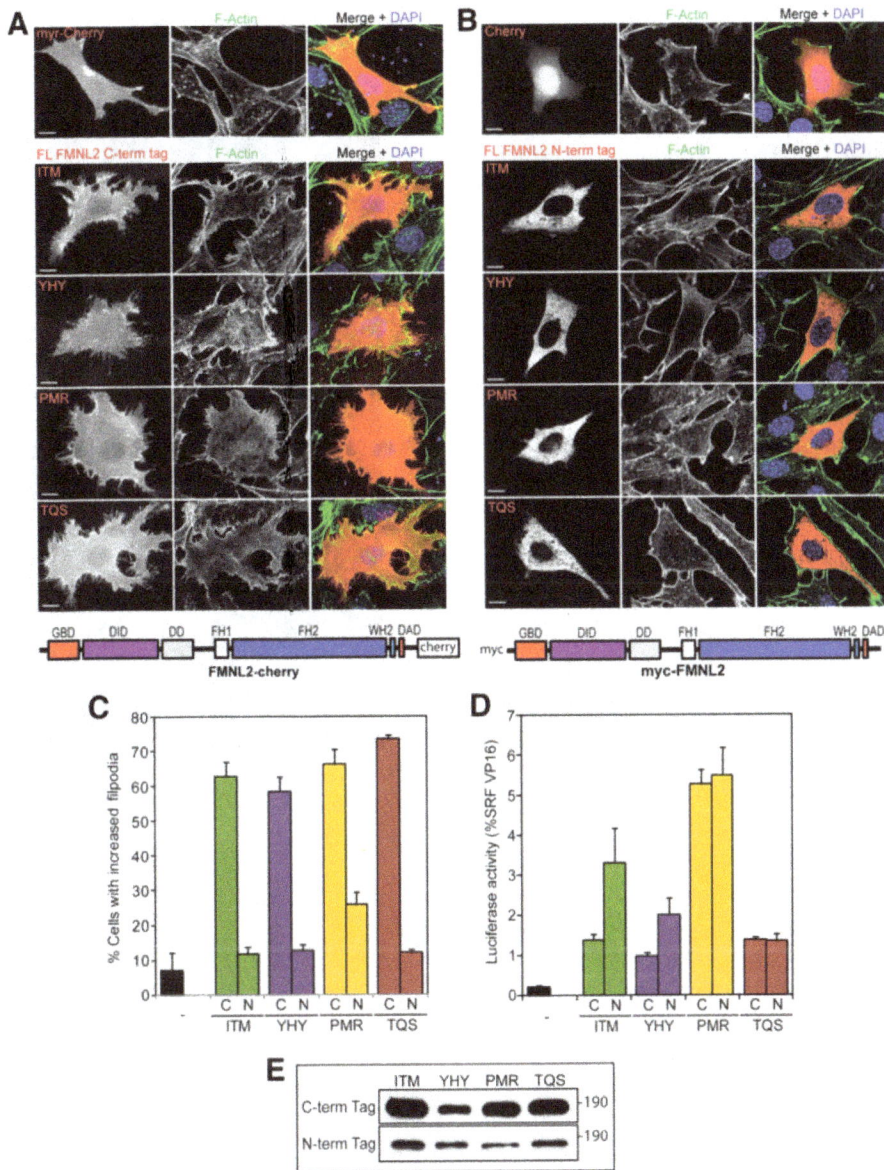

Fig. 2 FMNL2 expression induces filopodia formation. **a** C-terminally tagged full-length derivatives of each FMNL2 isoform were expressed in NIH 3T3 fibroblasts by transient transfection. F-actin (*green*) was detected with phalloidin and FMNL2 expression (*red*) was detected by virtue of the C-terminal mCherry tag. Nuclei were detected with DAPI (*blue*) in merged image. FMNL2 expression induced extensive filopodia formation in comparison to untransfected cells or cells expressing mCherryFP with an N-terminal myristoylation motif (*top panel*). **b** N-terminally tagged full-length derivatives of each of the FMNL2 isoforms were expressed in NIH 3T3 cells by transient transfection. F-actin (*green*) was detected with phalloidin and FMNL2 expression (*red*) was detected by virtue of the N-terminal myc tag. Expression of N-terminally tagged FMNL2 isoforms does not induce filopodia formation. mCherryFP was included as a control (*top panel*) **c** Quantification of data shown in (**a, b**). "C" indicates C-terminal mCherry tag, "N" indicates N-terminal myc tag. $N = 3$, >100 cells counted per sample, error bars = SEM. **d** NIH 3T3 fibroblasts were transiently transfected with an SRF luciferase reporter gene and the indicated full-length FMNL2 derivative. "C" indicates C-terminal mCherry tag, "N" indicates N-terminal myc tag. Reporter activation is expressed relative to an SRF-VP16 control fusion protein. $N = 3$, error bars = SEM. **e** Equivalent samples of transfected cell lysates were subjected to SDS-PAGE and immunoblotted. Expression of the indicated FMNL2 isoforms was detected with either anti-mCherry or anti-myc antibody

robust activation of the SRF reporter in this assay suggesting that, as expected, the full-length derivatives of each isoform are not constitutively active. Some differences, however, were apparent between the four isoforms, both in terms of basal activity and the effects of

the position of the tag. PMR showed the most activity, inducing activation of the SRF reporter gene to similar levels above background for both N-and C-terminal tagged derivatives (Fig. 2d). In contrast, the C-terminally tagged full-length derivatives of ITM, YHY and TQS were largely inactive, while the N-terminal tag on ITM had a modest effect on its autoinhibition as well.

The results of the SRF reporter gene assays suggest that the full-length FMNL2 isoforms are largely autoinhibited and that FMNL2-induced filopodia are likely a product of basal levels of FMNL2-induced F-actin formation on top of remodeling of pre-existing actin filaments. We therefore also wanted to compare the activity of constitutively active derivatives of the four isoforms. Deletion derivatives consisting of FH1, FH2 and the C-terminal tail (FH1FH2 + C) of each isoform were expressed in NIH 3T3 cells by transient transfection and the effects on cell morphology and actin polymerization were assessed by immunofluorescence (Fig. 3a, b). All four isoforms behaved similarly in this assay and induced a typical "formin" phenotype with extensive formation of thin stress fibers running from one end of the cell to the other. No obvious differences were apparent in the subcellular localization of each of these derivatives either. The SRF reporter gene assay was also used to compare the effects of each isoform on actin dynamics. Expression of the FH1FH2 + C derivative of each isoform was sufficient to induce robust SRF reporter gene activation consistent with previous results for ITM [11]. As with the formation of stress fibers, there was no obvious difference in the level of activation of the SRF reporter induced by expression of each isoform.

Our preliminary analysis of the full-length FMNL2 isoforms suggested that there are some functional differences between them (Fig. 2c, d). We wished to determine if this extends to their effects on cell behavior. Previous work has suggested that FMNL2 is required for both 2D and 3D cell motility in specific cell-types [6, 26]. Therefore we wanted to determine if FMNL2 is required for motility in A375 cells which express the TQS isoform almost exclusively. We used siRNA duplexes targeted against FMNL2 to knockdown its expression in A375 cells and the level of depletion was assessed by immunoblotting with an anti-FMNL2 antibody (Fig. 4b); cells transfected with non-specific duplex were used as a control. FMNL2 expression was knocked down very efficiently and the effects of FMNL2 depletion on cell motility were assessed using a variation on the scratch wound assay. In this assay, cells were seeded in a culture insert composed of two chambers separated by a divider and allowed to grow to confluence. Removal of the insert creates a standard "wound" and migration into the wound is followed by live-cell imaging. Control cells extensively infiltrated the wound by 20 h and filled the wound by 40 h (Fig. 4a, c). In contrast, FMNL2-depleted cells made little progress into the wound even after 40 h (Fig. 4a, c). This suggests that FMNL2 plays a role in 2D migration in A375 cells and encouraged us to investigate if the same holds true for 3D invasion.

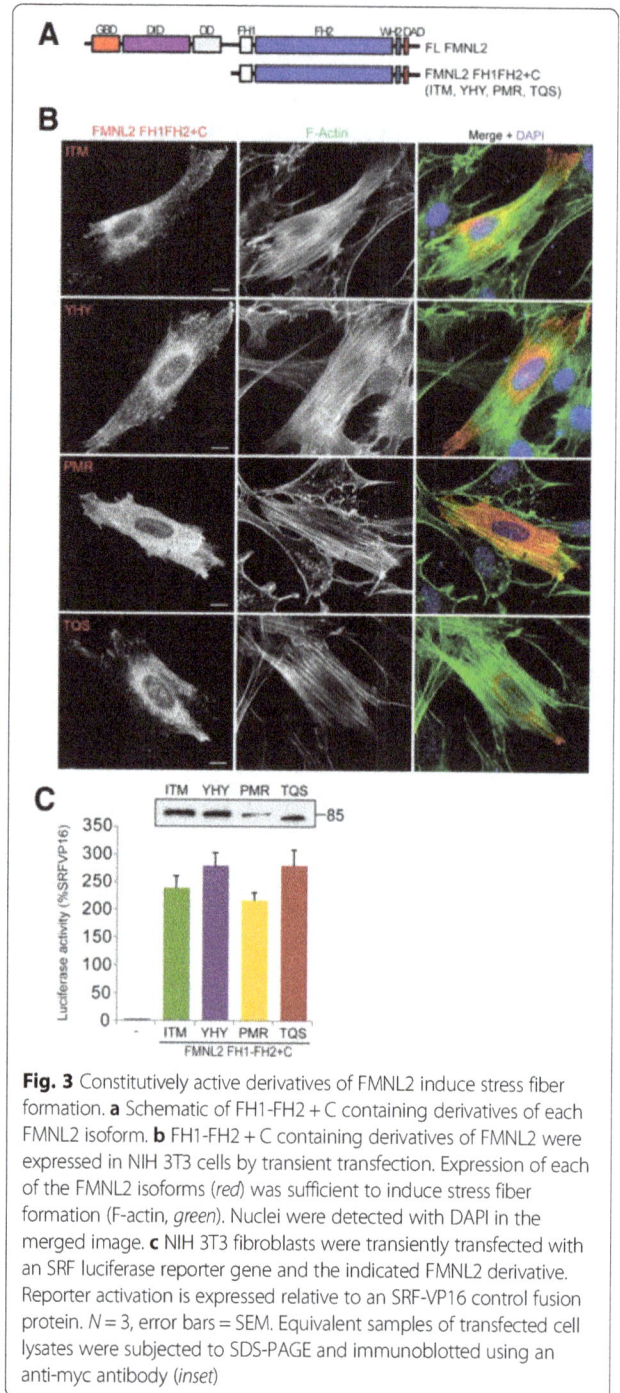

Fig. 3 Constitutively active derivatives of FMNL2 induce stress fiber formation. **a** Schematic of FH1-FH2 + C containing derivatives of each FMNL2 isoform. **b** FH1-FH2 + C containing derivatives of FMNL2 were expressed in NIH 3T3 cells by transient transfection. Expression of each of the FMNL2 isoforms (*red*) was sufficient to induce stress fiber formation (F-actin, *green*). Nuclei were detected with DAPI in the merged image. **c** NIH 3T3 fibroblasts were transiently transfected with an SRF luciferase reporter gene and the indicated FMNL2 derivative. Reporter activation is expressed relative to an SRF-VP16 control fusion protein. $N = 3$, error bars = SEM. Equivalent samples of transfected cell lysates were subjected to SDS-PAGE and immunoblotted using an anti-myc antibody (*inset*)

A375 and WM266.4 melanoma cells both invade using an amoeboid mode of 3D migration [3]. To determine if FMNL2 is required for 3D migration in these cells, we used a modified Boyden chamber assay to measure invasion [6]. In this assay cells are seeded on the underside of a transwell insert and their ability to invade upward into a matrigel matrix is assessed by confocal microscopy. siRNA duplexes were used to knockdown FMNL2 expression in both A375 and WM266.4 cells and FMNL2 was efficiently

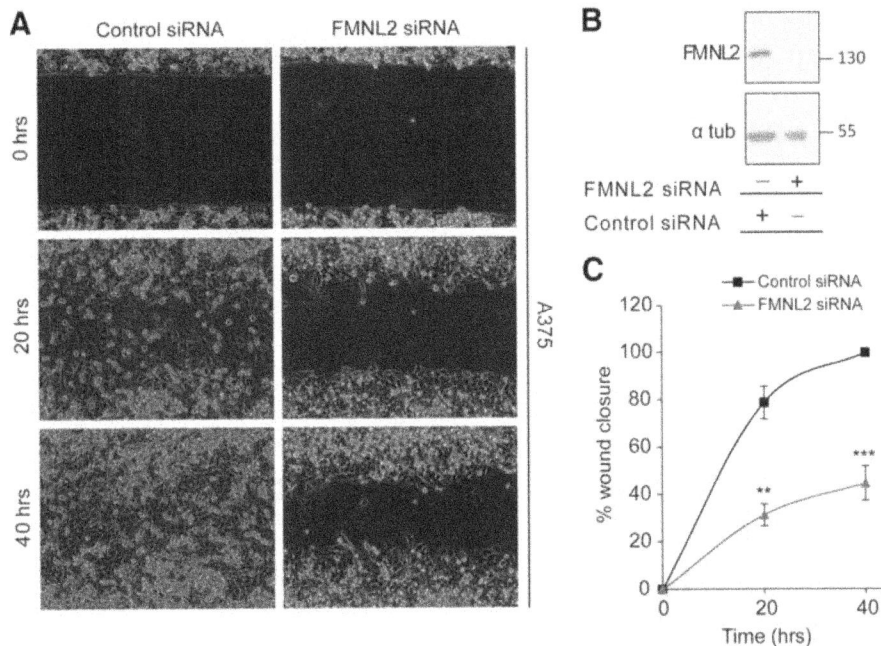

Fig. 4 FMNL2 is required for 2-D migration in A375 melanoma cells. **a** A375 cells were transfected with either control siRNA duplex or duplexes targeting all FMNL2 isoforms. The transfected cells were grown to confluence in scratch wound assay chambers (see Experimental Procedures). Wound closure was followed at the indicated times by phase contrast live-cell microscopy. **b** Equivalent samples of cell lysates from (**a**) were subjected to SDS-PAGE and immunoblotted to detect FMNL2 expression with anti-FMNL2 antibody. The pan-FMNL2 blot was stripped and re-probed with anti α-tubulin as a loading control. **c** Quantification of the data shown in (**a**). The gap in the cellular monolayer is represented as % wound closure. $N = 3$, error bars = SEM. ** indicates $P < 0.01$ and *** indicates $P < 0.001$

depleted in both cell-types as measured by immunoblotting (Fig. 5b, c). FMNL2 depletion did not affect expression levels of the related protein FMNL3 in A375 cells (Fig. 5b) and expression of FMNL1 was not detected (data not shown). The effect of FMNL2 depletion on the ability of A375 and WM266.4 cells to invade was compared to control cells transfected with non-specific siRNA duplexes. For both cell-types we found that FMNL2 depletion was sufficient to significantly inhibit invasion in the Transwell assay (Fig. 5a).

Having shown that FMNL2 is required for invasion in these cells, we wanted to compare the relative ability of each of the four FMNL2 isoforms to rescue invasion in knockdown cells. A375 cells were selected for this experiment as they were more invasive in our assay, FMNL2 was efficiently knocked down, and the effect of FMNL2 knockdown on invasion was more striking with these cells than with WM266.4 (Fig. 5a–c). Lentiviral rescue vectors were generated for ITM, YHY, PMR and TQS. This vector does not encode an epitope tag on the protein of interest, but does express mCherryFP as a separate protein via an IRES within the same mRNA. Empty vector expressing mCherry alone was used as a control. Using these vectors we were able to achieve nearly 100 % transduction efficiency in A375 cells. As before, transfection of control siRNA duplex had no

effect on invasion while knockdown of FMNL2 strongly inhibits it. Efficiency of knockdown, and inhibition of invasion, is not affected by infection with the control lentiviral vector in FMNL2-depleted cells (Fig. 5d–f). In contrast, expression of each of the individual FMNL2 isoforms was sufficient to rescue invasion. It should be noted, however, that each isoform did not rescue to the same extent. PMR was the least efficient, TQS was the most efficient, while YHY and ITM fell in between. Indeed, the enhanced invasion observed with TQS was significantly greater than both the parental cell-line and PMR-rescued cells (although not significantly different from ITM or YHY). This suggests that there are unique functional differences between FMNL2 isoforms when it comes to driving invasive cell migration.

Discussion

We provide here the initial description of two previously uncharacterized isoforms of FMNL2 (PMR and TQS) and show that at least four alternatively spliced FMNL2 isoforms are expressed in primary melanocytes. In contrast, the TQS isoform predominates in invasive melanoma and CRC cell-lines and we find that knockdown of FMNL2 expression in A375 and WM266.4 melanoma cells is sufficient to inhibit invasion in an in vitro assay. Taken with previous reports that FMNL2 is required for

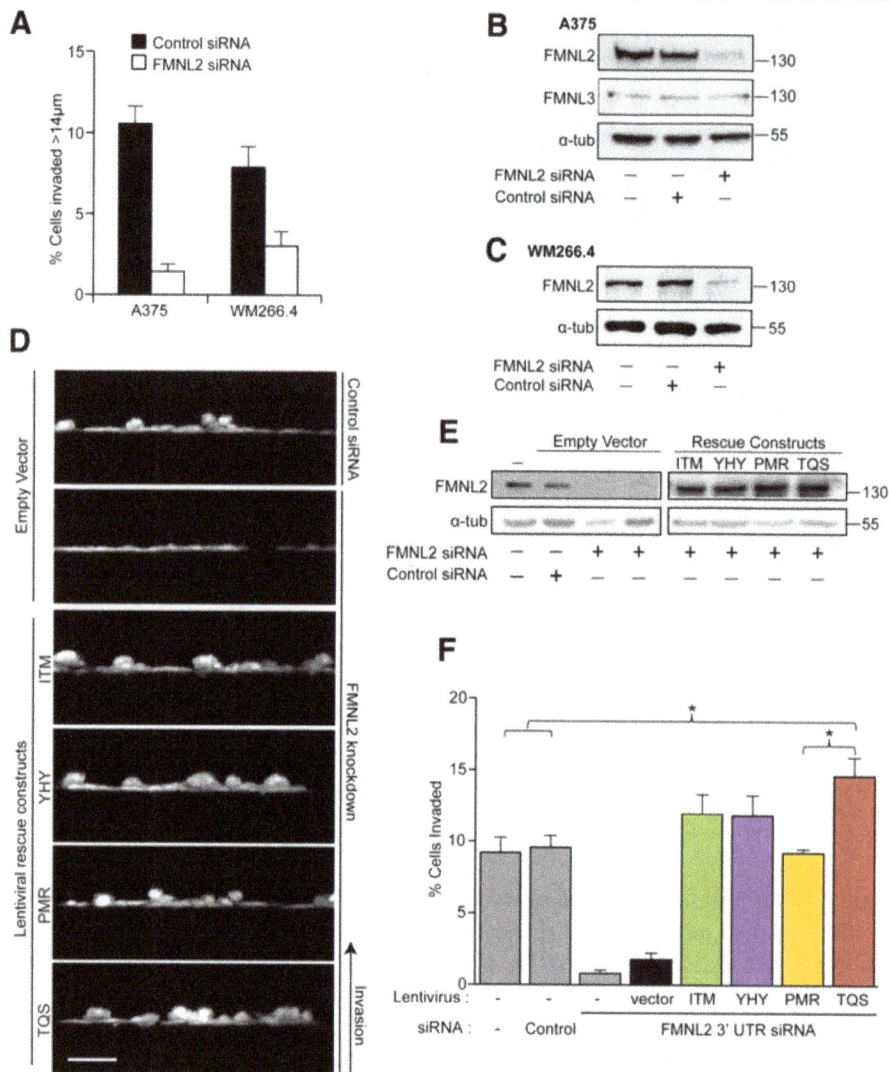

Fig. 5 FMNL2 is required for invasion in A375 and WM266.4 melanoma cells. **a** A375 and WM266.4 melanoma cells were transfected with control or FMNL2-specific siRNA duplexes. Control and knockdown cells were then assayed for their ability to invade in vitro in an inverted Boyden chamber assay (Kitzing et al. [6]). FMNL2 knockdown inhibited invasion in both A375 and WM266.4 cells. **b** Equivalent samples of A375 cell lysates from (**a**) were subjected to SDS-PAGE and immunoblotted to detect FMNL2 and FMNL3. The pan-FMNL2 blot was stripped and re-probed with anti α-tubulin as a loading control. **c** Equivalent samples of WM266.4 cell lysates from (**a**) were subjected to SDS-PAGE and immunoblotted to detect FMNL2. The pan-FMNL2 blot was stripped and re-probed with anti α-tubulin as a loading control. **d** A375 cells were transfected with control and FMNL2-targeted siRNA duplexes and then infected with control lentiviral vector or lentiviral vectors expressing the indicated untagged full-length FMNL2 isoforms. The cells were then assayed for their ability to invade as in (**a**). Images show a representative Z-stack projection from the invasion assays. **e** Equivalent samples of cell lysates from (**d**) were subjected to SDS-PAGE and immunoblotted to detect FMNL2. α-tubulin was used as a loading control. **f** Quantification of data shown in (**d**). $N = 3$, error bars = SEM. *indicates $P < 0.05$ where noted. There was no significant difference between TQS and ITM or TQS and YHY

migration in mouse B16 melanoma cells and for invasion in MDA-MB-435 [6, 26], our results suggest that FMNL2 is likely to be generally required for invasion in melanoma cells. Like MDA-MB-435 cells, both A375 and WM 266.4 cells use an amoeboid mode of migration for 3D invasion and these results suggest that this mode of invasion might also generally require FMNL2 or related activity [19].

At least four FMNL2 isoforms (ITM, YHY, PMR and TQS) are expressed in the assessed cell-types at both the transcript and protein level. Although we cannot discount that additional isoforms may be expressed in other cell-types, we did not find any evidence for additional isoforms in our assays. In addition, the four unique C-terminal tails we describe are the only C-termini predicted for all FMNL2 isoforms currently listed in the NCBI database. Thus, the antiserum directed against each unique FMNL2 C-terminal peptide should account for all predicted isoforms of FMNL2. In primary melanocytes all four splice forms were readily detected at the protein level, however, in WMM266.4, A375 and SW620 cells TQS is by far the predominant isoform. This raises the intriguing possibility that TQS represents an "invasive isoform" of FMNL2 similar to previous descriptions of invasive isoforms of Mena [43] and palladin [35]. In support of this hypothesis we find that in FMNL2-depleted A375 cells, TQS is able to rescue invasion significantly better than PMR and to a significantly higher level than control cells. It should be noted that under the conditions of the rescue experiment, the levels of exogenous FMNL2 expression were considerably higher than the endogenous protein and it is likely that the system is saturated for FMNL2 activity. If that is the case, it would suggest that the observed effects represent the maximal potentiation of invasion by each isoform in these cells.

Increased FMNL2 expression is reported to directly correlate with increased invasiveness in colorectal cancer cell-lines [23, 24]. Our results confirm this observation and show that FMNL2 expression is up-regulated at both the mRNA and protein level in SW620 CRC cells when compared to their poorly invasive SW480 progenitor cell-line. Indeed, the level of total FMNL2 expression in SW620 cells is similar to that observed in the melanoma cell-lines. Surprisingly, FMNL2 is expressed to comparable levels in non-transformed primary melanocytes. As part of their normal physiological role melanocytes themselves are somewhat "invasive" and deliver melanosomes via widespread formation of filopodia [44–46]. Given the ability of FMNL2 to support invasion and induce filopodia formation, it might therefore be expected that elevated levels of FMNL2 expression would also be required in these cells. A key difference between the primary melanocytes and melanoma cells, however, is the variety of FMNL2 isoforms expressed. As noted, the TQS isoform predominates in A375 and WM266.4 cells while at least four FMNL2 isoforms are expressed in primary melanocytes. This suggests an intriguing model where it is not the total level of FMNL2 expression that determines invasiveness in melanoma, but the specific isoform that is expressed.

The ITM, YHY, PMR and TQS isoforms of FMNL2 differ only at a relatively small region in their C-termini. Nevertheless, the sequences affected are suggested to be critical to the normal regulation and function of diaphanous-related formins (DRFs); the domain C-terminal to DAD is thought to influence both FH2 activity [42] and autoregulation [9, 42]. Consistent with this, we find that auto-inhibition of PMR is less efficient than the other three isoforms and that TQS is the isoform most efficient at promoting invasion. Beyond direct effects on FMNL2 activity, the DRF C-terminus may also serve as a target for multiple regulatory and accessory factors [10]. Indeed, a recent study investigating the role of FMNL2 in integrin trafficking found that the C-terminal tail of FMNL2.ITM is a target of

PKCα-dependent phosphorylation [28]. A comparison of the four FMNL2 isoforms on the Eukaryotic Linear Motif resource (http://elm.eu.org/index.html) reveals that the PKCα site is conserved in the ITM and YHY isoforms, but not in PMR or TQS. Instead the TQS C-terminal tail contains consensus PKB and proline-directed kinase sites and PMR contains no conserved kinase consensus sites at all. This comparison suggests that the C-terminal tail of each isoform is likely to be targeted by a variety of differing upstream signaling pathways. Determining how the prevalence of each isoform affects the integration of specific signaling pathways that regulate the effects of FMNL2 on cell behavior will be of great future interest.

Conclusions

We show here that FMNL2 expression is required for invasion in A375 and WM266.4 melanoma cells and, taken with previous results, this suggests that FMNL2 is likely to be generally required by melanoma cells for invasion. At least four alternatively-spliced FMNL2 isoforms are expressed in primary melanocytes as well as in specific cancer cell-lines and sequence analysis of the encoded proteins suggests they will be targeted by distinct regulatory pathways. Of the four isoforms, FMNL2.TQS is preferentially up-regulated in invasive CRC and melanoma cell-lines and is the most efficient in supporting cellular invasion. This raises the possibility that TQS could be considered as an invasive isoform of FMNL2, analogous to Mena[INV] and the invasive isoform of palladin. Regardless, our results suggest that future studies on the role of FMNL2 in metastasis should concentrate on TQS as the most relevant isoform.

Acknowledgements
The authors thank Laura Trinkle-Mulcahy, Stephen Lee and Jonathan Lee for insightful discussions on experimental design and interpretation of the data.

Funding
This work was initiated with support from grant G-13-0003059 from the Heart and Stroke Foundation of Canada and is now supported by grant 19078 to JC from the Cancer Research Society of Canada.

Authors' contributions
JC conceived the project, analyzed data and wrote the paper. CP performed and analyzed the experiments in Figs 1–5. AH assisted with the experiments in Figs 1, 4 and 5. MM assisted with the experiments in Figs 2 and 3. SC performed the experiment in Fig 1d, generated reagents, provided technical support and assisted with the preparation of figures. MM and CP prepared all figures. All authors read and approved the final manuscript.

Competing interests
The authors declare that they have no competing interests.

References

1. Nurnberg A, Kitzing T, Grosse R. Nucleating actin for invasion. Nat Rev Cancer. 2011;11:177–87.
2. Petrie RJ, Yamada KM. At the leading edge of three-dimensional cell migration. J Cell Sci. 2012;125:5917–26.
3. Sahai E, Marshall CJ. Differing modes of tumour cell invasion have distinct requirements for Rho/ROCK signalling and extracellular proteolysis. Nat Cell Biol. 2003;5:711–9.
4. Friedl P, Wolf K. Tumour-cell invasion and migration: diversity and escape mechanisms. Nat Rev Cancer. 2003;3:362–74.
5. Sanz-Moreno V, Marshall CJ. The plasticity of cytoskeletal dynamics underlying neoplastic cell migration. Curr Opin Cell Biol. 2010;22:690–6.
6. Kitzing TM, Wang Y, Pertz O, Copeland JW, Grosse R. Formin-like 2 drives amoeboid invasive cell motility downstream of RhoC. Oncogene. 2010;29:2441–8.
7. Faix J, Grosse R. Staying in shape with formins. Dev Cell. 2006;10:693–706.
8. Paul AS, Pollard TD. Review of the mechanism of processive actin filament elongation by formins. Cell Motil Cytoskeleton. 2009;66:606–17.
9. Schonichen A, Alexander M, Gasteier JE, Cuesta FE, Fackler OT, et al. Biochemical characterization of the diaphanous autoregulatory interaction in the formin homology protein FHOD1. J Biol Chem. 2006;281:5084–93.
10. Chesarone MA, DuPage AG, Goode BL. Unleashing formins to remodel the actin and microtubule cytoskeletons. Nat Rev Mol Cell Biol. 2010;11:62–74.
11. Vaillant DC, Copeland SJ, Davis C, Thurston SF, Abdennur N, et al. Interaction of the N- and C-terminal autoregulatory domains of FRL2 does not inhibit FRL2 activity. J Biol Chem. 2008;283:33750–62.
12. Thurston SF, Kulacz WA, Shaikh S, Lee JM, Copeland JW. The ability to induce microtubule acetylation is a general feature of formin proteins. PLoS One. 2012;7:e48041.
13. Bartolini F, Moseley JB, Schmoranzer J, Cassimeris L, Goode BL, et al. The formin mDia2 stabilizes microtubules independently of its actin nucleation activity. J Cell Biol. 2008;181:523–36.
14. Bartolini F, Gundersen GG. Formins and microtubules. Biochim Biophys Acta. 2010;1803:164–73.
15. Favaro P, Traina F, Machado-Neto JA, Lazarini M, Lopes MR, et al. FMNL1 promotes proliferation and migration of leukemia cells. J Leukoc Biol. 2013;94:503–12.
16. Favaro PM, de Souza Medina S, Traina F, Basseres DS, Costa FF, et al. Human leukocyte formin: a novel protein expressed in lymphoid malignancies and associated with Akt. Biochem Biophys Res Commun. 2003;311:365–71.
17. Favaro PM, Traina F, Vassallo J, Brousset P, Delsol G, et al. High expression of FMNL1 protein in T non-Hodgkin's lymphomas. Leuk Res. 2006;30:735–8.
18. Martin-Rufian M, Segura JA, Lobo C, Mates JM, Marquez J, et al. Identification of genes downregulated in tumor cells expressing antisense glutaminase mRNA by differential display. Cancer Biol Ther. 2006;5:54–8.
19. Vega FM, Fruhwirth G, Ng T, Ridley AJ. RhoA and RhoC have distinct roles in migration and invasion by acting through different targets. J Cell Biol. 2011;193:655–65.
20. Lynch J, Meehan MH, Crean J, Copeland J, Stallings RL, et al. Metastasis suppressor microRNA-335 targets the formin family of actin nucleators. PLoS One. 2013;8:e78428.
21. Zeng YF, Xiao YS, Lu MZ, Luo XJ, Hu GZ, et al. Increased expression of formin-like 3 contributes to metastasis and poor prognosis in colorectal carcinoma. Exp Mol Pathol. 2015;98:260–7.
22. Li Y, Zhu X, Zeng Y, Wang J, Zhang X, et al. FMNL2 enhances invasion of colorectal carcinoma by inducing epithelial-mesenchymal transition. Mol Cancer Res. 2010;8:1579–90.
23. Zhu XL, Liang L, Ding YQ. Overexpression of FMNL2 is closely related to metastasis of colorectal cancer. Int J Color Dis. 2008;23:1041–7.
24. Zhu XL, Zeng YF, Guan J, Li YF, Deng YJ, et al. FMNL2 is a positive regulator of cell motility and metastasis in colorectal carcinoma. J Pathol. 2011;224:377–88.
25. Ross DT, Scherf U, Eisen MB, Perou CM, Rees C, et al. Systematic variation in gene expression patterns in human cancer cell lines. Nat Genet. 2000;24:227–35.
26. Block J, Breitsprecher D, Kuhn S, Winterhoff M, Kage F, et al. FMNL2 drives actin-based protrusion and migration downstream of Cdc42. Curr Biol. 2012;22:1005–12.
27. Grikscheit K, Frank T, Wang Y, Grosse R. Junctional actin assembly is mediated by Formin-like 2 downstream of Rac1. J Cell Biol. 2015;209:367–76.
28. Wang Y, Arjonen A, Pouwels J, Ta H, Pausch P, et al. Formin-like 2 Promotes beta1-Integrin Trafficking and Invasive Motility Downstream of PKCalpha. Dev Cell. 2015;34:475–83.
29. Moriya K, Yamamoto T, Takamitsu E, Matsunaga Y, Kimoto M, et al. Protein N-myristoylation is required for cellular morphological changes induced by two formin family proteins, FMNL2 and FMNL3. Biosci Biotechnol Biochem. 2012;76:1201–9.
30. Gasman S, Kalaidzidis Y, Zerial M. RhoD regulates endosome dynamics through Diaphanous-related Formin and Src tyrosine kinase. Nat Cell Biol. 2003;5:195–204.
31. Iskratsch T, Lange S, Dwyer J, Kho AL, dos Remedios C, et al. Formin follows function: a muscle-specific isoform of FHOD3 is regulated by CK2 phosphorylation and promotes myofibril maintenance. J Cell Biol. 2010;191:1159–72.
32. Ramabhadran V, Korobova F, Rahme GJ, Higgs HN. Splice variant-specific cellular function of the formin INF2 in maintenance of Golgi architecture. Mol Biol Cell. 2011;22:4822–33.
33. Han Y, Eppinger E, Schuster IG, Weigand LU, Liang X, et al. Formin-like 1 (FMNL1) is regulated by N-terminal myristoylation and induces polarized membrane blebbing. J Biol Chem. 2009;284:33409–17.
34. Kobielak A, Pasolli HA, Fuchs E. Mammalian formin-1 participates in adherens junctions and polymerization of linear actin cables. Nat Cell Biol. 2004;6:21–30.
35. Goicoechea SM, Bednarski B, Stack C, Cowan DW, Volmar K, et al. Isoform-specific upregulation of palladin in human and murine pancreas tumors. PLoS One. 2010;5:e10347.
36. Goicoechea SM, Garcia-Mata R, Staub J, Valdivia A, Sharek L, et al. Palladin promotes invasion of pancreatic cancer cells by enhancing invadopodia formation in cancer-associated fibroblasts. Oncogene. 2014;33:1265–73.
37. Di Modugno F, Iapicca P, Boudreau A, Mottolese M, Terrenato I, et al. Splicing program of human MENA produces a previously undescribed isoform associated with invasive, mesenchymal-like breast tumors. Proc Natl Acad Sci U S A. 2012;109:19280–5.
38. Goswami S, Philippar U, Sun D, Patsialou A, Avraham J, et al. Identification of invasion specific splice variants of the cytoskeletal protein Mena present in mammary tumor cells during invasion in vivo. Clin Exp Metastasis. 2009;26:153–9.
39. Copeland JW, Treisman R. The diaphanous-related formin mDia1 controls serum response factor activity through its effects on actin polymerization. Mol Biol Cell. 2002;13:4088–99.
40. Young KG, Thurston SF, Copeland S, Smallwood C, Copeland JW. INF1 is a novel microtubule-associated formin. Mol Biol Cell. 2008;19:5168–80.
41. Hetheridge C, Scott AN, Swain RK, Copeland JW, Higgs HN, et al. The formin FMNL3 is a cytoskeletal regulator of angiogenesis. J Cell Sci. 2012;125:1420–8.
42. Gould CJ, Maiti S, Michelot A, Graziano BR, Blanchoin L, et al. The formin DAD domain plays dual roles in autoinhibition and actin nucleation. Curr Biol. 2011;21:384–90.
43. Philippar U, Roussos ET, Oser M, Yamaguchi H, Kim HD, et al. A Mena invasion isoform potentiates EGF-induced carcinoma cell invasion and metastasis. Dev Cell. 2008;15:813–28.
44. Beaumont KA, Hamilton NA, Moores MT, Brown DL, Ohbayashi N, et al. The recycling endosome protein Rab17 regulates melanocytic filopodia formation and melanosome trafficking. Traffic. 2011;12:627–43.
45. Yamaguchi Y, Hearing VJ. Melanocytes and their diseases. Cold Spring Harb Perspect Med. 2014;4:a017046.
46. Scott G, Leopardi S, Printup S, Madden BC. Filopodia are conduits for melanosome transfer to keratinocytes. J Cell Sci. 2002;115:1441–51.

13

Prolactin-induced PAK1 tyrosyl phosphorylation promotes FAK dephosphorylation, breast cancer cell motility, invasion and metastasis

Alan Hammer and Maria Diakonova*

Abstract

Background: The serine/threonine kinase PAK1 is an important regulator of cell motility. Both PAK1 and the hormone/cytokine prolactin (PRL) have been implicated in breast cancer cell motility, however, the exact mechanisms guiding PRL/PAK1 signaling in breast cancer cells have not been fully elucidated. Our lab has previously demonstrated that PRL-activated tyrosine kinase JAK2 phosphorylates PAK1 on tyrosines 153, 201, and 285, and that tyrosyl phosphorylated PAK1 (pTyr-PAK1) augments migration and invasion of breast cancer cells.

Results: Here we further investigate the mechanisms by which pTyr-PAK1 enhances breast cancer cell motility in response to PRL. We demonstrate a distinct reduction in PRL-induced FAK auto-phosphorylation in T47D and TMX2-28 breast cancer cells overexpressing wild-type PAK1 (PAK1 WT) when compared to cells overexpressing either GFP or phospho-tyrosine-deficient mutant PAK1 (PAK1 Y3F). Furthermore, pTyr-PAK1 phosphorylates MEK1 on Ser298 resulting in subsequent ERK1/2 activation. PRL-induced FAK auto-phosphorylation is rescued in PAK1 WT cells by inhibiting tyrosine phosphatases and tyrosine phosphatase inhibition abrogates cell motility and invasion in response to PRL. siRNA-mediated knockdown of the tyrosine phosphatase PTP-PEST rescues FAK auto-phosphorylation in PAK1 WT cells and reduces both cell motility and invasion. Finally, we provide evidence that PRL-induced pTyr-PAK1 stimulates tumor cell metastasis in vivo.

Conclusion: These data provide insight into the mechanisms guiding PRL-mediated breast cancer cell motility and invasion and highlight a significant role for pTyr-PAK1 in breast cancer metastasis.

Keywords: PAK1, FAK, Prolactin, Tyrosyl phosphorylation, Breast cancer cells

Background

Prolactin (PRL) is a peptide hormone/cytokine that is typically secreted from the anterior pituitary gland, and has been found to be locally produced in various other organs such as the prostate, uterus, and mammary gland (for review [1]). Upon PRL binding, PRL-receptor (PRLR) dimerizes resulting in activation of the non-receptor tyrosine kinase JAK2 (Janus kinase 2) and subsequent downstream signaling cascades including signal tranducers and activators of transcription (STATs), mitogen activated protein kinases (MAPKs), including ERK1/2, and phosphoinositol-3 kinase pathways (for review [2]). PRL signaling at both an endocrine and paracrine/autocrine levels regulates a variety of physiological processes in an eclectic range of tissues (for review [3]). There is mounting evidence that PRL plays a significant role in breast cancer. The PRLR has been found in the vast majority of human breast cancers and PRL signaling has been implicated in breast cancer cell proliferation, survival, motility and angiogenesis (for review [2]). Furthermore, elevated circulating PRL levels have been positively correlated with breast cancer metastasis and PRLR-deficient mice have prevention of neoplasia progression into invasive carcinoma [4–7]. Importantly, PRL has been noted as a chemoattractant for breast cancer cells and augments tumor

* Correspondence: mdiakon@utnet.utoledo.edu
Department of Biological Sciences, University of Toledo, 2801 W. Bancroft Street, Toledo 43606-3390, OH, USA

metastasis in nude mice [8, 9]. However, the exact mechanisms guiding PRL-induced cell migration and tumor metastasis are not fully understood.

We have implicated the serine/threonine kinase PAK1 (p21-activated kinase-1) as a substrate of PRL-activated JAK2 [10]. PAK1 has been associated with breast cancer progression (for review [11]). Aberrant expression/activation of PAK1 has been described in breast cancer as well as among several other cancers including brain, pancreas, colon, bladder, ovarian, hepatocellular, urinary tract, renal cell carcinoma, and thyroid cancers (for review [12]). The PAK1 gene lies within the 11q13 region and 11q13.5 → 11q14 amplifications involving the PAK1 locus are present in 17 % of breast cancers [13, 14]. PAK1 overexpression was observed in over half of observed breast tumor specimens [15] and PAK1 expression is correlated with tumor grade [16–18]. In transgenic mouse models, hyperactivation of PAK1 promotes mammary gland tumor formation [19]. Interestingly, overexpression of constitutively active PAK1 T423E in non-invasive breast cancer cells stimulates cell motility and anchorage independence [17], while expression of kinase dead PAK in highly invasive breast cancer cells significantly reduces cell invasiveness [20]. PAK1 kinase activity promotes directional cell motility and is a major regulator of the actin cytoskeleton (for review [11]). We have previously demonstrated that PRL-activated JAK2 directly phosphorylates PAK1 on tyrosines 153, 201, and 285 [10]. We have also demonstrated that tyrosyl phosphorylated PAK1 (pTyr-PAK1) enhances PRL-mediated cell invasion via MAPK activation and increased matrix metalloproteinase expression [21] as well as cell motility through increased phosphorylation of actin-crosslinking protein filamin A ([22]; reviewed in [23]). Additionally, PRL-induced pTyr-PAK1 is localized at small adhesion complexes at the cell periphery and regulates adhesion turnover in breast cancer cells, a process that is absolutely critical for cell motility [24].

Cell motility is essential in the regulation of many significant biological processes including embryogenesis, wound healing, and immune responses; however aberrant cell migration is present in malignant cancers and results in the establishment of tumors in distant tissues. Cell motility is a highly coordinated process that requires tight regulation of the actin cytoskeleton, cell-matrix adhesion turnover, and complex intracellular signaling cascades. The tyrosine kinase focal adhesion kinase (FAK) has been implicated as an important regulator of cell motility (for review [25]). FAK is localized to cell/matrix adhesions and is activated by integrin engagement to the extracellular matrix as well as by several other extracellular ligands (for review [26]). Auto-phosphorylation of FAK at tyrosine 397 (Y397) promotes FAK activation and recruits SH2- and SH3-domain containing proteins, most notably c-Src, leading to Src-mediated FAK activation and activation of

Src/FAK signaling pathways, including the ERK MAPK signaling cascade (for review [26]). FAK activation has been most well implicated in the positive regulation of cell motility (for review [26, 27]). However, recently more evidence has demonstrated a controversial role for FAK as a negative regulator of cancer cell migration [28–30].

Here we extend our knowledge on the role for pTyr-PAK1 in PRL-induced breast cancer cell motility and invasion. We use T47D and TMX2-28 breast cancer cells stably overexpressing GFP, PAK1 WT, or tyrosyl phosphorylation-deficient mutant of PAK1 in which the three JAK2 phosphorylation sites have been mutated to phenylalanine (PAK1 Y3F). These cells were previously characterized in [22, 24] and [21]. We demonstrate here that tyrosyl phosphorylation of PAK1 in response to PRL regulates PTP-PEST-dependent FAK dephosphorylation, resulting in augmented breast cancer cell migration and invasion and proposed the mechanism explaining these findings. Furthermore, we provide in vivo evidence that PRL-induced pTyr-PAK1 increases breast cancer cell metastasis. Taken together, these data suggest that PRL-mediated pTyr-PAK1 is important in regulating the dynamic activation of FAK and subsequent breast cancer cell migration and invasion.

Methods
Antibodies and reagents
Polyclonal αpY397-FAK (Abcam), monoclonal αFAK (EMD Millipore), polyclonal αpS298-MEK (Cell Signaling), monoclonal αMEK (GeneTex), monoclonal αphospho-ERK1/2 (pT202/Y204) and polyclonal αERK1/2 (Cell Signaling), monoclonal αmyc (9E10, Santa Cruz Biotechnology), and αγ-tubulin (Sigma-Aldrich) were used for immunoblotting. Na_3VO_4 was purchased from Sigma. siRNA and primers for PTP-PEST were purchased from Santa Cruz Biotechnology. Control nontargeting siRNA was purchased from Cell Signaling. Human PRL was purchased from the National Hormone and Peptide Program (Dr. Parlow, National Institute of Diabetes and Digestive and Kidney Diseases).

Cell culture
Prolactin receptor- and estrogen receptor-positive T47D cells stably overexpressing GFP, myc-tagged PAK1 WT, and myc-tagged PAK1 Y3F were described previously [22, 24]. T47D clones were maintained in RPMI 1640 medium (Corning Cellgro, Corning, Inc) supplemented with 10 % fetal bovine serum (FBS; Sigma-Aldrich) and insulin (Sigma-Aldrich). Prolactin receptor-positive but estrogen receptor-negative TMX2-28 cells (a variant of the MCF-7 breast cancer cell line [31]) and their clones stably overexpressing GFP, PAK1 WT or PAK1 Y3F were described previously [21] and maintained in DMEM supplemented with 10 % fetal bovine serum. The levels

of overexpressed PAK1 WT and PAK1 Y3F were roughly estimated to be around 20-fold over the level of endogenous PAK1 in both T47D cells and TMX2-28 cells. MCF-7 cells were kindly donated by Dr. Ethier (University of Michigan) and T47D cells were purchased from the ATCC. TMX2-28 cells were kindly donated by Dr. Eisenmann (University of Toledo, OH).

Assessing FAK, MEK, and ERK phosphorylation
T47D or TMX2-28 clones were seeded into 6-well dishes and deprived of serum for 72 h before treatment with or without PRL (200 ng/ml) for the indicated times. Cells were lysed and proteins were resolved by SDS-PAGE followed by immunoblotting with the indicated antibodies. Fold FAK, MEK, and ERK activation was assessed by densitometric analysis of αphospho-protein bands normalized to αtotal-protein bands using ImageJ software. To assess FAK activation in T47D clones in the absence of tyrosine phosphatase activity, cells were treated with 100 ng/ml of Na_3VO_4 for one hour before treatment with or without PRL (200 ng/ml) for the indicated times. Cells were lysed and proteins were resolved by SDS-PAGE followed by immunoblotting with the indicated antibodies. FAK activation was assessed by densitometric analysis of αpY397-FAK bands normalized to αFAK bands using ImageJ software.

PTP-PEST knockdown
PTP-PEST siRNA or control nontargeting siRNA were transfected into T47D or TMX2-28 cells using Lipofectamine RNAiMAX (Invitrogen) according to the manufacturer's instructions. The final concentration of the siRNA was 100 nM. Knockdown of PTP-PEST mRNA was assessed by RT-PCR method using PTP-PEST primers.

To assess PRL-induced FAK activation in the absence of PTP-PEST, T47D and TMX2-28 clones were transfected with PTP-PEST siRNA, deprived of serum for 48 h, and treated with or without PRL for the indicated times. Cells were lysed and proteins were resolved by SDS-PAGE followed by immunoblotting with the indicated antibodies.

Cell viability
To assess cell viability in the presence of 100 ng/ml Na_3VO_4 for 48 h, equal numbers of T47D cells were resuspended in deprivation media (RPMI 1640 medium supplemented with 1 % BSA) with or without PRL (200 ng/ml) and Na_3VO_4 (100 ng/ml) then seeded into a 96-well plate. After 48 h, cells were subjected to the Vybrant® MTT Cell Proliferation Assay (Molecular Probes) according to the manufacturer's instructions.

Cell migration and cell invasion assays
Cell migration and cell invasion assays were performed as we described previously [21, 22]. Equal cell numbers of

the T47D (1×10^6 cells/chamber) or TMX2-28 (0.5×10^6 cells/chamber) stable cell lines for each condition were placed in deprivation media with or without 100 ng/ml Na_3VO_4 in the upper chamber of a Boyden chamber (8.0 μm pores, Corning, Inc) (migration assay) or a Boyden chamber (8.0 μm pores), coated with Matrigel (BD Biosciences) (invasion assay). Deprivation media with or without 200 ng/ml PRL was placed in the lower chamber. Cells were allowed to migrate or invade for 48 h, after which the cells remaining in the upper chamber were removed from the upper chamber by a cotton swab. Cells from five separate fields that had migrated through the pores of the membrane to the underside of the filter were counted after fixation with 4 % formalin (Sigma) and staining with Differential Quik Stain (Polysciences, Inc). Brightfield images of migrated/invaded cells were acquired on an inverted Olympus IX81 microscope using LUCPlan FLN 40× objective lens and wide field WHN 10X eyepiece (Olympus, Tokyo, Japan).

To assess the effect of PTP-PEST knockdown on cell migration and invasion, T47D and TMX2-28 stable clones were transfected with PTP-PEST siRNA. After 24 h, cells were placed in deprivation media in the upper chamber of a Boyden chamber (migration assay) or a Boyden chamber coated with Matrigel (invasion assay). Cells were allowed to migrate/invade for 48 h and processed as described above.

In vivo metastasis
TMX2-28 clones stably overexpressing GFP, myc-PAK1 WT or myc-PAK1 Y3F were inoculated directly into mammary fat pad of NSG (NOD/SCID/ IL2Rgamma) female mice. hPRL (20 μg/100 μl) was injected subcutaneously every other day for 8 weeks and mice were terminated in 12 weeks. 8 mice were used for TMX2-28 PAK1 clone, 6 mice for TMX2-28 PAK1 Y3F clone and 4 mice for TMX2-28 GFP clone. Mouse experimental procedures were performed in the animal research core of Lerner Research Institute, Cleveland Clinic (Dr. Lindner), and were approved by the Institutional Animal Care and Use Committee, Cleveland Clinic. The first half of tumors and lungs from mice was frozen and kept at –80 °C. Before use, the tissues were homogenized in RIPA buffer with protease inhibitors (50 mM Tris-HCl, 150 mM NaCl, 2 nM EGTA, 1 % Triton X-100, aprotinin 10 μg/ml, leupeptin, 10 μg/ml, pH7.5; 500 μL per 10 mg tissue) at 4 °C. Homogenized tissues were rotated in RIPA buffer for 1 h at 4 °C to ensure cell lysis. Samples were centrifuged at 10,000 g to pellet debris and protein concentration in supernatant was determined by Bradford assay. Proteins were separated by SDS-PAGE and transferred to PVDF membrane. Lysates of TMX2-28 PAK1 WT cells were loaded as a control for PAK1-myc position in the gels. Membranes were probed with anti-myc to detect myc-PAK1 WT or Y3F in the

tissues and anti-tubulin for loading control. The second half of tumors and lungs was fixed with 10 % formalin and embedded in paraffin. Immunohistochemistry using paraffin-embedded sections was done as described previously [24]. Briefly, formalin-fixed, paraffin-embedded sections were boiled for 15 min in 0.01 M sodium citrate buffer (pH 6.0) to expose antigenic epitopes. Sections were blocked with 2.5 % normal horse serum for 30 min and then incubated overnight with anti-myc (1:100) or control pre-immune serum. The biotinylated secondary antibody was used followed by streptavidin horseradish peroxidase solution (R.T.U. Vectstatin universal quick kit, Vector Laboratories). The chromogen was 3,3' diaminobenzidine (ImmPACT DAB kit, Vector Laboratories). Staining with pre-immune serum was negligible (not shown).

Statistical analysis

Data from at least 3 separate experiments were pooled and analyzed using 1-way ANOVA plus Tukey's honest significant difference test. Differences were considered to be statistically significant at $P < 0.05$. Results are expressed as the mean ± SE.

Results

Tyrosyl phosphorylated PAK1 negatively regulates FAK auto-phosphorylation

We have previously demonstrated that PRL promotes breast cancer cell motility in a pTyr-PAK1-dependent manner [22]. In an attempt to understand the pTyr-PAK1-

dependent mechanism that regulates PRL-induced cell motility, we first examined the auto-phosphorylation of FAK in response to PRL, as FAK is an important regulator of cell motility (for review [25]). T47D GFP (control), PAK1 WT, or PAK1 Y3F (phospho-tyrosine-deficient mutant) clones were treated with PRL over a time-course and whole cell lysates (WCL) were analyzed for FAK auto-phosphorylation at Y397, which is critical for Src/FAK interaction and maximal FAK activation (reviewed in [32]). PRL treatment led to maximal FAK auto-phosphorylation in 15 min in control GFP cells (Fig. 1a, left blot; Fig. 1b, solid line). On the contrary, there was no significant Y397-FAK auto-phosphorylation in response to PRL in the PAK1 WT cells (Fig. 1a, middle blot; Fig. 1b, dashed line), suggesting that PRL-induced pTyr-PAK1 has a negative effect on FAK auto-phosphorylation. FAK was maximally auto-phosphorylated by PRL in 7.5 min in PAK1 Y3F cells (Fig. 1a, right blot; Fig. 1b, dotted line). Similar results were obtained in TMX2-28 (estrogen-receptor-negative sub-line of the MCF-7 breast cancer cells [31]) stably overexpressing GFP, myc-PAK1 WT, or myc-PAK1 Y3F (Fig. 2d, anti-pY397-FAK and anti-FAK blots) indicating that this finding was not restricted to T47D cells. It is important to note that PRL-dependent Y395-FAK phosphorylation was transient in both T47D GFP and T47D PAK1 Y3F clones. Our data suggest that tyrosyl phosphorylation of PAK1 in response to PRL promotes FAK dephosphorylation and tyrosines 153, 201 and 285 of PAK1 are responsible for this effect.

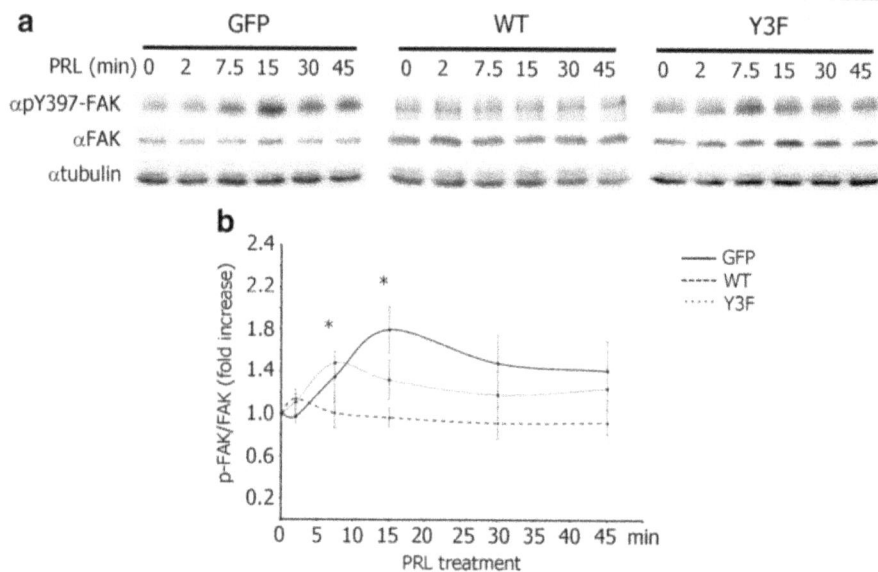

Fig. 1 Tyrosyl phosphorylation of PAK1 negatively regulates PRL-induced FAK auto-phosphorylation. **a** Whole cell lysates (WCL) of T47D cells stably overexpressing GFP, PAK1 WT, or PAK1 Y3F treated with PRL (200 ng/ml) for the indicated times were probed for FAK auto-phosphorylation using αpY397-FAK antibody. The expression levels of γtubulin were used as an internal loading control. **b** Graph represents the densitometric analysis of the bands obtained for pY397-FAK normalized to total FAK for at least 3 independent experiments. The solid line represents T47D GFP cells, the dashed line represents T47D PAK1 WT cells, and the dotted line represents T47D PAK1 Y3F cells. Bars represent mean ± SE. *$P < 0.05$ compared with the same cells not treated with PRL

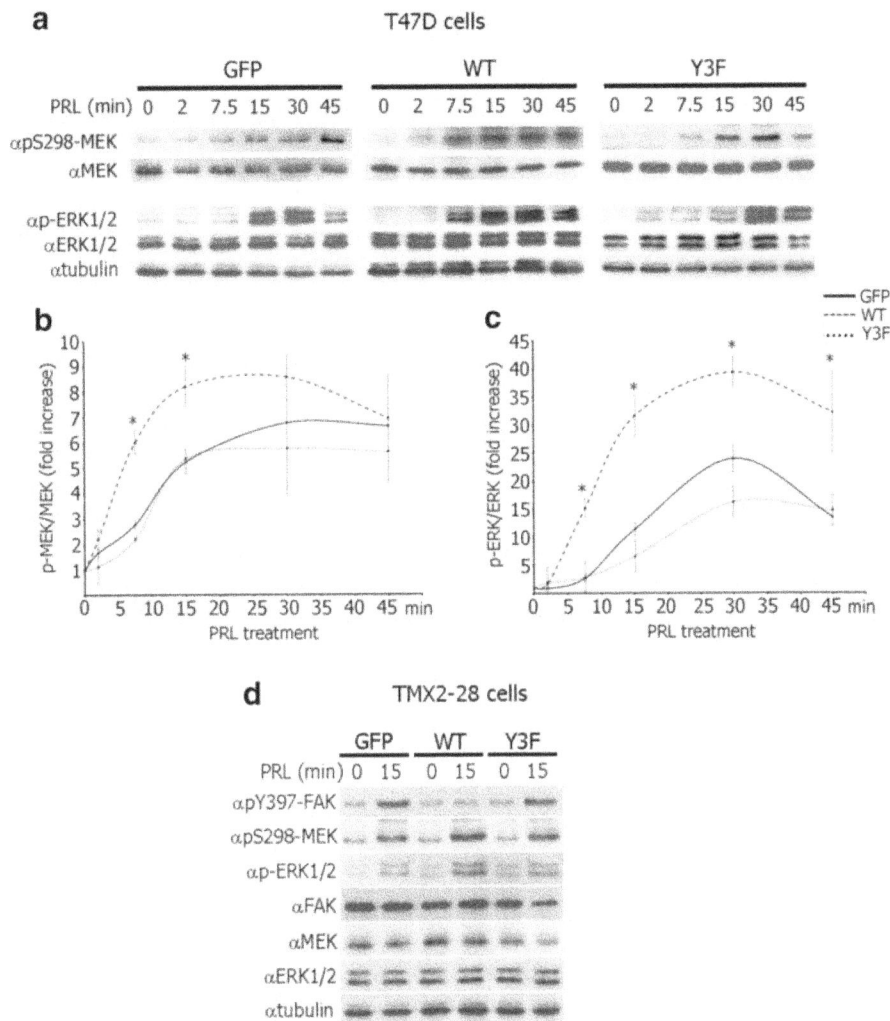

Fig. 2 Tyrosyl phosphorylation of PAK1 promotes S298-MEK1 phosphorylation and ERK activation in response to PRL. **a** WCL of T47D cells stably overexpressing GFP, PAK1 WT, or PAK1 Y3F treated with PRL (200 ng/ml) for the indicated times were probed for MEK phosphorylation using αpS298-MEK and ERK1/2 activation using αphospho-ERK1/2 (pT202/Y204) antibodies. **b, c** Graphs represent the densitometric analysis of the bands obtained for phospho-MEK (**b**) or phospho-ERK1/2 (**c**) normalized to total MEK or ERK1/2, respectively, for at least 3 independent experiments. Bars represent mean ± SE . *$P < 0.05$ compared with cells expressing GFP with the same treatment. **d** WCL of TMX2-28 cells stably overexpressing GFP, PAK1 WT, or PAK1 Y3F treated with PRL (200 ng/ml) for the indicated times were probed with the indicated antibodies. The expression levels of γtubulin were used as an internal loading control

Tyrosyl phosphorylation of PAK1 promotes S298-MEK1 phosphorylation and ERK activation in response to PRL

To uncover the mechanism by which pTyr-PAK1 may regulate FAK phosphorylation, we assessed S298-MEK phosphorylation and consequent ERK1/2 activation (dual phosphorylation of T202 and Y204 of ERK1/2 mediates ERK activity [33, 34]) in response to PRL because a PAK1/MEK/ERK signaling cascade has been implicated in Ras-mediated FAK dephosphorylation [30]. PRL promoted PAK1-dependent MEK phosphorylation 6-fold in as early as 7.5 min and maximal 8-fold MEK phosphorylation after 15 min in T47D PAK1 WT cells (Fig. 2a, middle blot, Fig. 2b). PRL also induced pS298-MEK signal in the T47D GFP and T47D PAK1 Y3F cells albeit slower and to a lesser extent when compared to the PAK1 WT cells (Fig. 2a, left and right blots, Fig. 2b). Subsequently, ERK1/2 was phosphorylated in response to PRL in all three T47D clones, however earlier and to a much greater extent in the PAK1 WT cells when compared to GFP and PAK1 Y3F cells (Fig. 2a and c). Similar results were obtained in TMX2-28 GFP, PAK1 WT and PAK1 Y3F cell clones (Fig. 2d, anti-pS298-MEK, anti-p-ERK1/2, anti-MEK and anti-ERK1/2 blots). These data suggest that PAK1 tyrosyl phosphorylation promotes PAK-dependent MEK phosphorylation and ERK activation in response to PRL.

Protein tyrosine phosphatase inhibition rescues PRL-mediated auto-phosphorylation of FAK

In order to determine whether tyrosine phosphatases are involved in the negative effect of pTyr-PAK1 on FAK auto-phosphorylation, we assessed Y397- FAK phosphorylation in response to PRL in the presence or absence of Na_3VO_4, a tyrosine phosphatase inhibitor. T47D PAK1 WT cells were treated with vehicle or Na_3VO_4 for one hour before PRL treatment and WCL were assessed for pY397-FAK (Fig. 3). As expected, there was no PRL-mediated increase in FAK tyrosyl phosphorylation in vehicle treated cells (Fig. 3a, lanes 1 and 2). However, phosphatase inhibition led to a significant increase in both basal and PRL-induced FAK auto-phosphorylation (Fig. 3a, lanes 3 vs. 1 and 4 vs. 2). To confirm that phosphatase activity is important for the pTyr-PAK1-dependent effect of PRL on FAK dephosphorylation, all three T47D clones were subjected to a PRL time-course in the presence of Na_3VO_4 and WCL were assessed for Y397 FAK auto-phosphorylation (Fig. 3b, c). In the presence of Na_3VO_4, PRL treatment activated FAK in all three cell lines regardless of the status of PAK1 tyrosyl phosphorylation (Fig. 3b, c). Furthermore, FAK remained phosphorylated until the end of the PRL time-course in all three clones in the presence of Na_3VO_4.

Next we aimed to determine whether tyrosine phosphatase PTP-PEST, which dephosphorylates FAK at Y397 [30], participates in PRL- and PAK1-dependent lack of FAK auto-phosphorylation. PTP-PEST silencing in T47D and TMX2-28 clones was confirmed by RT-PCR method (Fig. 4a). We performed siRNA-based silencing of PTP-PEST in T47D (Fig. 4b) and TMX2-28 (Fig. 4c) clones, treated the cells with or without PRL and assessed for Y397 FAK auto-phosphorylation. Indeed, PTP-PEST silencing rescued Y397-FAK phosphorylation in PAK1 WT cells to similar levels to that of GFP and PAK1 Y3F cells in response to PRL (Fig. 4b, c). On the contrary, there was no significant Y397-FAK auto-phosphorylation in response to PRL in the PAK1 WT clones transfected with control siRNA (Fig. 4b, c).

These data suggest that tyrosine phosphatase activity of PTP-PEST is responsible for the apparent lack of FAK auto-phosphorylation in response to PRL in PAK1 WT cells. Given the complexity of these signaling cascades, it is likely that additional signaling molecules are also involved in the modulation of FAK phosphorylation.

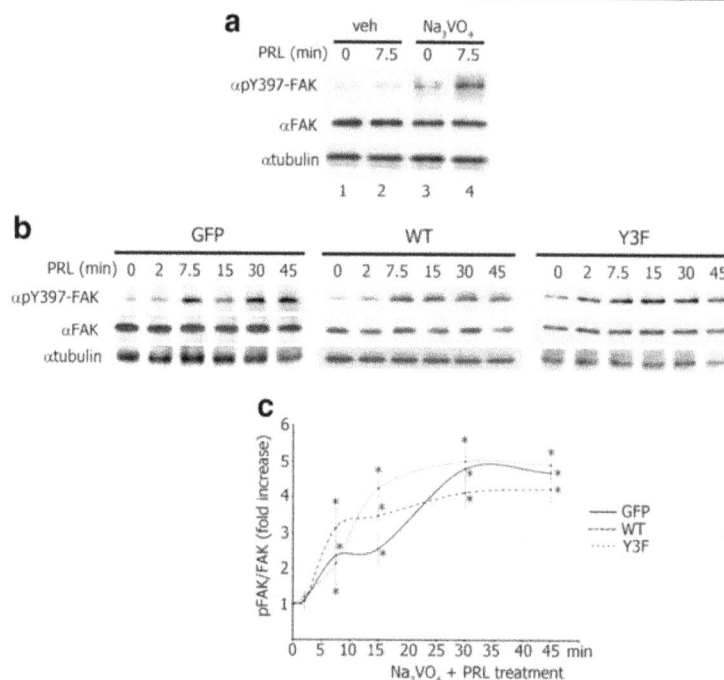

Fig. 3 Protein tyrosine phosphatase inhibition rescues PRL-mediated FAK auto-phosphorylation in T47D WT cells. **a** Tyrosine phosphatase inhibition by Na_3VO_4 permits PRL-induced FAK auto-phosphorylation in PAK1 WT cells. WCL of T47D PAK1 WT cells treated with either vehicle (veh) or Na_3VO_4 (100 ng/ml) for 1 h before PRL (200 ng/ml) treatment were probed for FAK auto-phosphorylation by αpY397-FAK antibody. **b** FAK is auto-phosphorylation in T47D GFP, PAK1 WT, and PAK1 Y3F cells in response to PRL in the presence of Na_3VO_4. The cells were treated with Na_3VO_4 as in A and with PRL (200 ng/ml) for the indicated times. FAK auto-phosphorylation was assessed as in A. The expression levels of γtubulin were used as an internal loading control. **c** Graph represents the densitometric analysis of the bands obtained for pY397-FAK normalized to total FAK for at least 3 independent experiments. Bars represent mean ± SE. *$P < 0.05$ compared with the same cells not treated with PRL

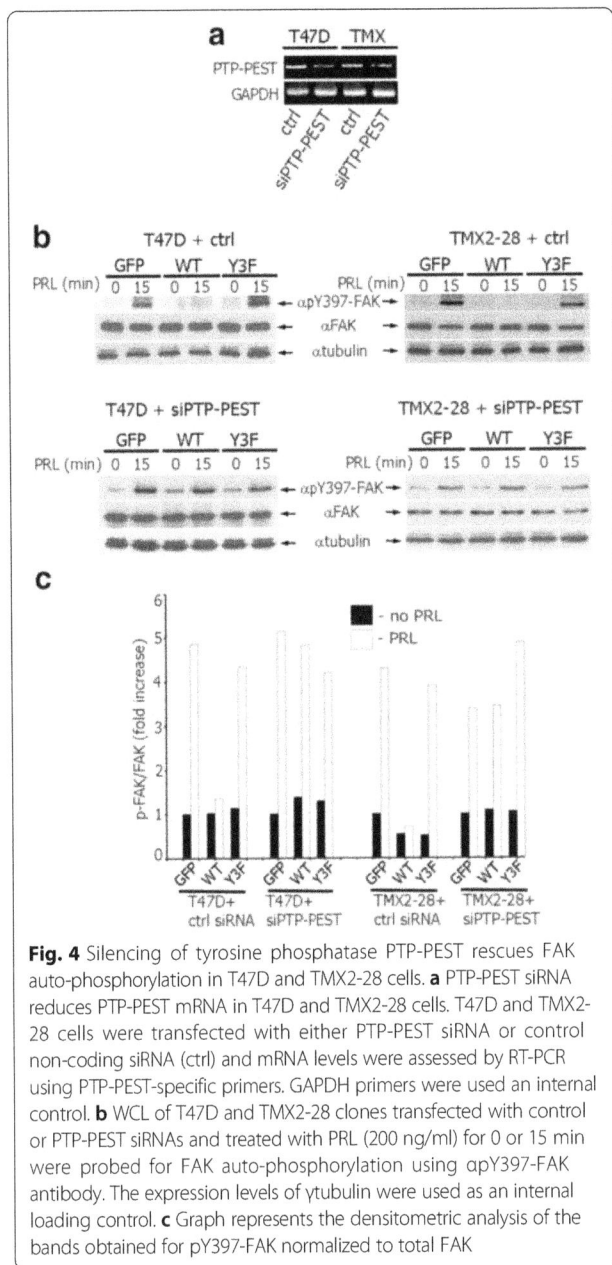

Fig. 4 Silencing of tyrosine phosphatase PTP-PEST rescues FAK auto-phosphorylation in T47D and TMX2-28 cells. **a** PTP-PEST siRNA reduces PTP-PEST mRNA in T47D and TMX2-28 cells. T47D and TMX2-28 cells were transfected with either PTP-PEST siRNA or control non-coding siRNA (ctrl) and mRNA levels were assessed by RT-PCR using PTP-PEST-specific primers. GAPDH primers were used an internal control. **b** WCL of T47D and TMX2-28 clones transfected with control or PTP-PEST siRNAs and treated with PRL (200 ng/ml) for 0 or 15 min were probed for FAK auto-phosphorylation using αpY397-FAK antibody. The expression levels of γtubulin were used as an internal loading control. **c** Graph represents the densitometric analysis of the bands obtained for pY397-FAK normalized to total FAK

Protein tyrosine phosphatase inhibition impedes PRL-mediated T47D and TMX2-28 cell migration and invasion

To investigate whether tyrosine phosphatases regulate PRL/pTyr-PAK1-dependent T47D breast cancer cell migration, we examined migration of T47D clones in the presence and absence of PRL and Na_3VO_4 for 48 h. As dynamic tyrosyl phosphorylation events are crucial to many cellular processes, it was important to test whether phosphatase inhibition for an extended period of 48 h had any cytotoxic effect. The cell viability was assessed in serum deprived T47D cells treated with or without PRL and Na_3VO_4 (Fig. 5c). Na_3VO_4 had no significant cytotoxic effect on any of the three stable cell lines in the presence or absence of PRL (Fig. 5c). Next, the effect of tyrosine phosphatase inhibition on PRL-mediated cell migration was assessed using a transwell migration assay. Equal numbers of T47D GFP, PAK1 WT and PAK1 Y3F cells were seeded into the upper part of a Boyden chamber with or without Na_3VO_4 and PRL or vehicle were added to the bottom part. The number cells that migrated through the chamber towards PRL were counted (Figs. 5a and d). As we demonstrated previously [22], PRL stimulated cell migration to a greater extent in PAK1 WT cells when compared to GFP and PAK1 Y3F cells in the absence of Na_3VO_4 (Fig. 5d, veh). However, phosphatase inhibition by Na_3VO_4 completely abolished cell migration in response to PRL in all T47D clones (Fig. 5d, Na_3VO_4). These data suggest that phosphatase activity is required for pTyr-PAK1-induced cell migration.

Cell migration is a key step in cell invasion so we decided to assess the effect of phosphatase inhibition on cell invasion. Equal numbers of T47D GFP, PAK1 WT and PAK1 Y3F cells were seeded into the upper chamber of a Boyden chamber coated with Matrigel, in the presence of either Na_3VO_4 or vehicle. Deprivation media with or without PRL (200 ng/ml) was added to the lower chamber of the Boyden chamber. The number of cells that invaded through the Matrigel towards PRL was counted. As we demonstrated previously [21], PRL stimulated cell invasion to a greater extent in PAK1 WT cells when compared to GFP and PAK1 Y3F cells in the absence of Na_3VO_4 (Fig. 5b and e, veh). However, Na_3VO_4-mediated tyrosine phosphatase inhibition abolished cell invasion in response to PRL in all T47D clones (Fig. 5e, Na_3VO_4).

To demonstrate that the role of PAK1 in PRL-mediated signaling is not limited to T47D cells, we assessed migration and invasion in the presence and absence of Na_3VO_4 in TMX2-28 clones. In TMX2-28 GFP and TMX2-28 WT cells PRL induced cell migration (Fig. 6a, veh) and invasion (Fig. 6b, veh) while in TMX2-28 Y3F cells did not (Fig. 6a and b, veh). However, Na_3VO_4 treatment abolished PRL-dependent cell migration and invasion in all TMX2-28 clones (Fig. 6a and b, Na_3VO_4). Silencing of PTP-PEST also abolished PRL-dependent cell migration and invasion of all T47D clones (Fig. 6c and d) and migration of TMX2-28 clones (Fig. 6e). PRL-induced invasion of TMX2-28 control (GFP) and WT cells was significantly decreased by silencing of PTP-PEST although not completely abolished (Fig. 6f) suggesting that, in addition to PTP-PEST, other tyrosine phosphatases may participate in the pTyr-PAK1-dependent invasion of TMX2-28 cells.

PAK1 tyrosyl phosphorylation stimulates PRL-induced tumor metastasis in vivo

Cell migration is critical for tumor cell metastasis. In order to assess whether PRL-induced tyrosyl phosphorylation of PAK1 has a physiological effect on breast cancer

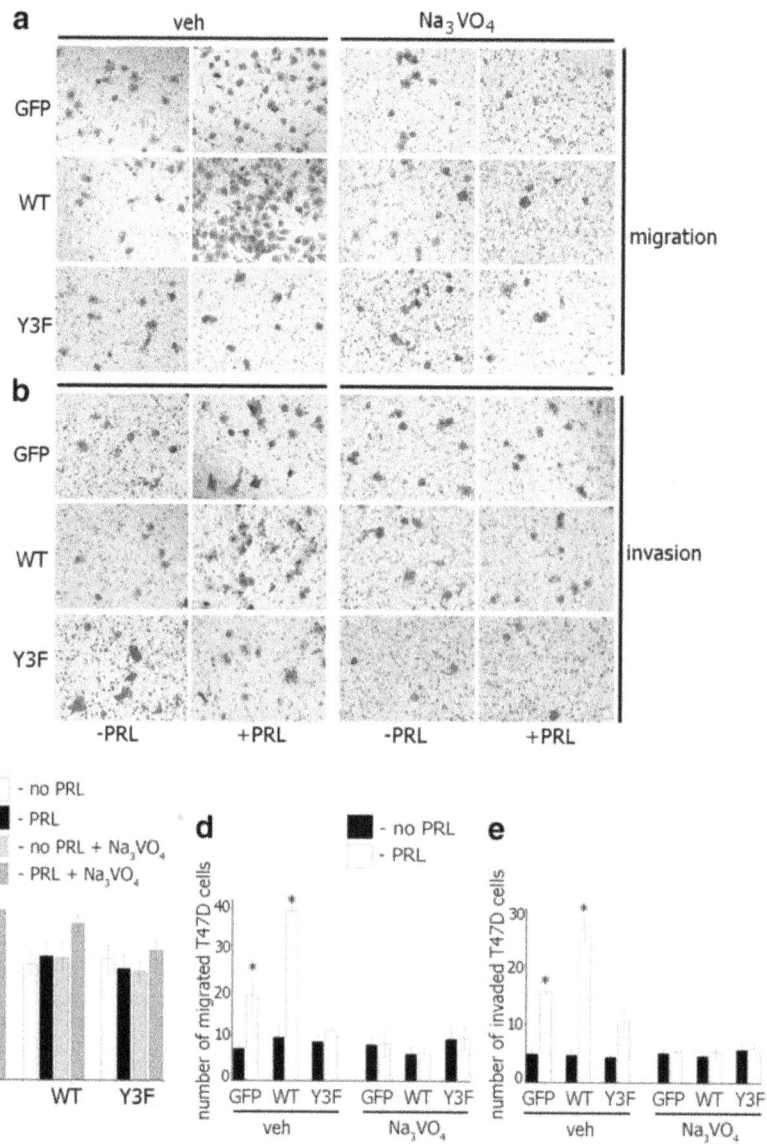

Fig. 5 Protein tyrosine phosphatase inhibition impedes PRL-mediated T47D cell migration and invasion. **a, b** Equal amounts of T47D GFP, PAK WT, or PAK1 Y3F cells were loaded into the upper part of the Boyden chamber uncovered (**a**) or covered with Matrigel (**b**) with or without Na_3VO_4 (100 ng/ml). PRL (200 ng/ml) was added to the lower part. Representative brightfield images of the cells migrated/invaded to the lower chamber were taken in 48 h. A LUCPlan FLN 40X objective lens and wide field WHN 10X eyepiece on an inverted Olympus IX81 microscope were used. (**c**) Na_3VO_4 (100 ng/ml) treatment on T47D cells for 48 h has no cytotoxic effect. **d, e** The number of cells that migrated to the lower surface of the chamber toward PRL (white bar) or vehicle (black bar) after 48 h was counted and plotted. Bars represent mean ± SE. *$P < 0.05$ compared with the same cells not treated with PRL

metastasis, TMX2-28 stably overexpressing GFP, myc-PAK1 WT, or myc-PAK1 Y3F were inoculated in mouse mammary fat pads and mice were treated with PRL for 8 weeks. Tumors and lungs were harvested and homogenized and proteins were separated by SDS-PAGE and analyzed for myc-tagged PAK1 to indicate metastasis of the primary tumor into distant tissues. We focused on the primary tumor and the lungs, as the lungs are one of the most common sites for secondary tumor in patients with metastatic breast cancer. As expected, each primary tumor

from all PAK1 WT and PAK1 Y3F mice was positive for myc-tagged PAK1 (Fig. 7a) while GFP cells do not produce tumors. Myc-tagged PAK1 was detected in 3 out of 8 lungs from the PAK1 WT mice while there was no detectable myc-PAK1 in any of the PAK1 Y3F or GFP mouse lungs (Fig. 7a). Tumors and lungs were also fixed and analyzed by immunocytochemistry (IHC) with anti-myc. Our IHC analysis revealed that anti-myc signal was detected in breast tumor (B) of myc-PAK1 WT- and myc-PAK1 Y3F-inoculated mice (C) as well as in lung of myc-PAK1 WT-

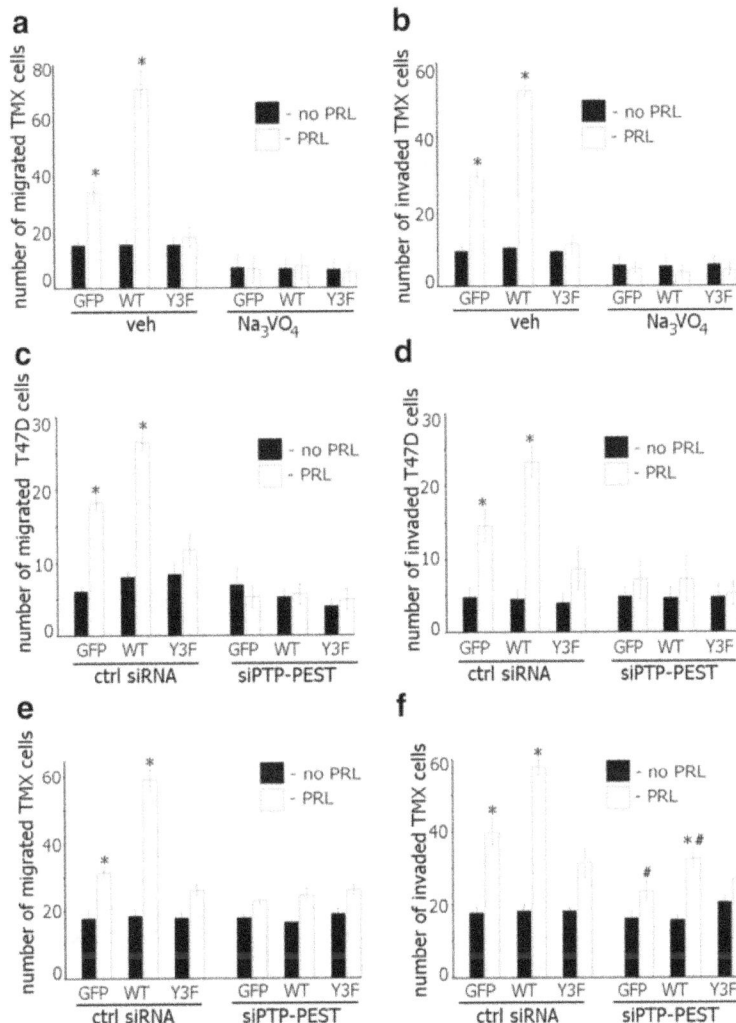

Fig. 6 Silencing of the tyrosine phosphatase PTP-PEST reduces PRL-mediated cell migration and invasion. **a** Tyrosine phosphatase inhibition abolishes PRL-induced TMX2-28 cell migration **a** and invasion **b**. TMX2-28 GFP, PAK WT, or PAK1 Y3F cells were assessed as in Fig. 5. **c–f** Equal amount of T47D (**c, d**) or TMX2-28 (**e, f**) clones were transfected with either control or PTP-PEST siRNA and loaded into the upper part of the Boyden chamber covered (**d, f**) or not (**c, e**) with Matrigel. The number of cells that migrated/invaded to the lower chamber toward PRL (white bar) or vehicle (black bar) after 48 h was counted and plotted. Bars represent mean ± SE. *$P < 0.05$ compared with the same cells not treated with PRL. #$P < 0.05$ compared with the same cells treated with PRL but transfected with control siRNA (**f**)

inoculated mice (D) but not in lungs of control GFP- (E) or PAK1 Y3F-inoculated mice (F). These data provide first in vivo evidence that tyrosyl phosphorylation of PAK1 plays a significant role in PRL-induced breast cancer cell motility and metastasis, as only cells overexpressing PAK1 WT, but not phospho-tyrosine-deficient PAK1 Y3F, were able to migrate from the primary tumor to the lungs. Here we provide new insight into the mechanisms regulating PRL-dependent breast cancer cell metastasis.

Discussion

The role of PAK1 in the regulation of cell motility is well documented (reviewed in [11]). The role of PAK1 in the regulation of cell adhesion is also well documented and at least one mechanism has been proposed ([35], reviewed in

[36]). According to this mechanism, PAK1 phosphorylates paxillin on Ser273, leading to increased paxillin-GIT1 binding and adhesion turnover [35]. We have previously implicated PRL/JAK2-dependent tyrosyl phosphorylation of PAK1 in regulation of cell motility and invasion [21, 22]. We have also implicated pTyr-PAK1 in the regulation of breast cancer cell adhesion and demonstrated that phosphorylation of tyrsines 153, 201 and 285 of PAK1 regulates cell adhesion, contribute to maximal PAK1 kinase activity and increased ability to bind βPIX and GIT1 [24]. Here we extend our findings and demonstrate that pTyr-PAK1 phosphorylates MEK1 on Ser289 resulting in subsequent ERK1/2 activation. We also show that PRL-induced FAK auto-phosphorylation on Tyr397 is inhibited by pTyr-PAK1 and can be rescued by inhibiting

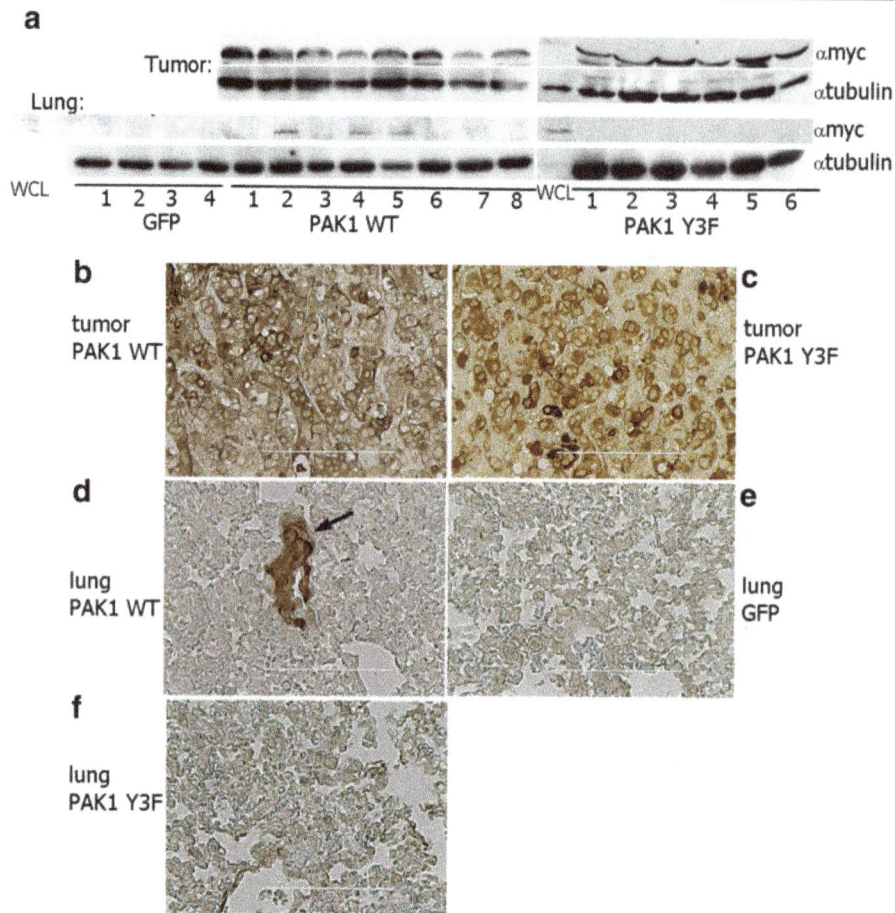

Fig. 7 PAK1 tyrosyl phosphorylation stimulates PRL-induced tumor metastasis in vivo. **a** myc-PAK1 was detected in all tumor lysates isolated from myc-PAK1 WT and myc-PAK1 Y3F inoculated mice. PRL-induced tyrosyl phosphorylation of PAK1 increased tumor metastasis, as 3 out of 8 lungs from WT mice contained myc-PAK1 (lanes 2, 4 and 5). No myc-PAK1 was detected in any of the lungs from the Y3F or GFP mice. Anti-tubulin antibody was used as a loading control. Whole cell lysate (WCL) of TMX2-28 PAK1 WT cells was loaded as a control for PAK1-myc position in the gels. **b–f** Representative images of myc-PAK1 detected with anti-myc in breast tumor (**b**, **c**) and lung tissue (**d–f**). myc-PAK1 was detected in breast tumors of myc-PAK1 WT- (**b**) and Y3F- (**c**) and lung of myc-PAK1 WT- inoculated mice (**d**) but not in lungs of GFP (**e**) or PAK1 Y3F-inoculated mice (**f**). The arrow highlights metastatic nodule. Counterstaining with hematoxylin was omitted. Scale bar is 200 μm in (**b**) and 100 μm in (**c–e**)

tyrosine phosphatases and silencing tyrosine phosphatase PTP-PEST. These tyrosine phosphatase inhibitions abrogate cell motility and invasion in response to PRL. We hypothesize that pSer910-FAK recruits tyrosine phosphatase PTP-PEST to dephosphorylate pTyr397-FAK and thereby promotes cell motility as shown previously [30].

Dynamic of FAK phosphorylation is significant for cell motility. Previously, FAK activation has been demonstrated to positively regulate cell motility (for review [25–27]) however it is becoming evident that the role of FAK activation in cell migration is more complex. Silencing FAK using siRNA enhanced HeLa cell migration on collagen, and FAK dephosphorylation on Y397 by the tyrosine phosphatase PTP-PEST promoted Ras-induced cell migration in transformed NIH3T3-v-H-Ras cells [29, 30]. Cells with reduced FAK dephosphorylation had diminished cell motility [37] and overexpression of the

tyrosine phosphatase LMR-PTP, which dephosphorylates FAK, enhanced cell motility [38]. Importantly, Zheng et al. implicated PAK1 in regulation of Ras-induced FAK dephosphorylation, as overexpression of constitutively active PAK1 T423E promoted FAK dephosphorylation while inhibition of PAK1 severely abolished FAK dephosphorylation at Y397 [30]. With agreement with these data, we have shown here that pTyr-PAK1 abolished PRL-dependent phosphorylation of Ser397-FAK.

We previously demonstrated that tyrosyl phosphorylation of PAK1 promotes both PAK1 kinase activity and protein-protein interaction capabilities (for review [23]). PAK1 directly binds to ERK in response to adhesion to fibronectin, and both PAK1 and ERK co-localize at nascent adhesions on the cell periphery [39]. Here, PAK1 can serve as a scaffold, bringing together Raf, MEK and ERK at cell/matrix adhesions and thereby stimulating ERK-dependent signal transduction [39]. In addition to PAK1 scaffolding activity, PAK1 promotes Raf activation by directly phosphorylated Raf on S338/339 [40, 41], and stimulates MEK/ERK binding and subsequent ERK activity by directly phosphorylating S298 on MEK1 [42–44]. Importantly, pS298-MEK has been shown to localize at peripheral adhesion complexes in response to cell adhesion to fibronectin [45]. Concurrently, we have demonstrated that tyrosyl phosphorylated PAK1 is localized at peripheral adhesion complexes in response to PRL and is responsible for proper adhesion turnover, an important process in cell migration [24]. This is important, as FAK is also localized at peripheral adhesion complexes and dynamic FAK localization and phosphorylation is important for proper adhesion turnover and cell migration (for review [25]). FAK localization to peripheral cell/matrix adhesions is dependent on its focal adhesion targeting (FAT) domain and binding to adhesion proteins paxillin and vinculin [46, 47]. Paxillin phosphorylation at Y31 and Y118 by FAK is necessary for cell migration and adhesion turnover [48–50], however, constitutive tyrosyl phosphorylation of paxillin impedes cell migration, and dephosphorylation of FAK by PTP-PEST is required for proper adhesion turnover in migrating cells [51]. Furthermore, overexpression of the dominant negative form of protein phosphatase LMR-PTP leads to FAK hyperphosphorylation and reduced cell motility [38] suggesting that complex regulation of FAK at adhesion complexes is necessary for proper cell migration. Phosphorylation of the FAK FAT domain on S910 and Y925 by ERK2 and Src, respectively, results in reduced FAK/paxillin binding and promotes adhesion turnover [52, 53]. Furthermore, pS298-MEK/ERK activation in NIH3T3 cells was shown to induce FAK dephosphorylation through ERK-mediated FAK S910 phosphorylation and resulting recruitment of tyrosine phosphatase PTP-PEST and thereby promote cell motility [30]. In this regard, PRL-induced tyrosyl phosphorylation of PAK1 and resulting adhesion localization could be creating localized PAK1/MAPK/FAK signaling at adhesion complexes and promoting adhesion turnover during cell migration.

In the present study we demonstrated that tyrosyl phosphorylation of PAK1 stimulates tumor cell metastasis in vivo. These data, combined with an animal study reporting prevention of neoplasia progression into invasive carcinoma in PRL receptor deficient mice [7], suggest that PRL is involved in the development of metastasis and tumor progression. Thus, our current data on pTyr-PAK1 regulation of FAK phosphorylation bring insight into the mechanism of PRL-stimulated motility of breast cancer cells.

Conclusions

Here we propose a mechanism by which PRL regulates motility of T47D and TMX2-28 cells through pTyr-

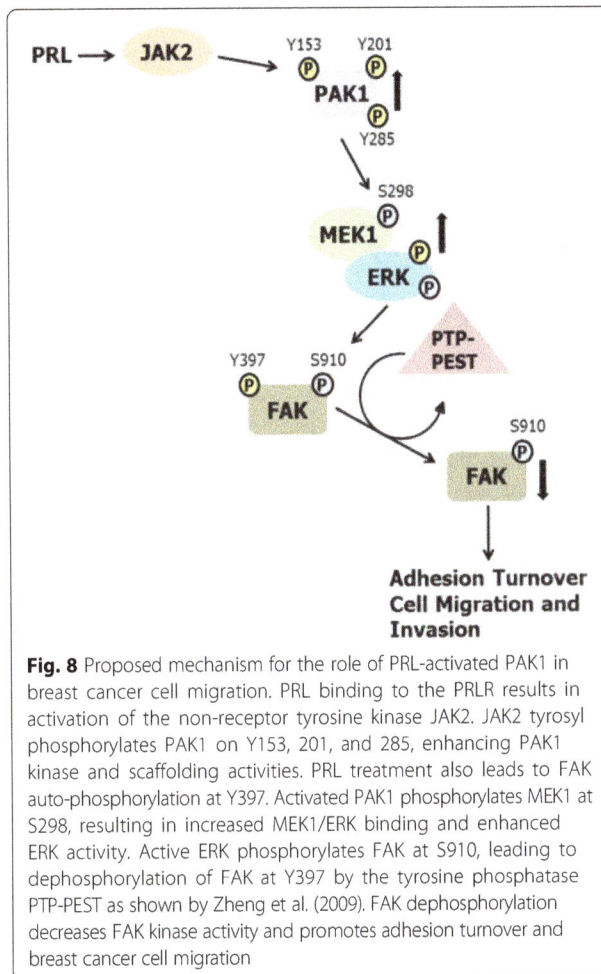

Fig. 8 Proposed mechanism for the role of PRL-activated PAK1 in breast cancer cell migration. PRL binding to the PRLR results in activation of the non-receptor tyrosine kinase JAK2. JAK2 tyrosyl phosphorylates PAK1 on Y153, 201, and 285, enhancing PAK1 kinase and scaffolding activities. PRL treatment also leads to FAK auto-phosphorylation at Y397. Activated PAK1 phosphorylates MEK1 at S298, resulting in increased MEK1/ERK binding and enhanced ERK activity. Active ERK phosphorylates FAK at S910, leading to dephosphorylation of FAK at Y397 by the tyrosine phosphatase PTP-PEST as shown by Zheng et al. (2009). FAK dephosphorylation decreases FAK kinase activity and promotes adhesion turnover and breast cancer cell migration

PAK1, MEK/ERK and FAK that integrates our findings with previous studies (Fig. 8). In response to PRL, FAK is auto-phosphorylated and PAK1 is tyrosyl phosphorylated by JAK2, stimulating PAK1 kinase activity and increasing PAK1 protein-protein binding abilities. We show that tyrosyl phosphorylated PAK1 phosphorylates MEK at serine 298, resulting in MEK-mediated ERK1/2 activation. Activated ERK phosphorylates FAK at S910, leading to subsequent recruitment of PTP-PEST and dephosphorylation of Y397-FAK [30]. PRL-dependent down-regulation of FAK activity may promote focal contact turnover thereby promoting cell migration. Finally, we demonstrate for the first time that tyrosyl phosphorylation of PAK1 by PRL increases breast cancer cell metastasis in vivo.

Abbreviations

ERK, extracellular signal-related kinase; FAK, focal adhesion kinase; JAK2, Janus kinase 2; MEK1, MAPK/ERK kinase 1; PAK1, p21-activated kinase 1; PRL, prolactin; WCL, whole cell lysate

Acknowledgments

We thank Dr. Eisenmann (University of Toledo, OH) for providing TMX2-28 cells. We thank Dr. Lindner and Ms. Parker (Lerner Research Institute, Cleveland

Clinic, OH) for in vivo experiments. We also thank Prabesh Khatiwada for help during the re-submission of the manuscript.

Funding
This work was supported by a grant from the National Institutes of Health (R01DK88127 to MD).

Authors' contributions
Experiments were designed by AH and MD, and performed by AH. Manuscript was written by AH. Both authors have read and approved the final version of the manuscript.

Competing interests
The authors declare that they have no competing interests.

References

1. Marano RJ, Ben-Jonathan N. Minireview: extrapituitary prolactin: an update on the distribution, regulation, and functions. Mol Endocrinol. 2014;28(5):622–33.
2. Clevenger CV, Furth PA, Hankinson SE, Schuler LA. The role of prolactin in mammary carcinoma. Endocr Rev. 2003;24(1):1–27.
3. Bernichtein S, Touraine P, Goffin V. New concepts in prolactin biology. J Endocrinol. 2010;206(1):1–11.
4. Holtkamp W, Nagel GA, Wander HE, Rauschecker HF, von Heyden D. Hyperprolactinemia is an indicator of progressive disease and poor prognosis in advanced breast cancer. Int J Cancer. 1984;34(3):323–8.
5. Bhatavdekar JM, Shah NG, Balar DB, Patel DD, Bhaduri A, Trivedi SN, et al. Plasma prolactin as an indicator of disease progression in advanced breast cancer. Cancer. 1990;65(9):2028–32.
6. Mujagic Z, Mujagic H. Importance of serum prolactin determination in metastatic breast cancer patients. Croat Med J. 2004;45(2):176–80.
7. Oakes SR, Robertson FG, Kench JG, Gardiner-Garden M, Wand MP, Green JE, et al. Loss of mammary epithelial prolactin receptor delays tumor formation by reducing cell proliferation in low-grade preinvasive lesions. Oncogene. 2007;26(4):543–53.
8. Maus MV, Reilly SC, Clevenger CV. Prolactin as a chemoattractant for human breast carcinoma. Endocrinology. 1999;140(11):5447–50.
9. Liby K, Neltner B, Mohamet L, Menchen L, Ben-Jonathan N. Prolactin overexpression by MDA-MB-435 human breast cancer cells accelerates tumor growth. Breast Cancer Res Treat. 2003;79(2):241–52.
10. Rider L, Shatrova A, Feener EP, Webb L, Diakonova M. JAK2 tyrosine kinase phosphorylates PAK1 and regulates PAK1 activity and functions. J Biol Chem. 2007;282(42):30985–96.
11. Molli PR, Li DQ, Murray BW, Rayala SK, Kumar R. PAK signaling in oncogenesis. Oncogene. 2009;28(28):2545–55.
12. Kumar R, Gururaj AE, Barnes CJ. p21-activated kinases in cancer. Nat Rev Cancer. 2006;6(6):459–71.
13. Bekri S, Adelaide J, Merscher S, Grosgeorge J, Caroli-Bosc F, Perucca-Lostanlen D, et al. Detailed map of a region commonly amplified at 11q13→q14 in human breast carcinoma. Cytogenet Cell Genet. 1997;79(1–2):125–31.
14. Ong CC, Jubb AM, Haverty PM, Zhou W, Tran V, Truong T, et al. Targeting p21-activated kinase 1 (PAK1) to induce apoptosis of tumor cells. Proc Natl Acad Sci U S A. 2011;108(17):7177–82.
15. Balasenthil S, Sahin AA, Barnes CJ, Wang RA, Pestell RG, Vadlamudi RK, et al. p21-activated kinase-1 signaling mediates cyclin D1 expression in mammary epithelial and cancer cells. J Biol Chem. 2004;279(2):1422–8.
16. Salh B, Marotta A, Wagey R, Sayed M, Pelech S. Dysregulation of phosphatidylinositol 3-kinase and downstream effectors in human breast cancer. Int J Cancer. 2002;98(1):148–54.
17. Vadlamudi RK, Adam L, Wang RA, Mandal M, Nguyen D, Sahin A, et al. Regulatable expression of p21-activated kinase-1 promotes anchorage-independent growth and abnormal organization of mitotic spindles in human epithelial breast cancer cells. J Biol Chem. 2000;275(46):36238–44.
18. Holm C, Rayala S, Jirstrom K, Stal O, Kumar R, Landberg G. Association between Pak1 expression and subcellular localization and tamoxifen resistance in breast cancer patients. J Natl Cancer Inst. 2006;98(10):671–80.
19. Wang RA, Zhang H, Balasenthil S, Medina D, Kumar R. PAK1 hyperactivation is sufficient for mammary gland tumor formation. Oncogene. 2006;25(20):2931–6.
20. Adam L, Vadlamudi R, Mandal M, Chernoff J, Kumar R. Regulation of microfilament reorganization and invasiveness of breast cancer cells by kinase dead p21-activated kinase-1. J Biol Chem. 2000;275(16):12041–50.
21. Rider L, Oladimeji P, Diakonova M. PAK1 regulates breast cancer cell invasion through secretion of matrix metalloproteinases in response to prolactin and three-dimensional collagen IV. Mol Endocrinol. 2013;27(7):1048–64.
22. Hammer A, Rider L, Oladimeji P, Cook L, Li Q, Mattingly RR, et al. Tyrosyl phosphorylated PAK1 regulates breast cancer cell motility in response to prolactin through filamin a. Mol Endocrinol. 2013;27(3):455–65.
23. Hammer A, Diakonova M. Tyrosyl phosphorylated serine-threonine kinase PAK1 is a novel regulator of prolactin-dependent breast cancer cell motility and invasion. Adv Exp Med Biol. 2015;846:97–137.
24. Hammer A, Oladimeji P, De Las Casas LE, Diakonova M. Phosphorylation of tyrosine 285 of PAK1 facilitates betaPIX/GIT1 binding and adhesion turnover. FASEB J. 2015;29(3):943–59.
25. Mitra SK, Hanson DA, Schlaepfer DD. Focal adhesion kinase: in command and control of cell motility. Nat Rev Mol Cell Biol. 2005;6(1):56–68.
26. Parsons JT. Focal adhesion kinase: the first 10 years. J Cell Sci. 2003;116(Pt 8):1409–16.
27. Hanks SK, Ryzhova L, Shin NY, Brabek J. Focal adhesion kinase signaling activities and their implications in the control of cell survival and motility. Front Biosci. 2003;8:d982–96.
28. Schaller MD. FAK and paxillin: regulators of N-cadherin adhesion and inhibitors of cell migration? J Cell Biol. 2004;166(2):157–9.
29. Yano H, Mazaki Y, Kurokawa K, Hanks SK, Matsuda M, Sabe H. Roles played by a subset of integrin signaling molecules in cadherin-based cell-cell adhesion. J Cell Biol. 2004;166(2):283–95.
30. Zheng Y, Xia Y, Hawke D, Halle M, Tremblay ML, Gao X, et al. FAK phosphorylation by ERK primes ras-induced tyrosine dephosphorylation of FAK mediated by PIN1 and PTP-PEST. Mol Cell. 2009;35(1):11–25.
31. Fasco MJ, Amin A, Pentecost BT, Yang Y, Gierthy JF. Phenotypic changes in MCF-7 cells during prolonged exposure to tamoxifen. Mol Cell Endocrinol. 2003;206(1–2):33–47.
32. Schlaepfer DD, Mitra SK, Ilic D. Control of motile and invasive cell phenotypes by focal adhesion kinase. Biochim Biophys Acta. 2004;1692(2–3):77–102.
33. Payne DM, Rossomando AJ, Martino P, Erickson AK, Her JH, Shabanowitz J, et al. Identification of the regulatory phosphorylation sites in pp 42/mitogen-activated protein kinase (MAP kinase). EMBO J. 1991;10(4):885–92.
34. Zhang J, Zhang F, Ebert D, Cobb MH, Goldsmith EJ. Activity of the MAP kinase ERK2 is controlled by a flexible surface loop. Structure. 1995;3(3):299–307.
35. Nayal A, Webb DJ, Brown CM, Schaefer EM, Vicente-Manzanares M, Horwitz AR. Paxillin phosphorylation at Ser273 localizes a GIT1-PIX-PAK complex and regulates adhesion and protrusion dynamics. J Cell Biol. 2006;173(4):587–9.
36. Parrini MC. Untangling the complexity of PAK1 dynamics: the future challenge. Cell Logist. 2012;2(2):78–83.
37. Yu DH, Qu CK, Henegariu O, Lu X, Feng GS. Protein-tyrosine phosphatase Shp-2 regulates cell spreading, migration, and focal adhesion. J Biol Chem. 1998;273(33):21125–31.
38. Rigacci S, Rovida E, Dello Sbarba P, Berti A. Low Mr phosphotyrosine protein phosphatase associates and dephosphorylates p125 focal adhesion kinase, interfering with cell motility and spreading. J Biol Chem. 2002;277(44):41631–6.
39. Sundberg-Smith LJ, Doherty JT, Mack CP, Taylor JM. Adhesion stimulates direct PAK1/ERK2 association and leads to ERK-dependent PAK1 Thr212 phosphorylation. J Biol Chem. 2005;280(3):2055–64.
40. Chaudhary A, King WG, Mattaliano MD, Frost JA, Diaz B, Morrison DK, et al. Phosphatidylinositol 3-kinase regulates Raf1 through Pak phosphorylation of serine 338. Curr Biol. 2000;10(9):551–4.
41. Zang M, Hayne C, Luo Z. Interaction between active Pak1 and Raf-1 is necessary for phosphorylation and activation of Raf-1. J Biol Chem. 2002;277(6):4395–405.
42. Frost JA, Steen H, Shapiro P, Lewis T, Ahn N, Shaw PE, et al. Cross-cascade activation of ERKs and ternary complex factors by Rho family proteins. Embo J. 1997;16(21):6426–38.
43. Coles LC, Shaw PE. PAK1 primes MEK1 for phosphorylation by Raf-1 kinase during cross-cascade activation of the ERK pathway. Oncogene. 2002;21(14):2236–44.
44. Park ER, Eblen ST, Catling AD. MEK1 activation by PAK: a novel mechanism. Cell Signal. 2007;19(7):1488–96.
45. Slack-Davis JK, Eblen ST, Zecevic M, Boerner SA, Tarcsafalvi A, Diaz HB, et al. PAK1 phosphorylation of MEK1 regulates fibronectin-stimulated MAPK activation. J Cell Biol. 2003;162(2):281–91.

46. Tachibana K, Sato T, D'Avirro N, Morimoto C. Direct association of pp125FAK with paxillin, the focal adhesion-targeting mechanism of pp125FAK. J Exp Med. 1995;182(4):1089–99.

47. Chen HC, Appeddu PA, Parsons JT, Hildebrand JD, Schaller MD, Guan JL. Interaction of focal adhesion kinase with cytoskeletal protein talin. J Biol Chem. 1995;270(28):16995–9.

48. Petit V, Boyer B, Lentz D, Turner CE, Thiery JP, Valles AM. Phosphorylation of tyrosine residues 31 and 118 on paxillin regulates cell migration through an association with CRK in NBT-II cells. J Cell Biol. 2000;148(5):957–70.

49. Webb DJ, Donais K, Whitmore LA, Thomas SM, Turner CE, Parsons JT, et al. FAK-Src signalling through paxillin, ERK and MLCK regulates adhesion disassembly. Nat Cell Biol. 2004;6(2):154–61.

50. Zaidel-Bar R, Milo R, Kam Z, Geiger B. A paxillin tyrosine phosphorylation switch regulates the assembly and form of cell-matrix adhesions. J Cell Sci. 2007;120(Pt 1):137–48.

51. Angers-Loustau A, Cote JF, Charest A, Dowbenko D, Spencer S, Lasky LA, et al. Protein tyrosine phosphatase-PEST regulates focal adhesion disassembly, migration, and cytokinesis in fibroblasts. J Cell Biol. 1999;144(5):1019–31.

52. Hunger-Glaser I, Fan RS, Perez-Salazar E, Rozengurt E. PDGF and FGF induce focal adhesion kinase (FAK) phosphorylation at Ser-910: dissociation from Tyr-397 phosphorylation and requirement for ERK activation. J Cell Physiol. 2004;200(2):213–22.

53. Katz BZ, Romer L, Miyamoto S, Volberg T, Matsumoto K, Cukierman E, et al. Targeting membrane-localized focal adhesion kinase to focal adhesions: roles of tyrosine phosphorylation and SRC family kinases. J Biol Chem. 2003;278(31):29115–20.

14

Specific localization of nesprin-1-α2, the short isoform of nesprin-1 with a KASH domain, in developing, fetal and regenerating muscle, using a new monoclonal antibody

Ian Holt[1,2]*, Nguyen Thuy Duong[1,3], Qiuping Zhang[4], Le Thanh Lam[1], Caroline A. Sewry[1,5], Kamel Mamchaoui[6], Catherine M. Shanahan[4] and Glenn E. Morris[1,2]

Abstract

Background: Nesprin-1-giant (1008kD) is a protein of the outer nuclear membrane that links nuclei to the actin cytoskeleton via amino-terminal calponin homology domains. The short nesprin-1 isoform, nesprin-1-α2, is present only in skeletal and cardiac muscle and several pathogenic mutations occur within it, but the functions of this short isoform without calponin homology domains are unclear. The aim of this study was to determine mRNA levels and protein localization of nesprin-1-α2 at different stages of muscle development in order to shed light on its functions.

Results: mRNA levels of all known nesprin-1 isoforms with a KASH domain were determined by quantitative PCR. The mRNA for the 111 kD muscle-specific short isoform, nesprin-1-α2, was not detected in pre-differentiation human myoblasts but was present at significant levels in multinucleate myotubes. We developed a monoclonal antibody against the unique amino-terminal sequence of nesprin-1-α2, enabling specific immunolocalization for the first time. Nesprin-1-α2 protein was undetectable in pre-differentiation myoblasts but appeared at the nuclear rim in post-mitotic, multinucleate myotubes and reached its highest levels in fetal muscle. In muscle from a Duchenne muscular dystrophy biopsy, nesprin-1-α2 protein was detected mainly in regenerating fibres expressing neonatal myosin. Nesprin-1-giant was present at all developmental stages, but was also highest in fetal and regenerating fibres. In fetal muscle, both isoforms were present in the cytoplasm, as well as at the nuclear rim. A pathogenic early stop codon (E7854X) in nesprin-1 caused reduced mRNA levels and loss of protein levels of both nesprin-1-giant and (unexpectedly) nesprin-1-α2, but did not affect myogenesis in vitro.

Conclusions: Nesprin-1-α2 mRNA and protein expression is switched on during myogenesis, alongside other known markers of muscle differentiation. The results show that nesprin-1-α2 is dynamically controlled and may be involved in some process occurring during early myofibre formation, such as re-positioning of nuclei.

Keywords: SYNE1, Nuclear membrane, Monoclonal antibody, Cardiomyopathy, Emery-Dreifuss muscular dystrophy, Lamin A/C

* Correspondence: ian.holt@rjah.nhs.uk
[1]Wolfson Centre for Inherited Neuromuscular Disease, RJAH Orthopaedic Hospital, Oswestry, SY10 7AG, UK
[2]Institute for Science and Technology in Medicine, Keele University, Keele ST5 5BG, UK
Full list of author information is available at the end of the article

Background

Nuclear envelope spectrin-repeat proteins (nesprins) form a structural link between the nuclear envelope and cytoskeletal filaments. The *SYNE1* gene, which encodes nesprin-1, was identified by yeast two-hybrid screening of fetal mouse post-synaptic membrane cDNA [1] and by differential screening of a rat vascular smooth muscle cell cDNA [2]. These early studies also identified a related gene, *SYNE2*, which encodes nesprin-2 [1, 2]. The protein products originally identified were, in fact, shorter C-terminal isoforms of larger proteins, nesprin-1-giant (1008kD) and nesprin-2-giant (792kD) [2–4]. The *SYNE1* gene on human chromosome 6q25 is also known as *MYNE1* [5] or *Enaptin* [6] and the *SYNE2* gene on human chromosome 14q23 is also known as *NUANCE* [3]. Both giant nesprins have calponin homology (CH) domains at their N-terminals that bind the actin cytoskeleton, and transmembrane Klarsicht-ANC-Syne-homology (KASH) domains at their C-terminals that bind to inner nuclear membrane SUN proteins in the luminal gap between inner and outer nuclear membranes [7, 8]. Additionally, on the nucleoplasmic side of the inner nuclear membrane, SUN proteins interact with A-type lamin components of the nuclear lamina. These linker of nucleoskeleton and cytoskeleton (LINC) complexes form a physical connection joining the cytoskeleton and the nucleus [9, 10]. Several human mutations in the C-terminal region of nesprin-1 are associated with Emery-Dreifuss muscular dystrophy (EDMD) and dilated cardiomyopathy [11–14]. Mutations in nesprin-interaction partners, emerin (*EMD*, [15] and lamin A/C (*LMNA*, [16]) account for about 50 % of EDMD cases (reviewed [17]).

SYNE1 and *SYNE2* have multiple internal promoters which may give rise to shorter nesprin isoforms which are truncated at the N-terminus but have a common C-terminal region. Three additional members of the nesprin family (nesprin-3, nesprin-4 and KASH5) are similar to the shorter nesprin isoforms in that they lack N-terminal CH domains. Nesprin-3 (112kD) contains a plectin-binding domain at the N-terminal which interacts with intermediate filaments [18], nesprin-4 (44kD) interacts with microtubules via kif5b [19] and KASH5 (63kD) links to chromosomes via dynein [20, 21]. These functional products of 3 separate, but related, genes suggest possible related functions for the similar short products of the *SYNE1/2* genes.

We recently showed by qPCR that mRNAs for nesprin short forms were present at only very low levels in most of the 20 human tissues studied, but were significantly expressed in specific cell types or stages of development [22]. Thus, nesprin-1-α2 was found in skeletal and cardiac muscle only and nesprin-2-epsilon-1 was associated with cells at very early stages of development, while nesprin-2-epsilon-2 was expressed in cardiac, but not skeletal, muscle [22].

In the present study, we extended our qPCR studies of adult human tissues to different stages of skeletal muscle development and, finding that the mRNA for nesprin-1-α2 appeared only after myogenic differentiation, we developed a new monoclonal antibody specific for the unique N-terminal amino-acid sequence of nesprin-1-α2 for immunolocalization studies.

Results

We have studied four stages of human muscle development: pre-differentiation, dividing myoblast cultures, differentiated myotube cultures expressing muscle-specific proteins, fetal muscle and adult muscle. We have also studied dystrophic muscle containing a significant proportion of immature fibres and muscle cultures with pathogenic mutations in nesprin-1 and lamin A/C.

Nesprin-1-α2 mRNA appears during myogenesis in vitro and nesprin-1-giant mRNA increases

mRNA levels in control human cells were determined by quantitative PCR (qPCR: Table 1) with appropriate standards and controls (see Methods). As expected, mRNA for the muscle differentiation marker, M-creatine kinase (CKM), was absent from dividing myoblasts, but increased markedly during muscle development (Fig. 1b). mRNA for the muscle cell marker protein, desmin, also increased; it was already present in the committed, though still dividing myogenic cells, as expected (Fig. 1b). Like CKM, nesprin-1-α2 mRNA was absent from myoblasts, but present in myotubes (Fig. 1a). Nesprin-1-giant mRNA was present in myoblasts, but also higher in myotubes (Fig. 1a). Other short isoforms of nesprin-1 did not make a major contribution to total nesprin mRNA content of cultured cells (Table 1). For these control myogenesis studies in vitro, five different transformed myoblast lines were used (Table 1), derived from control subjects of different ages and each individual cell line showed changes comparable to the mean values shown in Fig. 1a.

Effects of pathogenic mutations in nesprin-1 and lamin A/C on mRNA levels of nesprin-1 isoforms during myogenesis

Myoblast cell lines were also derived from a muscular dystrophy patient with an early termination mutation in the nesprin-1 gene [23], located just before the start of the short nesprin-1-α2 isoform. One would expect all nesprin-1 mRNAs to be synthesised normally, though nesprin-1-giant mRNA might be less stable than normal, because of "nonsense-mediated decay". In fact, consistent with reduced mRNA stability, the level of nesprin-1-giant mRNA was reduced by 78 % in both myoblasts and myotubes compared to the means of the 5 control

Table 1 Relative expression of mRNA of nesprin-1 isoforms and markers of differentiation in cultured human myoblasts and myotubes from control donors

	Control 5 days		Control 25 years		Control 41 years		Control 53 years		Control 79 years	
	Myoblast	Myotube	Myoblast	Myotube	Myoblast	Myotube	Myoblast	Myotube	Myoblast	Myotube
N1-Giant	61.1 ± 11.8 (5)	130.0 ± 12.8 (6)	88.3 ± 12.3 (6)	206.3 ± 17.3 (5)	47.7 ± 4.6 (4)	113.7 ± 19.3 (4)	47.4 ± 6.7 (5)	172.4 ± 22.1 (4)	60.5 ± 10.1 (4)	212.4 ± 12.1 (5)
	$P < 0.001$		$P < 0.001$		$P < 0.001$		$P < 0.001$		$P < 0.001$	
N1-β1	<1 (1)	<1 (1)	<1 (2)	<1 (2)	<1 (1)	<1 (1)	<1 (1)	<1 (1)	1.1 (1)	1.2 (1)
N1-β2	<1 (1)	1.2 (1)	<1 (3)	2.1 ± 0.8 (3)	<1 (1)	2.7 (1)	<1 (1)	<1 (1)	<1 (1)	1.2 (1)
N1-α1	nd (1)	nd (1)	nd (1)	nd (1)	nd (1)	nd (1)	nd (1)	nd (1)	nd (1)	nd (1)
N1-α2	<1 (5)	71.0 ± 9.5 (4)	<1 (3)	47.5 ± 9.3 (4)	<1 (5)	22.0 ± 1.3 (5)	<1 (5)	44.7 ± 9.1 (4)	<1 (3)	47.4 ± 4.3 (5)
	$P < 0.001$		$P < 0.001$		$P < 0.001$		$P < 0.001$		$P < 0.001$	
CKM	<1 (3)	1158 ± 116 (3)	<1 (3)	2847 ± 285 (3)	<1 (3)	740 ± 186 (3)	<1 (3)	1884 ± 104 (3)	2.1 ± 1.7 (3)	1362 ± 68 (3)
	$P < 0.001$		$P < 0.001$		$P < 0.005$		$P < 0.001$		$P < 0.001$	
Desmin	2773 ± 174 (3)	16434 ± 157 (3)	6931 ± 471 (3)	16966 ± 710 (3)	3784 ± 459 (3)	7103 ± 692 (3)	6532 ± 605 (3)	14799 ± 798 (3)	2484 ± 478 (3)	21973 ± 1803 (3)
	$P < 0.001$		$P < 0.001$		$P < 0.005$		$P < 0.001$		$P < 0.001$	

Results expressed as: Mean ± SD (n), "n" = number of repeat measurements. nd = Not detected

Fig. 1 Expression of nesprin-1 isoforms and muscle markers in cultured control myoblasts and myotubes. Quantitative PCR to show mRNA expression relative to the expression of two endogenous controls. **a** nesprin-1-giant and nesprin-1-α2 and **b** desmin (intermediate filament) and muscle creatine kinase (differentiation marker). Bar charts represent the mean relative expression ± SEM of 5 control cell lines, with values for the individual cell lines shown in Table 1

LMNA gene, which encodes lamin A/C. Among the 3 myoblast lines from patients with 3 different mutations in the *LMNA* gene [24, 25], there was no consistent effect on the mRNA levels for nesprin-1 shown by all 3 *LMNA* mutant cell lines (Table 2). Nesprin-1-giant mRNA levels were very high in one *LMNA* mutant (L380S), but it is not clear, without the availability of additional similar cell lines, whether this is attributable to the specific mutation.

New monoclonal antibodies against nesprin-1 isoforms

Having established the changes in mRNA levels, it was necessary to determine whether protein levels changed in the same way. Since antibodies specific for short isoforms were not available, we produced new monoclonal antibodies and these are shown in Table 3. Human nesprin-1-α2 has a unique 31aa sequence at its amino-terminus and a synthetic peptide corresponding to amino acids 2-17 was linked to KLH via its C-terminal cysteine. This peptide conjugate was used to produce a monoclonal antibody (mAb: N1alpha2-1H2) specific for this short isoform, as described in "Experimental Procedures". Another 16 amino acid peptide within exon 130 was used similarly to produce a second mAb, N1G-Ex130, which does not recognise nesprin-1-α2, but it does recognise nesprin-1-giant and would recognise beta isoforms. Beta isoforms are only expressed at low levels in skeletal muscle relative to the giant form (Table 3 and [22]), but to exclude them we also produced mAbs corresponding to exons 81-86 of the nesprin-1 gene (Table 3). These mAbs will recognise nesprin-1-giant but not the beta or alpha isoforms. The epitope locations of the new nesprin-1 mAbs, along with the MANNES1A and MANNES1E C-terminal mAbs (Randles et al., 2010 [26]), are illustrated in Fig. 2. All of the exon 81-86 mAbs recognised a band of nesprin-1-giant on western blots of a human skin fibroblast extract (Fig. 3a), the same band as recognised by a previously-established mAb, MANNES1E [26]. On western blot of human skeletal muscle extract, mAb N1G-Ex130, like MANNES1E, gave a band of nesprin-1-giant (Fig. 3b). For verification of the size of the nesprin-1-giant band, human skeletal muscle extract was run on a gradient gel along with a molecular weight marker with an upper band of 250 kDa (Fig. 3c). As an additional size marker, mAb MANDRA1 was used to show a band of dystrophin protein at 427 kDa. Nesprin-1 mAbs recognise bands that are larger than 427 kDa and are consistent with nesprin-1-giant at around 1008 kDa (Fig. 3c).

The mAb N1G-Ex130 and the N1G-Exon 81-86 mAbs against the giant isoform recognise nesprin-1-giant, but would not be expected to recognise the short α2 isoform, on a western blot of adult human muscle, whereas MAN-NES1E does recognise a band of the size expected for

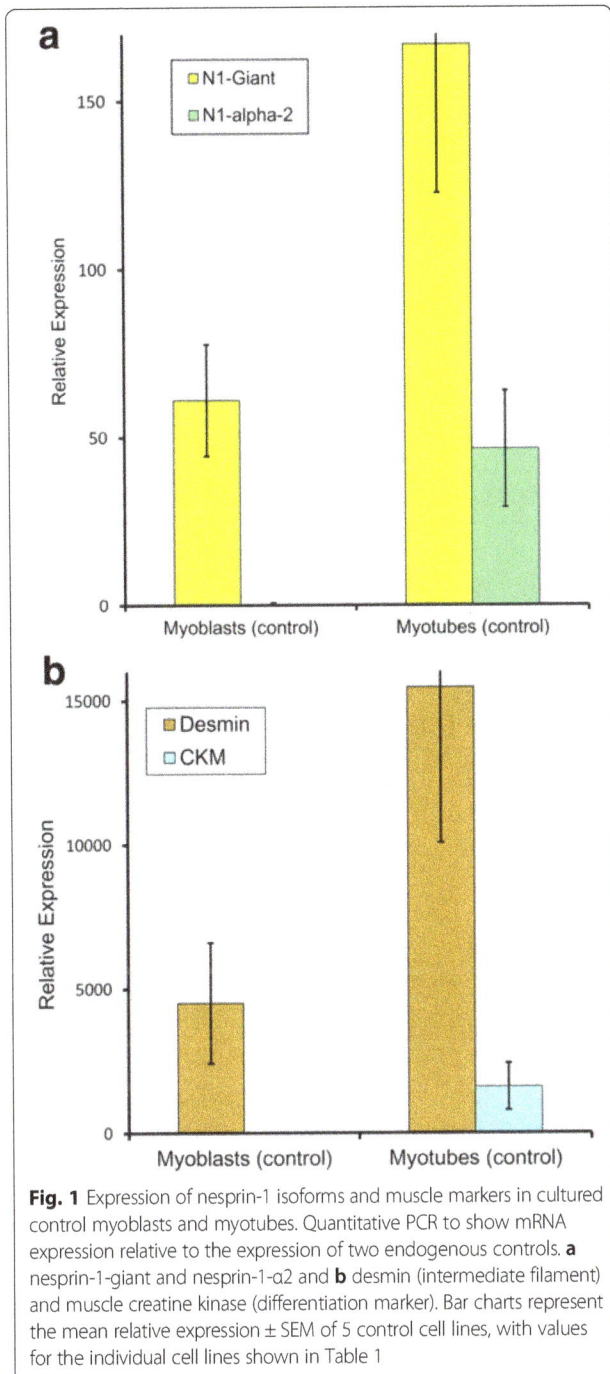

cell lines and was lower than the corresponding values in each of the 5 controls individually (Table 2). Surprisingly, the mRNA for nesprin-1-α2 was also reduced by 85 % in the mutant myotubes, even though the mutation lies 5' to its first exon. Levels of muscle marker mRNAs for CKM and desmin were unaffected by this mutation. Nesprins at the outer nuclear membrane are linked to lamin A/C in the nucleus via SUN proteins in the lumen of the nuclear membrane [10]. Therefore, we also studied myoblast cell lines with mutations in the

Table 2 Relative expression of mRNA of nesprin-1 isoforms and markers of differentiation in cultured human myoblasts and myotubes from patients with pathogenic mutations in Syne1 or LMNA

	SYNE-1 E7854X		LMNA L380S		LMNA del.K32		LMNA R249W	
	Myoblast	Myotube	Myoblast	Myotube	Myoblast	Myotube	Myoblast	Myotube
N1-Giant	13.9 ± 2.2 (6)	33.5 ± 5.0 (5)	102.1 ± 8.8 (5)	429.5 ± 42.4 (5)	48.3 ± 9.4 (4)	161.7 ± 33.6 (5)	53.4 ± 10.0 (4)	90.7 ± 19.0 (5)
	$P < 0.001$		$P < 0.001$		$P < 0.001$		$P < 0.01$	
N1-β1	<1 (2)	<1 (2)	1.0 (2)	<1 (2)	<1 (1)	<1 (1)	<1 (1)	<1 (1)
N1-β2	<1 (3)	1.8 ± 0.6 (4)	1.2 ± 0.1 (3)	3.0 (2)	<1 (1)	<1 (1)	<1 (1)	<1 (1)
N1-α1	nd (1)	nd (1)	nd (1)	nd (1)	nd (1)	nd (1)	nd (1)	nd (1)
N1-α2	<1 (3)	7.0 ± 4.1 (5)	<1 (4)	46.8 ± 6.2 (4)	<1 (3)	40.1 ± 7.8 (4)	<1 (4)	56.1 ± 11.2 (4)
	$P < 0.01$		$P < 0.001$		$P < 0.001$		$P < 0.001$	
CKM	3 ± 2 (3)	1290 ± 296 (3)	3 ± 1 (4)	1492 ± 206 (3)	<1 (4)	2346 ± 768 (3)	2 ± 1 (4)	1279 ± 320 (4)
	$P < 0.005$		$P < 0.001$		$P < 0.005$		$P < 0.001$	
Desmin	10331 ± 671 (3)	13038 ± 303 (3)	4064 ± 673 (6)	14334 ± 581 (4)	6084 ± 160 (4)	16701 ± 1321 (4)	7952 ± 441 (4)	11358 ± 231 (4)
	$P < 0.005$		$P < 0.001$		$P < 0.001$		$P < 0.001$	

Results expressed as: Mean ± SD (n), "n" = number of repeat measurements. nd = Not detected

nesprin-1-α2 (Fig. 3b and c). We cannot confirm that this band is nesprin-1-α2, because the α2-specific mAb, N1alpha2-1H2, does not work on western blots, but the specificity of the mAb is suggested by the absence of nuclear rim staining both in cells that do not express nesprin-1-α2, such as skin fibroblasts (not shown), and in myotubes with 85 % reduced α2 mRNA levels (*SYNE1* mutant cell line: Fig. 4b). The advantage of using N1G-Ex130 to detect nesprin-1-giant, instead of the exon 81-86mAbs, is that the exon 130 region does not encode any known N-terminal (KASH-less) isoforms of nesprin-1, whereas such isoforms containing the exon 81-86 region have been detected at the mRNA level [27]. There may be very few nesprin-1 sequences that are unique to the giant isoform.

Table 3 Nesprin-1 monoclonal antibodies used in this study

mAb name	Isotype	Epitope	Western blot	IMF
		Unique sequence preceding Exon 131 of N1-Giant		
N1alpha2-1H2	IgG1	VVAEDLSALRMAEDGC (aa 2-17 of unique N1-alpha-2 sequence)	-	++ (myotubes, immature and regenerating fibres)
		Exon 130		
N1G-Ex130	IgG1	SKASEIEYKLGKVNDRC	+++	+++
		Exons 81-84		
N1G-7D9	IgG1	QDKLPGSSA (Exon 82)	+++	+
N1G-8C8	IgG1	-	++	+
N1G-6F7	IgG1	-	+	+
N1G-4C11	IgG2b	EMIDQLQDKLP (Exon 82)	+ (cross reaction)	+ (cross reaction)
		Exons 84-86		
N1G-7C8	IgG1	-	++	++
N1G-5A6	IgG1	-	++	+
N1G-9G5	IgG1	LGLYTILPSELSL (Exon 84)	+++	+
N1G-6C9	IgG1	LKIRDQIQDK (Exon 84-85)	+ (cross reaction)	+
		Exons143-146		
MANNES1A	IgG1	-	++	++++
MANNES1E [25]	IgG1	-	+++	+++

Except for MANNES1A and MANNES1E (Randles et al., 2010 [26]), all other Nesprin-1 mAbs in the table are reported here for the first time
Exon numbering is based on the SYNE1 transcript variant 1 mRNA (accession: NM_182961.3) and is the same as that used by Rajgor et al., 2012 [27]

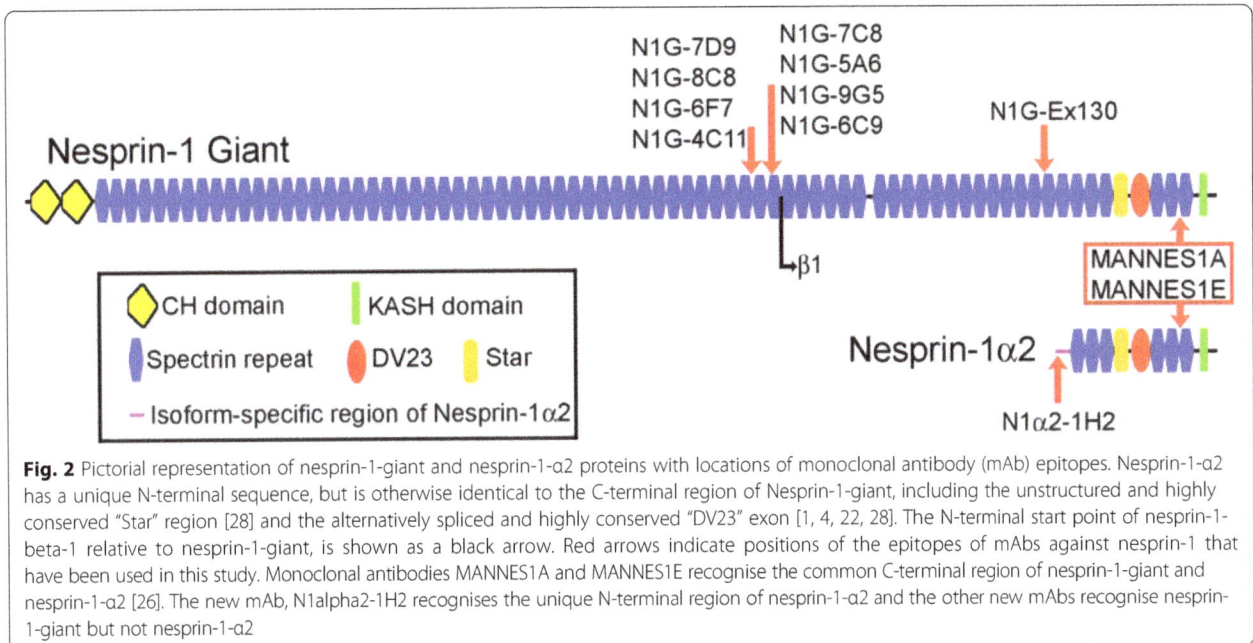

Fig. 2 Pictorial representation of nesprin-1-giant and nesprin-1-α2 proteins with locations of monoclonal antibody (mAb) epitopes. Nesprin-1-α2 has a unique N-terminal sequence, but is otherwise identical to the C-terminal region of Nesprin-1-giant, including the unstructured and highly conserved "Star" region [28] and the alternatively spliced and highly conserved "DV23" exon [1, 4, 22, 28]. The N-terminal start point of nesprin-1-beta-1 relative to nesprin-1-giant, is shown as a black arrow. Red arrows indicate positions of the epitopes of mAbs against nesprin-1 that have been used in this study. Monoclonal antibodies MANNES1A and MANNES1E recognise the common C-terminal region of nesprin-1-giant and nesprin-1-α2 [26]. The new mAb, N1alpha2-1H2 recognises the unique N-terminal region of nesprin-1-α2 and the other new mAbs recognise nesprin-1-giant but not nesprin-1-α2

Nesprin-1-α2 protein is located mainly at the nuclear rim in myotube nuclei and is absent from myoblast nuclei

N1alpha2-1H2 mAb does not stain myoblasts, but does stain the nuclear rim in control myotubes (Fig. 4a), consistent with mRNA data showing the first appearance of nesprin-1-α2 mRNA at the myotube stage (Fig. 1). Pre-incubation of N1alpha2-1H2 with the peptide epitope used for immunization abolished immunostaining of myotubes, whereas control pre-incubation with an unrelated peptide did not (Fig. 5). In contrast to N1alpha2-1H2, MANNES1A (against both giant and alpha isoforms) stains the nuclear rim both in dividing myoblasts and in post-mitotic multinucleate myotubes (Fig. 4c). There is also some cytoplasmic staining in myotubes with N1alpha2-1H2 (Fig. 4a) and to a lesser extent, with MANNES1A (Fig. 4c). Nesprin-1-giant-specific mAbs at both exon 130 (N1G-Ex130, Fig. 4a) and exons 84-86 (N1G-7C8, Fig. 4c) confirm that nesprin-1-giant is present at the nuclear rim throughout myogenesis in vitro.

Myotube cultures of the early termination nesprin-1 mutant were negative for all nesprin-1 mAbs, including the N1alpha2-1H2 mAb (Fig. 4b). This may seem surprising since the stop codon is 5' to the nesprin-1-α2 sequence, but qPCR shows that nesprin-1-α2 mRNA levels are only 15 % of control levels (Tables 1 and 2) and this may result in protein levels below the level of detection. The more surprising observation, therefore, is that the nesprin-1-α2 mRNA is so much reduced. There is the possibility of a truncated form of nesprin-1-giant protein, but qPCR (Tables 1 and 2) shows that this would probably exist at only 20-25 % of normal levels, even

assuming it were as stable as the full-length protein. In fact, neither N1G-Ex130 (Fig. 4b) nor the nesprin-1-specific exon 81-86 mAbs (not shown) detected any nesprin-1-giant in the nesprin-1 mutant myotubes, confirming that truncated mutant protein was present at only very low levels, if at all.

Nesprin-1-α2 protein is weakly expressed in mature muscle fibres, but is strongly expressed, alongside nesprin-1-giant in both fetal muscle and immature (regenerating) fibres of Duchenne muscular dystrophy muscle

Nuclei of human fetal skeletal muscle were clearly labelled with the N1alpha2-1H2 mAb (Fig. 6a) and low level cytoplasmic staining was also seen with this mAb. The large central nuclei are typical of immature muscle fibres, whereas mature muscle nuclei are compressed and peripheral. We were surprised to find that the N1alpha2-1H2 mAb only gave a weak, low level stain in many mature muscle fibres (Fig. 6b), suggesting that the short isoform is either present at low levels and/or the mAb has reduced accessibility to the epitope. Rarely (<0.1 %), normal human adult muscle nuclei were observed that were strongly positive for nesprin-1-α2 (not shown). Nesprin-1-giant was found at the nuclear envelope of both fetal (Fig. 6c) and adult (Fig. 6d) skeletal muscle, with additional low level cytoplasmic stain in fetal muscle. Figure 6 shows representative images of N1alpha2-1H2 and N1G-Ex130 immunofluorescent staining of a total of 3 fetal donors and 4 adult donors. Furthermore, when we studied a muscle section from a Duchenne muscular dystrophy patient, we found that regenerating fibres (identified by staining serial sections

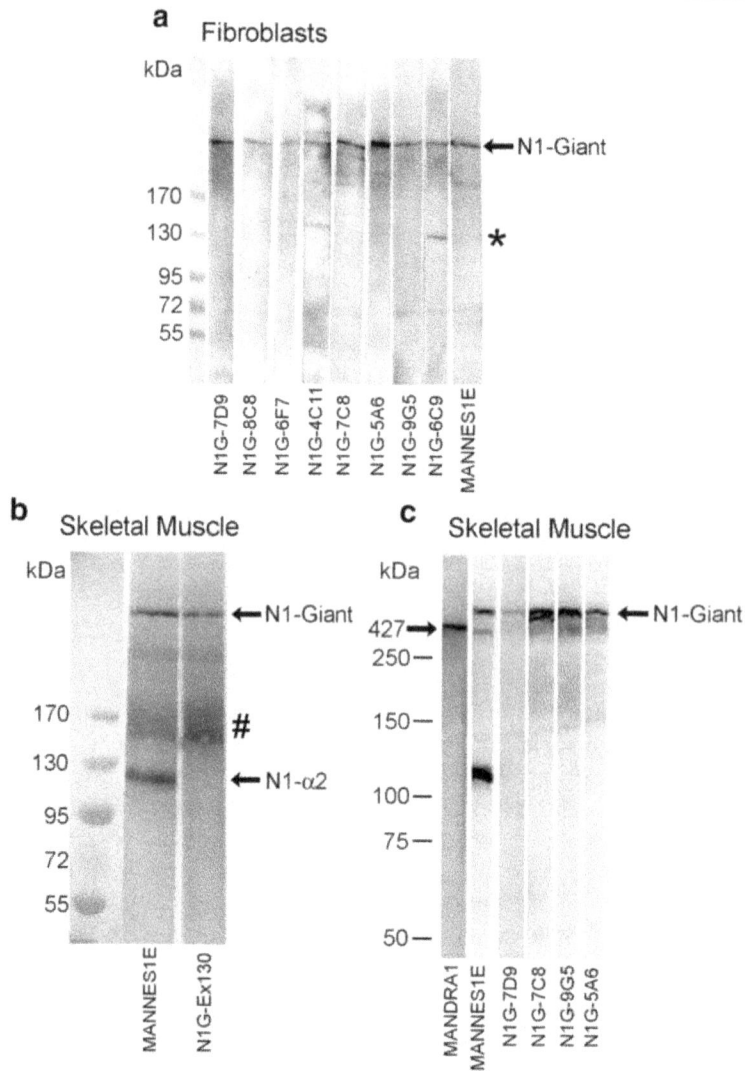

Fig. 3 Western blots with nesprin-1 mAbs. Extracts of **a** cultured human dermal fibroblasts probed with the new mAbs against exons 81 to 86 of nesprin-1-giant and **b** normal adult human skeletal muscle probed with the new mAb N1G-Ex130 against exon 130 of nesprin-1-giant. The molecular weight marker used for blots (**a**) and (**b**) was: EZ-Run Pre-Stained Rec Protein Ladder (Fisher; Cat No: BP3603), upper molecular weight 170 kDa. Blot (**c**) is human skeletal muscle extract alongside Precision Plus Protein Standards (BioRad; Cat No: 161-0374) as molecular weight markers, upper molecular weight 250 kDa. Strips of blot (**c**) were probed with MANDRA1 to show dystrophin as a size marker (427 kDa). Bands with nesprin-1 mAbs were higher than the 427 kDa marker and consistent with nesprin-1-giant (1008 kDa). All blots include the C-terminal mAb against nesprin-1, MANNES1E [26]. * = likely non-specific bands; # = likely nesprin degradation products

with neonatal myosin antibody specific for immature fibres) had high levels of nesprin-1-α2 protein (Fig. 7a). Although mainly at the nuclear rim, N1alpha2-1H2 mAb also showed some staining of the cytoplasm of regenerating fibres, with mature muscle fibres being much weaker (Fig. 7a). Using N1G-Ex130 mAb to locate nesprin-1-giant, we found that this isoform was also higher in regenerating fibres, compared with mature fibres in the same section, and also showed faint cytoplasmic staining (Fig. 7b). Nesprin-1-giant was present at the nuclear rim of all nuclei in the muscle biopsy sections.

Nesprin mRNA levels in fetal and adult human muscle

Both commercial and local sources were used for human fetal and adult skeletal muscle total RNAs, and total cDNAs were prepared by reverse transcription for qPCR quantification. Table 4 shows that nesprin isoform mRNA levels were highly variable between sources, although M-creatine kinase and desmin mRNA levels were significantly higher in adult, compared with fetal, muscle, as expected. Although Fetal 1 and Adult 1 (from the same commercial source) in Table 4 showed changes in nesprin mRNA consistent

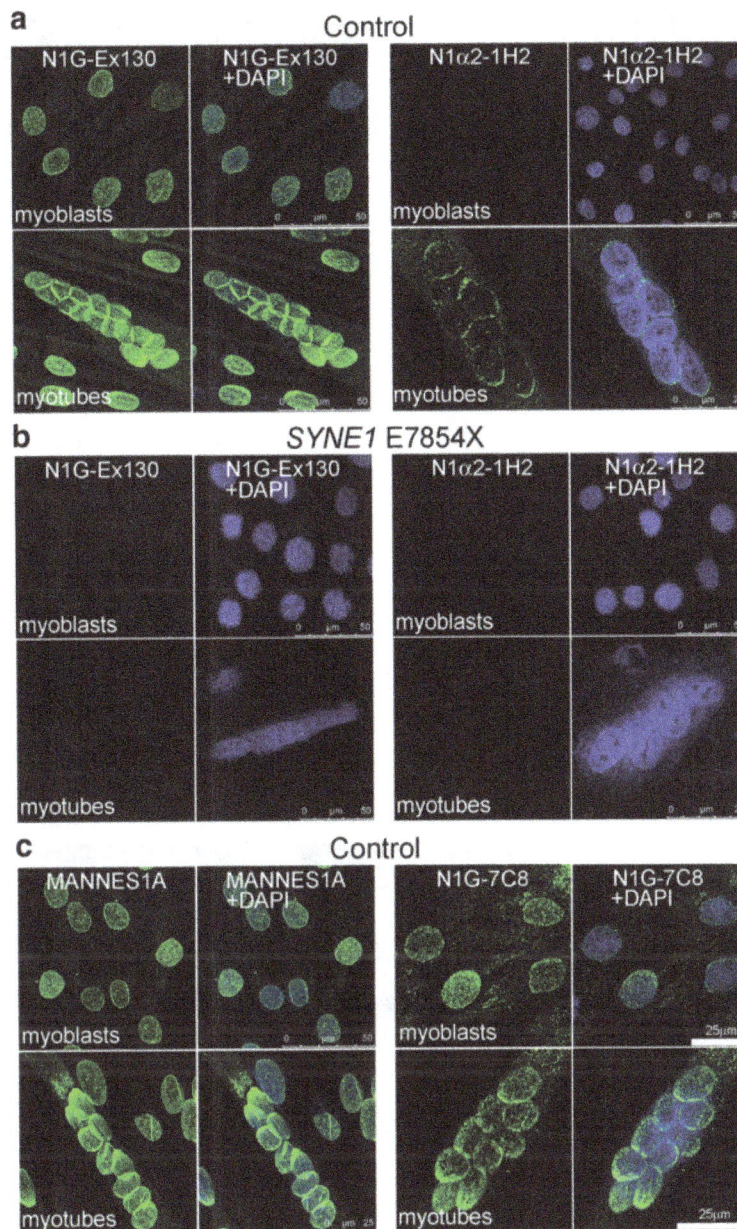

Fig. 4 Immunofluorescent staining for nesprin-1 in cultured myoblasts and myotubes. Individual microscopic fields are shown as paired images. The left image of each pair shows nesprin-1 antibody stain (green) and the right image of each pair shows the merged antibody (green) and DAPI (blue) nuclear stain. **a** mAb N1G-Ex130 stained the nuclei of control myoblasts and myotubes whereas mAb N1alpha2-1H2 stained the nuclear envelope of control myotubes but not myoblasts. **b** The mAbs did not stain nesprin-1 mutant myoblasts or myotubes. **c** Similar to the image seen with N1G-Ex130, mAb MANNES1A and also the mAbs targeted against exons 81-86 (N1G-7C8 shown), stained the nuclear rim of control myoblasts and myotubes

with the protein data (Figs. 4 and 6), they also expressed beta isoform mRNAs, not found in myotubes.

Human muscle tissue used for mRNA isolation will contain non-muscle cells, such as connective tissue, whereas the cultured cell lines are clonal and contain only desmin-positive myogenic cells. It is unclear, therefore, whether the beta isoform mRNAs in fetal and adult muscle are in myofibres (and therefore appear at some post-myotube stage) or in non-muscle components present in variable proportions in different muscle samples used for RNA extraction; this question can only be resolved with antibodies specific for the beta isoforms and these are not yet available. When we expanded our sampling to three fetal and five adult RNA preparations (Table 4), the results for all the nesprin-1 isoforms were much less consistent. It is possible that nesprin-1 mRNA

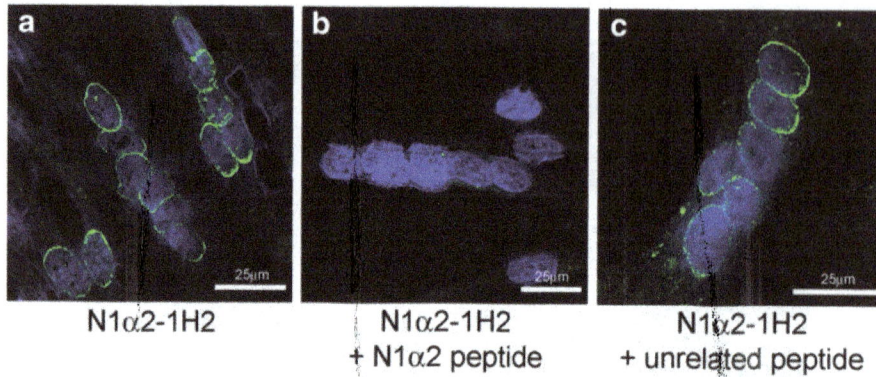

Fig. 5 Peptide competition experiment. Localization of mAb N1alpha2-1H2 to the nuclear envelope of cultured myotubes (**a**) was neutralized when the mAb was blocked by pre-incubation with the immunizing peptide (**b**). Pre-incubation with an unrelated peptide did not prevent normal localization (**c**)

content is variably affected by tissue storage and methods of RNA extraction, since the qPCRs themselves were quite reproducible and gave good dissociation curves. Nesprin-1-α2 mRNA is barely detectable in dividing myoblasts, but accounts for up to one-third of all the nesprin-1 mRNA in adult muscle (Table 4), a proportion consistent with the western blot in Fig. 3b and c.

Discussion

N1alpha2-1H2 is the first anti-nesprin-1 antibody that is specific for a single isoform since it recognises a sequence encoded by the unique first exon of nesprin-1-α2. Antibodies against the N-terminal region of nesprin-1-giant will not recognize any other KASH+ isoforms but a number of shorter N-terminal isoforms are now

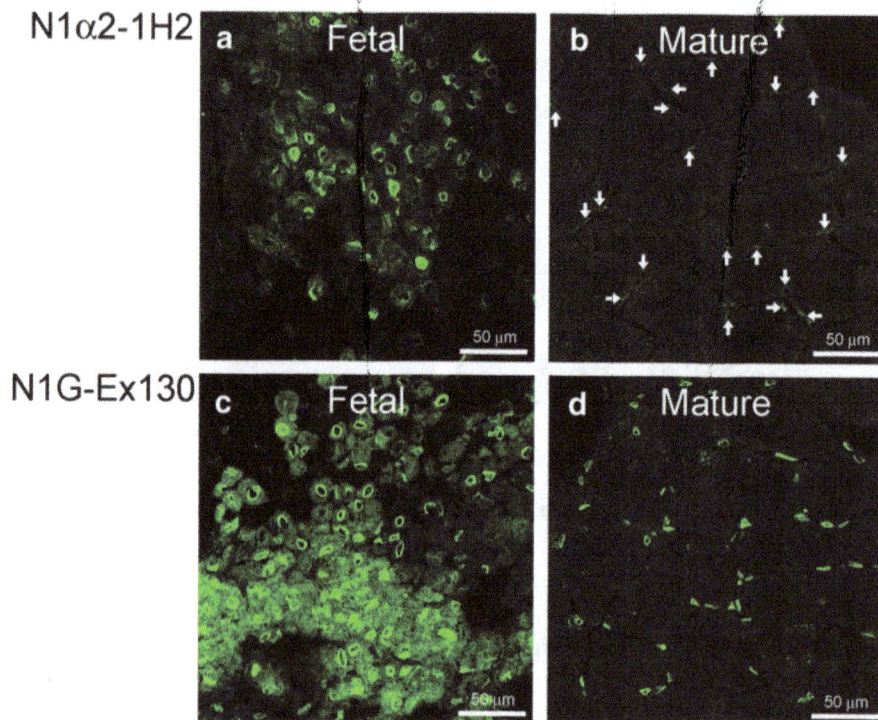

Fig. 6 Immunofluorescent staining of fetal and mature human muscle for nesprin-1-α2 and nesprin-1-giant. **a** mAb N1alpha2-1H2 stained nuclei at the nuclear rim and gave less intense staining of cytoplasm in fetal muscle. **b** Nesprin-1-α2 was weakly stained in nuclei (white arrows) of mature muscle. Nesprin-1-α2 in mature muscle was much less obvious than that seen in fetal muscle. **c** Nesprin-1-giant (mAb N1G-Ex130) was found at the nuclear rim, with low level cytoplasmic stain, in fetal muscle. Nesprin-1-giant was also seen at the nuclear rim in adult skeletal muscle (**d**). Figure shows representative images of total of 3 fetal donors and 4 mature donors

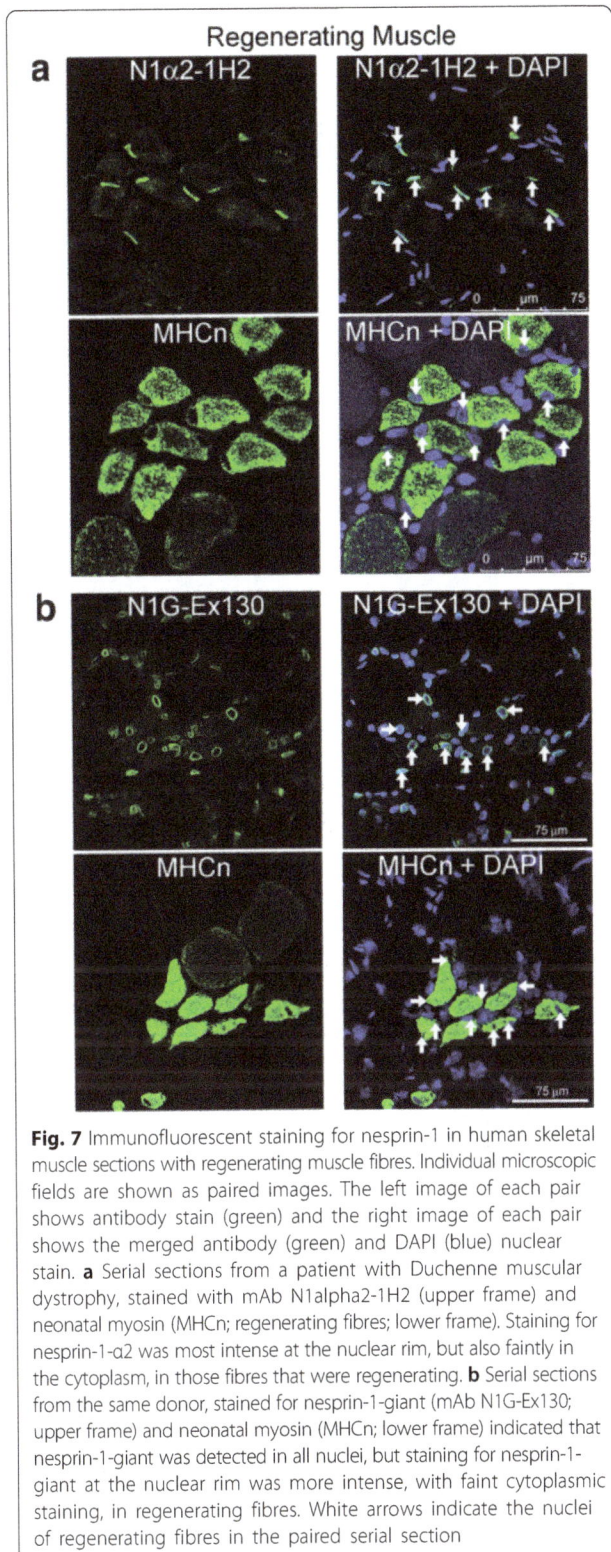

Fig. 7 Immunofluorescent staining for nesprin-1 in human skeletal muscle sections with regenerating muscle fibres. Individual microscopic fields are shown as paired images. The left image of each pair shows antibody stain (green) and the right image of each pair shows the merged antibody (green) and DAPI (blue) nuclear stain. **a** Serial sections from a patient with Duchenne muscular dystrophy, stained with mAb N1alpha2-1H2 (upper frame) and neonatal myosin (MHCn; regenerating fibres; lower frame). Staining for nesprin-1-α2 was most intense at the nuclear rim, but also faintly in the cytoplasm, in those fibres that were regenerating. **b** Serial sections from the same donor, stained for nesprin-1-giant (mAb N1G-Ex130; upper frame) and neonatal myosin (MHCn; lower frame) indicated that nesprin-1-giant was detected in all nuclei, but staining for nesprin-1-giant at the nuclear rim was more intense, with faint cytoplasmic staining, in regenerating fibres. White arrows indicate the nuclei of regenerating fibres in the paired serial section

abundance and tissue distribution has not been precisely established. For this reason, we have extended the repertoire of available antibodies by producing a new mAb, N1G-Ex130, which recognises only giant and beta isoforms of nesprin-1, plus a panel of new mAbs that recognises nesprin-1-giant and NOT beta isoforms (though they would also recognize some N-terminal KASH-less isoforms, if they were present).

In addition to N-terminal isoforms lacking large C-terminal regions, there are also known isoforms that lose only a short KASH sequence by alternative splicing (e.g. the last 99 amino-acids; [32]) and are otherwise similar to the known KASH+ isoforms. There are currently no antibodies that distinguish these KASH-less forms from their KASH+ equivalents. Such KASH-less forms may have a putative nuclear localization signal (over 800 amino-acids from the nesprin-1 C-terminus [2]) and the KASH-less form of nesprin-1-giant in cerebellum has been shown to locate to the cytoplasm [32]. In embryonic stem cells, however, KASH-less forms of the related protein, nesprin-2, located to the nucleoplasm [22].

Using the new mAb tools described here for the first time, we have shown that nesprin-1-α2, the principal short form of nesprin-1 in cardiac and skeletal muscle [1, 22], appears after myoblast differentiation into early myotubes, remains at high levels in immature muscle fibres and becomes very weakly present by immunolocalization in most mature, adult muscle fibres (Figs. 6 & 7). Nesprin-1-giant, the main product of the *SYNE1* gene in most, if not all, tissues, was present at all stages of muscle development (Figs. 4 and 6), though its levels were also somewhat elevated in immature fibres relative to mature fibres (Fig. 7b). We have previously reported a decrease in nesprin-1 and an increase in nesprin-2 during the transition from immature to mature muscle fibres [26].

The clonal myoblast cell lines used in this study are a valuable resource because, unlike primary myoblast cultures, they do not contain variable proportions of non-muscle cells, such as fibroblasts. Immortalization does not seem to have affected their differentiation potential, we did not observe any effect of donor age on nesprin mRNA expression (Table 1) and the mRNA data from qPCR were consistent with the protein data from antibody studies (Fig. 4).

Fetal and adult muscle RNA samples, however, whether obtained commercially as extracted total RNA or locally as tissue stored for RNA extraction, gave very variable results in qPCR (Table 4). It seems likely that variability between samples, both in their "non-muscle" tissue content and in the storage and extraction conditions, might affect the levels of some mRNA species. Indeed, Table 4 suggests that low abundance mRNAs, like those for nesprins, give less consistent qPCR results than abundant species, like creatine kinase and

known [6, 28], including CPG2 (109kD) [29], GSRP-56 (56 kD) [30] and Drop1 [31]. More recently, Rajgor et al, [10, 27] have shown numerous isoforms that cover much of the N-terminal half of the molecule, although their

Table 4 Relative expression of mRNA of nesprin-1 isoforms and markers of differentiation in normal human skeletal muscle

	Fetal Skeletal Muscle			Adult Skeletal Muscle				
	Fetal 1	Fetal 2	Fetal 3	Adult 1 (30y)	Adult 2 (18y)	Adult 3 (67y)	Adult 4 (85y)	Adult 5
N1-Giant	269.0 ± 127.7 (8)	290.3 ± 36.6 (4)	5.3 ± 3.3 (5)	172.4 ± 27 (5)	113.6 ± 16.5 (4)	119.2 ± 30.4 (6)	47.5 ± 9.3 (5)	86.8 ± 25.9 (3)
N1-β1	2.4 ± 1.9 (5)	1 (2)	<1 (2)	21.3 ± 2.9 (4)	<1 (2)	<1 (2)	8.2 ± 3.0 (2)	9.8 ± 2.6 (3)
N1-β2	8.0 ± 1.9 (4)	124.4 ± 43.2 (8)	<1 (3)	4.9 ± 1.7 (4)	<1 (2)	1 (2)	<1 (3)	1 (2)
N1-α1	nd (1)	nd (1)	nd (1)	nd (1)	nd (1)	nd (1)	nd (1)	nd (1)
N1-α2	120.8 ± 33.9 (4)	66.2 ± 18.1 (5)	10.3 ± 2.3 (5)	17.8 ± 5.8 (4)	48.6 ± 12.7 (6)	140.3 ± 25.4 (5)	30.1 ± 5.3 (5)	55.4 ± 18.2 (3)
CKM	4237 ± 95 (4)	1611 ± 295 (4)	3404 ± 971 (5)	33117 ± 2320 (3)	89214 ± 10651 (4)	117928 ± 25917 (4)	20372 ± 1290 (4)	19033 ± 1851 (4)
Desmin	11928 ± 802 (3)	3260 ± 1034 (6)	14542 ± 3431 (5)	38904 ± 1567 (3)	84550 ± 20608 (4)	82022 ± 23108 (5)	37307 ± 617 (5)	27197 ± 4613 (4)

Results expressed as: Mean $\pm$ SD (n). "n" = number of repeat measurements. nd = Not detected desmin mRNAs. However, the mRNA for nesprin-1-α2 was detected by qPCR in adult muscle (Table 4) in amounts, relative to nesprin-1-giant, consistent with the 111kD band in western blot (Fig. 3b and c). In other words, the mRNA was certainly present in adult muscle, although nesprin-1-α2 protein was detected weakly by immunofluorescence microscopy in normal adult muscle. One possibility is that the epitope recognized by our mAb at the N-terminus of nesprin-1-α2 becomes masked in mature fibres in situ; it is not possible to test this hypothesis using different nesprin-1-α2-specific mAbs because the unique N-terminal immunogen sequence for this isoform is very short. This seems the simplest way to reconcile qPCR, western blot and immunofluorescence data at the present time.

A less likely possibility is that not all of the western blot band at around 111kD detected by mAb MANNES1E (Fig. 3b and c) is actually nesprin-1-α2; in a previous study, we showed that some prominent protein bands on western blots of skeletal muscle extracts, previously interpreted as known isoforms, were likely degradation products of the giant isoforms, since the corresponding mRNAs were barely detectable by qPCR [22]. It is also possible, but very unlikely, that larger muscle samples used for RNA extraction contain nesprin-1-α2-positive cells that are not seen in our small biopsy samples for immunofluorescence microscopy.

Missense mutations in the 3'-region of *SYNE1* usually result in autosomal-dominant Emery-Dreifuss muscular dystrophy or inherited cardiomyopathy, with nesprin-1 function impaired, though perhaps not abolished [11–14]. The homozygous *SYNE1* early termination mutation produces a severe, congenital muscular dystrophy phenotype, affecting several tissues [23], probably because expression of KASH+ nesprin-1 is massively reduced. If the expression of nesprin-1-α2 had been unaffected, as predicted from the position of the early stop codon, it might have thrown light on roles for nesprin-1-giant that cannot be replaced by the short isoform, nesprin-1-α2. However, both the mRNA and protein levels of nesprin-1-α2 were as much affected as was nesprin-1-giant by the mutation, although there does not appear to be any simple explanation for the reduced levels of nesprin-1-α2 mRNA. It is interesting that the mutation did not seem to affect in vitro myogenesis, measured by either myotube formation (Fig. 4b) or creatine kinase mRNA expression (Table 2). Nuclear re-positioning to the fibre periphery, however, does not occur until later stages of muscle development.

The detection of cytoplasmic nesprin-1 isoforms (both giant and alpha-2) in fetal muscle fibres, and to a lesser extent in myotubes, in addition to their principal location at the nuclear membrane, is not altogether surprising, since non-nuclear rim localizations in some circumstances have been reported previously [22, 26, 33]. The possibility that these are KASH-less splice variants, unable to attach to the nuclear membrane, cannot be excluded.

Although we have no evidence that nesprin-1-α2 is involved in movement of nuclei during muscle formation, there is circumstantial evidence to suggest that this is one possibility. Nuclear positioning is disrupted in nesprin-1 knockout mice [34, 35] and nesprin-1-giant is known to link the nuclear rim to the actin cytoskeleton [2, 36]. Although nesprin-1-α2 is unable to bind to actin via CH domains, it does share with the giant form, in its highly-conserved STAR domain [28], a LEWD motif that appears to be involved in binding kinesins [37]. Kinesins act as a linker to the microtubule system, which is involved in active movement of nuclei [38], including the movement of myonuclei along the longitudinal axis of the developing myotube [39–41]. This LEWD motif is the only part of the STAR domain shared by nesprin-4, a 44kD KASH-domain protein also known to bind microtubules via kinesins, but with a distribution that may be limited to tissues with secretory epithelial cells [19]. KASH5 lacks a LEWD motif and appears to bind to microtubules via dynein; like nesprin-4, it also exhibits a limited range of expression in spermatocytes, bone marrow and fetal liver [21]. It is possible that KASH proteins in the size range of 44-111kD, such as nesprin-1-α2, carry out similar functions related to nuclear movements, but in specific tissues or at specific stages of development.

Conclusions

Expression of mRNA and protein of nesprin-1-α2, the skeletal and cardiac muscle-specific isoform of nesprin-1, is switched on during myogenesis, mirroring the expression of muscle-creatine kinase. The dynamic profile of nesprin-1-α2 expression suggests that it may be involved in some process occurring during early myofibre formation, such as re-positioning of nuclei. If the weak staining of nesprin-1-α2 in mature fibres is due to epitope masking, we might speculate that nesprin-1-α2 is carrying out one function in immature fibres, when it is accessible to antibody, and a different function in mature fibres, when antibody access is restricted.

Methods
Muscle cell culture

Clonal immortalized myoblast cell lines (Table 5) were from five human control donors without neuromuscular disease (aged 5 days and 25, 41, 53 and 79 years), from a congenital muscular dystrophy patient with a homozygous premature stop mutation in the nesprin-1 gene (Nucleotide: 23560 gaa to taa; Protein: E7854X; [23] and

Table 5 Immortalized human myoblasts

Name	Donor	Donor Muscle
C5d	Newborn (5.5 day), female	Quadriceps
C25yr	25 years, male	Semitendinosus
C41yr	41 years, male	Pectoralis Major
C53yr	53 years, male	Quadriceps
C79yr	79 years, female	Quadriceps
Syne-1 E7854X	16 years. *SYNE1* c.23560 G > T, p.Glu7854* Homozygous	Paravertebral
LMNA L380S	12 years, male. *LMNA* c.1139T > C, p.Leu380Ser Heterozygous	Paravertebral
LMNA del.K32	5 years, female. *LMNA* c.94_69delAAG, p.Lys32del Heterozygous	Gastrocnemius
LMNA R249W	3 years, male. *LMNA* c.745C > T, p.Arg249Trp Heterozygous	Deltoid

Control, Syne-1 mutant [23] and LMNA mutant myoblasts [24, 25], immortalised as described [42]

from three congenital muscular dystrophy patients with mutations in the lamin A/C gene (Protein: L380S, del.K32 and R249W; [24, 25]. They were immortalized by transduction with human telomerase reverse transcriptase (hTERT) and cyclin-dependent kinase-4 (Cdk4) containing retroviral vectors, at the Institut de Myologie, Paris, as described previously [42]. They were cultured in skeletal muscle cell growth medium (Cat No: C-23060; PromoCell GmbH, Heidelberg, Germany) containing supplement mix (Cat No: C-39365; PromoCell) with 20 % Fetal Bovine Serum (Cat No: 10270; Gibco; ThermoFisher Scientific, Paisley, UK). Differentiation was induced at 80 % confluency by washing the adherent myoblasts in medium lacking serum and then culturing in DMEM (Cat No: 31966-021; Gibco; ThermoFisher Scientific) supplemented with Insulin (1721nM), Transferrin (68.7nM), Selenium (38.7nM) (ITS-X; Cat No: 51500-056; Gibco; ThermoFisher Scientific) and Penicillin-Streptomycin (Cat No: DE17-603E; Lonza; Verviers, Belgium). After a further 4 days of cell culture, over 80 % of the cells had fused into myotubes.

RT-PCR and qPCR

Total RNA was prepared from cultured cells and from skeletal muscle samples using RNeasy Plus Mini Kit (Qiagen) and quantified with a NanoDrop ND-1000 spectrophotometer. Human skeletal muscle total RNA were obtained from the following sources: Fetal 1 (19 week female fetus; cat no: T5595-7587; lot no: L14020669) and Adult 1 (30 year female; cat no: T5595-7379; lot no: L11033012) were both purchased from United States Biological (Swampscott, MA 01907). Fetal 3 (18 week female fetus; cat no: 540181; lot no: 0006260887) and Adult 4 (85 year female; cat no: 540029; lot no: 0006167155) were both purchased from Agilent Technologies (Stockport, Cheshire, SK8 3GR, UK). Adult 5 from the First Choice Human Total RNA Survey Panel (cat no: AM6000; lot no: 1004067; [22]) was purchased from Ambion Inc (Austin, TX 78744). Fetal 2

muscle was obtained from the MRC Centre for Neuro-muscular Disease Biobank, London. Adult 2 muscle (18 year male) and Adult 3 muscle (67 year female) were obtained during routine surgery at RJAH Orthopaedic Hospital, Oswestry. Total RNA (maximum of 2.5μg in a 20μL reaction) was reverse transcribed (SuperScript VILO cDNA Synthesis Kit; Applied Biosystems) and then diluted with sterile water, in order to achieve the cDNA equivalent of 10ng total RNA for each 20μL reaction in the qPCR plate.

Forward primers for the short isoforms of nesprin-1 were designed to recognise unique sequences in the 5' UTR of each isoform [22]. Primer sequences for M-creatine kinase (accession: NM_001824) were F: GCTCGTCCGAAGTAGAACAGGTG and R: GGTTG GAACTCTGGTTGAAACTG (282 bp product size) and for desmin (accession: NM_001927) were F: GC TCAACGTGAAGATGGCCCT and R: CTGCTGCTG TGTGGCCTCACT (223 bp product size). Primer pairs were tested by conventional PCR and products were confirmed by DNA sequencing [22]. Primer pairs were chosen which gave dissociation curves with single peaks.

Relative quantitative PCR was performed with an ABI 7500 Real Time PCR system (Applied Biosystems) using SYBR green detection, as previously described [22]. For each preparation of cDNA, two endogenous controls (Beta-actin and GAPDH) were amplified along with each target sequence. The $2^{-\Delta CT}$ method [43, 44] was used to calculate the quantity of target transcripts relative to the two endogenous reference transcripts. The C_T (cycle threshold) value was plotted against the log cDNA dilution in order to calculate the efficiency of the primer pairs [44].

Hybridoma production
Peptide immunogens
The peptide VVAEDLSALRMAEDGC, with Keyhole Limpet Hemocyanin (KLH) conjugated to its C-terminus

(Davids Biotechnologie GmbH, Regensburg, Germany), was used to immunize CD1 mice for production of mAbs against the unique N-terminal sequence of Nesprin-1-α2. The peptide SKASEIEYKLGKVNDRC, with KLH conjugated to its C-terminus (AltaBioscience, Birmingham UK), was used to immunize Balb/C mice for production of mAbs against a sequence within exon 130 of Nesprin-1-giant. For hybridoma production (described in detail previously: Nguyen thi Man and Morris, 2010 [45]), spleen cells from immunized mice were fused with mouse myeloma cell line Sp2/O and 960 wells were screened by ELISA against unconjugated peptides. ELISA-positives were screened by immunofluorescence microscopy on unfixed human muscle sections and western blotting with human skeletal muscle extracts (for exon 130 mAbs), before cloning by limiting dilution.

Recombinant fusion protein immunogen

Nesprin-1-giant cDNA (corresponding to amino acids 5209-5570) in pGEX vector was used to transform *E. coli* BL21(DE3) which were induced by IPTG to give a GST-fusion protein. After incubation the cells were washed with TNE buffer and sonicated sequentially with TNE, 2M urea, 4M urea, 6M urea and 8M urea. Fusion protein in the 6M urea fraction was used in the protocol for immunization of Balb/C mice for production of mAbs [45]. ELISA plates were coated with either the GST-nesprin-1 amino acids 5209-5570 fusion protein used for immunization or with an unrelated GST-fusion protein (GST-MSH3), in order to eliminate those mAbs reacting with the GST fusion tag. For further refinement of the epitope positions, ELISAs were also performed with plates coated with recombinant protein GST-nesprin-1-giant amino acids 5377-5570 (nesprin-1 isoform p23, Accession No: JQ754364), in order to identify those mAbs with epitopes at the C-terminal end of the original immunogen. ELISA-positives were screened by immunofluorescence microscopy and western blotting, before cloning by limiting dilution. Further epitope mapping was performed by testing the mAbs for immunolocalization with COS-7 cells transfected with pCMV/nesprin-1-beta-1 (amino acids 1-97). These mAbs were therefore classed as reacting with nesprin-1-giant exons 81-84 (amino acids 5209-5376) or nesprin-1-giant exons 84-86 (amino acids 5377-5476), which does not include any sequence from nesprin-1-beta-1.

Other mAbs used in this study were: MANNES1A (for immunofluorescence microscopy) and MANNES1E (for western blot), both of which recognise the C-terminal region of nesprin-1 [26].

SDS-polyacrylamide gel electrophoresis and Western blotting

Cell lysis buffer (125 mM Tris pH 6.8, 2 % SDS, 5 % 2-beta mercaptoethanol, 5 % glycerol with protease inhibitors: Sigma P8340 and 1mM PMSF) was used for the extraction of cultured cells and tissue lysis buffer (50mM Tris pH 6.8, 1 % EDTA, 10 % SDS, 5 % beta mercaptoethanol, 10 % glycerol with protease inhibitors) was used for the extraction of tissue samples (250 mg/ml). Bromophenol blue was added to the samples which were then boiled and separated by SDS-PAGE using 4 to 12 % polyacrylamide gels (Ref: NW04125BOX; ThermoFisher Scientific) and then electroblotted onto nitrocellulose membranes (Protan BA85, Whatman). Non-specific sites were blocked with 5 % skimmed milk protein and the membranes then incubated with monoclonal antibody supernatants (diluted: 1/10, except MANNES1E: 1/50 and MANDRA1: 1/100), followed by washing and incubation with secondary antibody (peroxidase-labelled rabbit anti-mouse immunoglobulins; 1/1000, Dako, Denmark). Antibody reacting bands were detected with SuperSignal West Femto chemiluminescent reagent (Cat No: 34094; ThermoFisher Scientific) and visualized with a ChemiDoc Touch imaging system (BioRad Ltd.).

Immunofluorescence microscopy

Immunohistochemistry was performed on cultured cells on coverslips that had been fixed with 50 % acetone, 50 % methanol and then washed with PBS and also performed on unfixed cryostat sections of human muscle. Monoclonal antibodies in culture supernatants were diluted and incubated on specimens for 1 h. N1alpha2-1H2 was diluted 1:1 to 1:3 in PBS and all other nesprin-1 mAbs were diluted 1:3 in PBS. Following incubation with the monoclonal antibody, specimens were washed with PBS and then incubated with goat anti-mouse ALEXA 488 (Molecular Probes, Eugene, Oregon, USA) secondary antibody (diluted to 5μg/ml in PBS containing 1 % horse serum, 1 % fetal bovine serum, 0.1 % BSA). After 50 min, DAPI (diamidino phenylindole at 200 ng/ml) was also added to specimens to counterstain nuclei. After incubation for 10 min with DAPI, specimens were washed with PBS and mounted in Hydromount (Merck). Images were acquired by sequential scanning with a Leica TCS SP5 spectral confocal microscope (Leica Microsystems, Milton Keynes, UK). For peptide competition experiments, mAb N1alpha2-1H2 was pre-incubated with 1mg/ml unconjugated peptide (VVAEDLSALRMAEDGC) or unrelated control peptide (IFSHQQVKKLKETFAFIQQLC) for 1 h at room temperature.

Abbreviations

CH, Calponin homology; CKM, Creatine kinase, muscle; EDMD, Emery-Dreifuss muscular dystrophy; KASH, Klarsicht, ANC-1 and Syne homology; KLH, Keyhole limpet hemocyanin; LINC, Linker of Nucleoskeleton and Cytoskeleton; SUN, Sad-1, UNC-84

Acknowledgments

The Platform for Immortalization of Human Cells, Institute of Myology, Paris, for the immortalization of human myoblasts.

Funding

This study was supported by British Heart Foundation grant (PG/11/71/29091) to IH and GEM; RJAH Orthopaedic Hospital Institute of Orthopaedics (RPG 149) to IH and GEM; The Vietnam National Foundation for Science and Technology Development (NAFOSTED) (106.06-2010.62) to NTD; Muscular Dystrophy Association Research Infrastructure grant (173057) to GEM; Association Française contre les Myopathies to KM.

Authors' contributions

Conceived and designed the experiments: IH, NTD, QZ, CMS, GEM; Hybridoma production: IH, GEM; Quantitative PCR: IH, NTD, LTL; Immunofluorescence microscopy: IH, LTL, CAS; Myoblast immortalization: KM; Analysed the data: IH, NTD, QZ, LTL, CAS, CMS, GEM; Wrote the manuscript: IH, QZ, CMS, GEM. Revised the manuscript: IH, GEM. All authors read and approved the final manuscript.

Competing interests

The authors declare that they have no competing interests.

Author details

[1]Wolfson Centre for Inherited Neuromuscular Disease, RJAH Orthopaedic Hospital, Oswestry, SY10 7AG, UK. [2]Institute for Science and Technology in Medicine, Keele University, Keele ST5 5BG, UK. [3]Institute of Genome Research (IGR), Vietnam Academy of Science and Technology (VAST), Hanoi, Vietnam. [4]Cardiovascular Division, James Black Centre, King's College, London SE5 9NU, UK. [5]Dubowitz Neuromuscular Centre, Institute for Child Health and Great Ormond Street Hospital, London WC1 1EH, UK. [6]Sorbonne Universités, UPMC Univ Paris 06, INSERM UMRS974, CNRS FRE3617, Center for Research in Myology, 47 Boulevard de l'hôpital, 75013 Paris, France.

References

1. Apel ED, Lewis RM, Grady RM, Sanesi JR. Syne-1, a dystrophin- and klarsicht-related protein associated with synaptic nuclei at the neuromuscular junction. J Biol Chem. 2000;275:31986–95.
2. Zhang Q, Skepper JN, Yang F, Davies JD, Hegyi L, Roberts RG, et al. Nesprins: a novel family of spectrin-repeat-containing proteins that localize to the nuclear membrane in multiple tissues. J Cell Sci. 2011;114:4485–98.
3. Zhen YY, Libotte T, Munck M, Noegel AA, Korenbaum E. NUANCE, a giant protein connecting the nucleus and actin cytoskeleton. J Cell Sci. 2002;115: 3207–22.
4. Zhang Q, Ragnauth CD, Skepper JN, Worth NF, Warren DT, Roberts RG, et al. Nesprin-2 is a multi-isomeric protein that binds lamin and emerin at the nuclear envelope and forms a subcellular network in skeletal muscle. J Cell Sci. 2005;118:673–87.
5. Mislow JM, Kim MS, Davis DB, McNally EM. Myne-1, a spectrin repeat transmembrane protein of the myocyte inner nuclear membrane, interacts with lamin A/C. J Cell Sci. 2002;115:61–70.
6. Padmakumar VC, Abraham S, Braune S, Noegel AA, Tunggal B, Karakesisoglou I, Korenbaum E. Enaptin, a giant actin-binding protein, is an element of the nuclear membrane and the actin cytoskeleton. Exp Cell Res. 2004;295:330–9.
7. Sosa BA, Rothballer A, Kutay U, Schwartz TU. LINC complexes form by binding of three KASH peptides to domain interfaces of trimeric SUN proteins. Cell. 2012;149:1035–47.
8. Zhou Z, Du X, Cai Z, Song X, Zhang H, Mizuno T, et al. Structure of Sad1-UNC84 homology (SUN) domain defines features of molecular bridge in nuclear envelope. J Biol Chem. 2012;287:5317–26.
9. Lombardi ML, Lammerding J. Keeping the LINC: the importance of nucleocytoskeletal coupling in intracellular force transmission and cellular function. Biochem Soc Trans. 2011;39:1729–34.
10. Rajgor D, Shanahan CM. Nesprins: from the nuclear envelope and beyond. Expert Rev Mol Med. 2013;15:e5.
11. Zhang Q, Bethmann C, Worth NF, Davies JD, Wasner C, Feuer A, et al. Nesprin-1 and -2 are involved in the pathogenesis of Emery Dreifuss muscular dystrophy and are critical for nuclear envelope integrity. Hum Mol Genet. 2007;16:2816–33.

12. Puckelwartz MJ, Kessler EJ, Kim G, Dewitt MM, Zhang Y, Earley JU, et al. Nesprin-1 mutations in human and murine cardiomyopathy. J Mol Cell Cardiol. 2010;48:600–8.
13. Li C, Zhou B, Rao L, Shanahan CM, Zhang Q. Novel nesprin-1 mutations in human dilated cardiomyopathy. Nuclear envelope disease and chromatin organization meeting, Cambridge, 2011; Abstract P028.
14. Fanin M, Savarese M, Nascimbeni AC, Di Fruscio G, Pastorello E, Tasca E, et al. Dominant muscular dystrophy with a novel SYNE1 gene mutation. Muscle Nerve. 2015;51:145–7.
15. Bione S, Maestrini E, Rivella S, Mancini M, Regis S, Romeo G, Toniolo D. Identification of a novel X-linked gene responsible for Emery-Dreifuss muscular dystrophy. Nat Genet. 1994;8:323–7.
16. Bonne G, Di Barletta MR, Varnous S, Bécane HM, Hammouda EH, Merlini L, et al. Mutations in the gene encoding lamin A/C cause autosomal dominant Emery-Dreifuss muscular dystrophy. Nat Genet. 1999;21:285–8.
17. Meinke P, Nguyen TD, Wehnert MS. The LINC complex and human disease. Biochem Soc Trans. 2011;39:1693–7.
18. Wilhelmsen K, Litjens SH, Kuikman I, Tshimbalanga N, Janssen H, van den Bout I, et al. Nesprin-3, a novel outer nuclear membrane protein, associates with the cytoskeletal linker protein plectin. J Cell Biol. 2005;171:799–810.
19. Roux KJ, Crisp ML, Liu Q, Kim D, Kozlov S, Stewart CL, Burke B. Nesprin 4 is an outer nuclear membrane protein that can induce kinesin-mediated cell polarization. Proc Natl Acad Sci U S A. 2009;106:2194–9.
20. Morimoto A, Shibuya H, Zhu X, Kim J, Ishiguro K, Han M, Watanabe Y. A conserved KASH domain protein associates with telomeres, SUN1, and dynactin during mammalian meiosis. J Cell Biol. 2012;198:165–72.
21. Horn HF, Kim DI, Wright GD, Wong ES, Stewart CL, Burke B, Roux KJ. A mammalian KASH domain protein coupling meiotic chromosomes to the cytoskeleton. J Cell Biol. 2013;202:1023–39.
22. Duong NT, Morris GE, Lam le T, Zhang Q, Sewry CA, Shanahan CM, Holt I. Nesprins: tissue-specific expression of epsilon and other short isoforms. PLoS One. 2014;9:e94380.
23. Voit T, Cirak S, Abraham S, Karakesisoglou I, Parano E, Pavone P, et al. Congenital muscular dystrophy with adducted thumbs, mental retardation, cerebellar hypoplasia and cataracts is caused by mutation of Enaptin (Nesprin-1): The third nuclear envelopathy with muscular dystrophy (Abstract C.O.4). Neuromuscul Disord. 2007;17:833–834.24.
24. Quijano-Roy S, Mbieleu B, Bönnemann CG, Jeannet PY, Colomer J, Clarke NF, et al. De novo LMNA mutations cause a new form of congenital muscular dystrophy. Ann Neurol. 2008;64:177–86.
25. Bertrand AT, Ziaei S, Ehret C, Duchemin H, Mamchaoui K, Bigot A, et al. Cellular microenvironments reveal defective mechanosensing responses and elevated YAP signaling in LMNA-mutated muscle precursors. J Cell Sci. 2014;127:2873–84.
26. Randles KN, Lam LT, Sewry CA, Puckelwartz M, Furling D, Wehnert M, et al. Nesprins, but not sun proteins, switch isoforms at the nuclear envelope during muscle development. Dev Dyn. 2010;239:998–1009.
27. Rajgor D, Mellad JA, Autore F, Zhang Q, Shanahan CM. Multiple novel nesprin-1 and nesprin-2 variants act as versatile tissue-specific intracellular scaffolds. PLoS One. 2012;7:e40098.
28. Simpson JG, Roberts RG. Patterns of evolutionary conservation in the nesprin genes highlight probable functionally important protein domains and isoforms. Biochem Soc Trans. 2008;36:1359–67.
29. Cottrell JR, Borok E, Horvath TL, Nedivi E. CPG2: a brain- and synapse-specific protein that regulates the endocytosis of glutamate receptors. Neuron. 2004;44:677–90.
30. Kobayashi Y, Katanosaka Y, Iwata Y, Matsuoka M, Shigekawa M, Wakabayashi S. Identification and characterization of GSRP-56, a novel Golgi-localized spectrin repeat-containing protein. Exp Cell Res. 2006;312:3152–64.
31. Marmé A, Zimmermann HP, Moldenhauer G, Schorpp-Kistner M, Müller C, Keberlein O, et al. Loss of Drop1 expression already at early tumor stages in a wide range of human carcinomas. Int J Cancer. 2008;123:2048–56.
32. Razafsky D, Hodzic D. A variant of Nesprin1 giant devoid of KASH domain underlies the molecular etiology of autosomal recessive cerebellar ataxia type I. Neurobiol Dis. 2015;78:57–67.
33. Rajgor D, Mellad JA, Soong D, Rattner JB, Fritzler MJ, Shanahan CM. Mammalian microtubule P-body dynamics are mediated by nesprin-1. J Cell Biol. 2014;205:457–75.
34. Zhang X, Xu R, Zhu B, Yang X, Ding X, Duan S, et al. Syne-1 and Syne-2 play crucial roles in myonuclear anchorage and motor neuron innervation. Development. 2007;134:901–8.

35. Zhang J, Felder A, Liu Y, Guo LT, Lange S, Dalton ND, et al. Nesprin 1 is critical for nuclear positioning and anchorage. Hum Mol Genet. 2010;19: 329–41.

36. Zhang Q, Ragnauth C, Greener MJ, Shanahan CM, Roberts RG. The nesprins are giant actin-binding proteins, orthologous to Drosophila melanogaster muscle protein MSP-300. Genomics. 2002;80:473–81.

37. Wilson MH, Holzbaur EL. Nesprins anchor kinesin-1 motors to the nucleus to drive nuclear distribution in muscle cells. Development. 2015;142:218–28.

38. Fridolfsson HN, Starr DA. Kinesin-1 and dynein at the nuclear envelope mediate the bidirectional migrations of nuclei. J Cell Biol. 2010;191:115–28.

39. Metzger T, Gache V, Xu M, Cadot B, Folker ES, Richardson BE, et al. MAP and kinesin-dependent nuclear positioning is required for skeletal muscle function. Nature. 2012;484:120–4.

40. Wilson MH, Holzbaur ELF. Opposing microtubule motors drive robust nuclear dynamics in developing muscle cells. J Cell Sci. 2012;125:4158–69.

41. Cadot B, Gache V, Vasyutina E, Falcone S, Birchmeier C, Gomes ER. Nuclear movement during myotube formation is microtubule and dynein dependent and is regulated by Cdc42, Par6 and Par3. EMBO Rep. 2012;13: 741–9.

42. Mamchaoui K, Trollet C, Bigot A, Negroni E, Chaouch S, Wolff A, et al. Immortalized pathological human myoblasts: towards a universal tool for the study of neuromuscular disorders. Skelet Muscle. 2011;1:34.

43. Livak KJ, Schmittgen TD. Analysis of relative gene expression data using real-time quantitative PCR and the 2(-Delta Delta C(T)) Method. Methods. 2001;25:402–8.

44. Schmittgen TD, Livak KJ. Analyzing real-time PCR data by the comparative C(T) method. Nat Protoc. 2008;3:1101–8.

45. Nguyen TM, Morris GE. A rapid method for generating large numbers of high-affinity monoclonal antibodies from a single mouse. In: Walker JM, editor. The Protein Protocols Handbook. 3rd ed. Totowa NJ: Humana Press; 2010. p. 1961–74.

PERMISSIONS

All chapters in this book were first published in BMCCB, by BioMed Central; hereby published with permission under the Creative Commons Attribution License or equivalent. Every chapter published in this book has been scrutinized by our experts. Their significance has been extensively debated. The topics covered herein carry significant findings which will fuel the growth of the discipline. They may even be implemented as practical applications or may be referred to as a beginning point for another development.

The contributors of this book come from diverse backgrounds, making this book a truly international effort. This book will bring forth new frontiers with its revolutionizing research information and detailed analysis of the nascent developments around the world.

We would like to thank all the contributing authors for lending their expertise to make the book truly unique. They have played a crucial role in the development of this book. Without their invaluable contributions this book wouldn't have been possible. They have made vital efforts to compile up to date information on the varied aspects of this subject to make this book a valuable addition to the collection of many professionals and students.

This book was conceptualized with the vision of imparting up-to-date information and advanced data in this field. To ensure the same, a matchless editorial board was set up. Every individual on the board went through rigorous rounds of assessment to prove their worth. After which they invested a large part of their time researching and compiling the most relevant data for our readers.

The editorial board has been involved in producing this book since its inception. They have spent rigorous hours researching and exploring the diverse topics which have resulted in the successful publishing of this book. They have passed on their knowledge of decades through this book. To expedite this challenging task, the publisher supported the team at every step. A small team of assistant editors was also appointed to further simplify the editing procedure and attain best results for the readers.

Apart from the editorial board, the designing team has also invested a significant amount of their time in understanding the subject and creating the most relevant covers. They scrutinized every image to scout for the most suitable representation of the subject and create an appropriate cover for the book.

The publishing team has been an ardent support to the editorial, designing and production team. Their endless efforts to recruit the best for this project, has resulted in the accomplishment of this book. They are a veteran in the field of academics and their pool of knowledge is as vast as their experience in printing. Their expertise and guidance has proved useful at every step. Their uncompromising quality standards have made this book an exceptional effort. Their encouragement from time to time has been an inspiration for everyone.

The publisher and the editorial board hope that this book will prove to be a valuable piece of knowledge for researchers, students, practitioners and scholars across the globe.

LIST OF CONTRIBUTORS

Clemens Cammann, Luca Simeoni and Luca Simeoni
Institute of Molecular and Clinical Immunology, Otto-von-Guericke-University, Magdeburg, Germany

Alexander Rath and Udo Reichl
Max-Planck-Institute for Dynamics of Complex Technical Systems, Magdeburg, Germany

Holger Lingel and Monika Brunner-Weinzierl
Department of Experimental Pediatrics, Otto-von-Guericke-University, Magdeburg, Germany

Burkhart Schraven
Institute of Molecular and Clinical Immunology, Otto-von-Guericke-University, Magdeburg, Germany
Department of Immune Control, Helmholtz Centre for Infection Research, Braunschweig, Germany

Jonathan A. Lindquist
Institute of Molecular and Clinical Immunology, Otto-von-Guericke-University, Magdeburg, Germany
Department of Nephrology and Hypertension, Diabetes and Endocrinology, Otto-von-Guericke University, Magdeburg, Germany

David Dickerson, Marek Gierliński, Vijender Singh, Etsushi Kitamura, Graeme Ball and Tomoyuki U. Tanaka
Centre for Gene Regulation and Expression, College of Life Sciences, University of Dundee, Dundee DD1 5EH, UK

Tom Owen-Hughes
Centre for Gene Regulation and Expression, College of Life Sciences, University of Dundee, Dundee DD1 5EH, UK
Wellcome Trust Building, University of Dundee, Dow Street, Dundee DD1 5EH, UK

Toni M. Yeasky, Elizabeth R. Smith and Xiang-Xi Xu
Sylvester Comprehensive Cancer Center/University of Miami, Miami, Florida 33136, USA
Department of Cell Biology, University of Miami Miller School of Medicine, Miami, FL 33136, USA

Callinice D. Capo-chichi
Sylvester Comprehensive Cancer Center/University of Miami, Miami, Florida 33136, USA
Department of Cell Biology, University of Miami Miller School of Medicine, Miami, FL 33136, USA
Institute of Biomedical Sciences, Laboratory of Biochemistry and Molecular Biology, University of Abomey-Calavi, Abomey Calavi, Benin

Raquel Núñez-Toldrà, Ester Martínez-Sarrà, Carlos Gil-Recio, Sheyla Montori
Regenerative Medicine Research Institute, Universitat Internacional de Catalunya, Barcelona, Spain
Chair of Regenerative Implantology MIS-UIC, Barcelona, Spain

Maher Atari
Regenerative Medicine Research Institute, Universitat Internacional de Catalunya, Barcelona, Spain
Chair of Regenerative Implantology MIS-UIC, Barcelona, Spain
Surgery and Oral Implantology Department, Universitat Internacional de Catalunya, Barcelona, Spain

Miguel Ángel Carrasco
Area of Pathology, Universitat Internacional de Catalunya, Barcelona, Spain

Ashraf Al Madhoun
Research Division, Dasman Diabetes Institute, Dasman, Kuwait

Yu Meng and Bo Hu
Department of Nephrology, the First Hospital Affiliated to Jinan University, No. 613 Huangpu West Road, Guangzhou 510630, China

Changzhen Shi
Department of Radiology, the First Hospital Affiliated to Jinan University, No. 613 Huangpu West Road, Guangzhou 510630, China

Jian Gong, Xing Zhong, Xueyin Lin and Hao Xu
Department of Nuclear Medicine, the First Hospital Affiliated to Jinan University, No. 613 Huangpu West Road, Guangzhou 510630, China

Xinju Zhang, Jun Liu and Cong Liu
Shenzhen Engineering Laboratory for Genomics-Assisted Animal Breeding, BGI-Shenzhen, Shenzhen 518083, China

Mytre Koul, Ashok Kumar, Jasvinder Singh, Parduman Raj Sharma and Shashank Singh
Cancer Pharmacology Division, CSIR-Indian Institute of Integrative Medicine, Jammu, India
Academy of Scientific & Innovative Research (AcSIR), CSIR, New Delhi, India

Ramesh Deshidi and Bhahwal Ali Shah
Natural Product Chemistry, CSIR-Indian Institute of Integrative Medicine, Jammu, India
Academy of Scientific & Innovative Research (AcSIR), CSIR, New Delhi, India

Vishal Sharma and Sundeep Jaglan
Microbial Biotechnology Division, CSIR-Indian Institute of Integrative Medicine, Jammu, India
Academy of Scientific & Innovative Research (AcSIR), CSIR, New Delhi, India

Rachna D. Singh
Department of Conservative Dentistry & Endodontics, Indira Gandhi Govt. Dental College and Hospital, Jammu, India

Wildriss Viranaicken
UMR PIMIT, Processus Infectieux en Milieu Insulaire Tropical, Université de la Réunion, 97490 Sainte Clotilde, La Réunion, France

Rémy Beaujois, Elizabeth Ottoni, Xin Zhang, Christina Gagnon, Sami HSine, Stéphanie Mollet and Luc DesGroseillers
Département de biochimie et médecine moléculaire, Faculté de médecine, Université de Montréal, 2900 Edouard Montpetit, Montréal, QC H3T 1J4, Canada

Midori Mukai, Norihiko Suruga, Noritaka Saeki and Kazushige Ogawa
Laboratory of Veterinary Anatomy, Graduate School of Life and Environmental Sciences, Osaka Prefecture University, 1-58 Rinku-Ourai-Kita, Izumisano, Osaka 598-8531, Japan

Hao Wu, Zhanhai Yin, Feng Li and Yusheng Qiu
Department of Orthopaedics, The First Affiliated Hospital, College of Medicine, Xi'an Jiaotong University, Xi'an 710061, People's Republic of China

Ling Wang
Center for Biomedical Engineering and Regenerative Medicine, Frontier Institute of Science and Technology, Xi'an Jiaotong University, Xi'an 710049, People's Republic of China

Eva Latorre, Amy Hooper, David Melzer and Lorna W. Harries
Institute of Biomedical and Clinical Sciences, University of Exeter Medical School, University of Exeter, Barrack Road, Exeter, Devon EX2 5DW, UK

Vishal C. Birar, Angela N. Sheerin, Richard G. A. Faragher and Elizabeth L. Ostler
School of Pharmacy and Biomolecular Sciences, University of Brighton, Cockcroft Building, Moulsecoomb, Brighton BN2 4GJ, UK

J. Charles C. Jeynes
Centre for Biomedical Modelling and Analysis, University of Exeter, Exeter, Devon EX2 5DW, UK

Helen R. Dawe
College of Life and Environmental Sciences, University of Exeter, Exeter, Devon EX4 4QD, UK

Lynne S. Cox
Department of Biochemistry, University of Oxford, Oxford OX1 3QU, UK

Katerina Koloustroubis, Rong Wang and Ann O. Sperry
Anatomy and Cell Biology, East Carolina University, Brody School of Medicine, Greenville, NC, USA

Nicole DeVaul
Laboratory of Biochemistry and Genetics, National Institute of Diabetics and Digestive and Kidney Diseases, National Institutes of Health, Bethesda, MD, USA

Christine Péladeau, Allan Heibein, Melissa T. Maltez, Sarah J. Copeland and John W. Copeland
Department of Cellular and Molecular Medicine, Faculty of Medicine, University of Ottawa, 451 Smyth Road, Ottawa, ON K1H 8M5, Canada

Alan Hammer and Maria Diakonova
Department of Biological Sciences, University of Toledo, 2801 W. Bancroft Street, Toledo 43606-3390, OH, USA

Le Thanh Lam
Wolfson Centre for Inherited Neuromuscular Disease, RJAH Orthopaedic Hospital, Oswestry, SY10 7AG, UK

Ian Holt and Glenn E. Morris
Wolfson Centre for Inherited Neuromuscular Disease, RJAH Orthopaedic Hospital, Oswestry, SY10 7AG, UK
Institute for Science and Technology in Medicine, Keele University, Keele ST5 5BG, UK

Nguyen Thuy Duong
Wolfson Centre for Inherited Neuromuscular Disease, RJAH Orthopaedic Hospital, Oswestry, SY10 7AG, UK
Institute of Genome Research (IGR), Vietnam Academy of Science and Technology (VAST), Hanoi, Vietnam

Caroline A. Sewry
Wolfson Centre for Inherited Neuromuscular Disease, RJAH Orthopaedic Hospital, Oswestry, SY10 7AG, UK
Dubowitz Neuromuscular Centre, Institute for Child Health and Great Ormond Street Hospital, London WC1 1EH, UK

Qiuping Zhang and Catherine M. Shanahan
Cardiovascular Division, James Black Centre, King's College, London SE5 9NU, UK

Kamel Mamchaoui
Sorbonne Universités, UPMC Univ Paris 06, INSERM UMRS974, CNRS FRE3617, Center for Research in Myology, 47 Boulevard de l'hôpital, 75013 Paris, France

Index

9 781682 867440